P9-APS-419

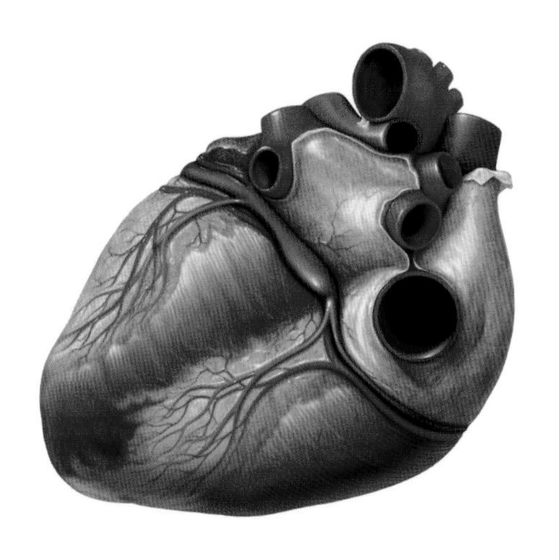

The HUMAN BODY IDENTIFICATION MANUAL

YOUR BODY & HOW IT WORKS

The HUMAN BODY IDENTIFICATION MANUAL

YOUR BODY & HOW IT WORKS

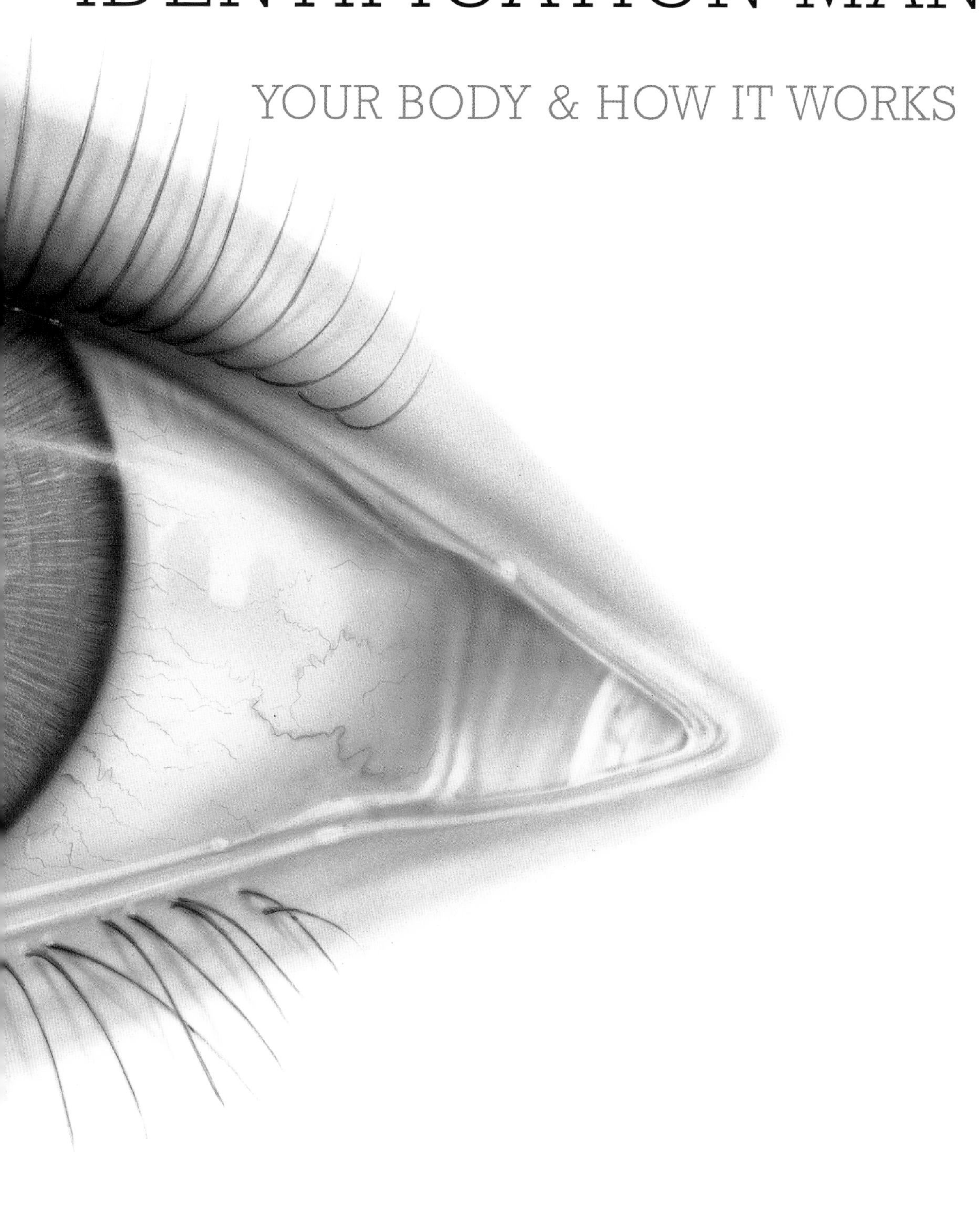

CHARTWELL
BOOKS

This edition published in 2016 by
CHARTWELL BOOKS
an imprint of Book Sales
a division of Quarto Publishing Group USA Inc.
142 West 36th Street, 4th Floor
New York, New York 10018
USA

First published in 2010 by
Global Book Publishing Pty Ltd
10.2 Level 10, 8 West Street,
North Sydney, 2060, Australia

ISBN 978-0-7858-3182-2

Printed and Bound in China

Conceived, designed and produced by Global Book Publishing Pty Ltd

Latin terminology consultant: Professor Ian Whitmore

Chief consultant on 2010 edition: Ken W. S. Ashwell

Design: Kate Haynes, Stan Lamond, Kylie Mulquin

Illustrations:
David Carroll, Peter Child, Deborah Clarke, Geoff Cook, Marcus Cremonese, Beth Croce, Hans De Haas, Wendy de Paauw, Levant Efe, Mike Golding, Mike Gorman, Jeff Lang, Alex Lavroff, Ulrich Lehmann, Ruth Lindsay, Richard McKenna, Annabel Milne, Tony Pyrzakowski, Oliver Rennert, Caroline Rodrigues, Otto Schmidinger, Bob Seal, Vicky Short, Graeme Tavendale, Jonathan Tidball, Paul Tresnan, Valentin Varetsa, Glen Vause, Spike Wademan, Trevor Weekes, Paul Williams, David Wood

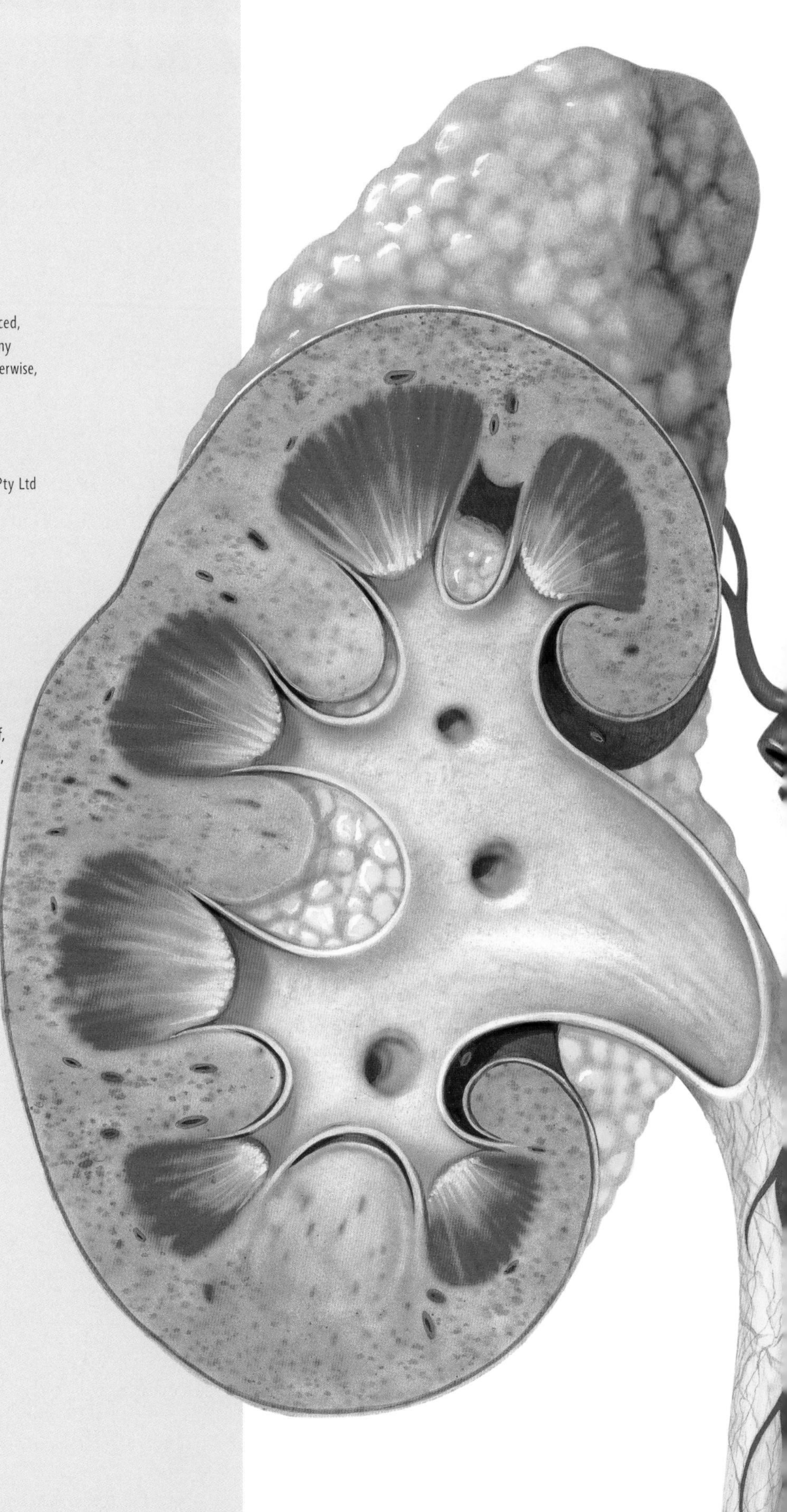

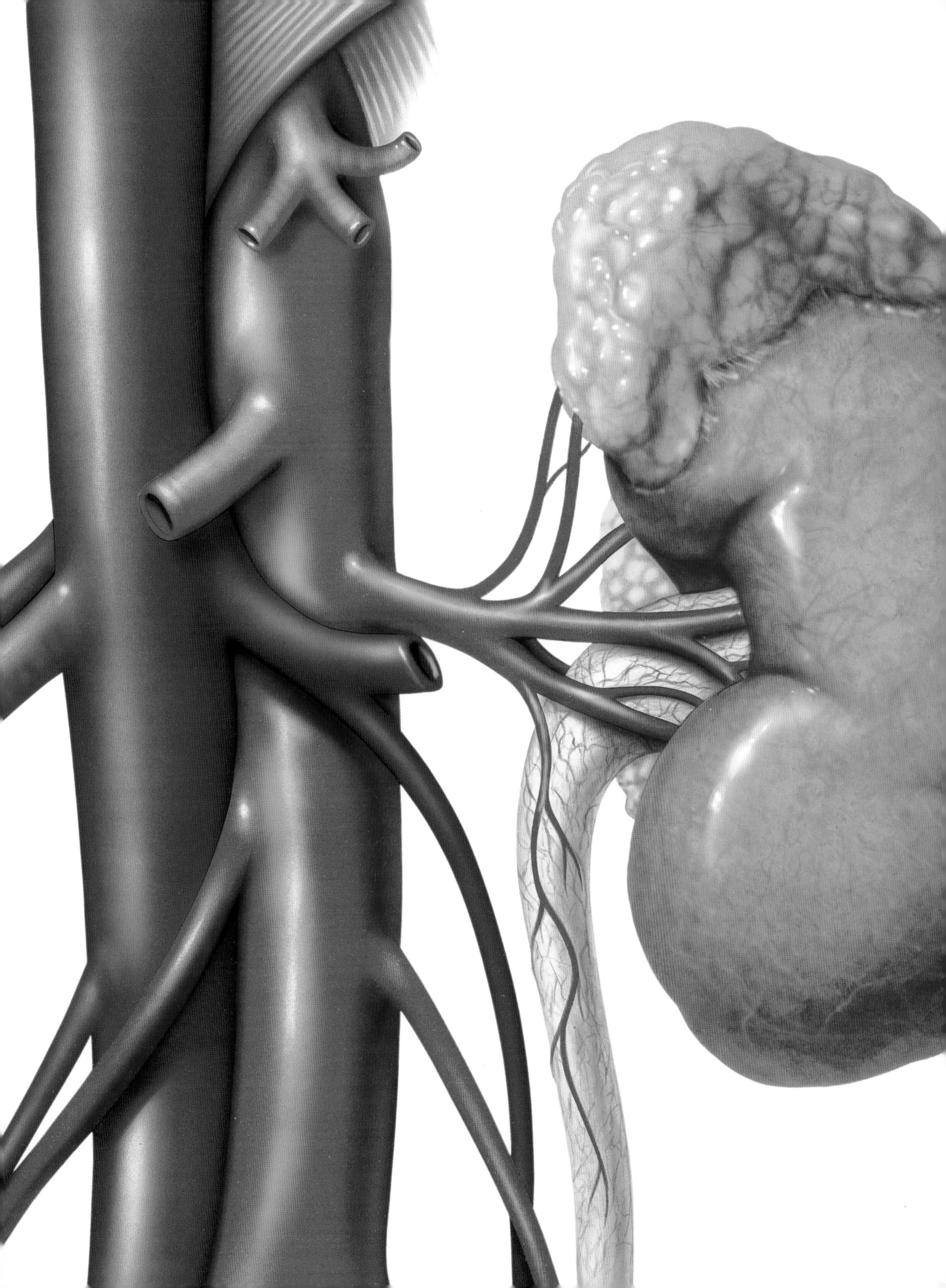

c o n t

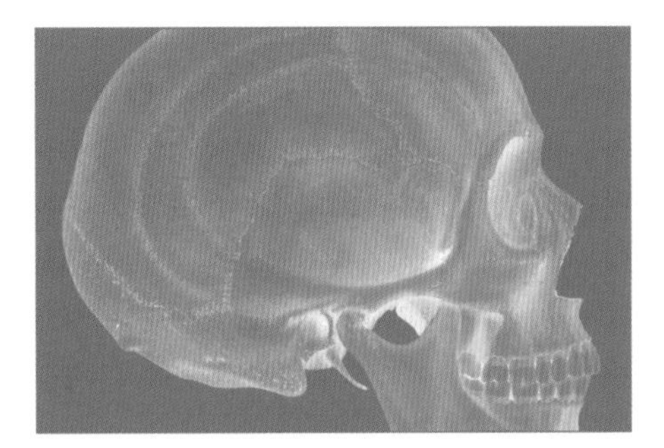

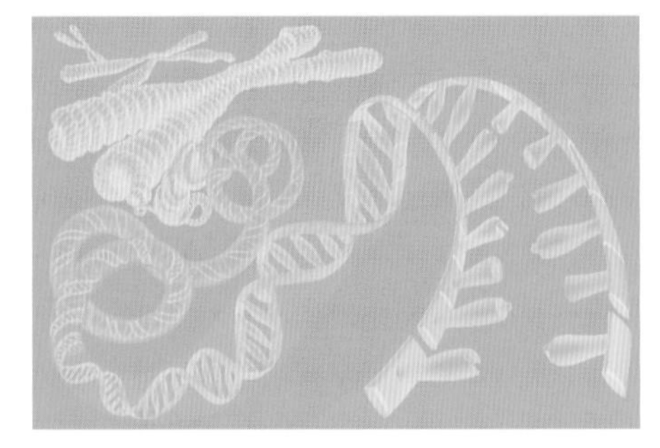

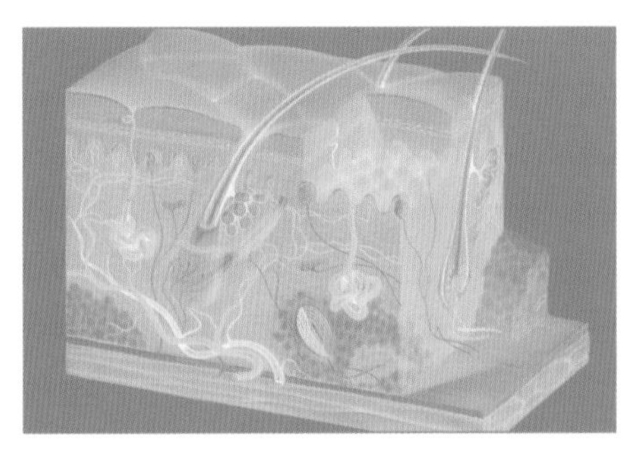

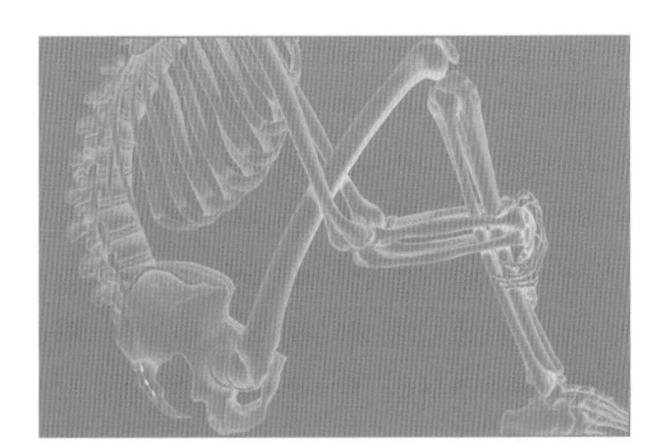

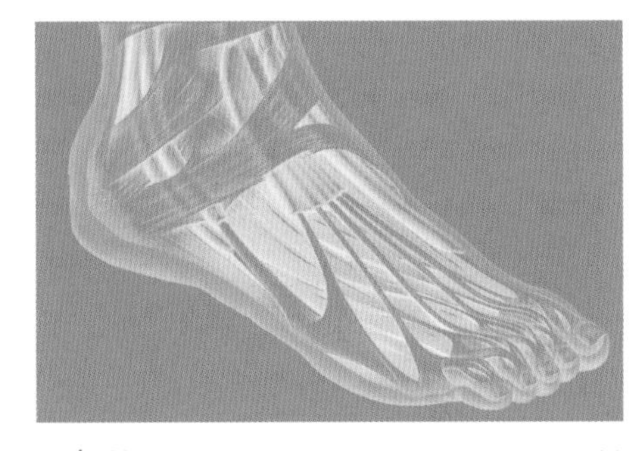

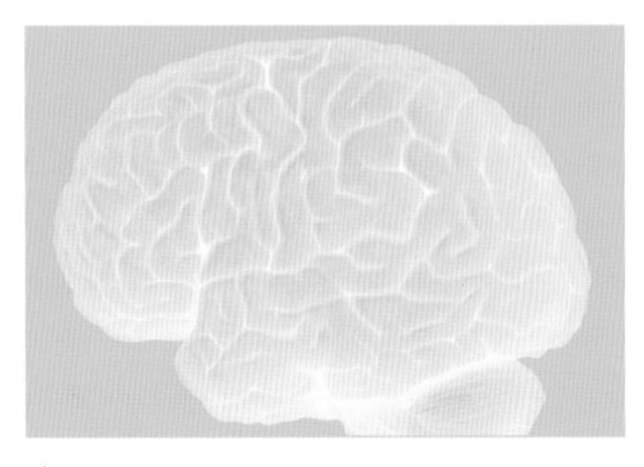

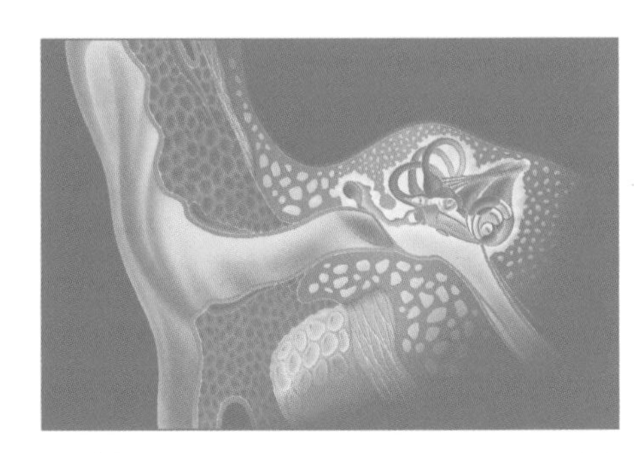

e n t s

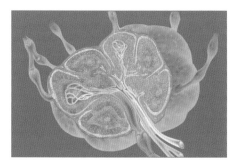

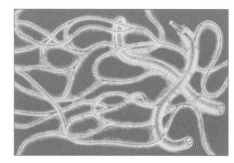

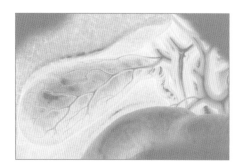

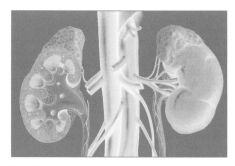

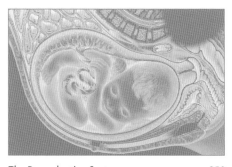

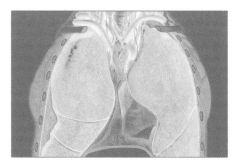

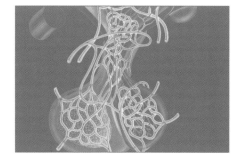

How This Book Works

The Human Body Identification Manual is a stunning pictorial handbook that reveals the marvels of the human body. By concentrating on graphic elements, this comprehensive book allows people learning about human anatomy to visualize the various parts of the body and their special links to each other quickly and easily.

Divided into 15 chapters, the book begins with an overview of the body systems and regions, followed by an introduction to cells and tissues. Subsequent chapters identify the structures and functions of the major body systems, from the skeletal and muscular to the endocrine and reproductive.

The book features over 500 anatomically correct illustrations with clear and informative labels showing both the English and formal medical terminology. Short but instructive captions explain physiological processes, microscopic structures, and other difficult anatomical concepts. As well as showing the location and name of hundreds of body parts, the color illustrations also reveal the unique composition of the human form.

In some instances, the illustrator has removed one part of the body so that another may be viewed more clearly: for example, in some of the illustrations of the abdominal organs, the liver has been peeled back to show the gallbladder. This is also true of some of the lung illustrations.

Many illustrations are supplemented by a locator diagram, which indicates where the organ is in relation to the rest of the body or shows the position and orientation of a cross-section illustration. The appearance of an organ often depends on the angle from which it is viewed, and for this reason some of the illustration names include an orientation such as Front View, Side View, or Rear View.

For ready reference, each chapter features an individually colored border along the side margins of its pages. The chapter name appears within the colored border on the left-hand page, while the body region or part appears within the colored border on the right-hand page. This helps readers to find the body system and region or part that they are looking for with ease.

Illustration headings
Illustration headings give the name of the body part. The orientation is included if necessary, and right or left limbs are also identified here.

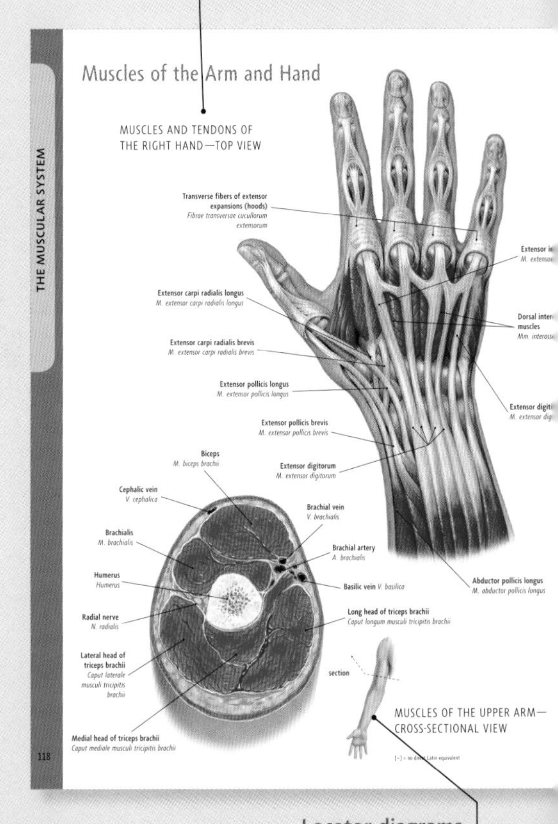
THE MUSCULAR SYSTEM
Muscles of the Arm and Hand
MUSCLES AND TENDONS OF THE RIGHT HAND—TOP VIEW
Transverse fibers of extensor expansions (hoods)
Fibrae transversae cucullorum extensorum
Extensor carpi radialis longus
M. extensor carpi radialis longus
Extensor carpi radialis brevis
M. extensor carpi radialis brevis
Extensor pollicis longus
M. extensor pollicis longus
Extensor pollicis brevis
M. extensor pollicis brevis
Extensor digitorum
M. extensor digitorum
Abductor pollicis longus
M. abductor pollicis longus
Biceps
M. biceps brachii
Cephalic vein
V. cephalica
Brachial vein
V. brachialis
Brachialis
M. brachialis
Brachial artery
A. brachialis
Humerus
Humerus
Basilic vein V. basilica
Radial nerve
N. radialis
Long head of triceps brachii
Caput longum musculi tricipitis brachii
Lateral head of triceps brachii
Caput laterale musculi tricipitis brachii
section
Medial head of triceps brachii
Caput mediale musculi tricipitis brachii
MUSCLES OF THE UPPER ARM—CROSS-SECTIONAL VIEW
118

Locator diagrams
Locator diagrams are included to show where in the body the particular organ or part is found, or to establish the body region from which a cross-section illustration has been drawn.

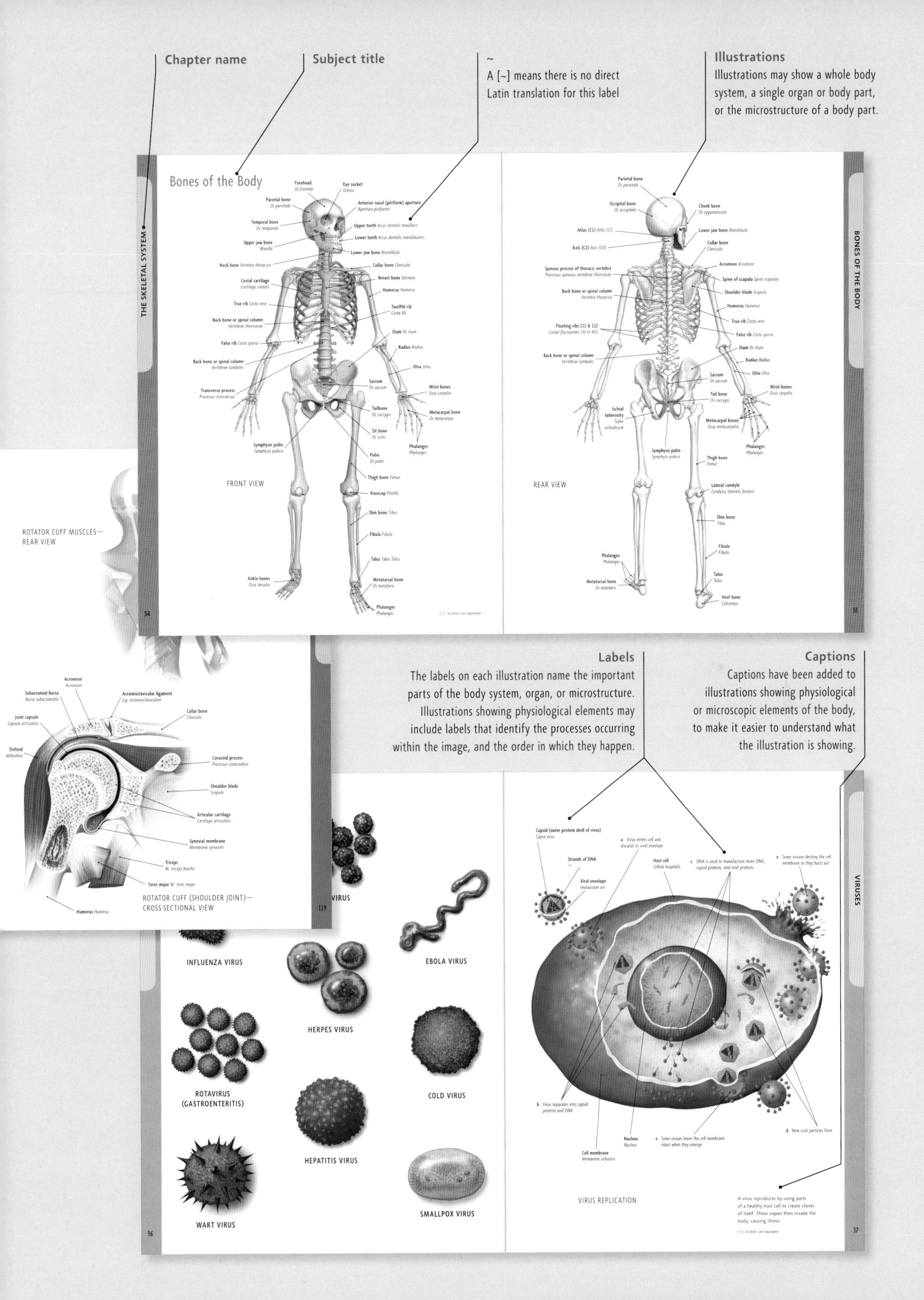

Chapter name
Subject title
~
A [~] means there is no direct Latin translation for this label
Illustrations
Illustrations may show a whole body system, a single organ or body part, or the microstructure of a body part.
Bones of the Body
THE SKELETAL SYSTEM
BONES OF THE BODY
FRONT VIEW
REAR VIEW
ROTATOR CUFF MUSCLES—REAR VIEW
ROTATOR CUFF (SHOULDER JOINT)—CROSS-SECTIONAL VIEW
Labels
The labels on each illustration name the important parts of the body system, organ, or microstructure. Illustrations showing physiological elements may include labels that identify the processes occurring within the image, and the order in which they happen.
Captions
Captions have been added to illustrations showing physiological or microscopic elements of the body, to make it easier to understand what the illustration is showing.
VIRUSES
INFLUENZA VIRUS
EBOLA VIRUS
HERPES VIRUS
ROTAVIRUS (GASTROENTERITIS)
COLD VIRUS
HEPATITIS VIRUS
WART VIRUS
SMALLPOX VIRUS
VIRUS REPLICATION
A virus reproduces by using parts of a healthy host cell to create clones of itself. These copies then invade the body, causing illness.

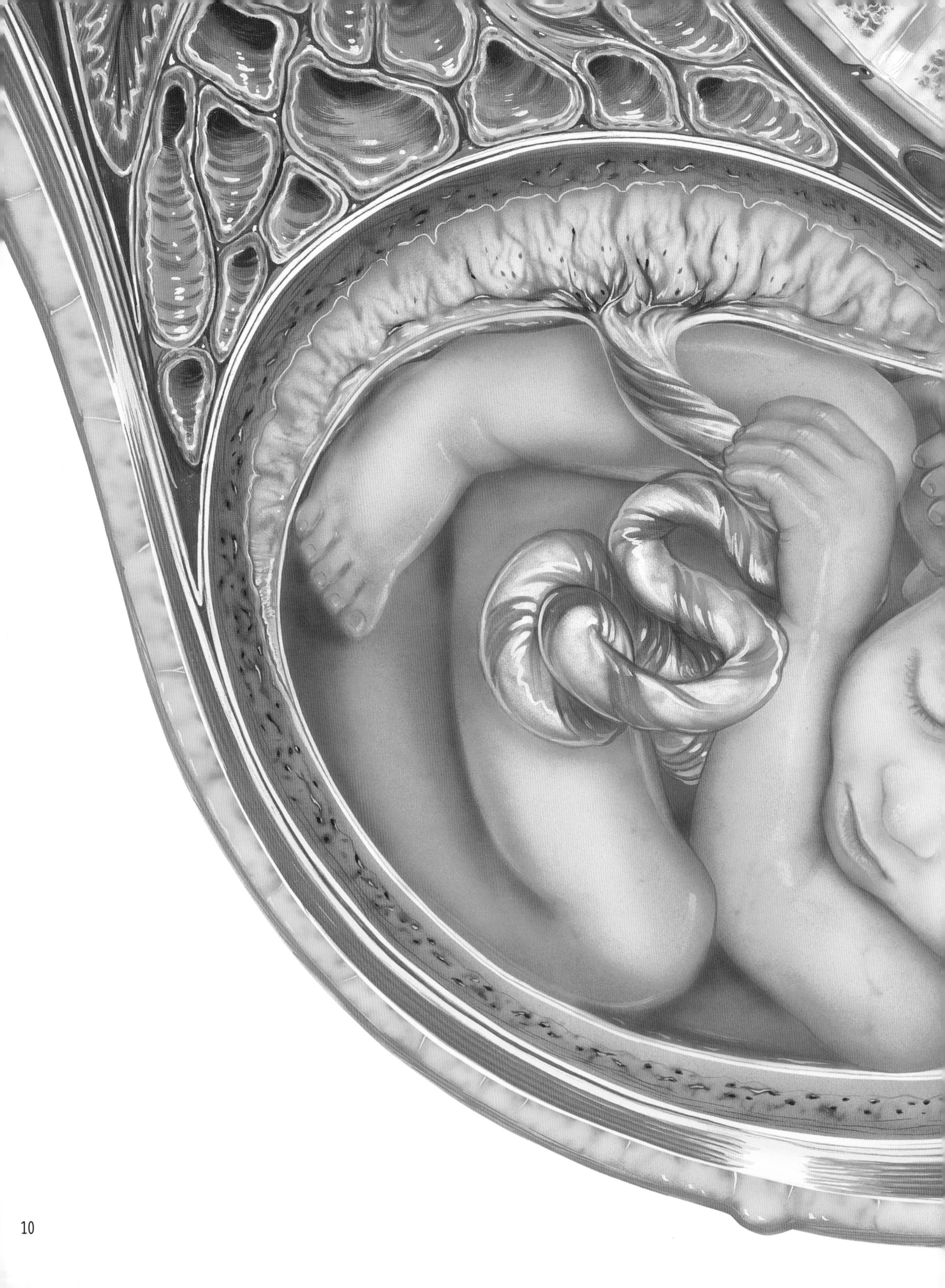

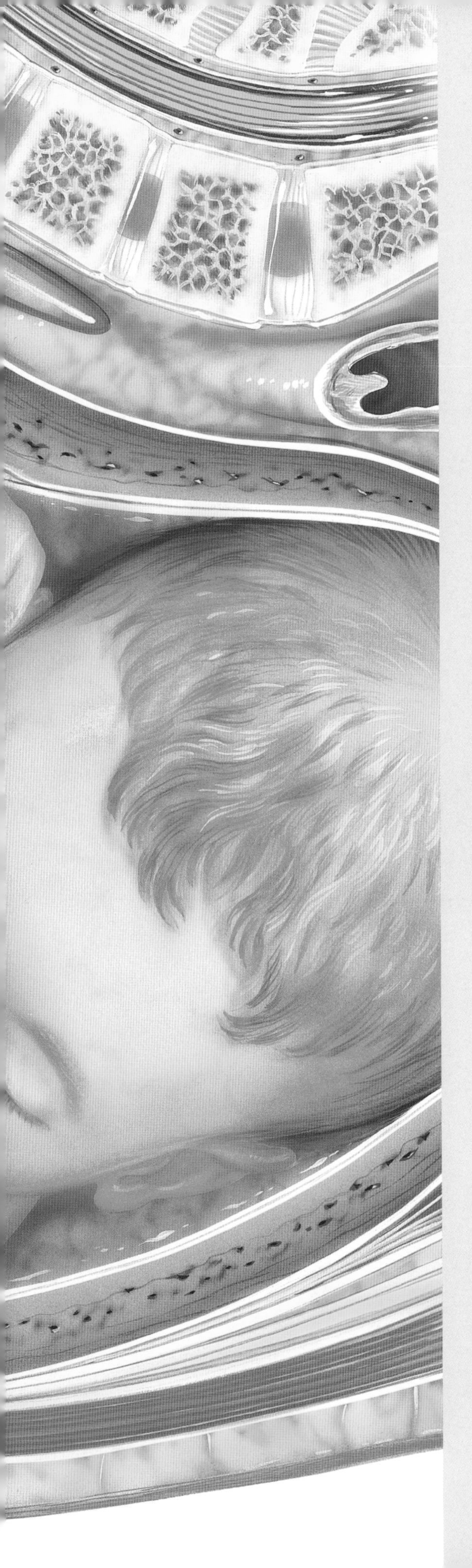

Introduction

Beneath the surface of our skin lies a complex system of muscles, internal organs, nerves, arteries, veins, lymph nodes, joints, and bones; a world of exquisite detail that we all rely on for our health and vitality, but one to which few of us ever give more than a passing thought. Although some of these internal structures can be felt through ridges and elevations on the skin's surface, most people would have only a vague idea of what lies just below their skin.

In this book, the intricate world of human internal structure is brought to life in beautiful, realistic, full-color illustrations that range from the molecular, through the microscopic, to the naked-eye level of magnification. The large-scale panoramic diagrams of the whole body in realistic poses allow the reader to quickly identify key anatomical features and their relationships to each other, while more detailed images focused on discrete body regions provide a seamless progression from the large to the small scale.

Hundreds of muscles, bony features, arteries, veins, and nerves have been illustrated and labeled on full-page, full-body images, so the reader can relate these structures to the external features of the body. The figure legends provide pertinent detail on the orientation and focus of each illustration.

This book will not only be useful for students of biology, physical education, osteopathy, chiropractic, podiatry, massage therapy, nursing, physiotherapy, and medicine, but will also be a valuable addition to the home library of those interested in the internal structure of the human form. The accessible, easy-to-understand nature of the images makes the book suitable for anyone who wishes to discover the wondrous workings of their own body.

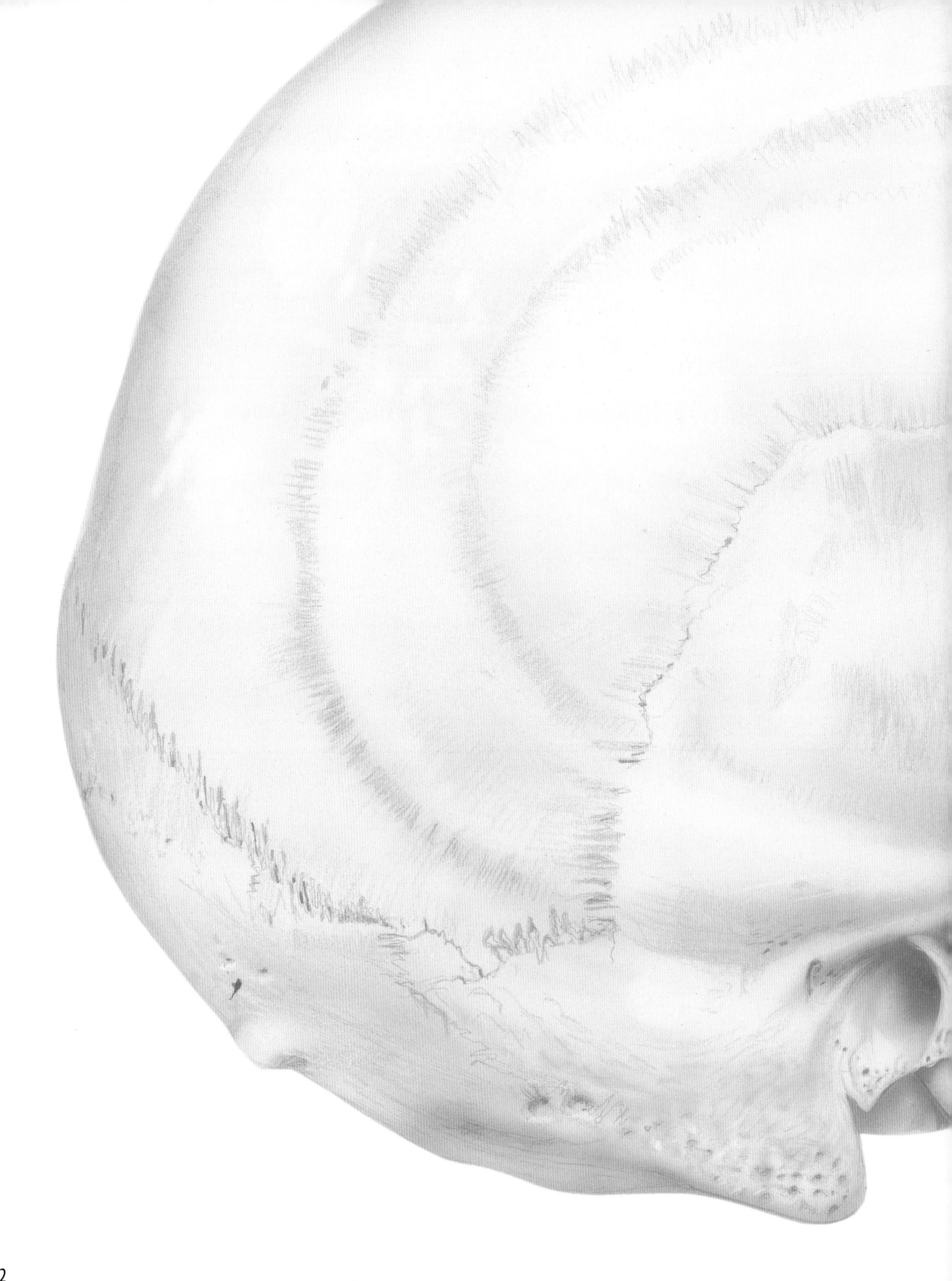

Body Overview

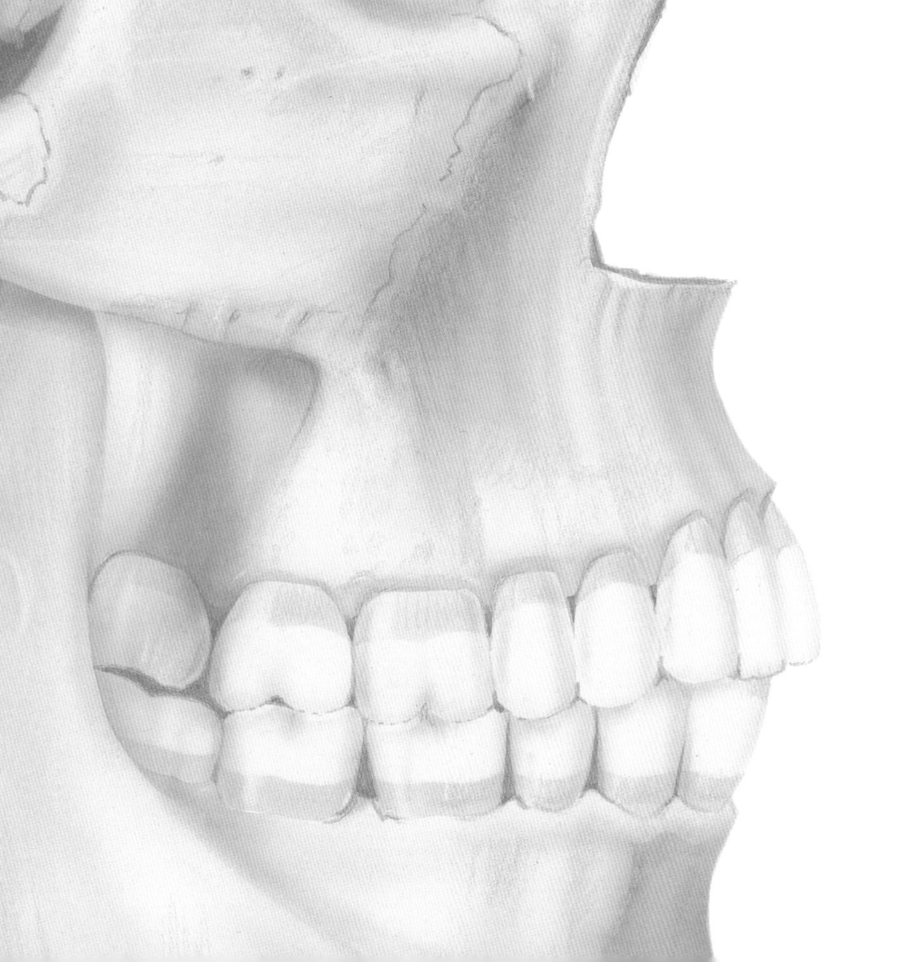

Overview of the Body Systems

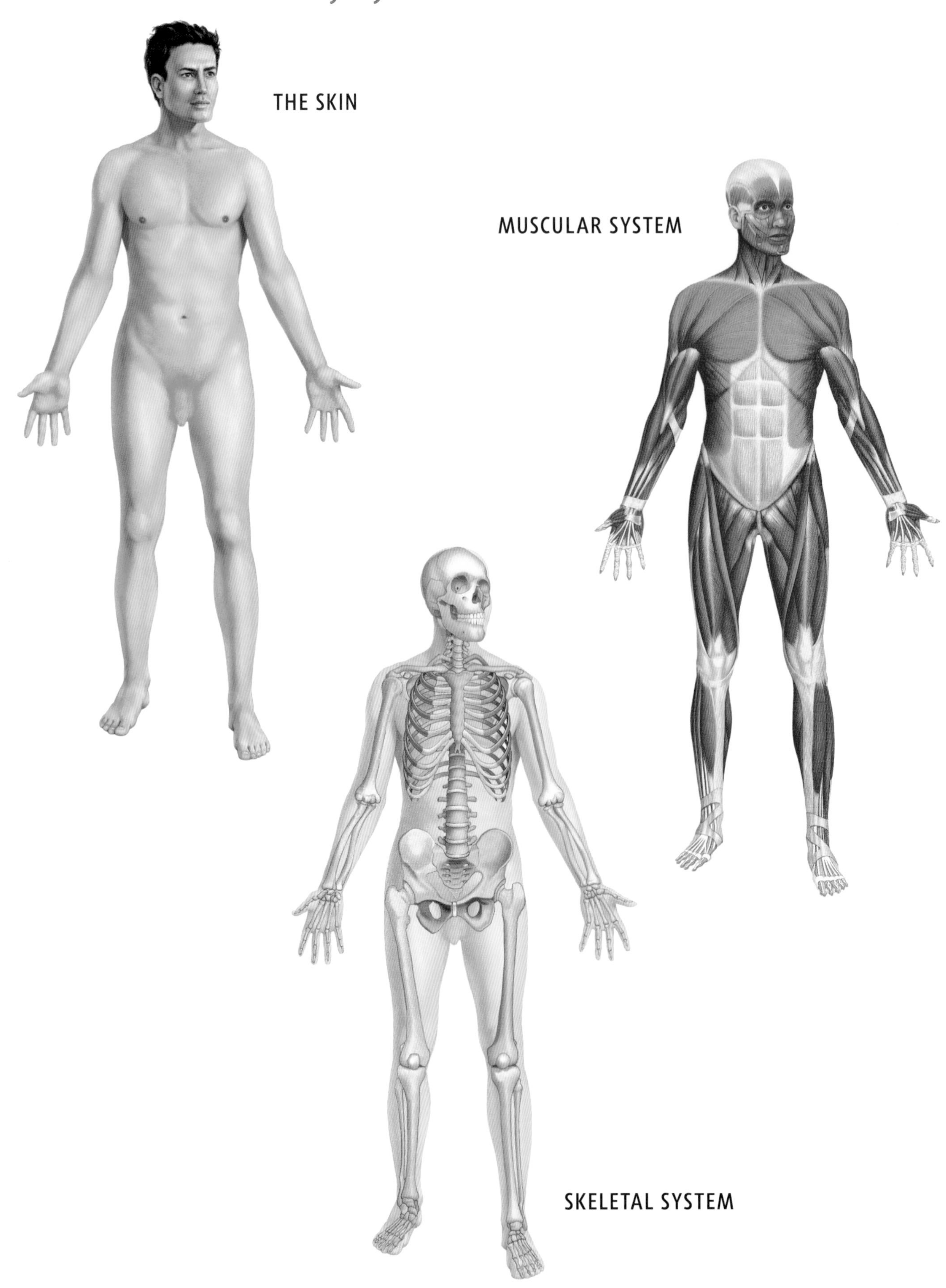
THE SKIN
MUSCULAR SYSTEM
SKELETAL SYSTEM

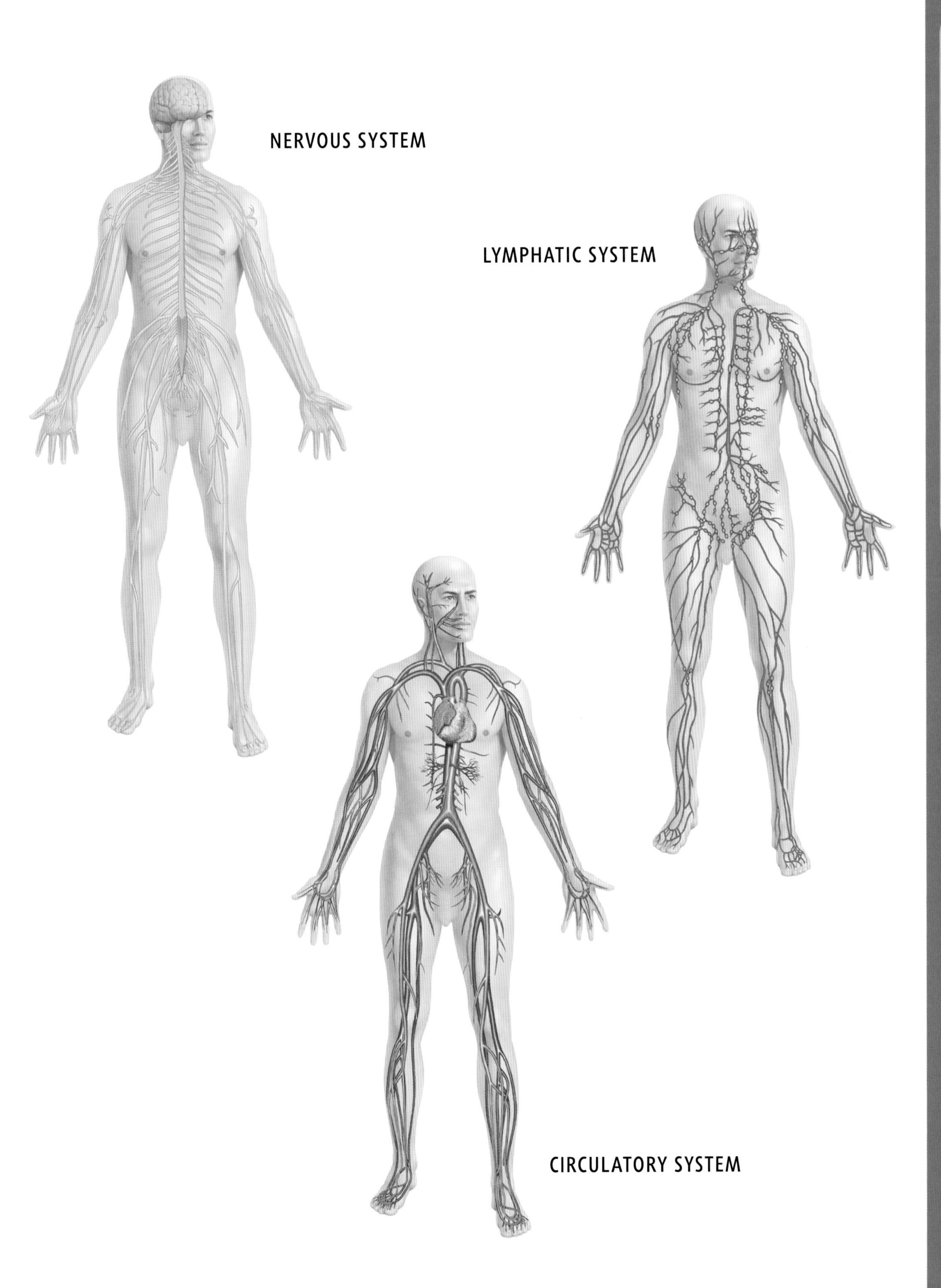
NERVOUS SYSTEM
LYMPHATIC SYSTEM
CIRCULATORY SYSTEM

Overview of the Body Systems

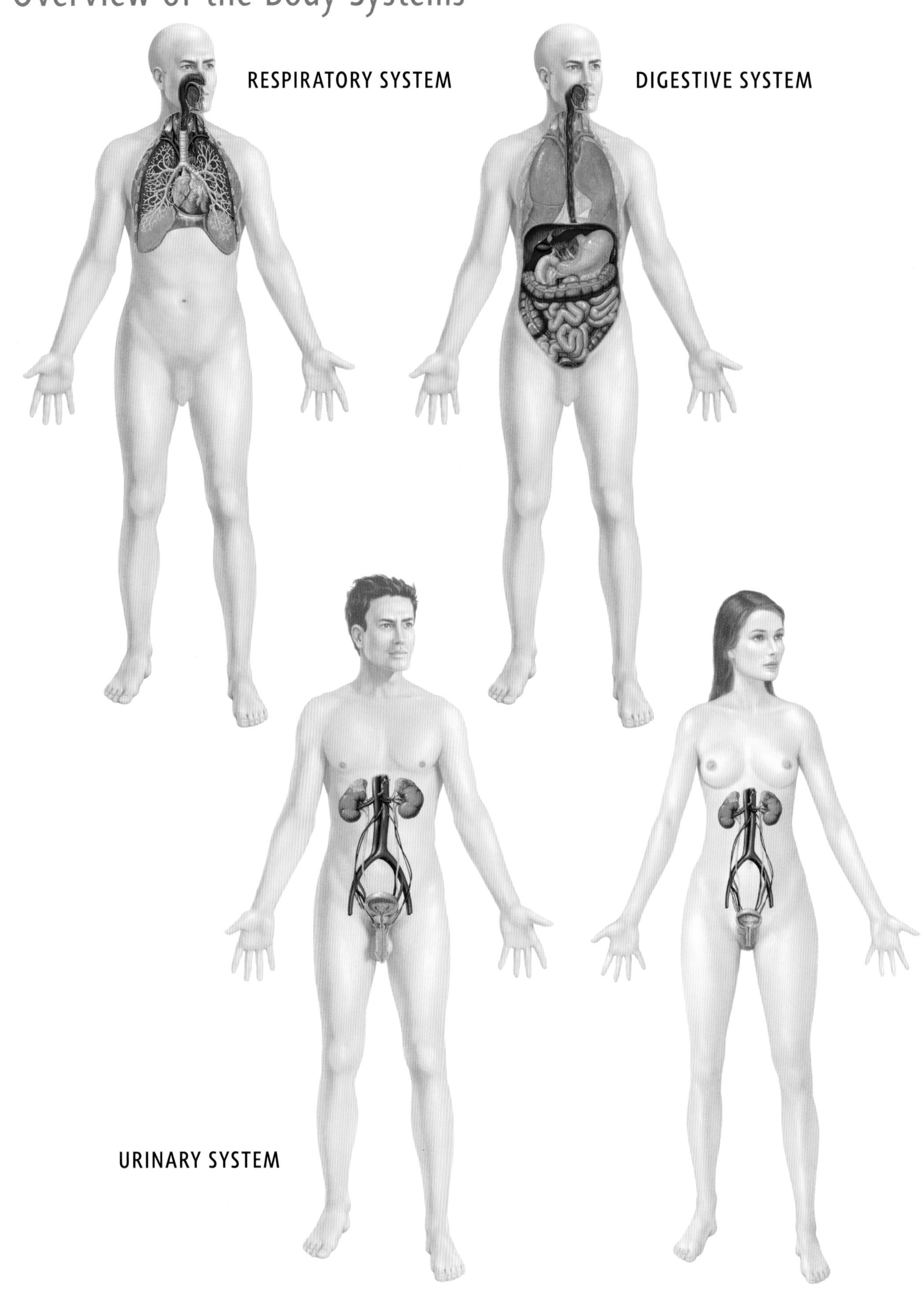
RESPIRATORY SYSTEM
DIGESTIVE SYSTEM
URINARY SYSTEM

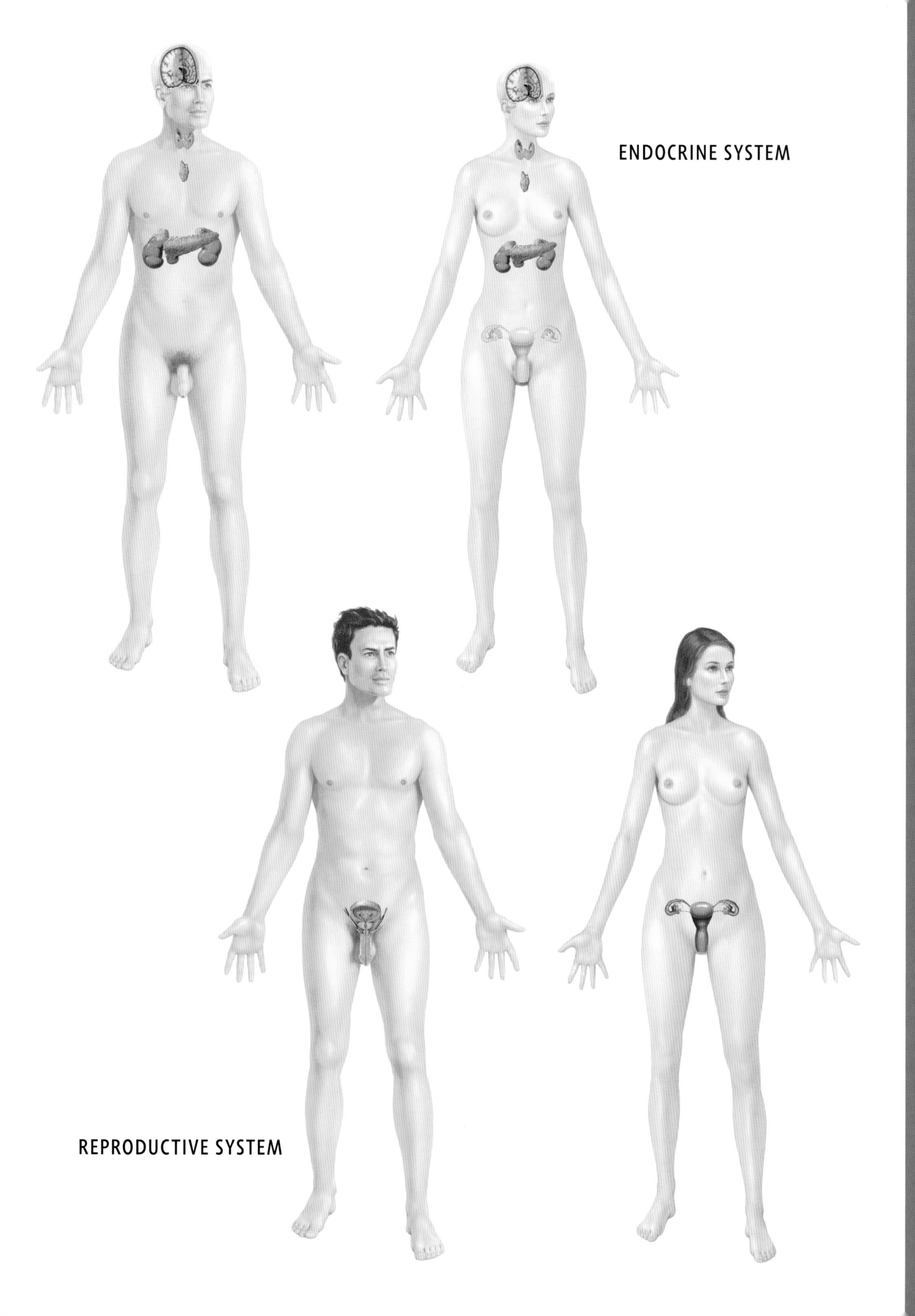
ENDOCRINE SYSTEM
REPRODUCTIVE SYSTEM

Body Regions

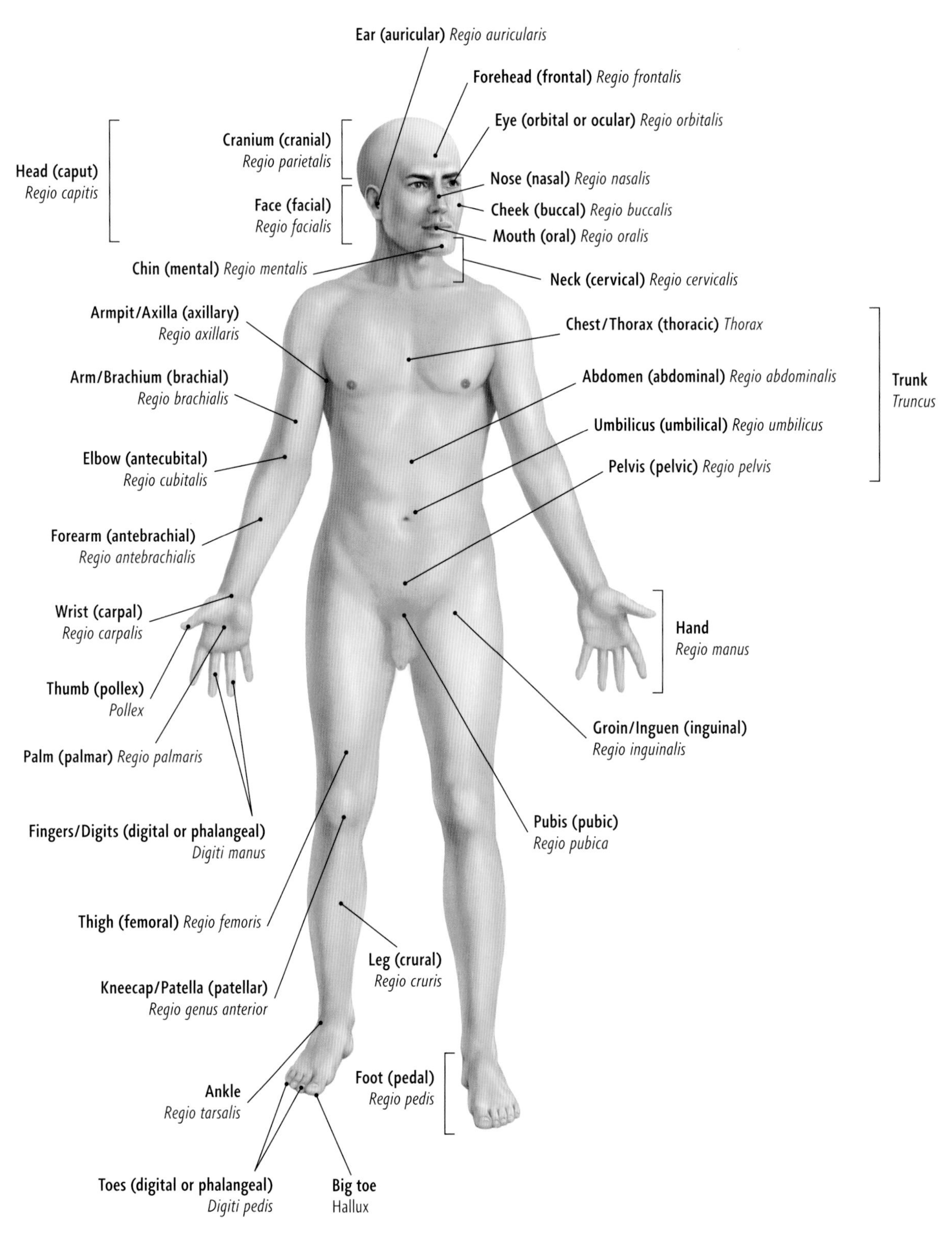

Ear (auricular) Regio auricularis
Forehead (frontal) Regio frontalis
Eye (orbital or ocular) Regio orbitalis
Head (caput)
Regio capitis
Cranium (cranial)
Regio parietalis
Face (facial)
Regio facialis
Nose (nasal) Regio nasalis
Cheek (buccal) Regio buccalis
Mouth (oral) Regio oralis
Chin (mental) Regio mentalis
Neck (cervical) Regio cervicalis
Armpit/Axilla (axillary)
Regio axillaris
Chest/Thorax (thoracic) Thorax
Arm/Brachium (brachial)
Regio brachialis
Abdomen (abdominal) Regio abdominalis
Trunk
Truncus
Umbilicus (umbilical) Regio umbilicus
Elbow (antecubital)
Regio cubitalis
Pelvis (pelvic) Regio pelvis
Forearm (antebrachial)
Regio antebrachialis
Wrist (carpal)
Regio carpalis
Hand
Regio manus
Thumb (pollex)
Pollex
Groin/Inguen (inguinal)
Regio inguinalis
Palm (palmar) Regio palmaris
Fingers/Digits (digital or phalangeal)
Digiti manus
Pubis (pubic)
Regio pubica
Thigh (femoral) Regio femoris
Leg (crural)
Regio cruris
Kneecap/Patella (patellar)
Regio genus anterior
Foot (pedal)
Regio pedis
Ankle
Regio tarsalis
Toes (digital or phalangeal)
Digiti pedis
Big toe
Hallux

FRONT VIEW

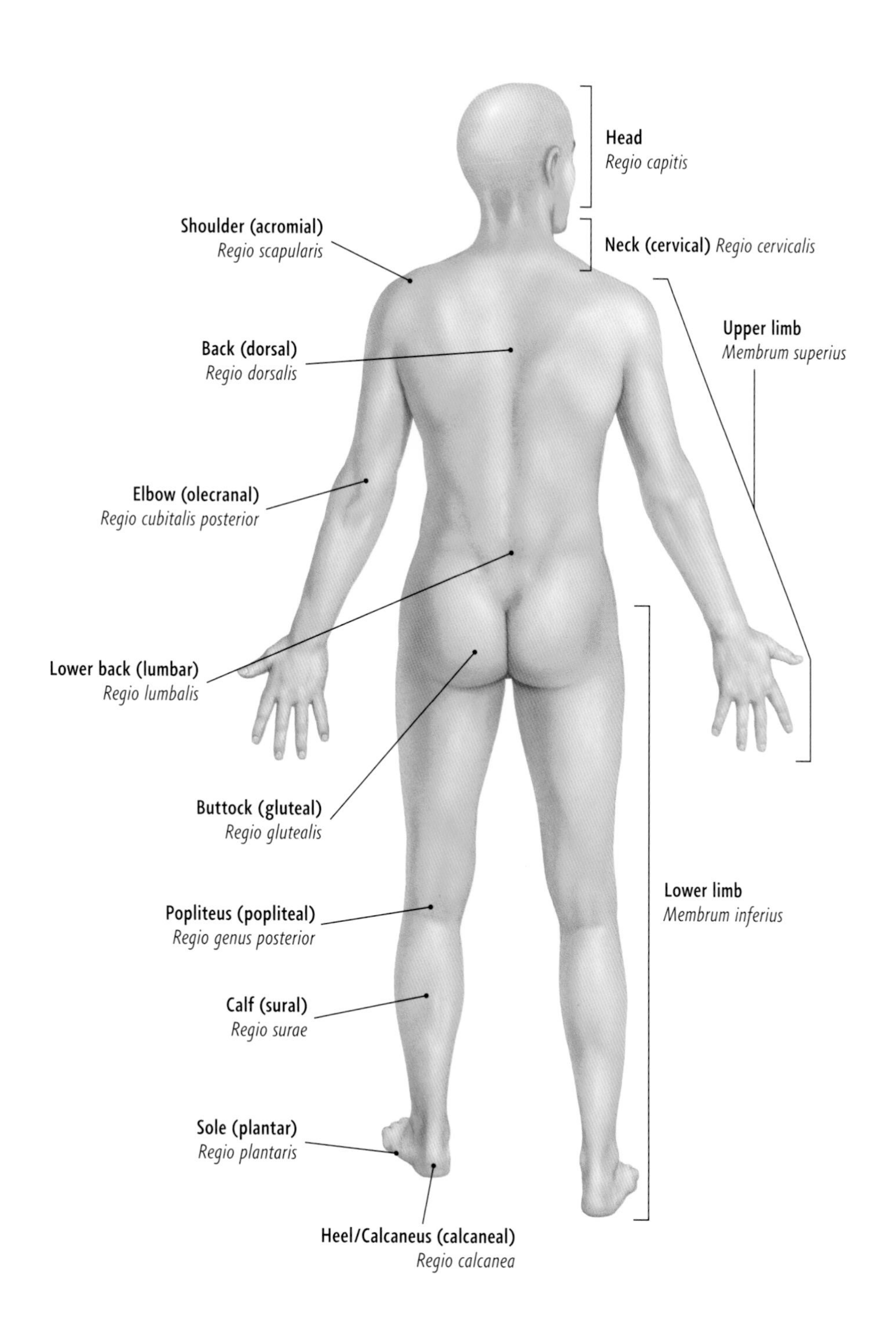
Head
Regio capitis
Shoulder (acromial)
Regio scapularis
Neck (cervical) Regio cervicalis
Upper limb
Membrum superius
Back (dorsal)
Regio dorsalis
Elbow (olecranal)
Regio cubitalis posterior
Lower back (lumbar)
Regio lumbalis
Buttock (gluteal)
Regio glutealis
Lower limb
Membrum inferius
Popliteus (popliteal)
Regio genus posterior
Calf (sural)
Regio surae
Sole (plantar)
Regio plantaris
Heel/Calcaneus (calcaneal)
Regio calcanea

REAR VIEW

Body Regions

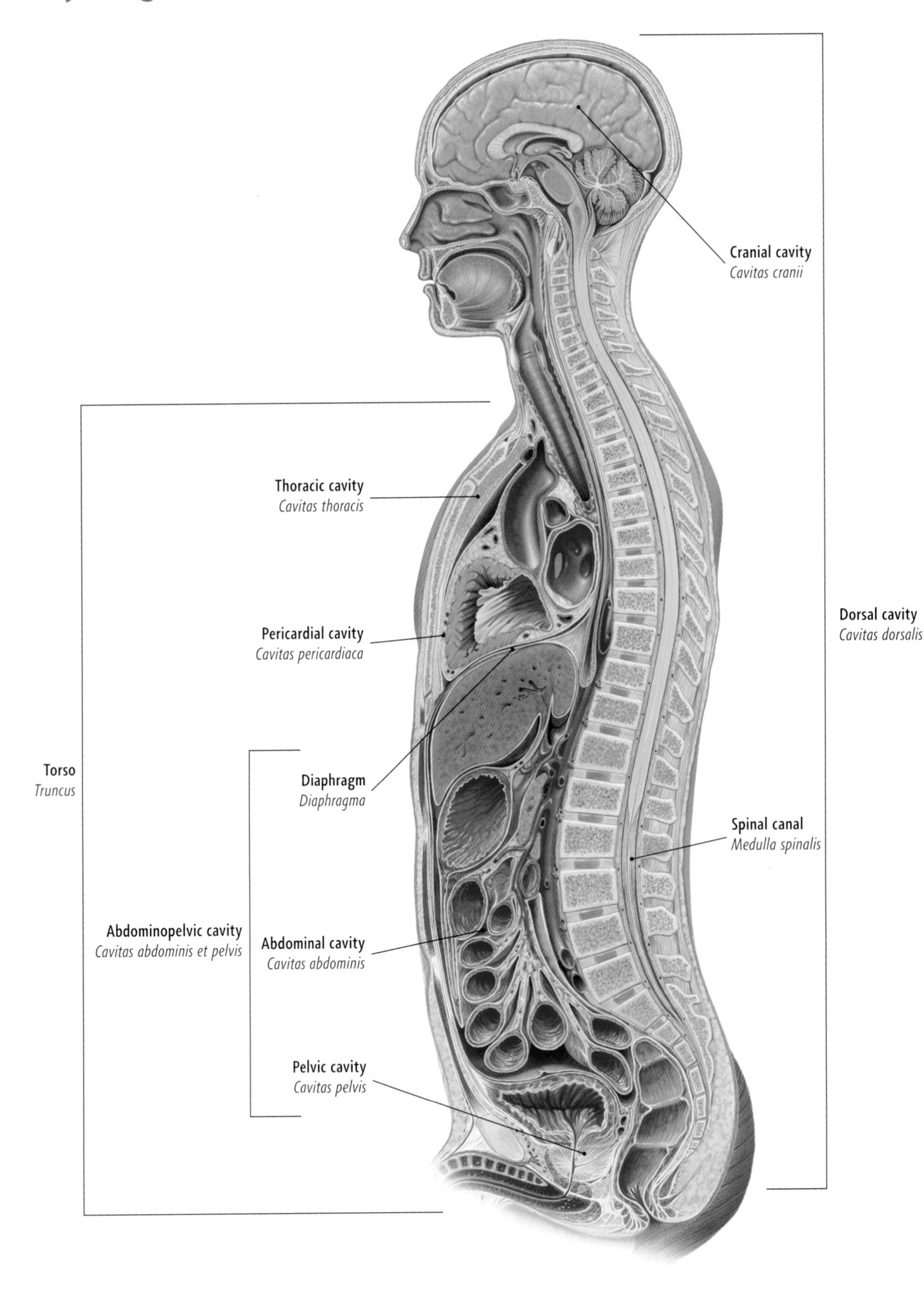
Cranial cavity
Cavitas cranii
Thoracic cavity
Cavitas thoracis
Pericardial cavity
Cavitas pericardiaca
Dorsal cavity
Cavitas dorsalis
Torso
Truncus
Diaphragm
Diaphragma
Spinal canal
Medulla spinalis
Abdominopelvic cavity
Cavitas abdominis et pelvis
Abdominal cavity
Cavitas abdominis
Pelvic cavity
Cavitas pelvis

BODY CAVITIES—CROSS-SECTIONAL VIEW

Body Shapes

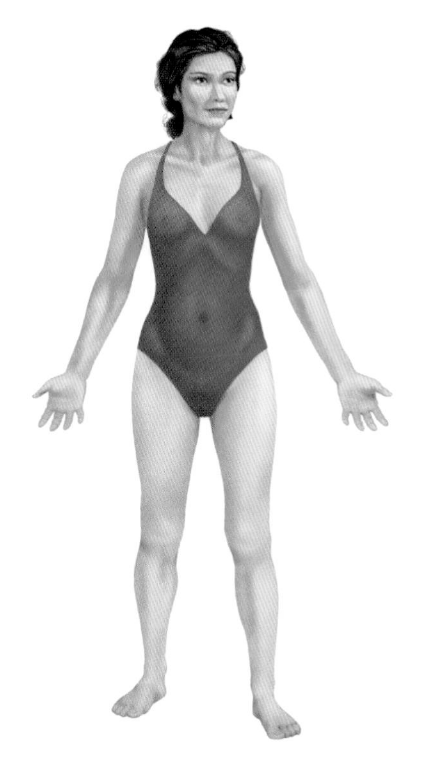

ECTOMORPH FEMALE

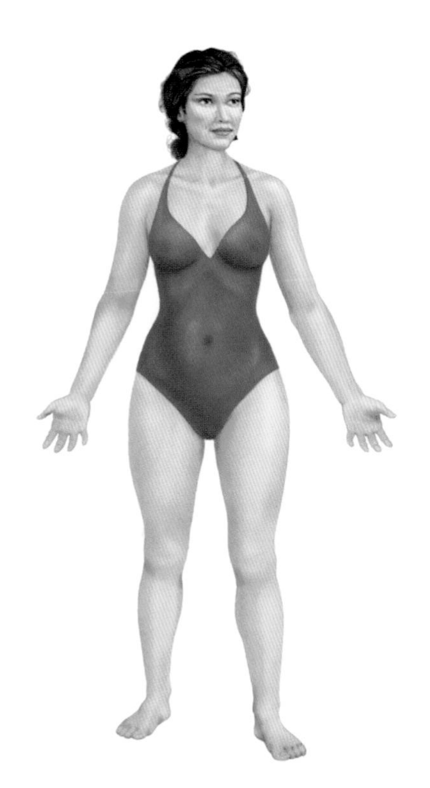

MESOMORPH FEMALE

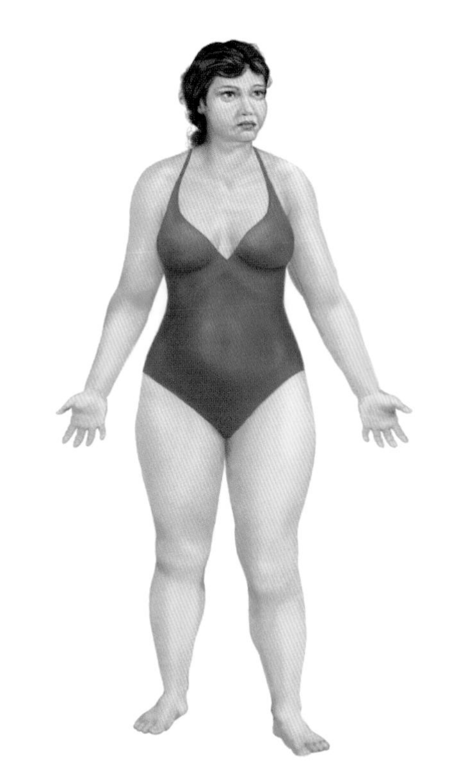

ENDOMORPH FEMALE

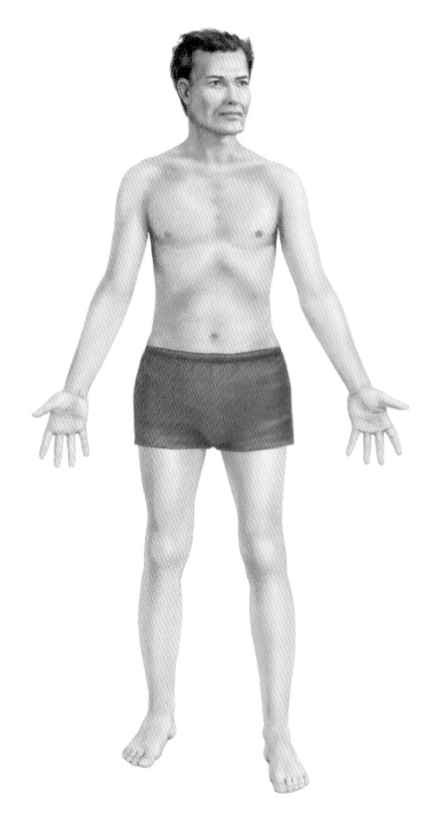

ECTOMORPH MALE

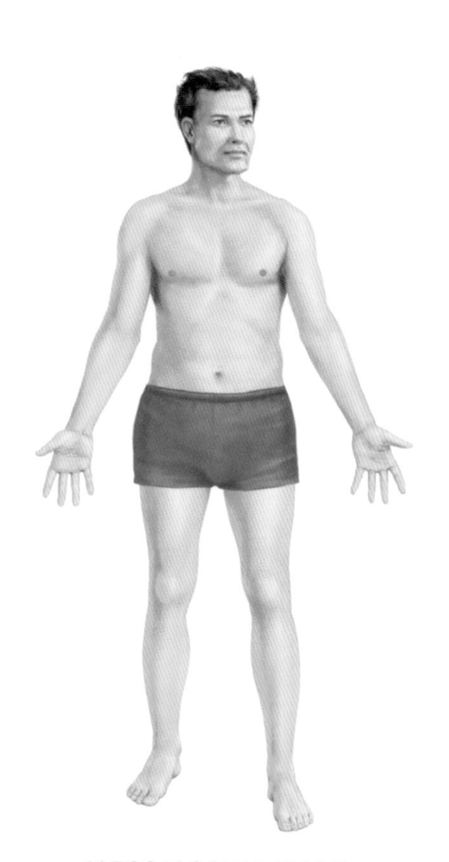

MESOMORPH MALE

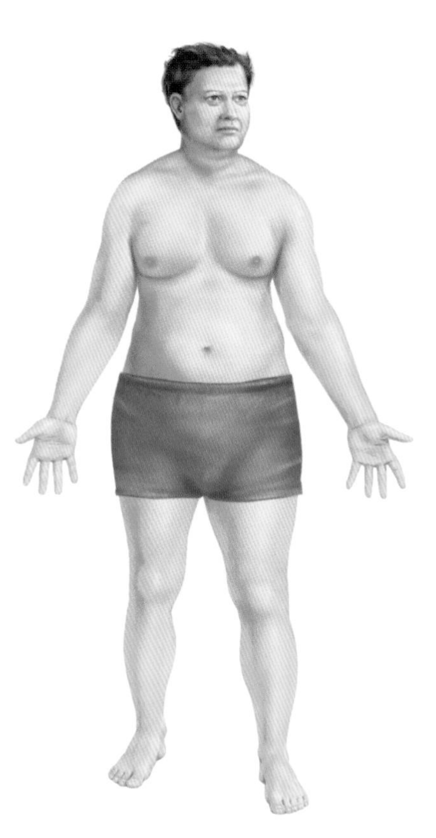

ENDOMORPH MALE

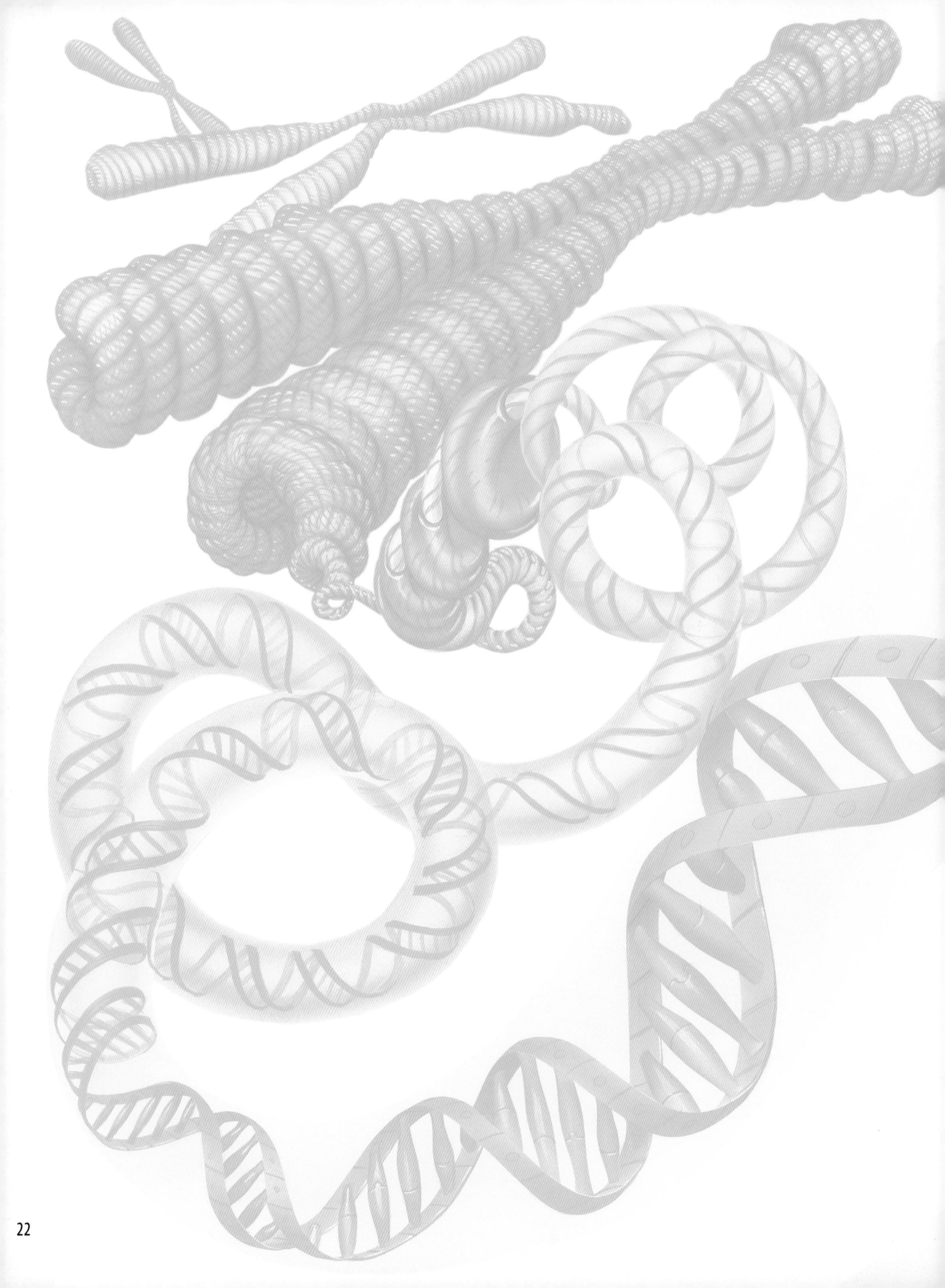

Cells and Tissues

Cell Structure and Major Cell Types

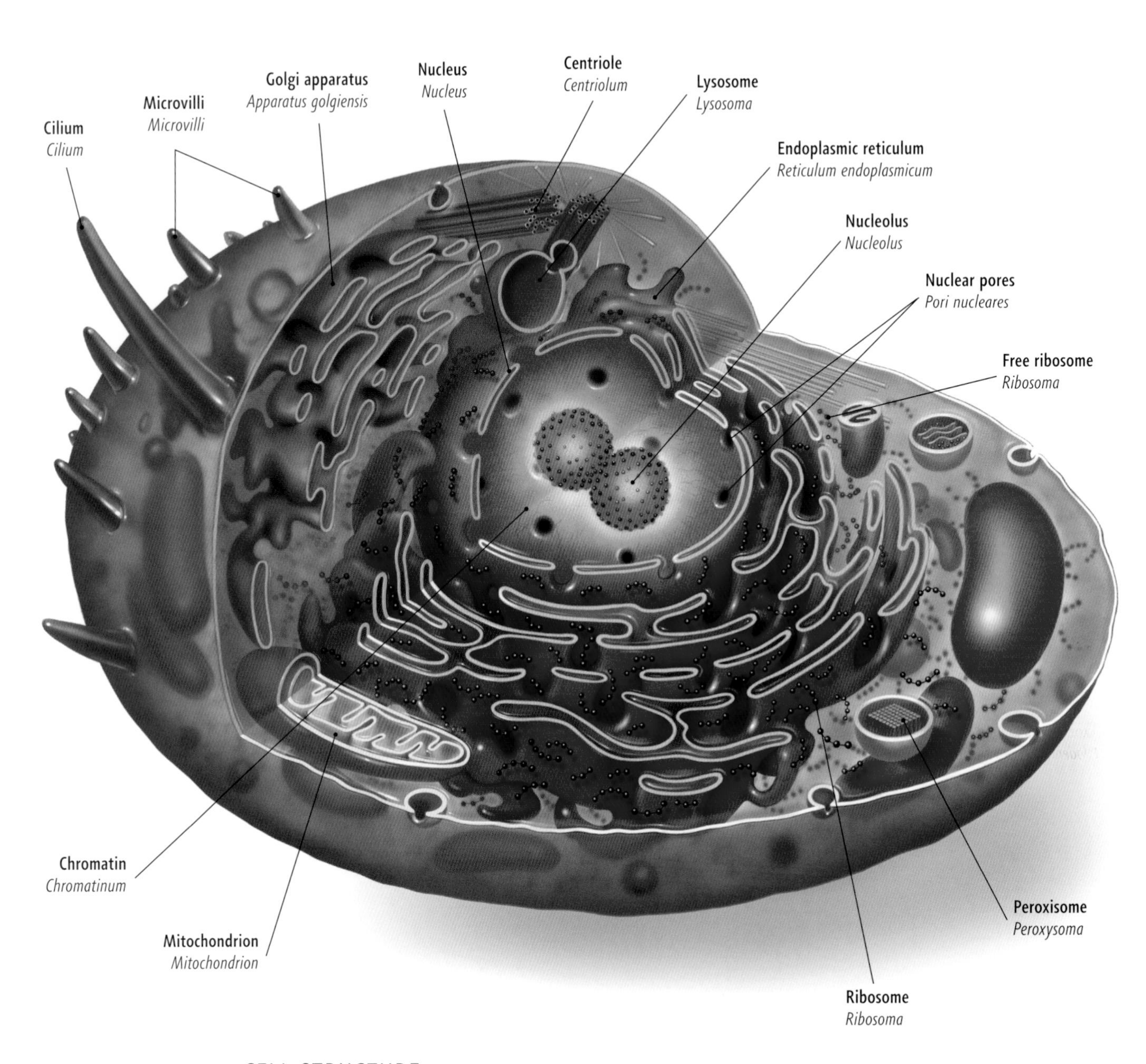
Cilium
Cilium
Microvilli
Microvilli
Golgi apparatus
Apparatus golgiensis
Nucleus
Nucleus
Centriole
Centriolum
Lysosome
Lysosoma
Endoplasmic reticulum
Reticulum endoplasmicum
Nucleolus
Nucleolus
Nuclear pores
Pori nucleares
Free ribosome
Ribosoma
Chromatin
Chromatinum
Mitochondrion
Mitochondrion
Peroxisome
Peroxysoma
Ribosome
Ribosoma

CELL STRUCTURE

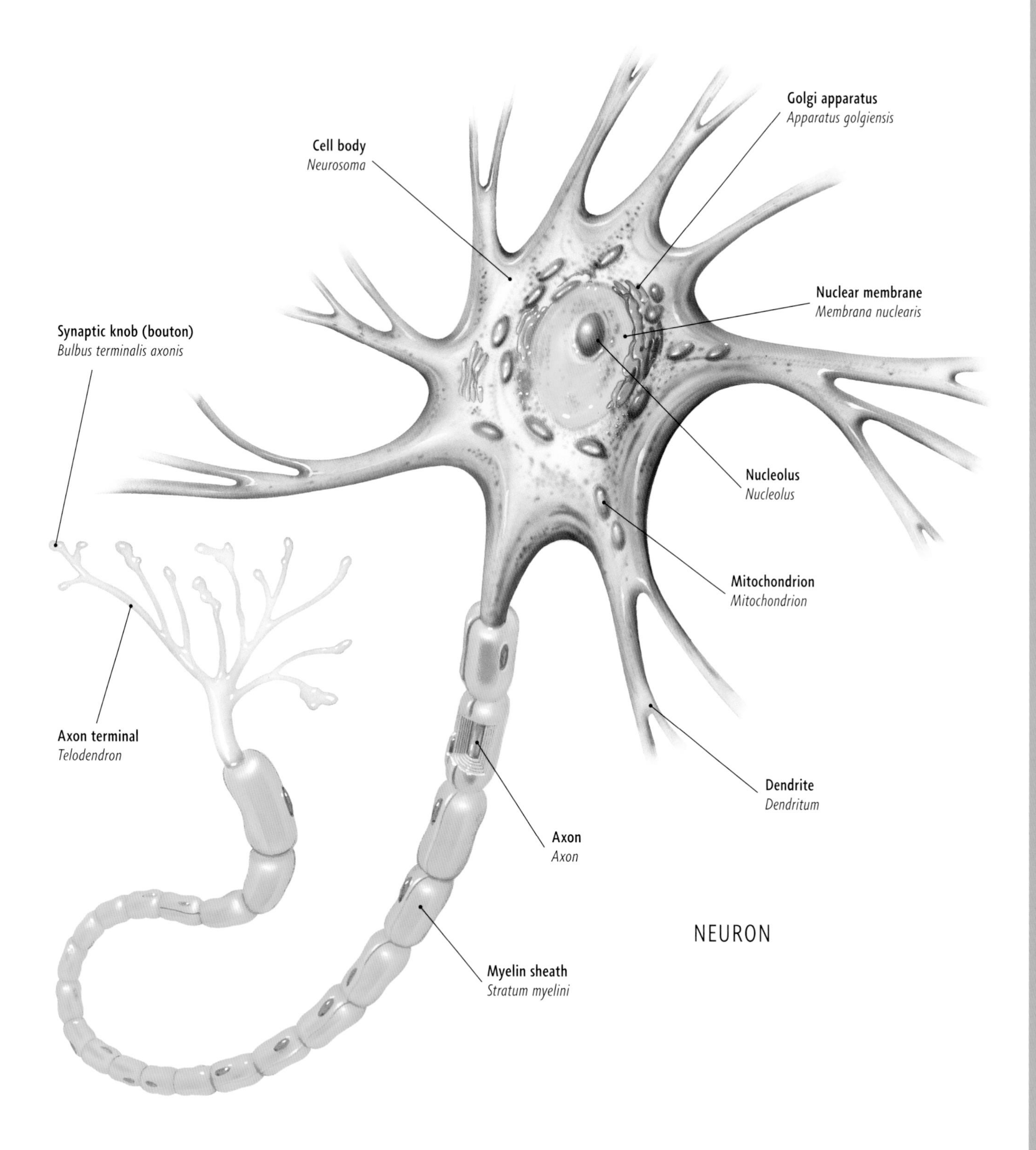
Cell body
Neurosoma
Golgi apparatus
Apparatus golgiensis
Nuclear membrane
Membrana nuclearis
Synaptic knob (bouton)
Bulbus terminalis axonis
Nucleolus
Nucleolus
Mitochondrion
Mitochondrion
Axon terminal
Telodendron
Dendrite
Dendritum
Axon
Axon
Myelin sheath
Stratum myelini

NEURON

Blood Cells

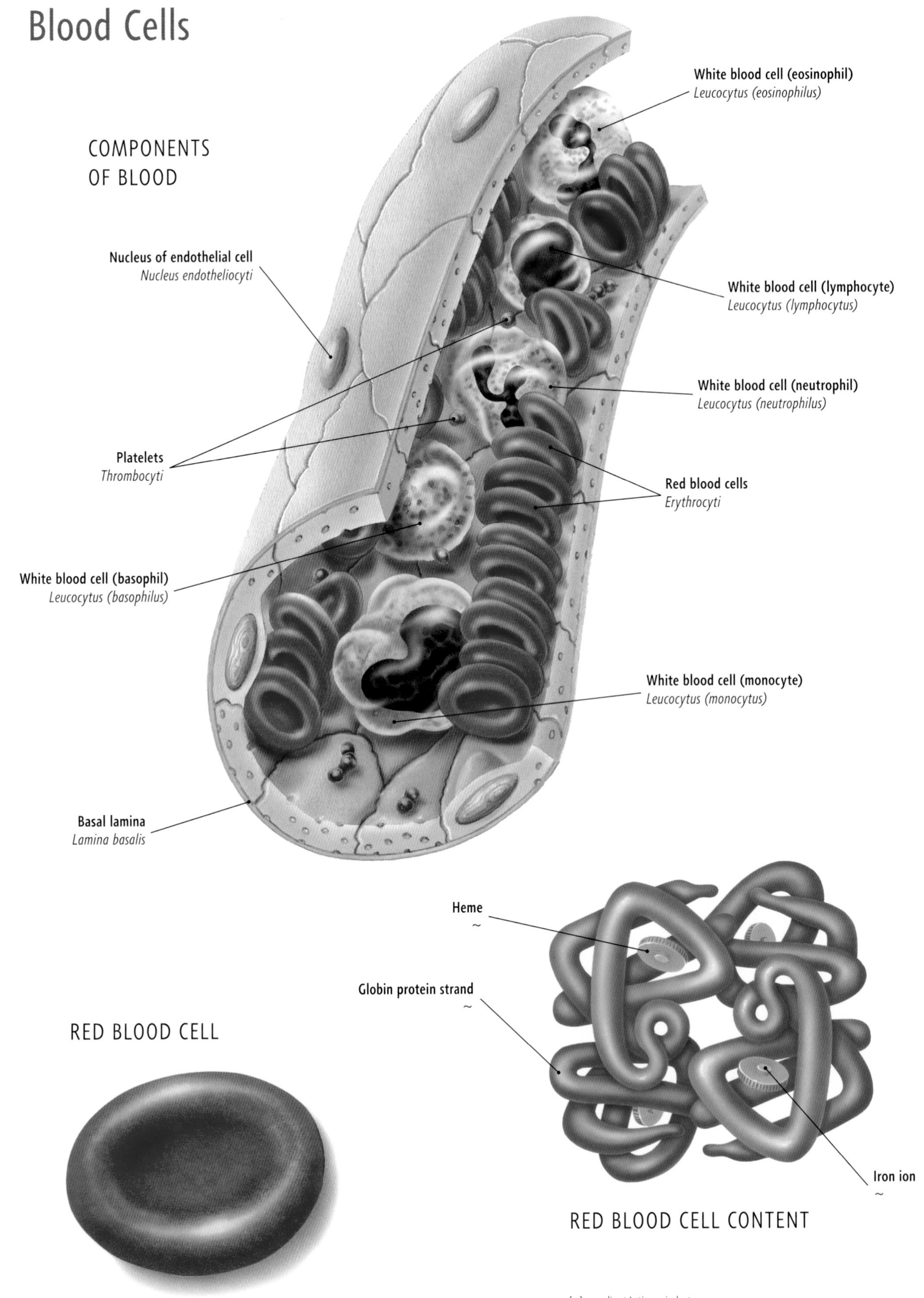
COMPONENTS OF BLOOD
White blood cell (eosinophil)
Leucocytus (eosinophilus)
Nucleus of endothelial cell
Nucleus endotheliocyti
White blood cell (lymphocyte)
Leucocytus (lymphocytus)
White blood cell (neutrophil)
Leucocytus (neutrophilus)
Platelets
Thrombocyti
Red blood cells
Erythrocyti
White blood cell (basophil)
Leucocytus (basophilus)
White blood cell (monocyte)
Leucocytus (monocytus)
Basal lamina
Lamina basalis
Heme
~
Globin protein strand
~
RED BLOOD CELL
Iron ion
~
RED BLOOD CELL CONTENT

[~] = no direct Latin equivalent

Monocyte
Monocytus

Macrophage
Macrophagocytus

Neutrophil
Neutrophilus

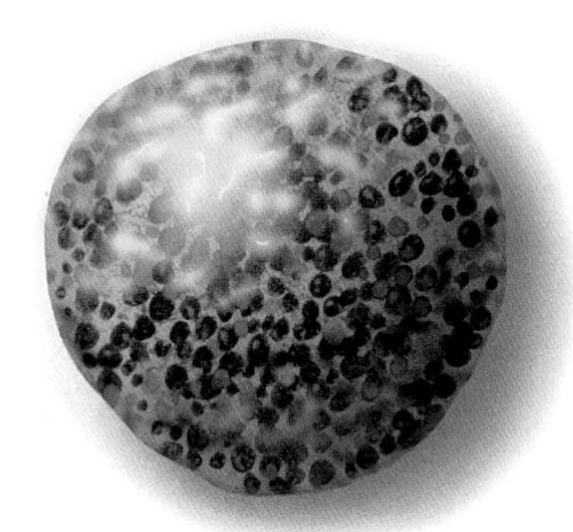

Basophil
Basophilus

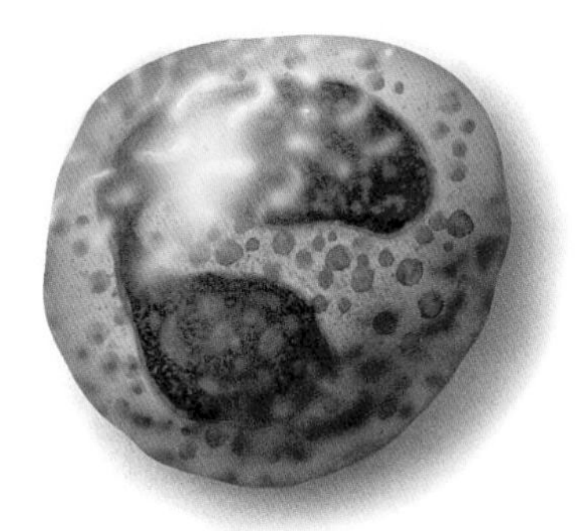

Eosinophil
Eosinophilus

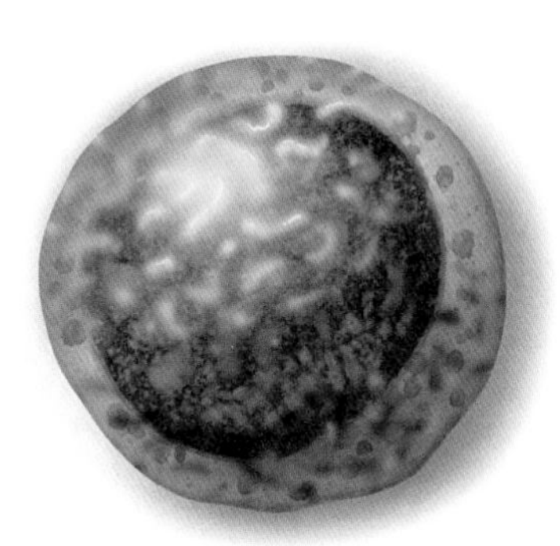

Lymphocyte
Lymphocytus

WHITE BLOOD CELLS

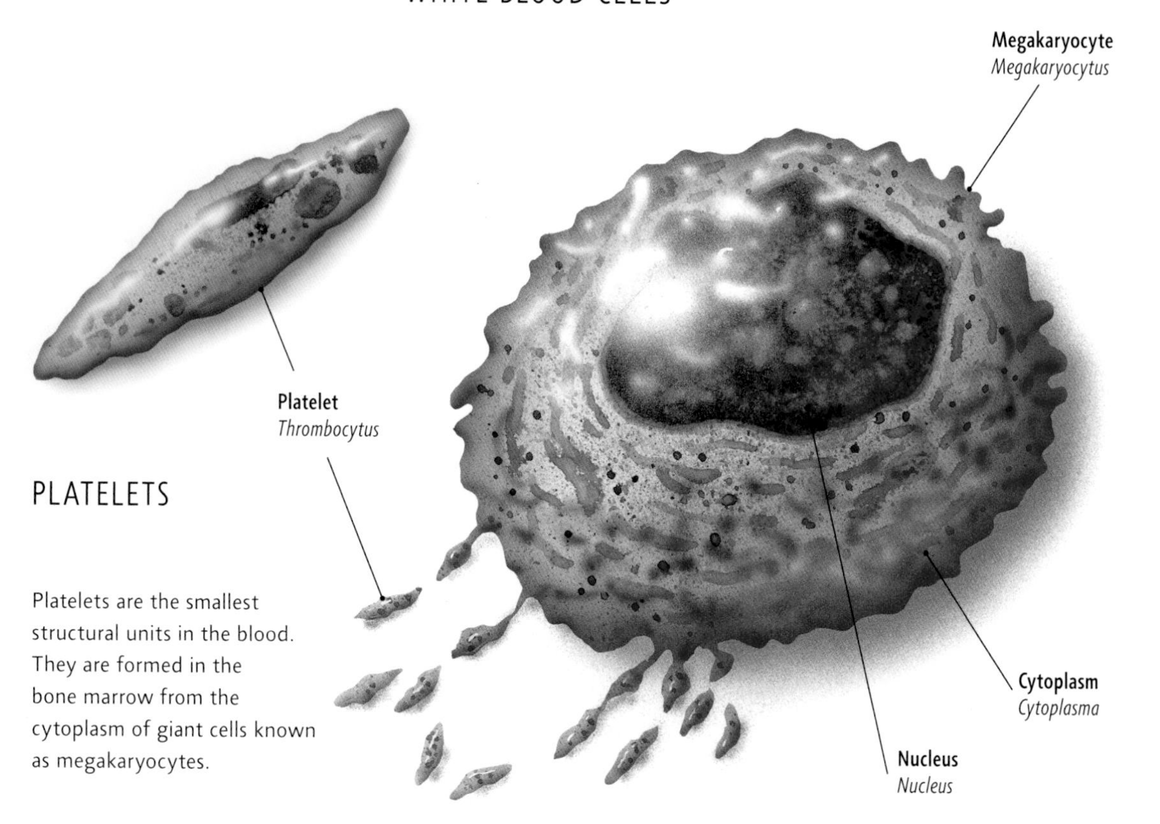
Megakaryocyte
Megakaryocytus
Platelet
Thrombocytus
Cytoplasm
Cytoplasma
Nucleus
Nucleus

PLATELETS

Platelets are the smallest structural units in the blood. They are formed in the bone marrow from the cytoplasm of giant cells known as megakaryocytes.

Healing

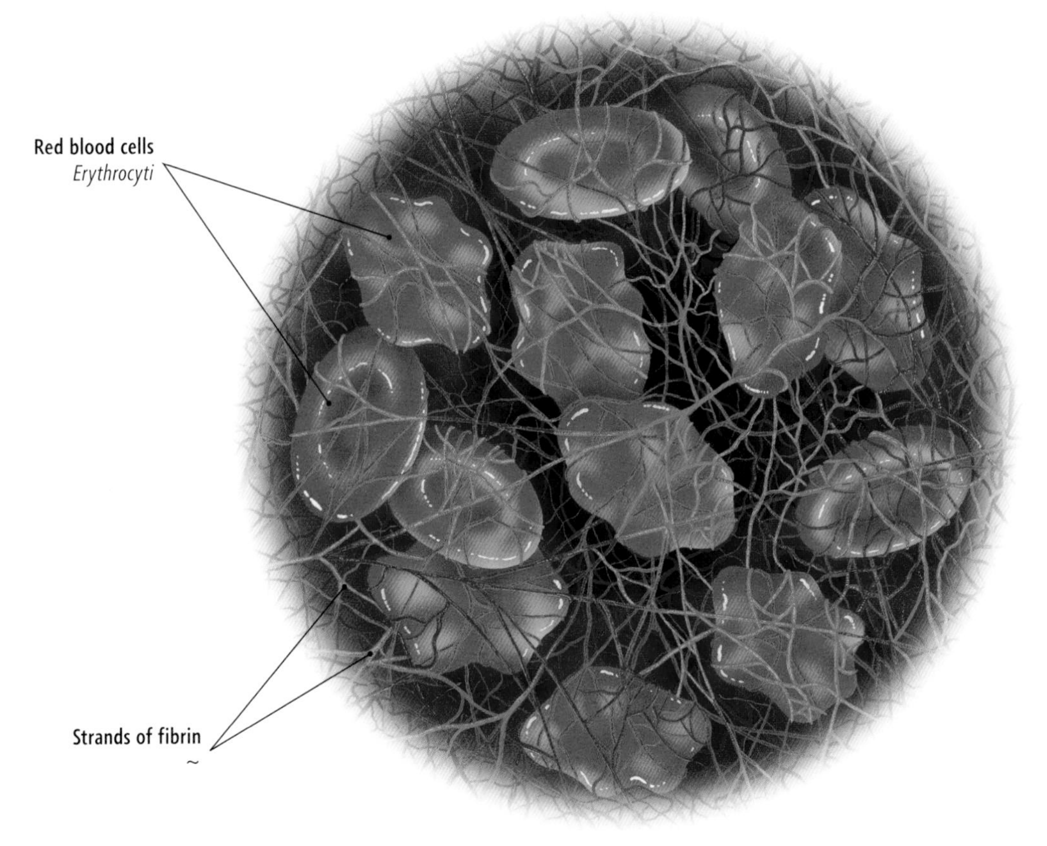
Red blood cells
Erythrocyti
Strands of fibrin
~

BLOOD CLOT

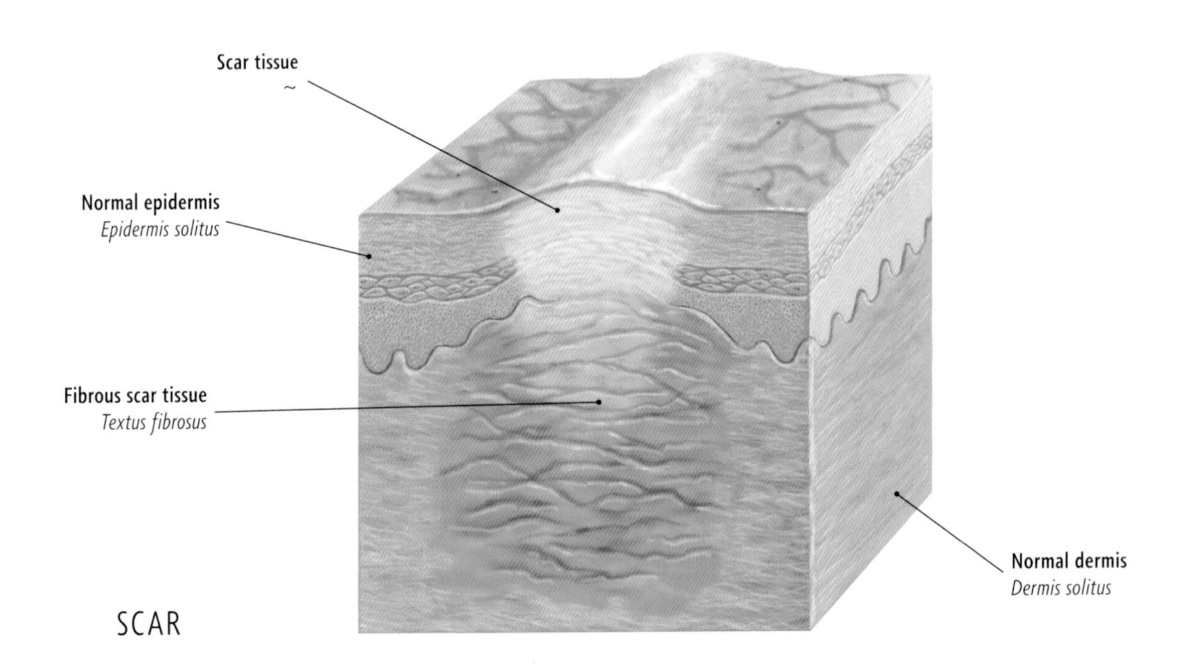
Scar tissue
~
Normal epidermis
Epidermis solitus
Fibrous scar tissue
Textus fibrosus
Normal dermis
Dermis solitus

SCAR

[~] = no direct Latin equivalent

BLOOD CLOTTING

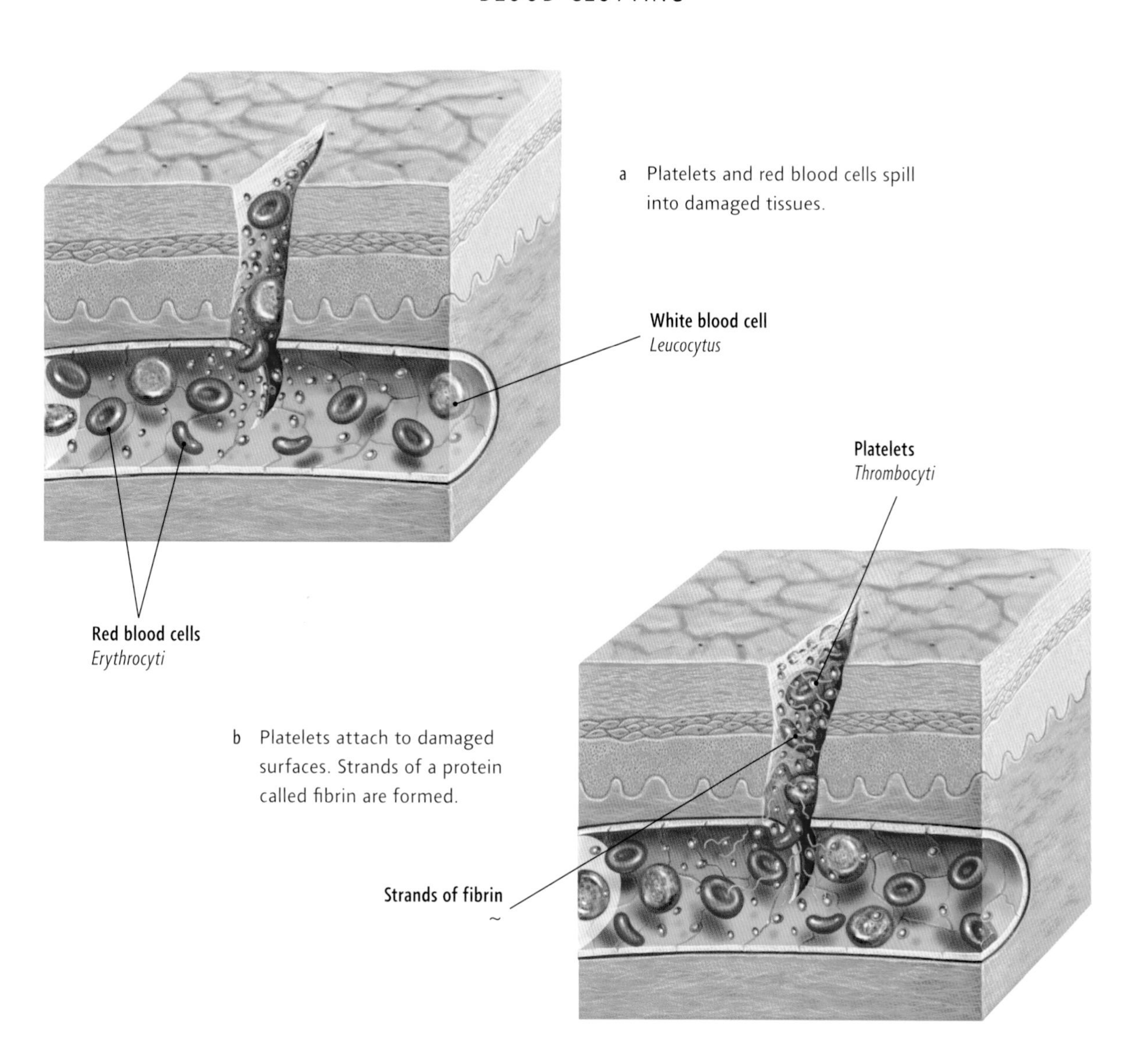
White blood cell
Leucocytus
Red blood cells
Erythrocyti
Platelets
Thrombocyti
Strands of fibrin
~

a Platelets and red blood cells spill into damaged tissues.

b Platelets attach to damaged surfaces. Strands of a protein called fibrin are formed.

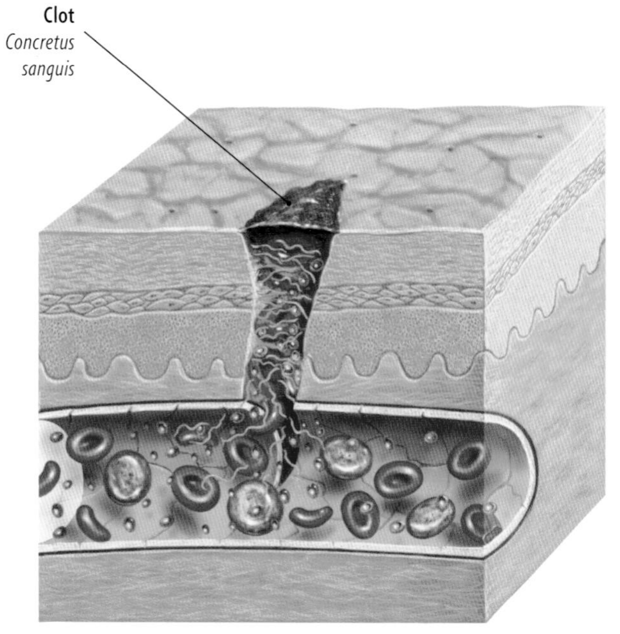
Clot
Concretus sanguis

c Blood cells, platelets, and fibrin strands become entangled and form a clot.

[~] = no direct Latin equivalent

DNA

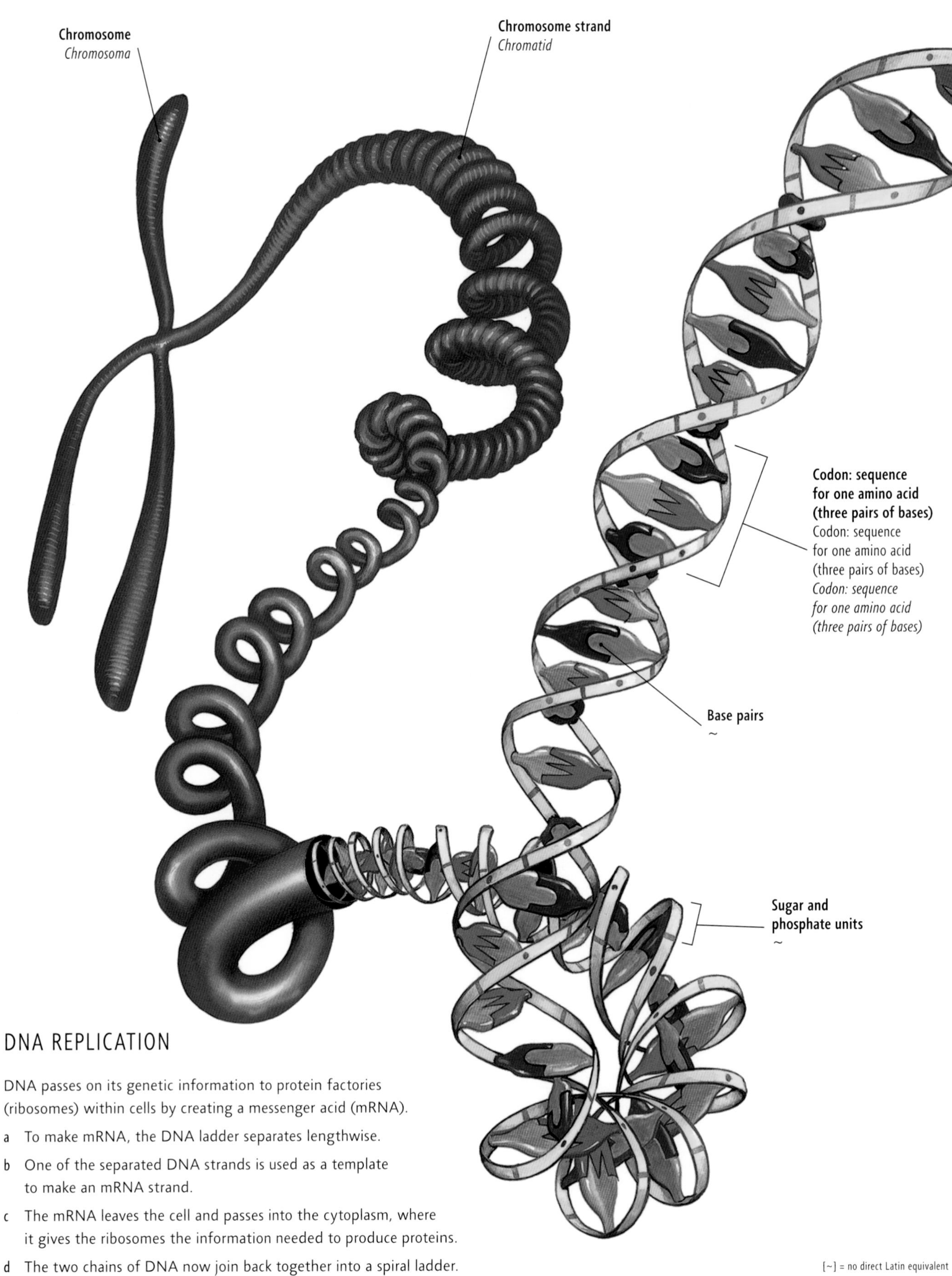
Chromosome
Chromosoma
Chromosome strand
Chromatid
Codon: sequence for one amino acid (three pairs of bases)
Codon: sequence for one amino acid (three pairs of bases)
Codon: sequence for one amino acid (three pairs of bases)
Base pairs
~
Sugar and phosphate units
~

DNA REPLICATION

DNA passes on its genetic information to protein factories (ribosomes) within cells by creating a messenger acid (mRNA).

a To make mRNA, the DNA ladder separates lengthwise.

b One of the separated DNA strands is used as a template to make an mRNA strand.

c The mRNA leaves the cell and passes into the cytoplasm, where it gives the ribosomes the information needed to produce proteins.

d The two chains of DNA now join back together into a spiral ladder.

[~] = no direct Latin equivalent

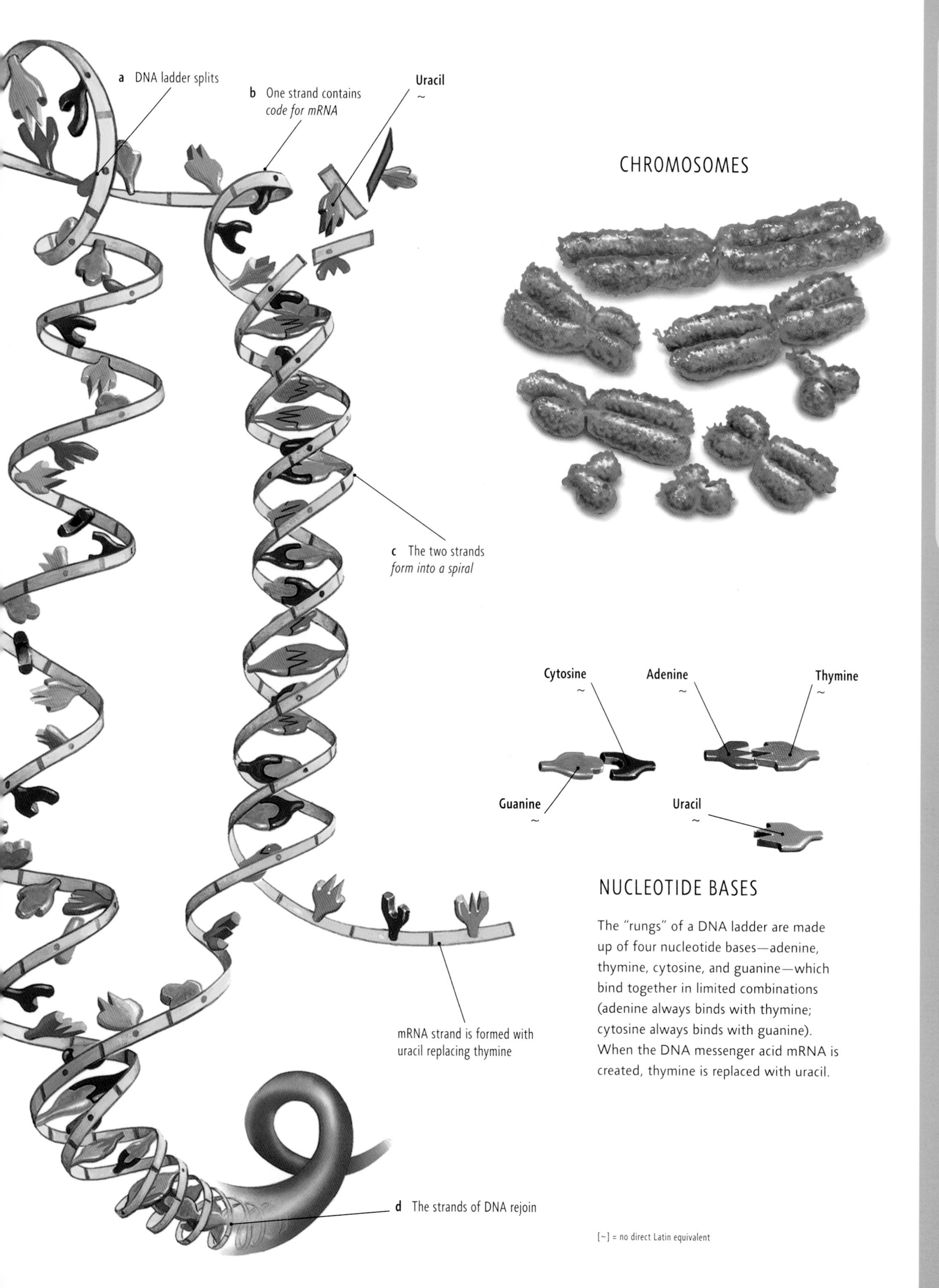
a DNA ladder splits
b One strand contains *code for mRNA*
Uracil
~
c The two strands *form into a spiral*
mRNA strand is formed with uracil replacing thymine
d The strands of DNA rejoin
CHROMOSOMES
Cytosine
~
Adenine
~
Thymine
~
Guanine
~
Uracil
~

NUCLEOTIDE BASES

The "rungs" of a DNA ladder are made up of four nucleotide bases—adenine, thymine, cytosine, and guanine—which bind together in limited combinations (adenine always binds with thymine; cytosine always binds with guanine). When the DNA messenger acid mRNA is created, thymine is replaced with uracil.

[~] = no direct Latin equivalent

Genes and Heredity

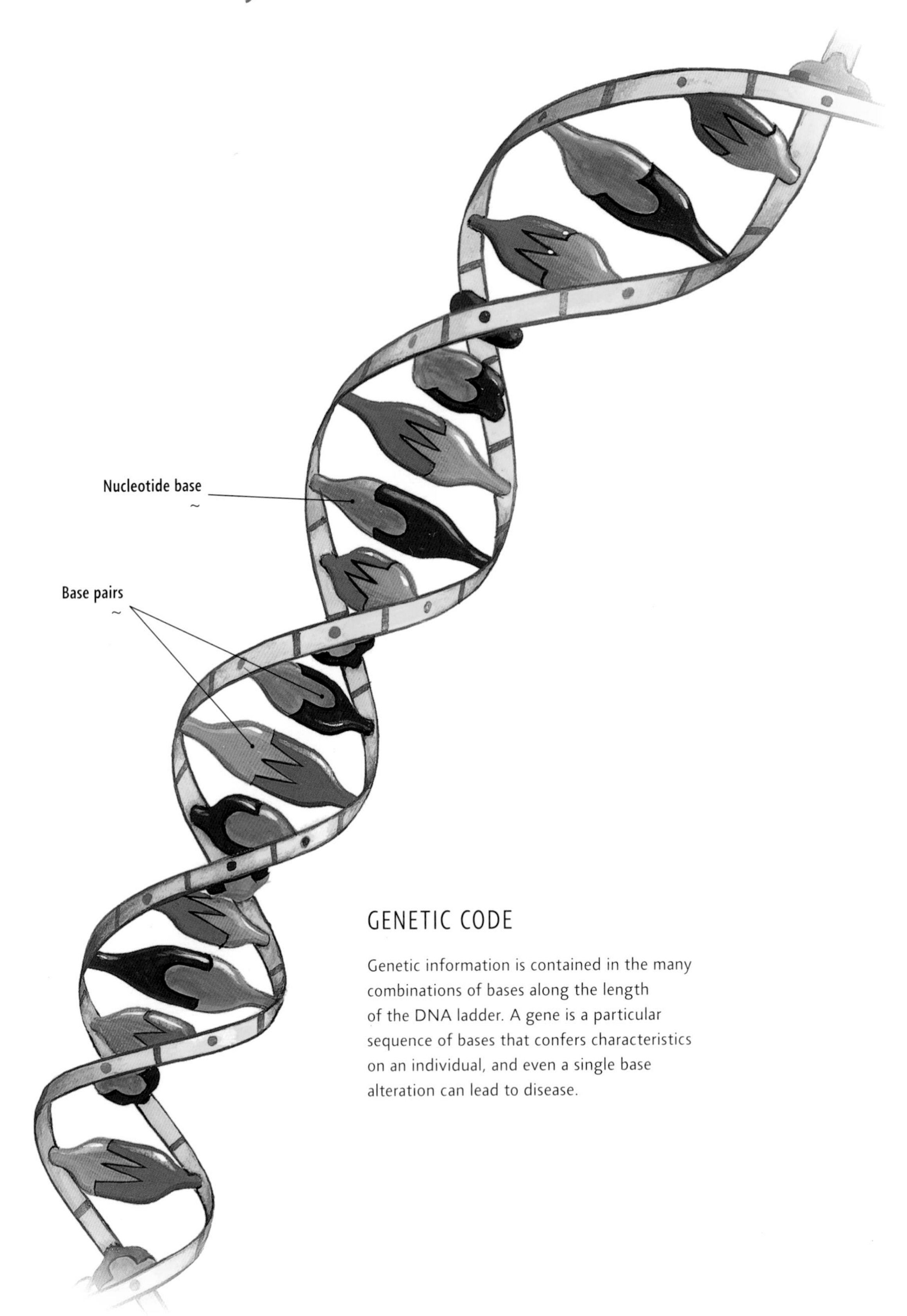
Nucleotide base
~
Base pairs
~

GENETIC CODE

Genetic information is contained in the many combinations of bases along the length of the DNA ladder. A gene is a particular sequence of bases that confers characteristics on an individual, and even a single base alteration can lead to disease.

[~] = no direct Latin equivalent

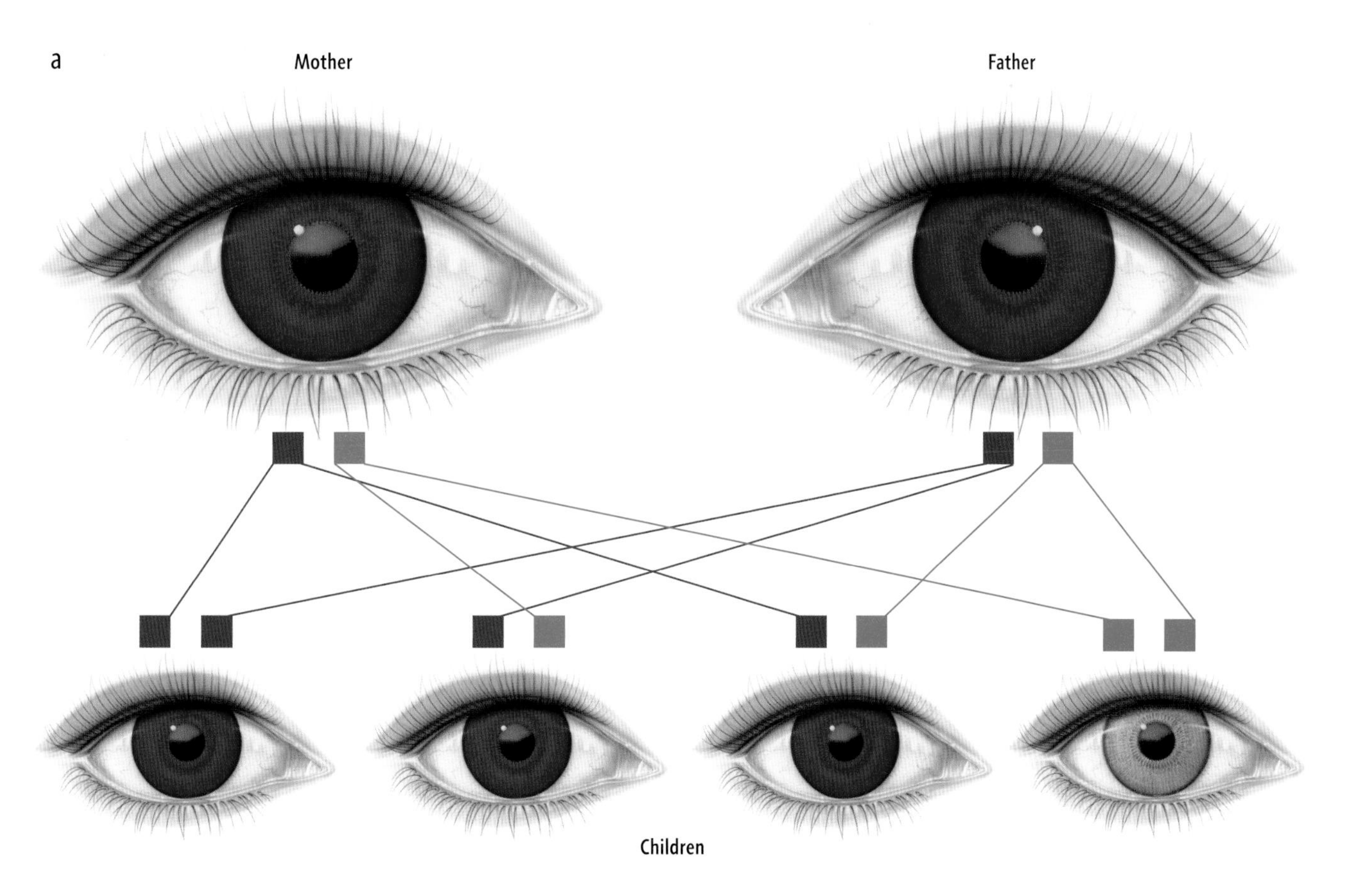
a
Mother
Father
Children

DOMINANT AND RECESSIVE GENES

Some features, such as eye color, are determined by a single gene. The gene for brown eyes is dominant over the gene for blue eyes. Two parents with brown eyes can only have a blue-eyed child if the child inherits a recessive blue gene from both parents (a). If one parent with brown eyes has two dominant brown-eye genes, all children will inherit at least one dominant gene and will have brown eyes (b).

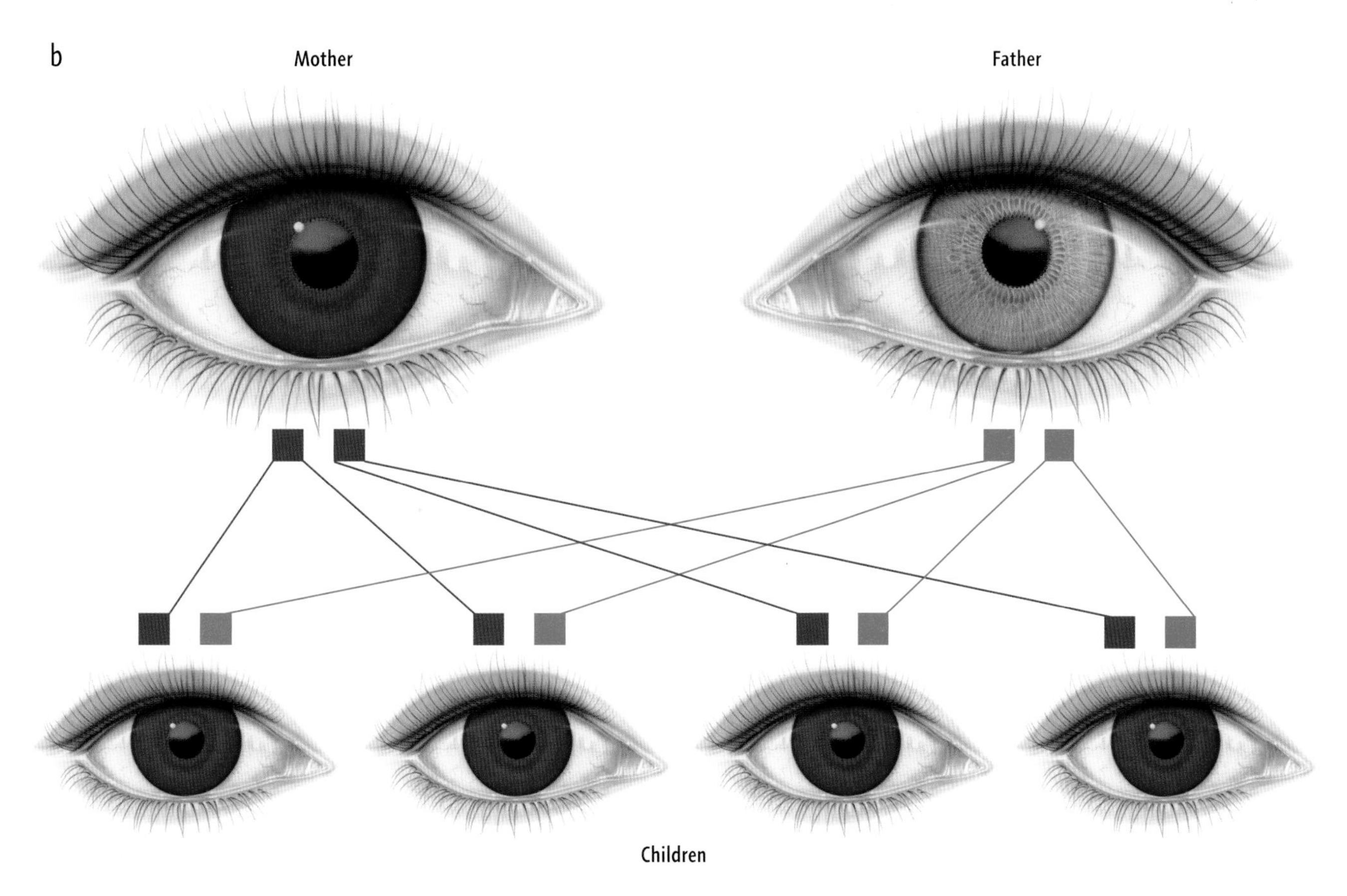
b
Mother
Father
Children

Bacteria

Bacteria are simple organisms of microscopic size. Many are beneficial and live in harmony with humans. Some are harmful and can cause and spread infections such as cholera, syphilis, and food poisoning.

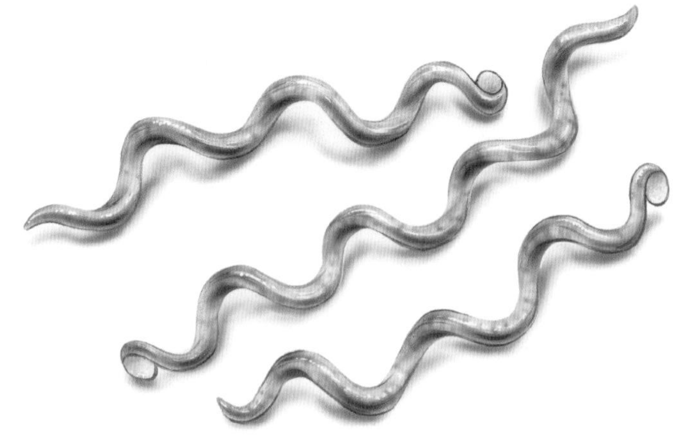

TREPONEMA PALLIDUM
(SYPHILIS)

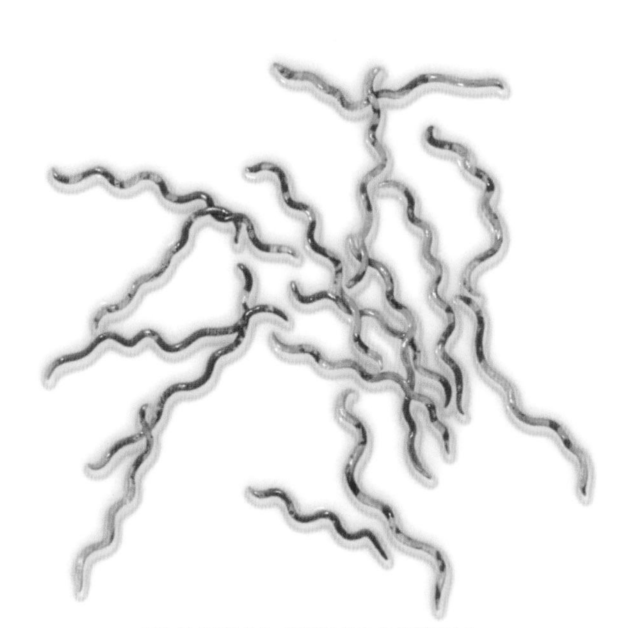

BORRELIA BURGDORFERI
(LYME DISEASE)

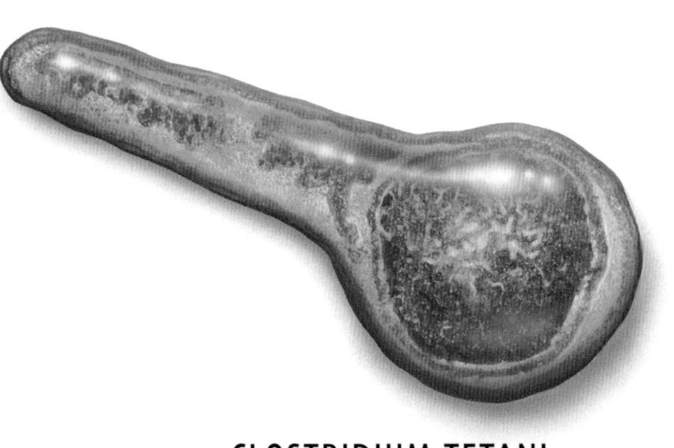

CLOSTRIDIUM TETANI
(TETANUS)

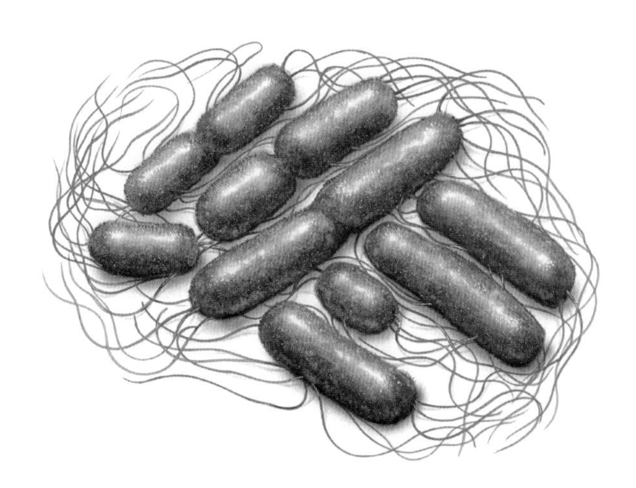

SALMONELLA
(FOOD POISONING)

E. COLI
(FOOD POISONING)

STREPTOCOCCUS
(INFECTION)

NEISSERIA MENINGITIDIS
(MENINGOCOCCAL DISEASE)

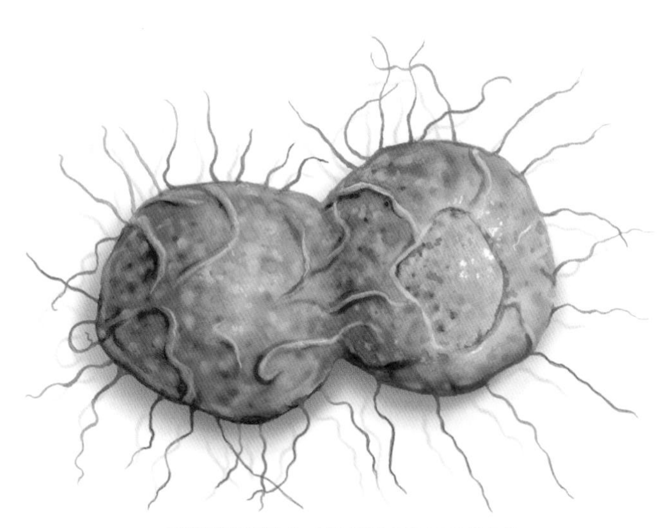

NEISSERIA GONORRHOEAE
(GONORRHEA)

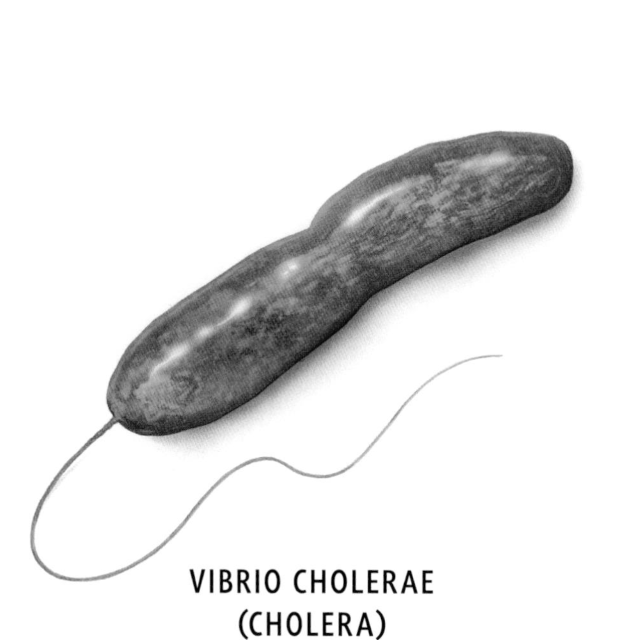

VIBRIO CHOLERAE
(CHOLERA)

CHLAMYDIA TRACHOMATIS
(CHLAMYDIA)

Viruses

These tiny organisms are much smaller than bacteria and vary considerably in shape and structure. Some viruses cause acute disease lasting for only a short time and others cause recurring or chronic disease, while others do not cause any disease.

POLIO VIRUS

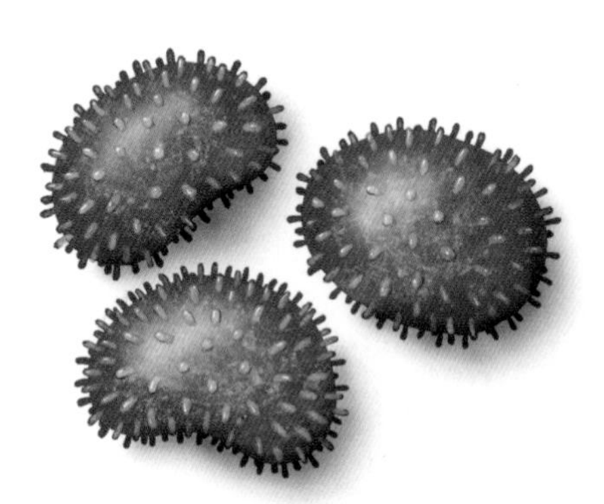

INFLUENZA VIRUS

EBOLA VIRUS

HERPES VIRUS

ROTAVIRUS (GASTROENTERITIS)

COLD VIRUS

HEPATITIS VIRUS

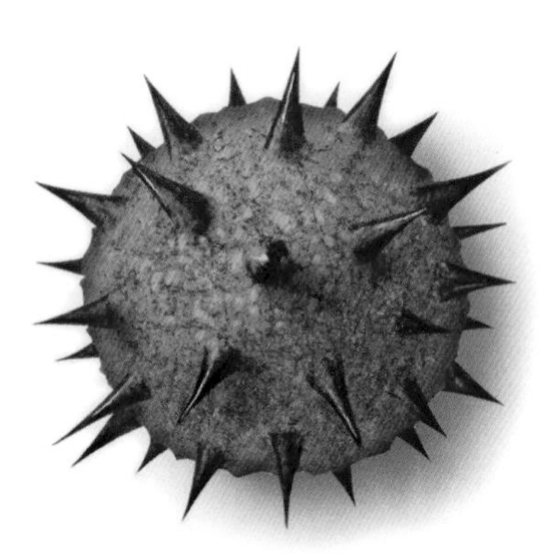

WART VIRUS

SMALLPOX VIRUS

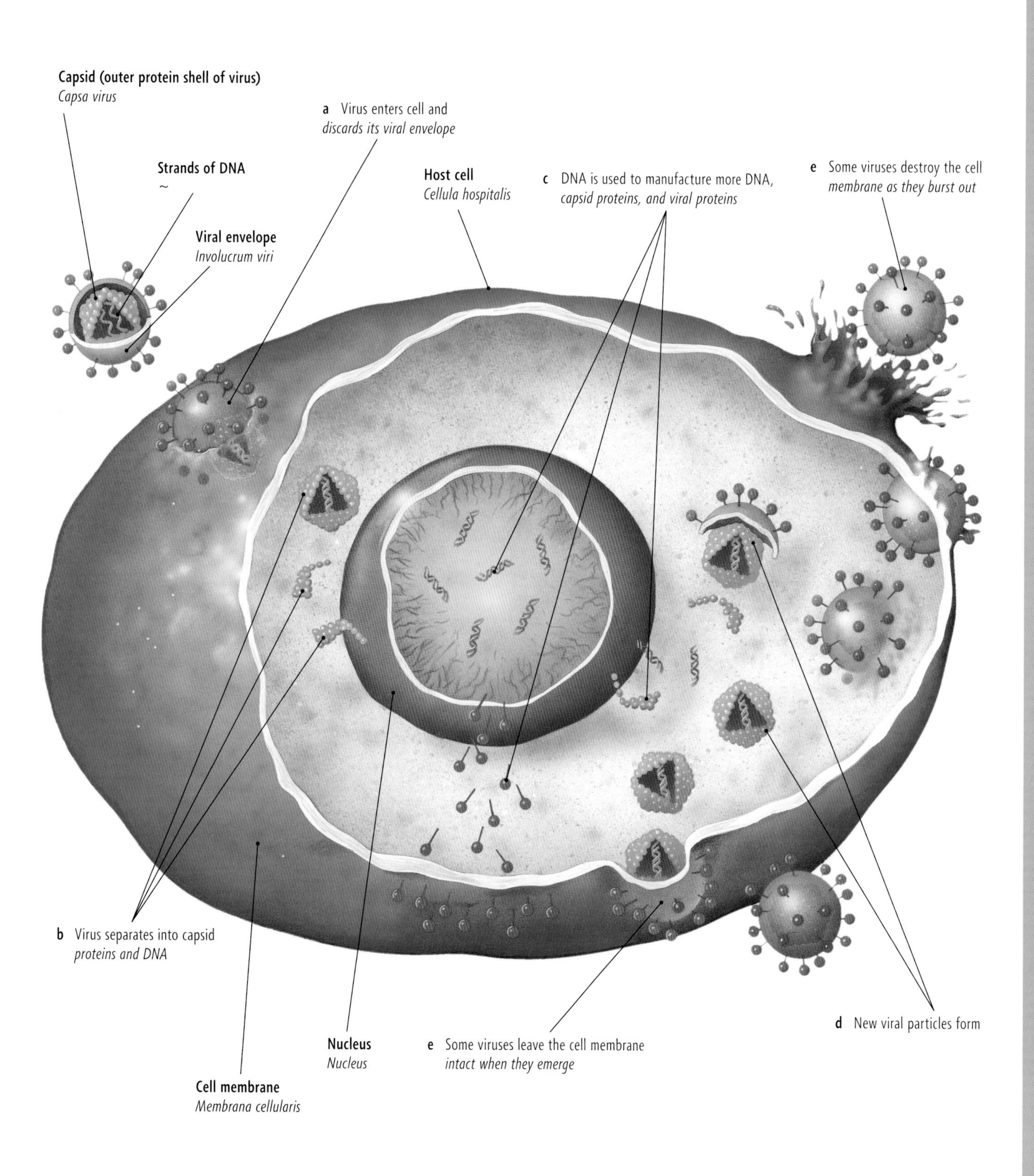
Capsid (outer protein shell of virus)
Capsa virus
a Virus enters cell and
discards its viral envelope
Strands of DNA
~
Host cell
Cellula hospitalis
c DNA is used to manufacture more DNA,
capsid proteins, and viral proteins
e Some viruses destroy the cell
membrane as they burst out
Viral envelope
Involucrum viri
b Virus separates into capsid
proteins and DNA
d New viral particles form
Nucleus
Nucleus
e Some viruses leave the cell membrane
intact when they emerge
Cell membrane
Membrana cellularis

VIRUS REPLICATION

A virus reproduces by using parts of a healthy host cell to create clones of itself. These copies then invade the body, causing illness.

[~] = no direct Latin equivalent

Immunity

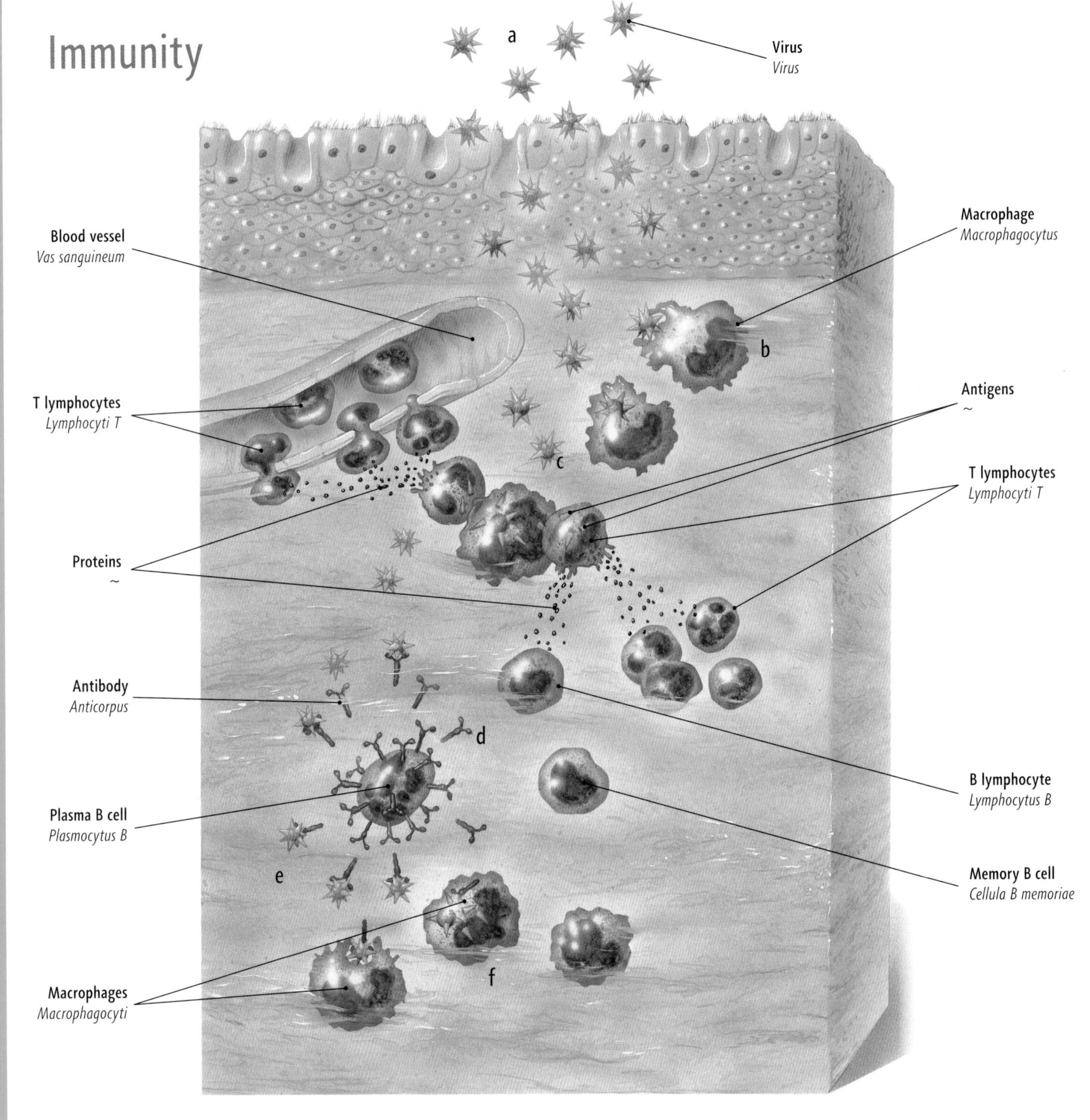
a
Virus
Virus
Blood vessel
Vas sanguineum
Macrophage
Macrophagocytus
b
T lymphocytes
Lymphocyti T
Antigens
~
c
T lymphocytes
Lymphocyti T
Proteins
~
Antibody
Anticorpus
d
B lymphocyte
Lymphocytus B
Plasma B cell
Plasmocytus B
Memory B cell
Cellula B memoriae
e
f
Macrophages
Macrophagocyti

HUMORAL IMMUNE RESPONSE

B lymphocytes (white blood cells) produce antibodies to help identify and eliminate invading antigens (carried by bacteria or viruses). They are helped in the body's defenses by circulating T lymphocytes and macrophages (scavenging white blood cells).

- a Virus particles invade tissue through surface cells and multiply.
- b Virus particles are consumed by macrophages.
- c Macrophages break down the virus and present antigens to circulating T lymphocytes, which release proteins that attract more T and B lymphocytes.
- d B lymphocytes divide into memory B cells (which remember the virus for future attacks) and plasma B cells (which make virus-specific antibodies).
- e Circulating antibodies attach to the virus particles.
- f Macrophages primed to recognize the antibody consume the virus and break it down, saving the body from infection.

[~] = no direct Latin equivalent

ALLERGIC REACTION

The body's exposure to allergens leads to the release of histamine, which causes symptoms such as sneezing and rash.

a On first exposure to allergens, plasma B cells produce antibodies.

b Antibodies attach to mast cells in the body's tissues.

c Subsequent exposure sees allergens captured by antibodies.

d Mast cells respond by releasing histamine.

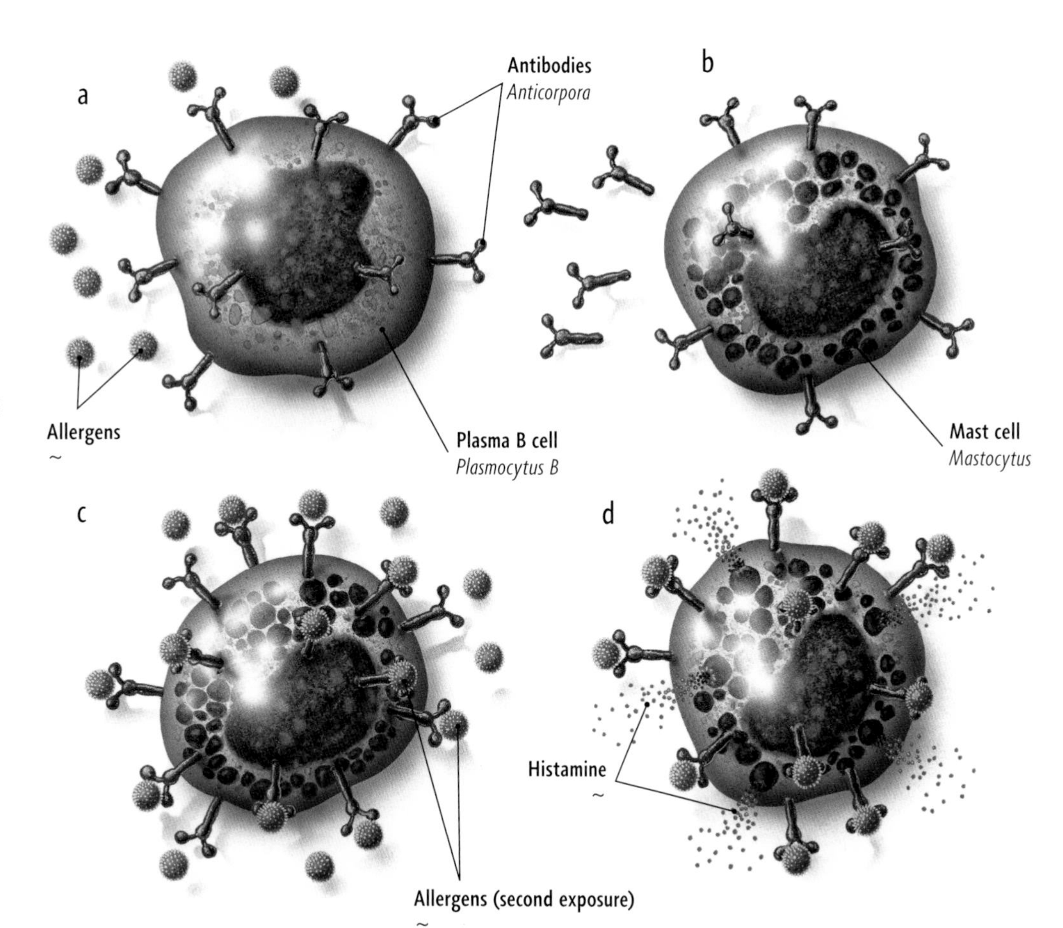
a
b
c
d
Antibodies
Anticorpora
Allergens
~
Plasma B cell
Plasmocytus B
Mast cell
Mastocytus
Histamine
~
Allergens (second exposure)
~

CELL-MEDIATED IMMUNE RESPONSE

T lymphocytes (a type of white blood cell) are responsible for the delayed action of the cell-mediated immune response.

a Circulating macrophages ingest the invading virus.

b Macrophages process the virus and present antigens to T cells.

c T cells produce clones that each play a special role in the immune response: memory T cells remember the invading antigen for future attacks; helper T cells recruit B and T cells to the site of the antigen attack; suppressor T cells inhibit the action of B and T cells; and killer T cells attach to invading antigens and destroy them.

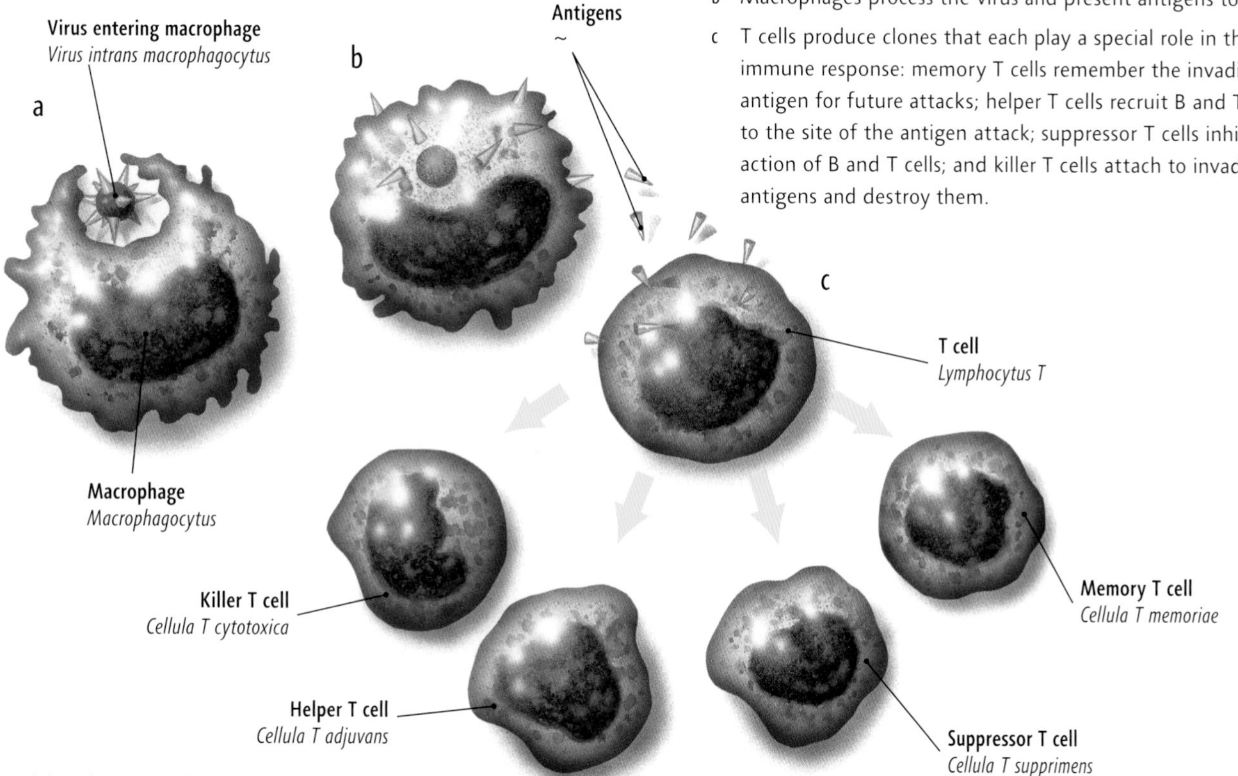
Virus entering macrophage
Virus intrans macrophagocytus
a
b
Antigens
~
c
T cell
Lymphocytus T
Macrophage
Macrophagocytus
Killer T cell
Cellula T cytotoxica
Memory T cell
Cellula T memoriae
Helper T cell
Cellula T adjuvans
Suppressor T cell
Cellula T supprimens

[~] = no direct Latin equivalent

Tissues

LOOSE CONNECTIVE TISSUE

Connective tissue is the framework that supports, connects, and fills out body structures.

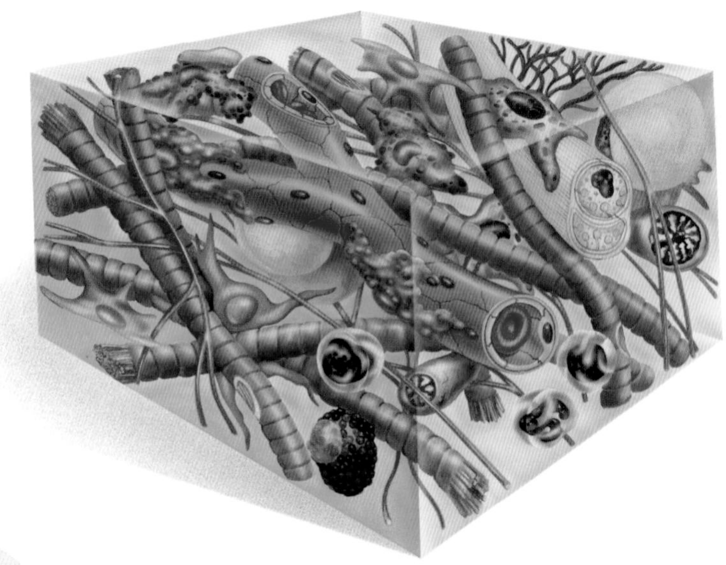

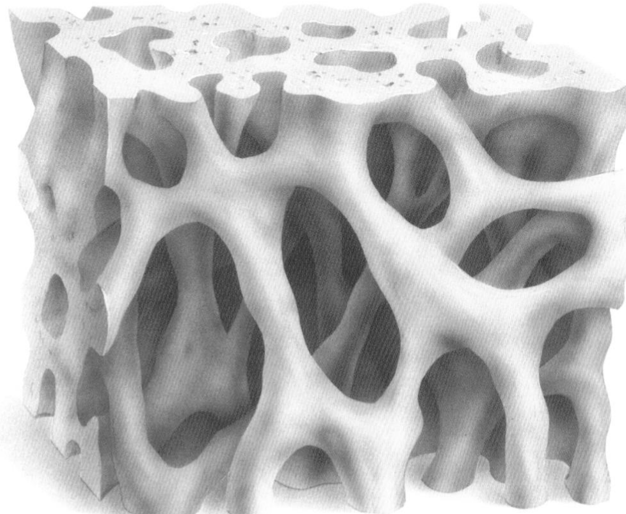

BONE TISSUE

Spongy or cancellous bone consists of a lattice-work system of bony spikes called trabeculae, arranged in different directions.

ADIPOSE TISSUE

Adipose tissue is a specialized connective tissue that stores fat.

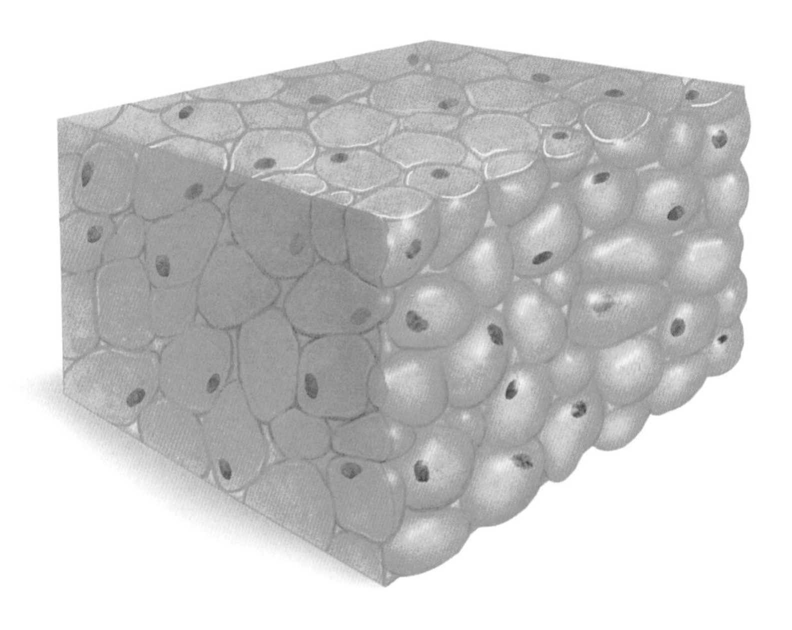

EPITHELIAL TISSUE

Formed from cells that are packed closely together, epithelial tissue is found in the outermost layer of the skin and in some internal organs.

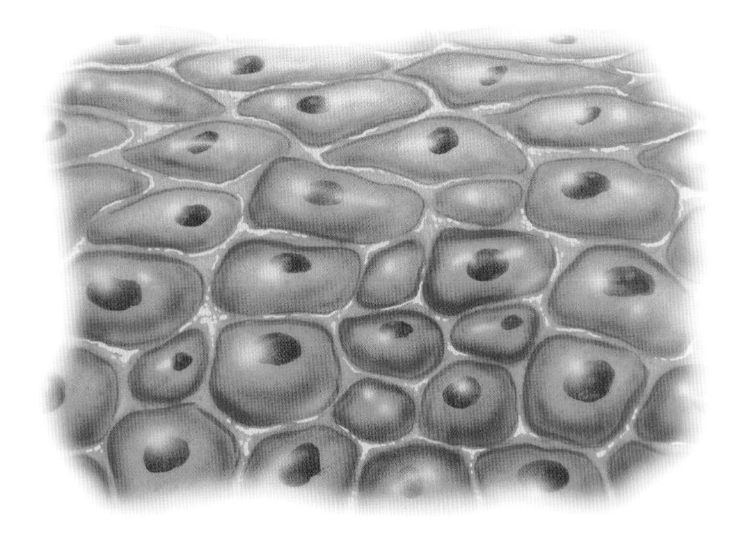

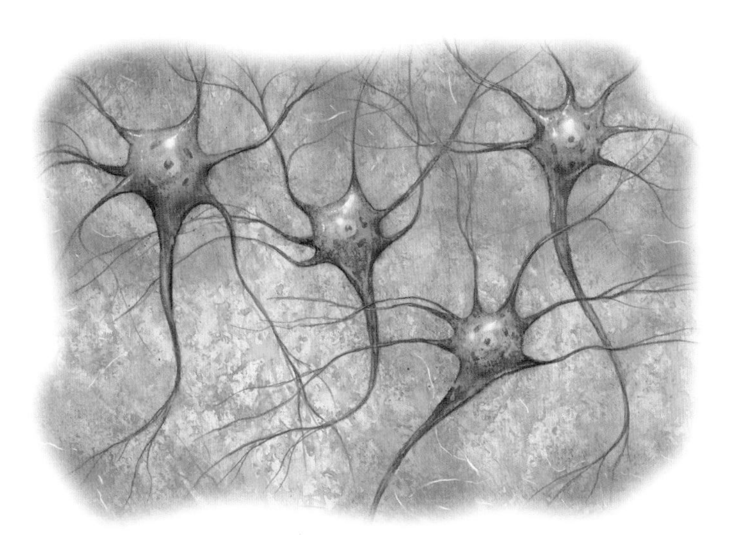

NEURAL TISSUE

The nervous system is made up of neural tissue, which transmits messages to and from the brain.

LYMPHATIC TISSUE

Found at the entrances to the respiratory system and in the digestive and urogenital tracts, lymphatic tissue acts as a first line of defense against infection.

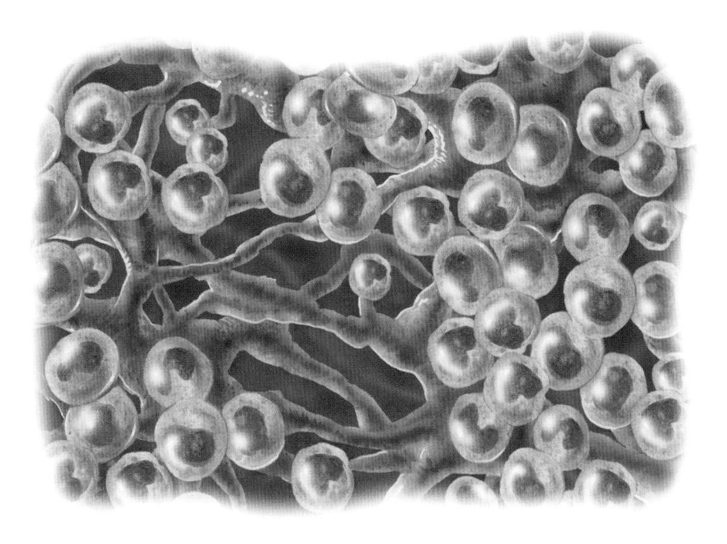

Tissues

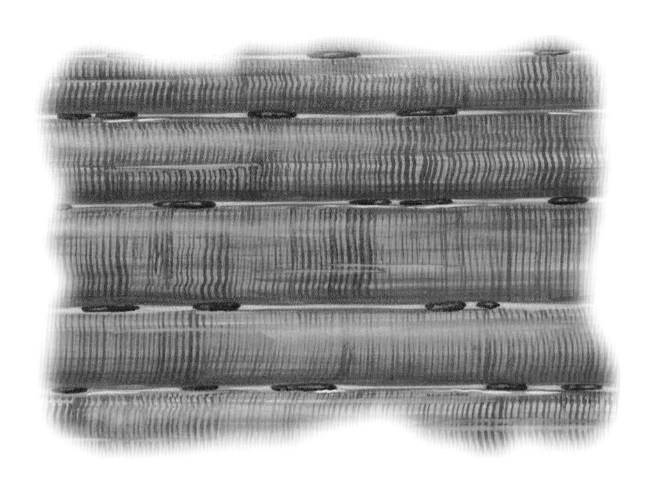

MUSCLE TISSUE: SKELETAL MUSCLE

Featuring long, cylindrical muscle fibers, skeletal muscle is usually attached to bones via tendons.

MUSCLE TISSUE: SMOOTH MUSCLE

Smooth muscle is found in the walls of blood vessels, in airways, and inside the eye.

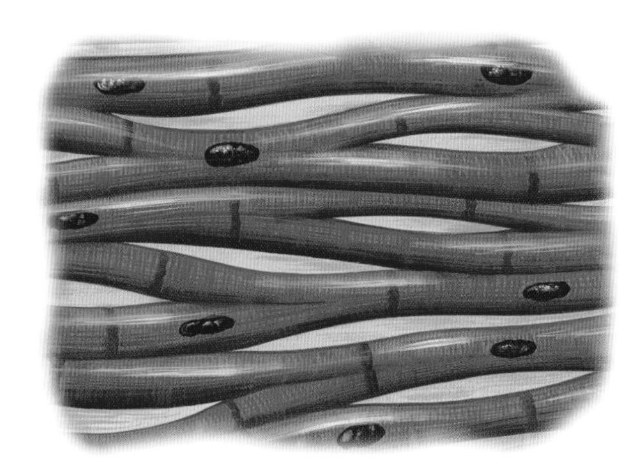

MUSCLE TISSUE: CARDIAC MUSCLE

Found only in the heart, cardiac muscle cannot regenerate after being destroyed.

TENDON TISSUE (RELAXED)

Tendons are constructed primarily of collagen fibers arranged in a regular formation, which provide the strength to attach muscles to bones.

CARTILAGE TISSUE: HYALINE CARTILAGE

Found in many places in the body, hyaline cartilage forms the skeleton in the embryo, the end of the nose and ribs, and the rings around the windpipe.

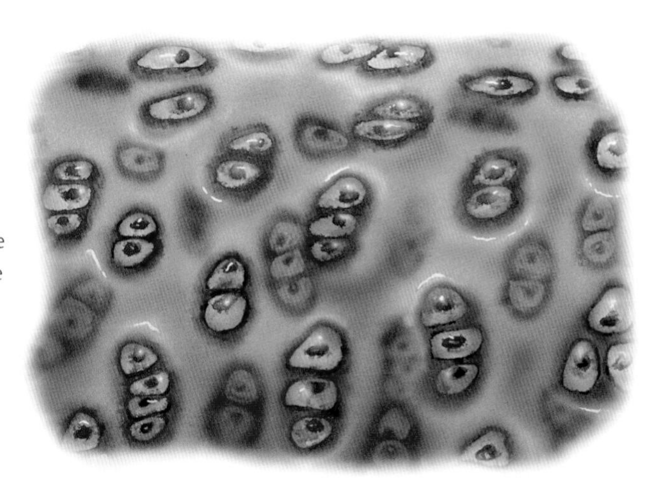

CARTILAGE TISSUE: FIBROCARTILAGE

Resilient and able to withstand compression, fibrocartilage is located between the bones of the spinal column, hip, and pelvis.

CARTILAGE TISSUE: ELASTIC CARTILAGE

Elastic cartilage is strong but supple, and makes up the epiglottis and the springy part of the outer ear.

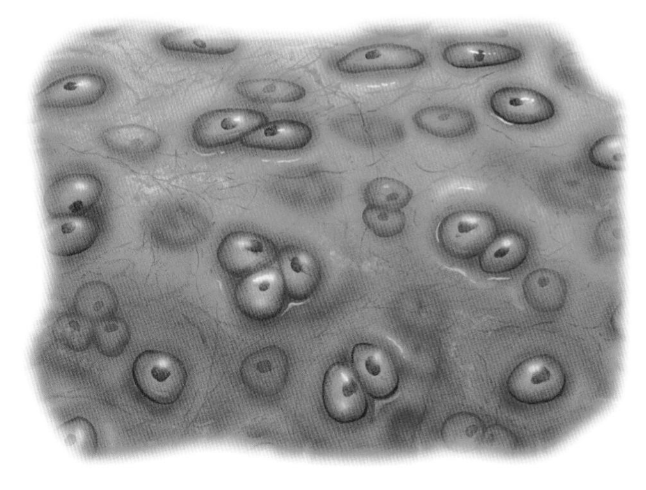

LIGAMENT TISSUE (TIGHT)

Ligaments are tough, white, fibrous, slightly elastic tissues that mainly support and strengthen joints.

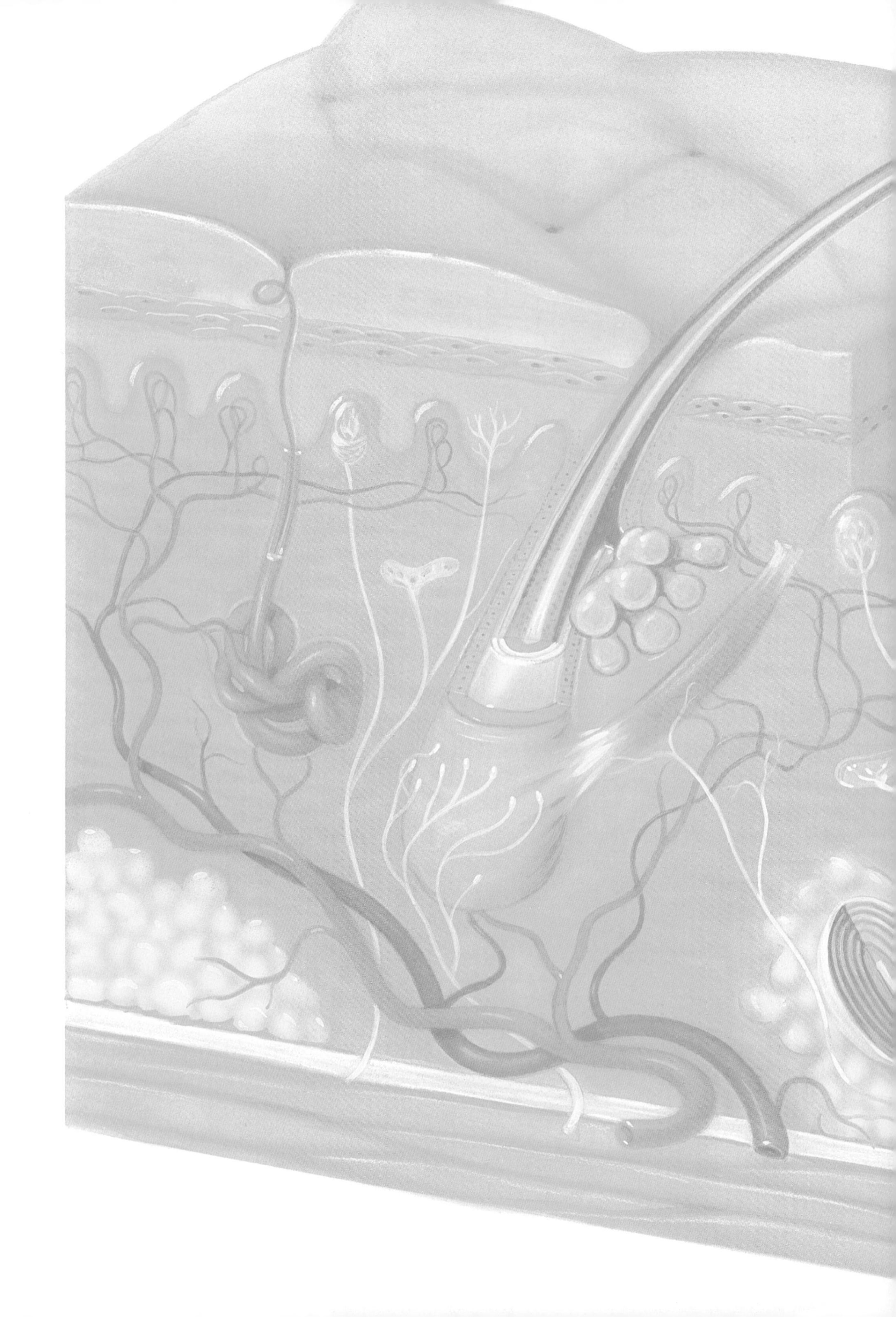

The Skin, Nails, and Hair

The Skin

Horny layer
Stratum corneum

Hair
Pilus

Free nerve ending
Terminationes nervorum liberum

Meissner corpuscles
Corpusculi tactile

Ruffini endings
Corpusculi sensorii fusiformia

Stratum granulosum
Stratum granulosum

Epidermis
Epidermis

Stratum spinosum
Stratum spinosum

Sebaceous gland
Glandula sebacea

Dermis
Dermis

Subcutaneous fat
Panniculus adiposus

Deep fascia
Fascia profunda

Subcutaneous fat
Panniculus adiposus

Sweat gland
Glandula sudorifera

Nerve endings
Terminationes nervorum

Hair follicle
Folliculus pili

Pacinian corpuscle
Corpusculum lamellosum

Krause bulb
C. bulboideum

Dermal papilla
Papilla dermis

SKIN—CROSS-SECTIONAL VIEW

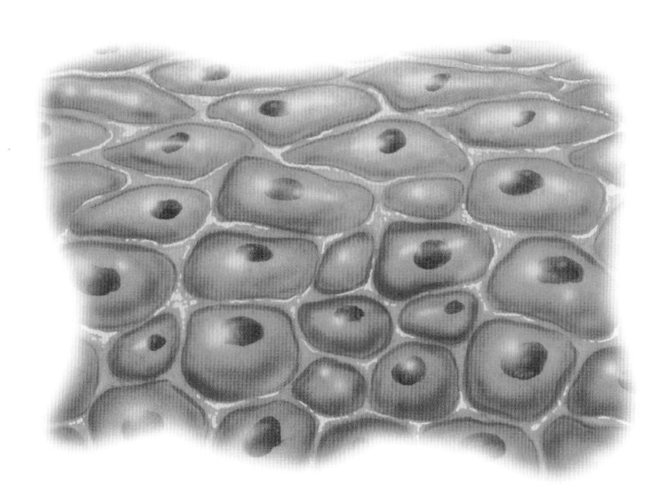

STRATIFIED SQUAMOUS SKIN CELLS

Near the surface of the skin (the horny layer), cells are flattened. The arrangement of cell layers provides a protective shield and prevents dehydration.

SWEAT GLANDS

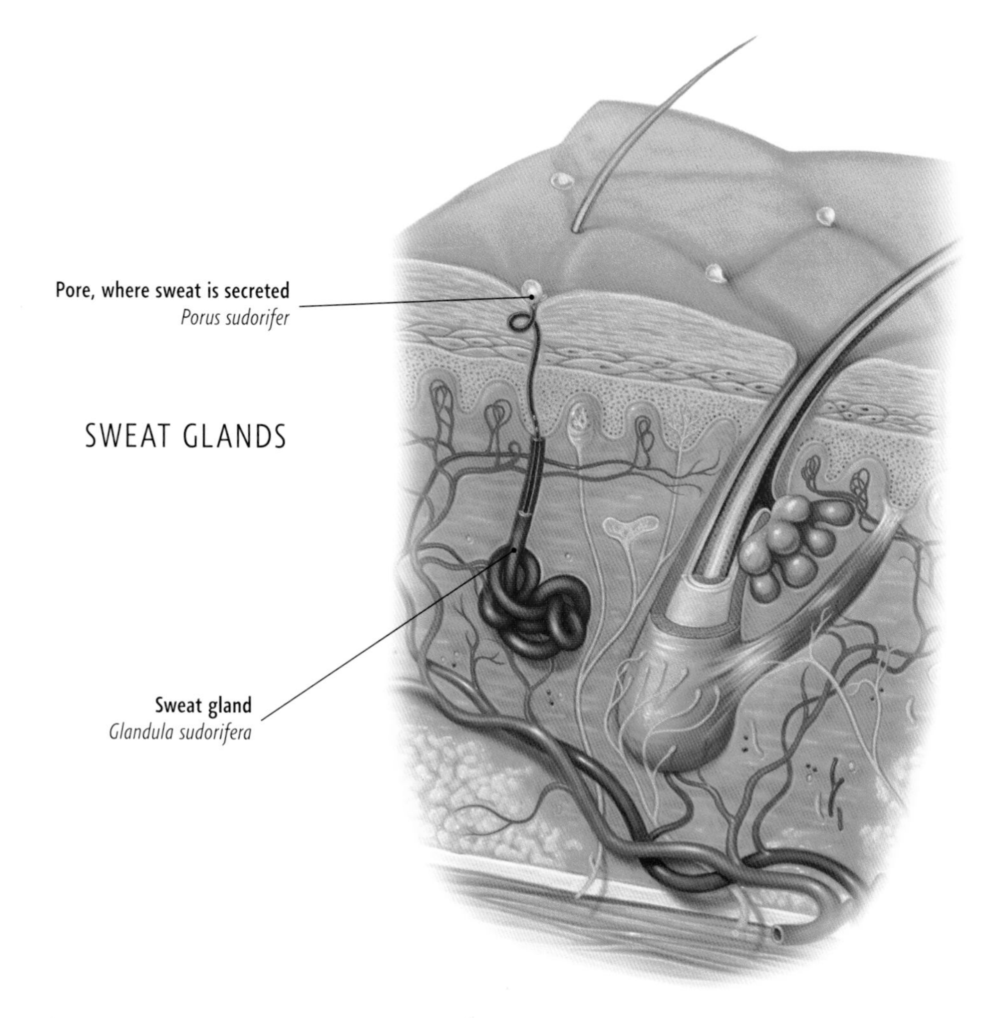

Pore, where sweat is secreted
Porus sudorifer
Sweat gland
Glandula sudorifera

SEBACEOUS GLANDS

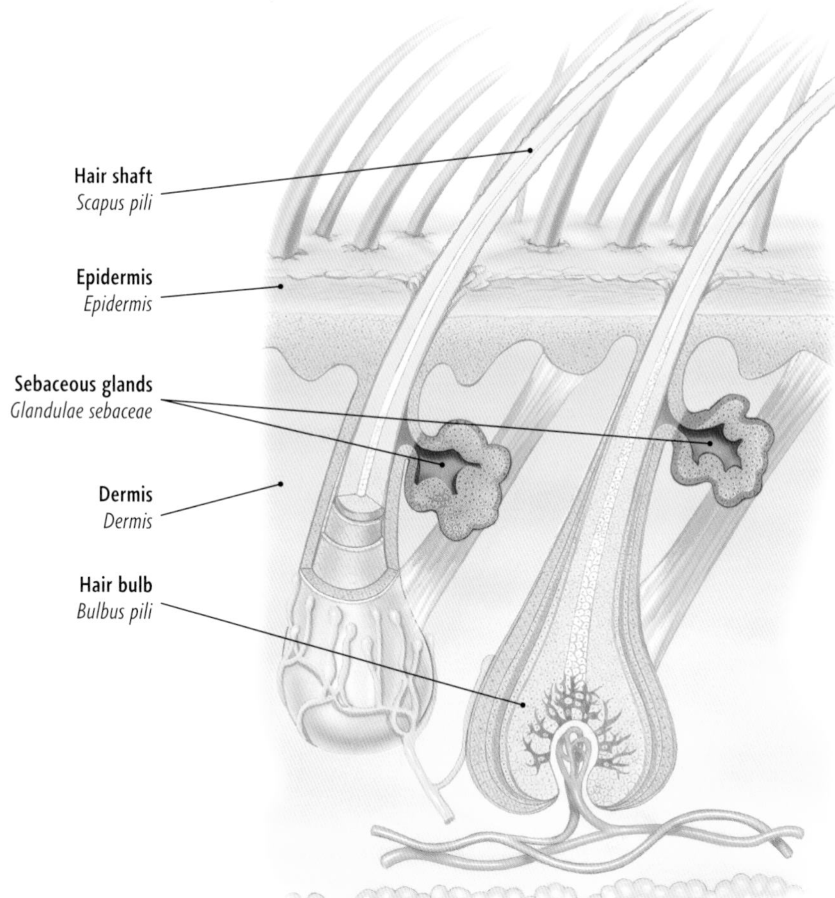

Hair shaft
Scapus pili
Epidermis
Epidermis
Sebaceous glands
Glandulae sebaceae
Dermis
Dermis
Hair bulb
Bulbus pili

[~] = no direct Latin equivalent

Skin and Temperature

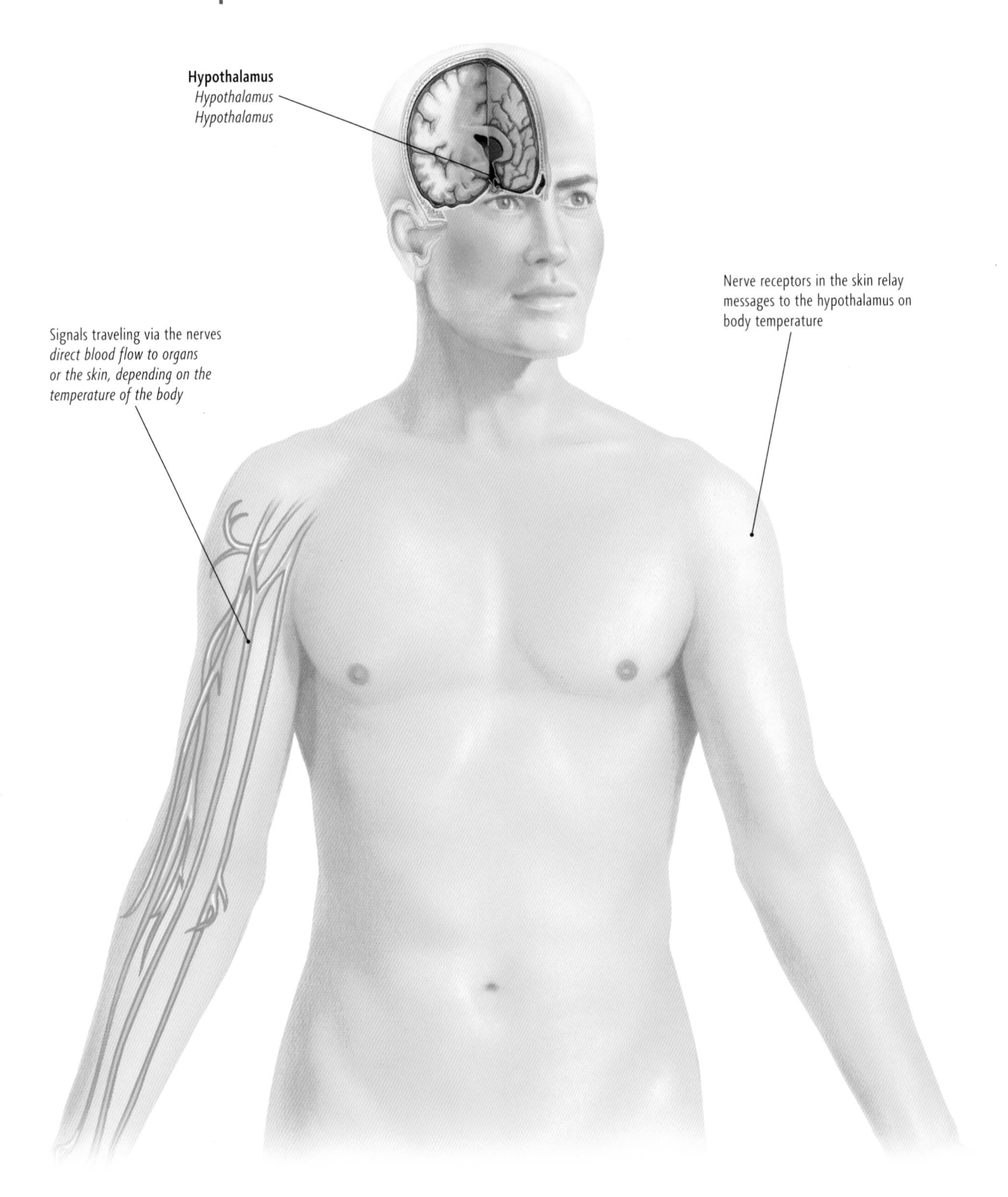
Hypothalamus
Hypothalamus
Hypothalamus
Nerve receptors in the skin relay messages to the hypothalamus on body temperature
Signals traveling via the nerves *direct blood flow to organs or the skin, depending on the temperature of the body*

TEMPERATURE REGULATION

The body has a built-in mechanism regulated by the hypothalamus for maintaining a stable temperature. Nerve endings in the skin relay temperature changes to the hypothalamus. If the body is cold, the hypothalamus increases heat production in the body by increasing the metabolic rate. If the body is hot, the hypothalamus sends blood to the skin, where heat can be lost through radiation, conduction, convection, and evaporation.

SKIN REACTION TO COLD

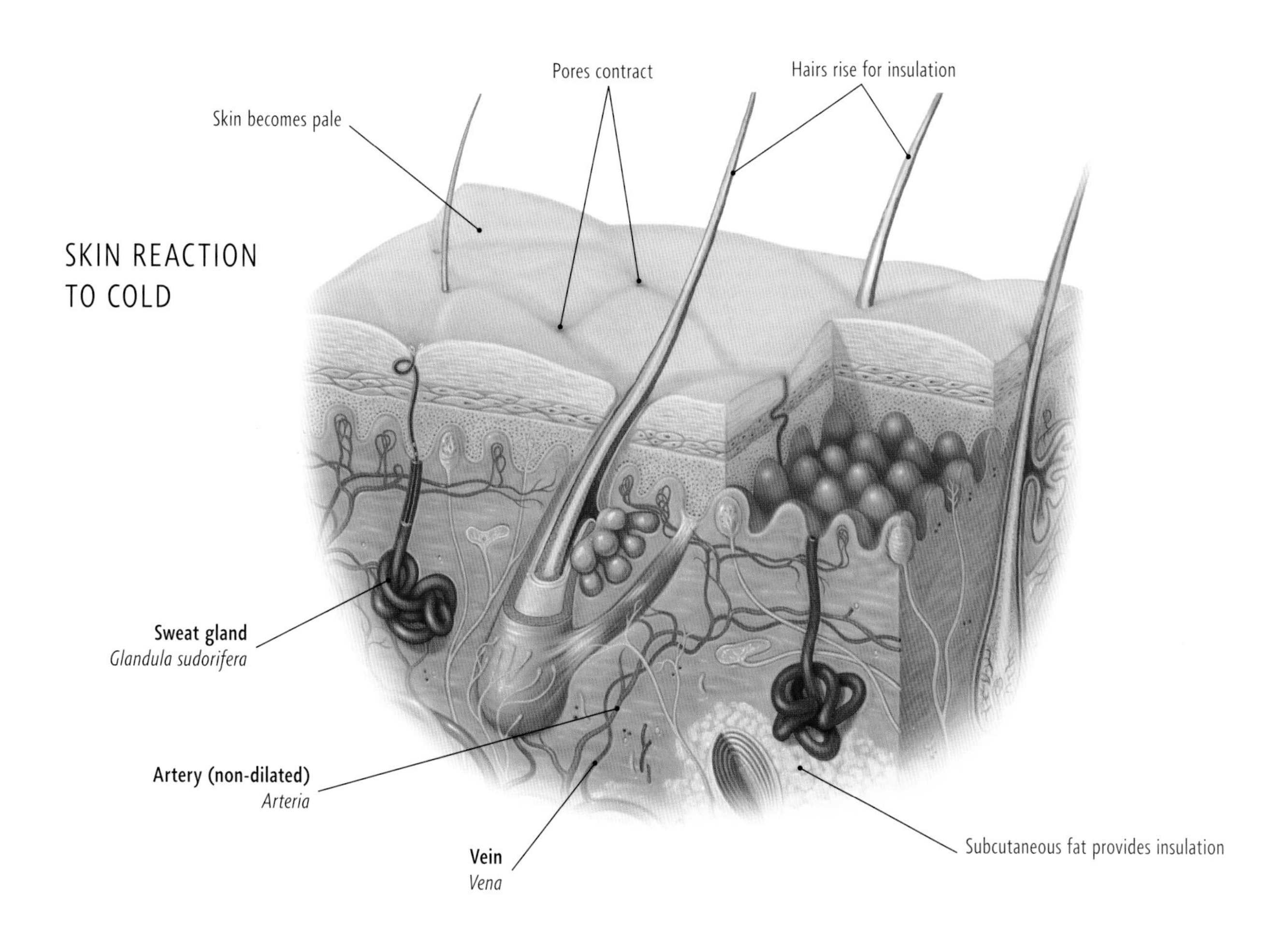
Pores contract
Hairs rise for insulation
Skin becomes pale
Sweat gland
Glandula sudorifera
Artery (non-dilated)
Arteria
Vein
Vena
Subcutaneous fat provides insulation

SKIN REACTION TO HEAT

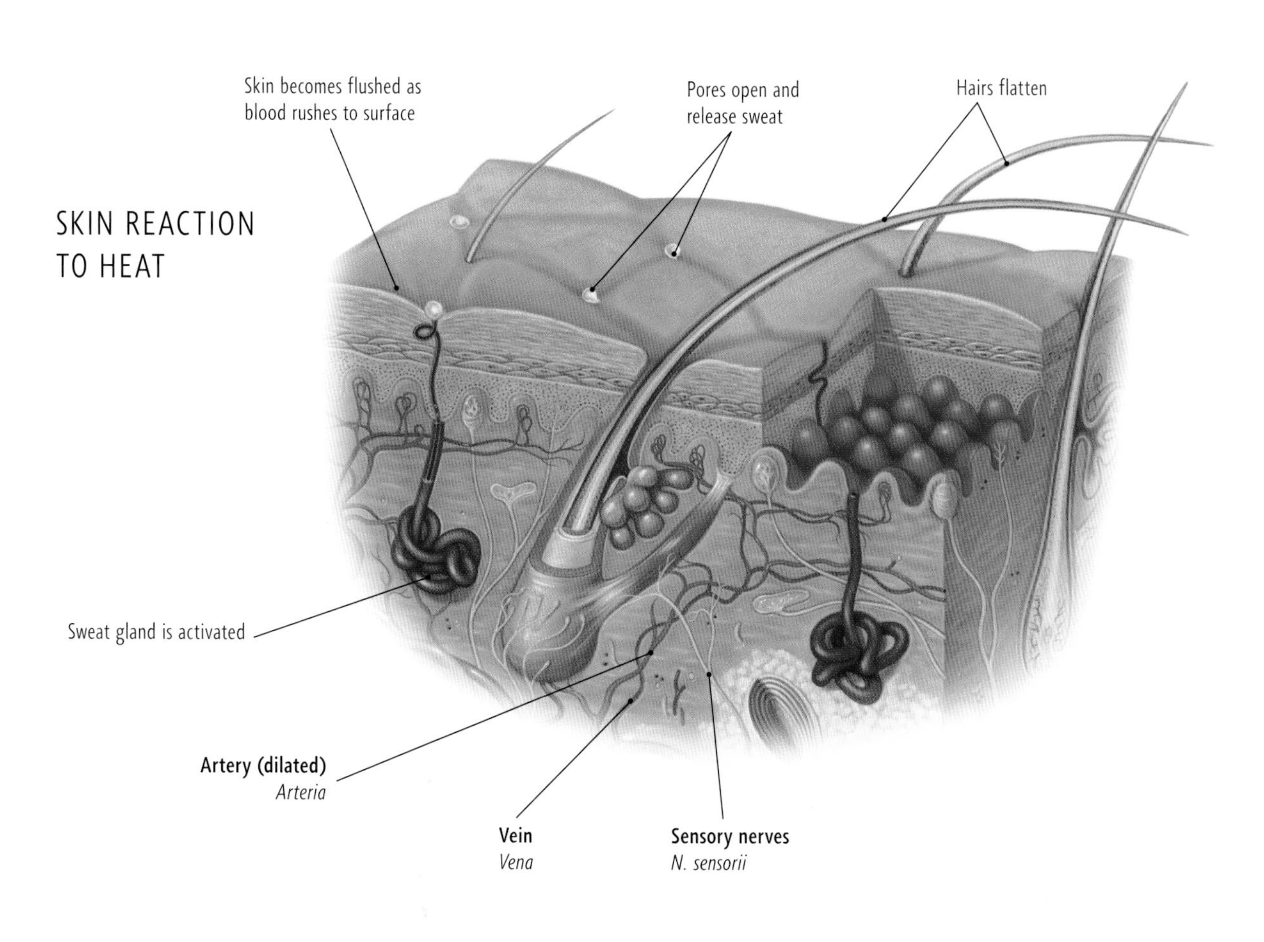
Skin becomes flushed as blood rushes to surface
Pores open and release sweat
Hairs flatten
Sweat gland is activated
Artery (dilated)
Arteria
Vein
Vena
Sensory nerves
N. sensorii

The Hair and Nails

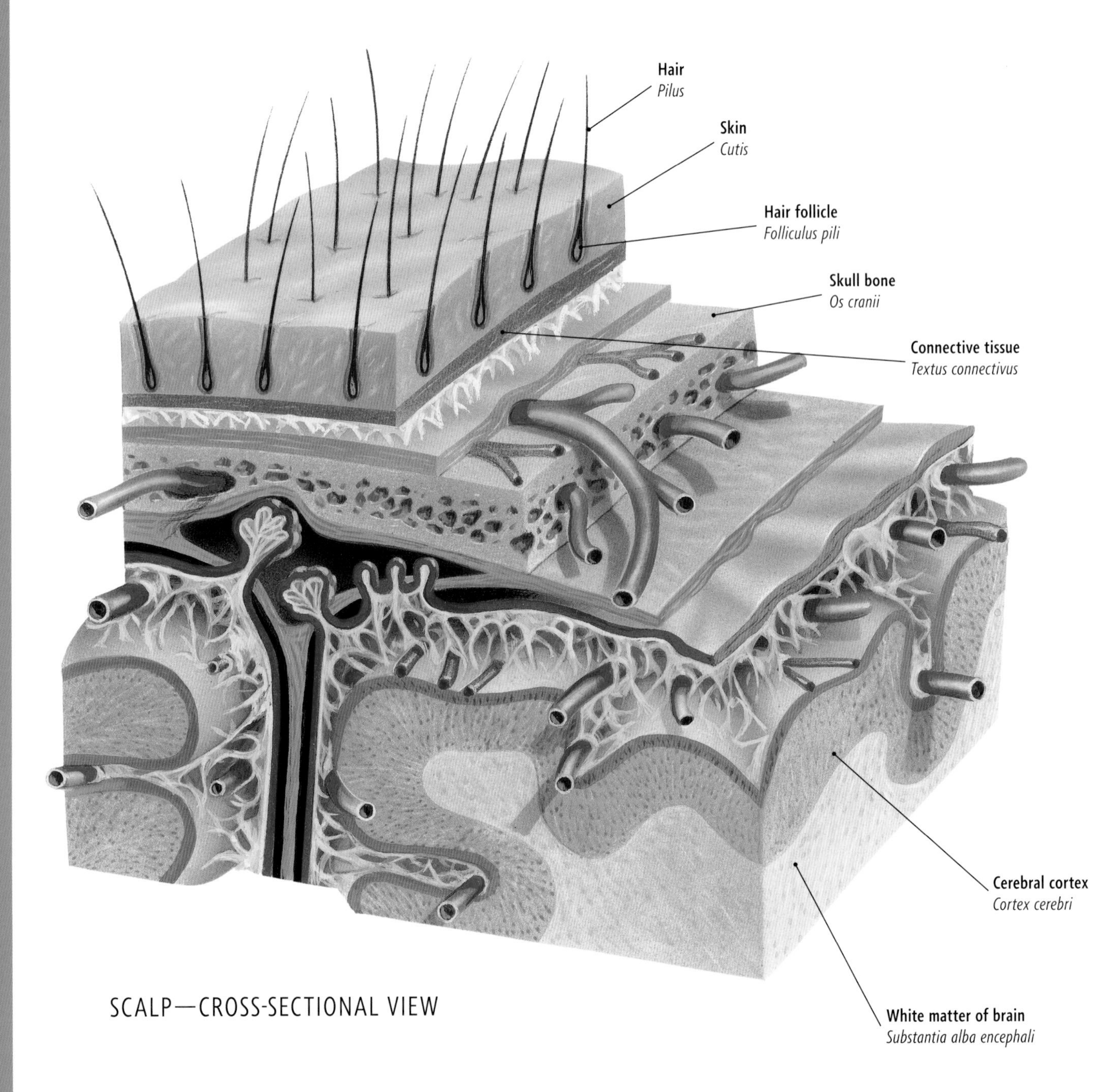
Hair
Pilus
Skin
Cutis
Hair follicle
Folliculus pili
Skull bone
Os cranii
Connective tissue
Textus connectivus
Cerebral cortex
Cortex cerebri
White matter of brain
Substantia alba encephali

SCALP—CROSS-SECTIONAL VIEW

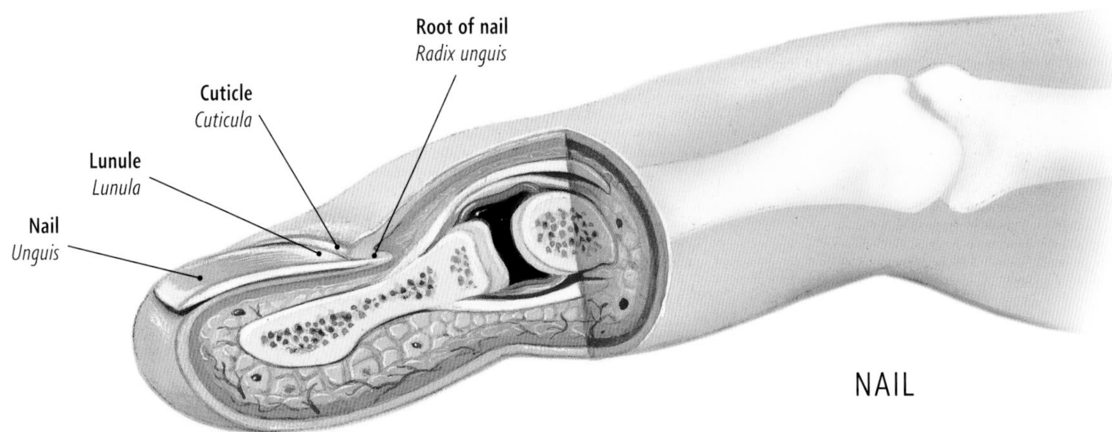
Root of nail
Radix unguis
Cuticle
Cuticula
Lunule
Lunula
Nail
Unguis

NAIL

HAIR FOLLICLE

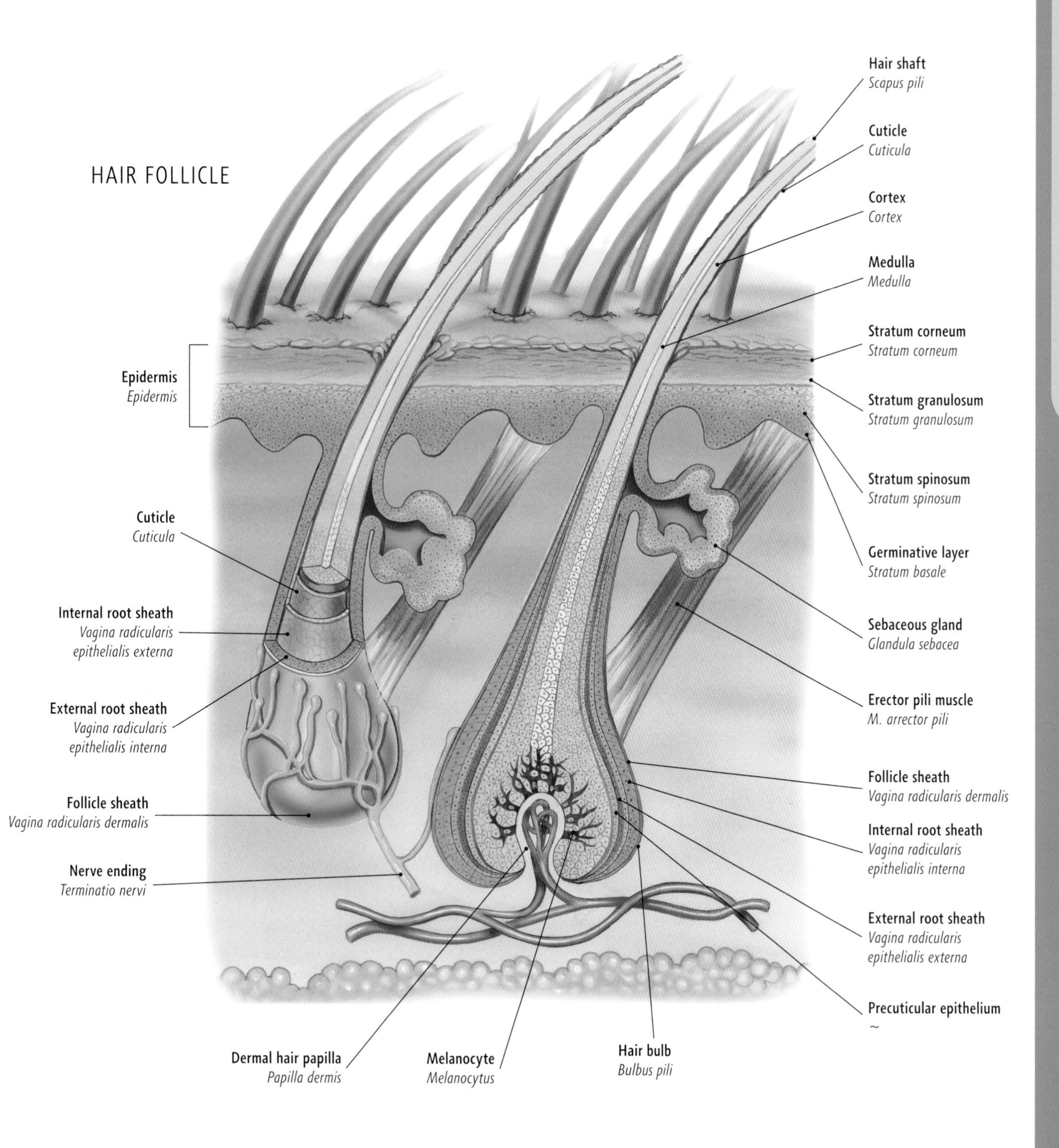
Hair shaft
Scapus pili
Cuticle
Cuticula
Cortex
Cortex
Medulla
Medulla
Stratum corneum
Stratum corneum
Stratum granulosum
Stratum granulosum
Stratum spinosum
Stratum spinosum
Germinative layer
Stratum basale
Sebaceous gland
Glandula sebacea
Erector pili muscle
M. arrector pili
Follicle sheath
Vagina radicularis dermalis
Internal root sheath
Vagina radicularis epithelialis interna
External root sheath
Vagina radicularis epithelialis externa
Precuticular epithelium
~
Epidermis
Epidermis
Cuticle
Cuticula
Internal root sheath
Vagina radicularis epithelialis externa
External root sheath
Vagina radicularis epithelialis interna
Follicle sheath
Vagina radicularis dermalis
Nerve ending
Terminatio nervi
Dermal hair papilla
Papilla dermis
Melanocyte
Melanocytus
Hair bulb
Bulbus pili

[~] = no direct Latin equivalent

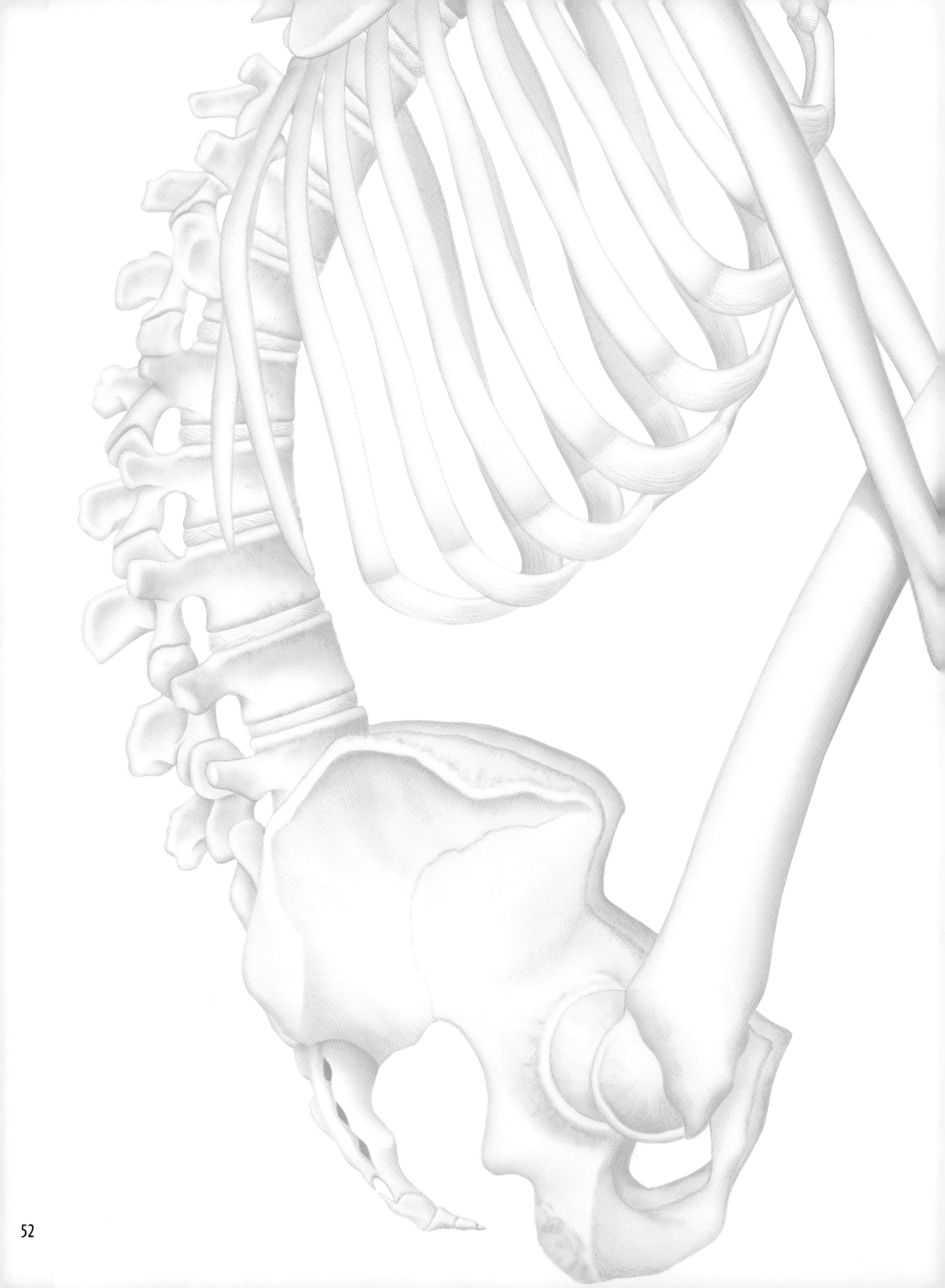

The Skeletal System

Bones of the Body

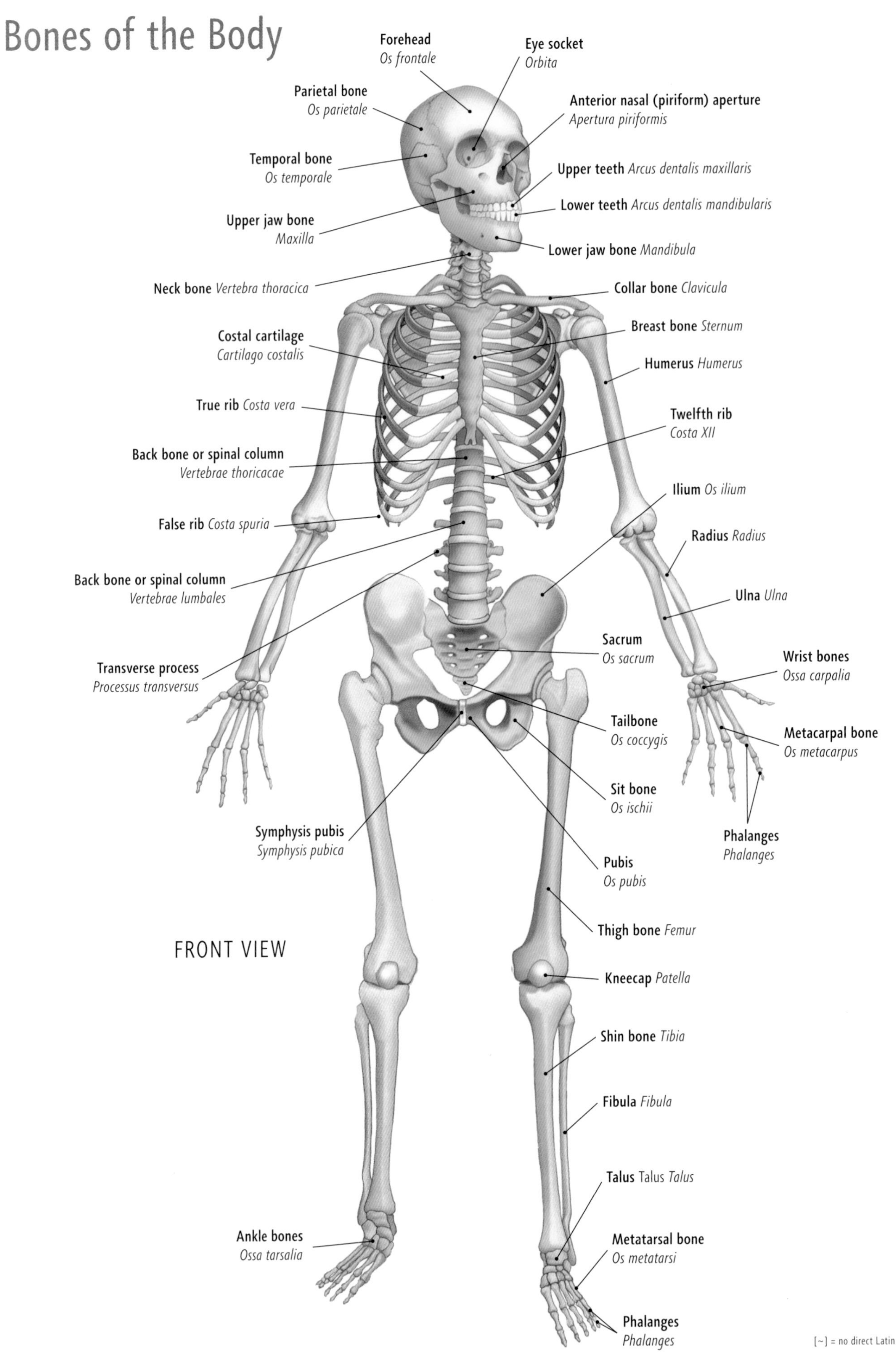
Forehead
Os frontale
Eye socket
Orbita
Parietal bone
Os parietale
Anterior nasal (piriform) aperture
Apertura piriformis
Temporal bone
Os temporale
Upper teeth Arcus dentalis maxillaris
Lower teeth Arcus dentalis mandibularis
Upper jaw bone
Maxilla
Lower jaw bone Mandibula
Neck bone Vertebra thoracica
Collar bone Clavicula
Breast bone Sternum
Costal cartilage
Cartilago costalis
Humerus Humerus
True rib Costa vera
Twelfth rib
Costa XII
Back bone or spinal column
Vertebrae thoricacae
Ilium Os ilium
False rib Costa spuria
Radius Radius
Back bone or spinal column
Vertebrae lumbales
Ulna Ulna
Sacrum
Os sacrum
Wrist bones
Ossa carpalia
Transverse process
Processus transversus
Tailbone
Os coccygis
Metacarpal bone
Os metacarpus
Sit bone
Os ischii
Symphysis pubis
Symphysis pubica
Phalanges
Phalanges
Pubis
Os pubis
Thigh bone Femur
FRONT VIEW
Kneecap Patella
Shin bone Tibia
Fibula Fibula
Talus Talus Talus
Ankle bones
Ossa tarsalia
Metatarsal bone
Os metatarsi
Phalanges
Phalanges

[~] = no direct Latin equivalent

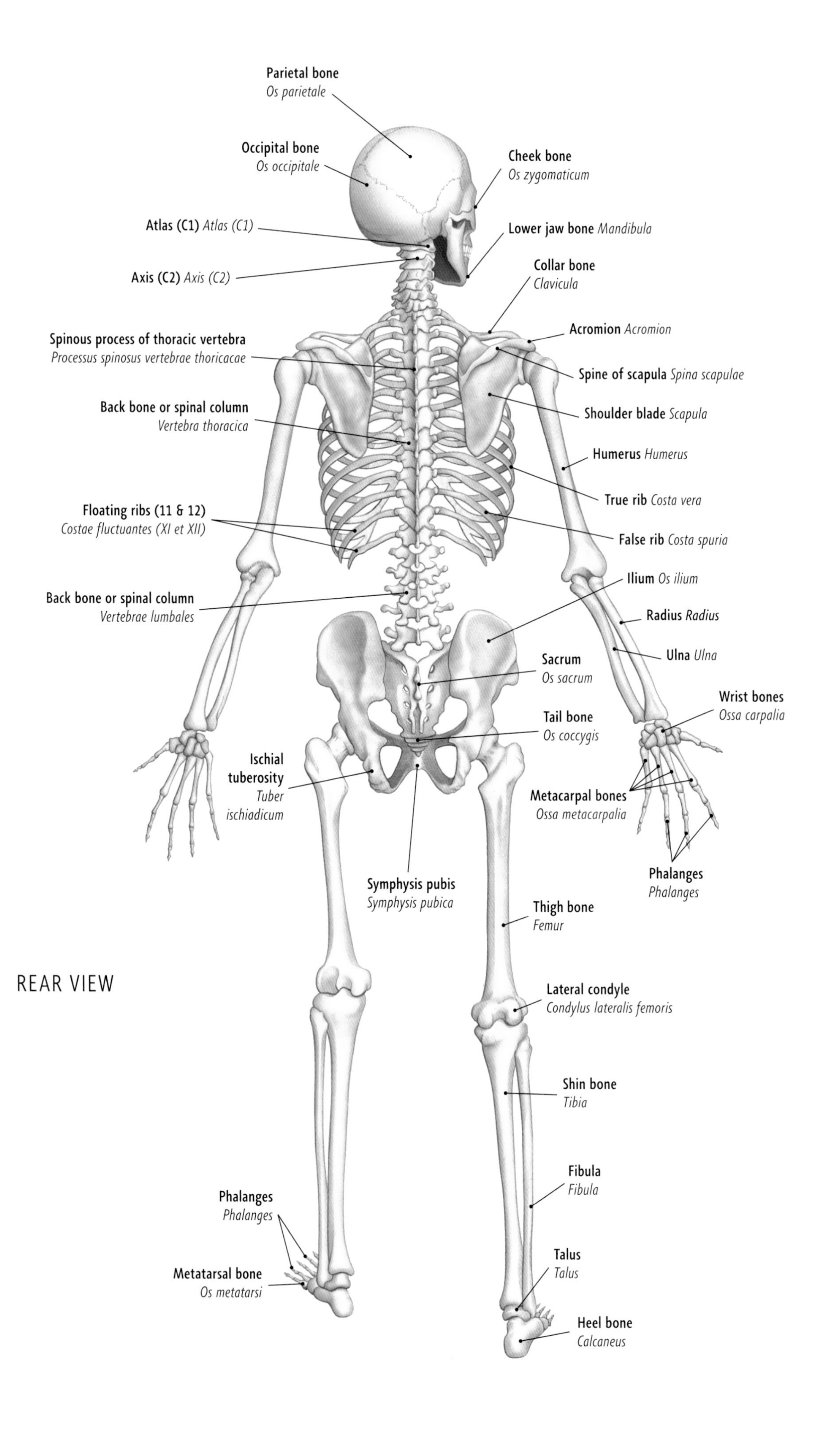
Parietal bone
Os parietale
Occipital bone
Os occipitale
Cheek bone
Os zygomaticum
Atlas (C1) Atlas (C1)
Lower jaw bone Mandibula
Axis (C2) Axis (C2)
Collar bone
Clavicula
Spinous process of thoracic vertebra
Processus spinosus vertebrae thoricacae
Acromion Acromion
Spine of scapula Spina scapulae
Back bone or spinal column
Vertebra thoracica
Shoulder blade Scapula
Humerus Humerus
True rib Costa vera
Floating ribs (11 & 12)
Costae fluctuantes (XI et XII)
False rib Costa spuria
Ilium Os ilium
Back bone or spinal column
Vertebrae lumbales
Radius Radius
Ulna Ulna
Sacrum
Os sacrum
Wrist bones
Ossa carpalia
Tail bone
Os coccygis
Ischial
tuberosity
Tuber
ischiadicum
Metacarpal bones
Ossa metacarpalia
Phalanges
Phalanges
Symphysis pubis
Symphysis pubica
Thigh bone
Femur
Lateral condyle
Condylus lateralis femoris
Shin bone
Tibia
Fibula
Fibula
Phalanges
Phalanges
Talus
Talus
Metatarsal bone
Os metatarsi
Heel bone
Calcaneus

REAR VIEW

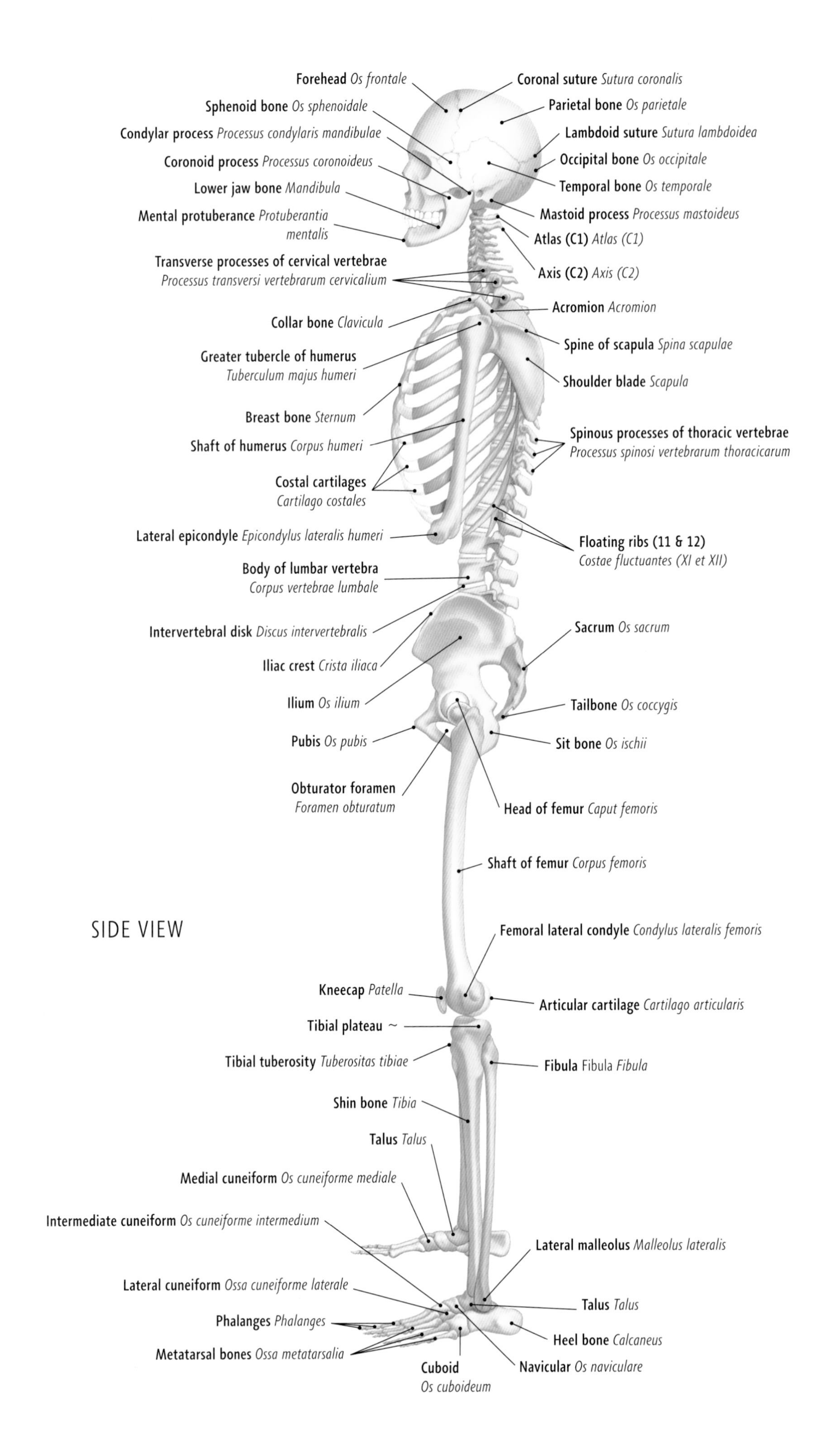
Forehead Os frontale
Coronal suture Sutura coronalis
Sphenoid bone Os sphenoidale
Parietal bone Os parietale
Condylar process Processus condylaris mandibulae
Lambdoid suture Sutura lambdoidea
Coronoid process Processus coronoideus
Occipital bone Os occipitale
Lower jaw bone Mandibula
Temporal bone Os temporale
Mental protuberance Protuberantia mentalis
Mastoid process Processus mastoideus
Atlas (C1) Atlas (C1)
Transverse processes of cervical vertebrae Processus transversi vertebrarum cervicalium
Axis (C2) Axis (C2)
Acromion Acromion
Collar bone Clavicula
Spine of scapula Spina scapulae
Greater tubercle of humerus Tuberculum majus humeri
Shoulder blade Scapula
Breast bone Sternum
Spinous processes of thoracic vertebrae Processus spinosi vertebrarum thoracicarum
Shaft of humerus Corpus humeri
Costal cartilages Cartilago costales
Lateral epicondyle Epicondylus lateralis humeri
Floating ribs (11 & 12) Costae fluctuantes (XI et XII)
Body of lumbar vertebra Corpus vertebrae lumbale
Intervertebral disk Discus intervertebralis
Sacrum Os sacrum
Iliac crest Crista iliaca
Ilium Os ilium
Tailbone Os coccygis
Pubis Os pubis
Sit bone Os ischii
Obturator foramen Foramen obturatum
Head of femur Caput femoris
Shaft of femur Corpus femoris
SIDE VIEW
Femoral lateral condyle Condylus lateralis femoris
Kneecap Patella
Articular cartilage Cartilago articularis
Tibial plateau ~
Tibial tuberosity Tuberositas tibiae
Fibula Fibula Fibula
Shin bone Tibia
Talus Talus
Medial cuneiform Os cuneiforme mediale
Intermediate cuneiform Os cuneiforme intermedium
Lateral malleolus Malleolus lateralis
Lateral cuneiform Ossa cuneiforme laterale
Talus Talus
Phalanges Phalanges
Heel bone Calcaneus
Metatarsal bones Ossa metatarsalia
Cuboid Os cuboideum
Navicular Os naviculare

Bone

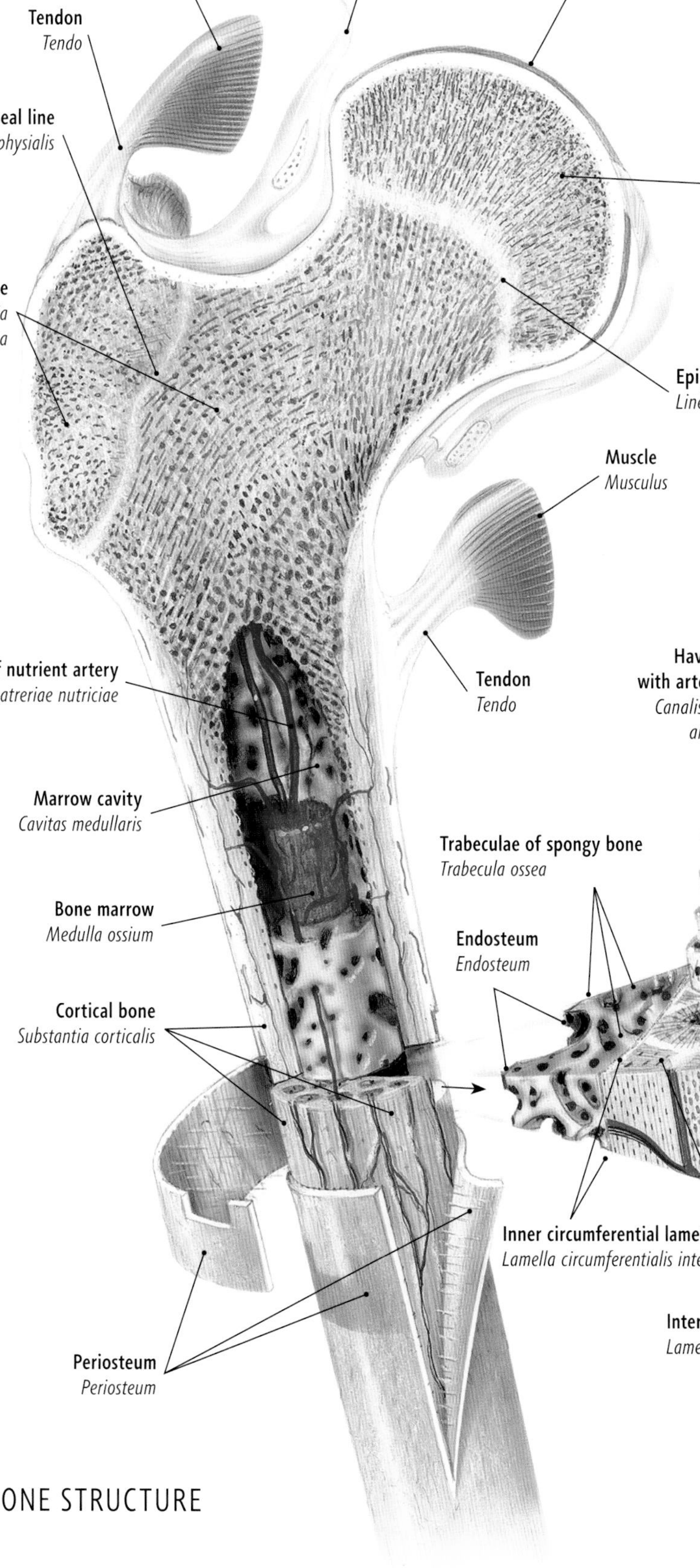

Ligament
Ligamenta

Muscle
Musculus

Articular cartilage on articular surface
Cartilago articularis in facii articularis

Tendon
Tendo

Epiphyseal line
Linea epiphysialis

Spongy bone
Substantia spongiosa

pongy bone
Substantia spongiosa

Epiphyseal line
Linea epiphysialis

Muscle
Musculus

Branch of nutrient artery
Ramus atreriae nutriciae

Tendon
Tendo

Marrow cavity
Cavitas medullaris

Bone marrow
Medulla ossium

Cortical bone
Substantia corticalis

Periosteum
Periosteum

BONE STRUCTURE

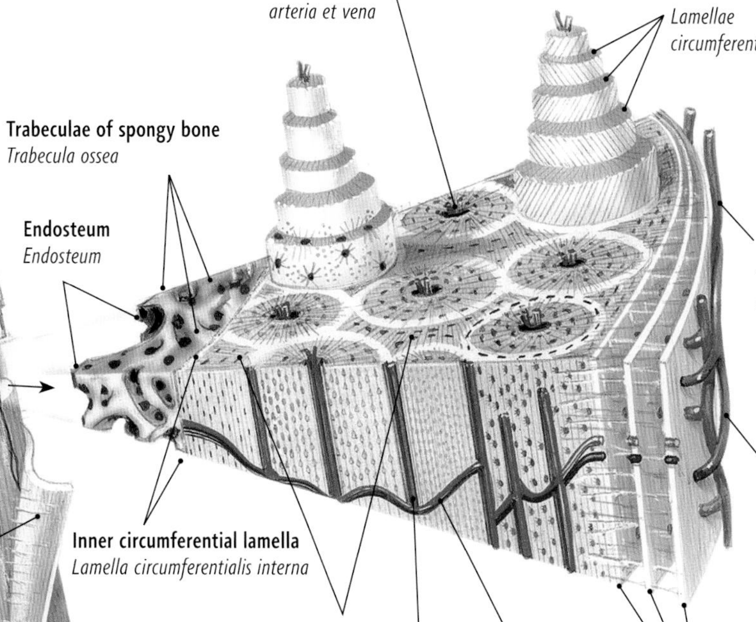

BONE DETAIL

Haversian canal with artery and vein
Canalis osteoni cum arteria et vena

Concentric lamellae
Lamellae circumferentiales

Trabeculae of spongy bone
Trabecula ossea

Endosteum
Endosteum

Periosteal artery
Arteria periostei

Periosteal vein
Vena periostei

Inner circumferential lamella
Lamella circumferentialis interna

Interstitial lamellae
Lamellae interstitiales

Volkmann's canal
Canalis transversus

Haversian canal with artery and vein
Canalis osteoni cum arteria et vena

Outer circumferential lamellae
Lamella circumferentialis externa

Bones of the Skull

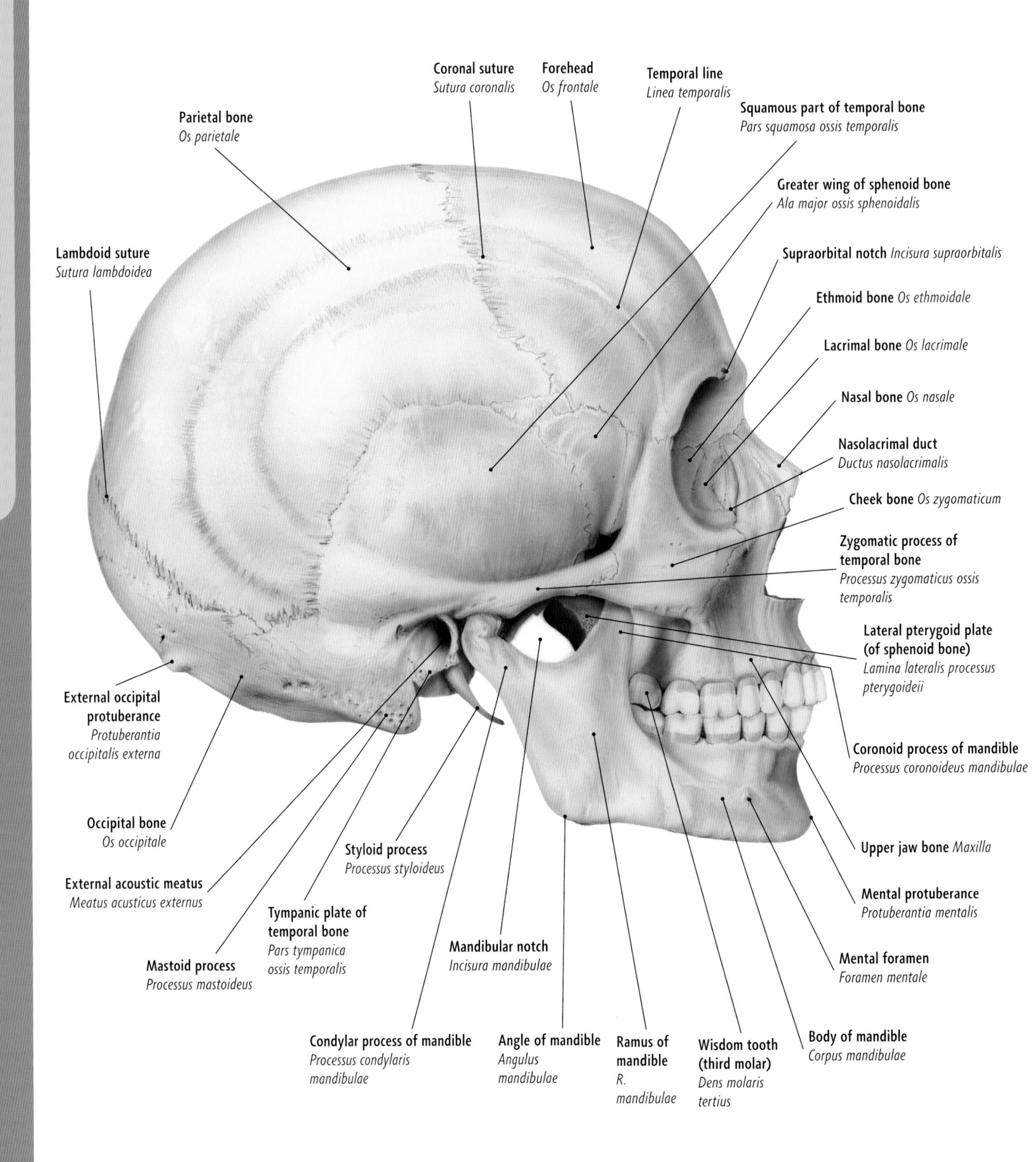
Coronal suture
Sutura coronalis
Forehead
Os frontale
Temporal line
Linea temporalis
Parietal bone
Os parietale
Squamous part of temporal bone
Pars squamosa ossis temporalis
Greater wing of sphenoid bone
Ala major ossis sphenoidalis
Lambdoid suture
Sutura lambdoidea
Supraorbital notch Incisura supraorbitalis
Ethmoid bone Os ethmoidale
Lacrimal bone Os lacrimale
Nasal bone Os nasale
Nasolacrimal duct
Ductus nasolacrimalis
Cheek bone Os zygomaticum
Zygomatic process of temporal bone
Processus zygomaticus ossis temporalis
Lateral pterygoid plate (of sphenoid bone)
Lamina lateralis processus pterygoideii
External occipital protuberance
Protuberantia occipitalis externa
Coronoid process of mandible
Processus coronoideus mandibulae
Occipital bone
Os occipitale
Upper jaw bone Maxilla
Styloid process
Processus styloideus
External acoustic meatus
Meatus acusticus externus
Mental protuberance
Protuberantia mentalis
Tympanic plate of temporal bone
Pars tympanica ossis temporalis
Mandibular notch
Incisura mandibulae
Mental foramen
Foramen mentale
Mastoid process
Processus mastoideus
Condylar process of mandible
Processus condylaris mandibulae
Angle of mandible
Angulus mandibulae
Ramus of mandible
R. mandibulae
Wisdom tooth (third molar)
Dens molaris tertius
Body of mandible
Corpus mandibulae

SIDE VIEW

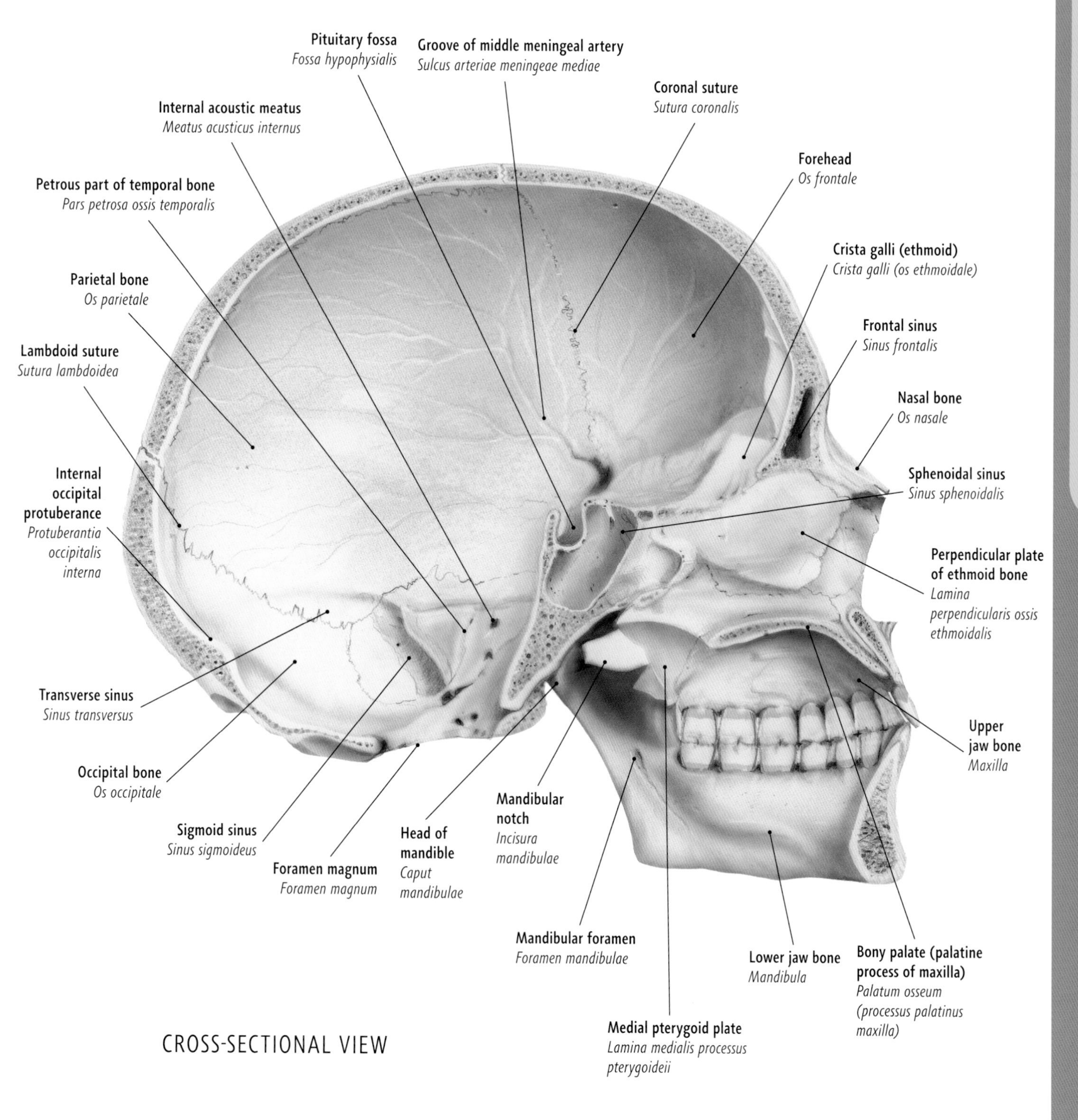

Pituitary fossa
Fossa hypophysialis
Groove of middle meningeal artery
Sulcus arteriae meningeae mediae
Coronal suture
Sutura coronalis
Internal acoustic meatus
Meatus acusticus internus
Forehead
Os frontale
Petrous part of temporal bone
Pars petrosa ossis temporalis
Crista galli (ethmoid)
Crista galli (os ethmoidale)
Parietal bone
Os parietale
Frontal sinus
Sinus frontalis
Lambdoid suture
Sutura lambdoidea
Nasal bone
Os nasale
Internal occipital protuberance
Protuberantia occipitalis interna
Sphenoidal sinus
Sinus sphenoidalis
Perpendicular plate of ethmoid bone
Lamina perpendicularis ossis ethmoidalis
Transverse sinus
Sinus transversus
Upper jaw bone
Maxilla
Occipital bone
Os occipitale
Mandibular notch
Incisura mandibulae
Sigmoid sinus
Sinus sigmoideus
Head of mandible
Caput mandibulae
Foramen magnum
Foramen magnum
Mandibular foramen
Foramen mandibulae
Lower jaw bone
Mandibula
Bony palate (palatine process of maxilla)
Palatum osseum (processus palatinus maxilla)
Medial pterygoid plate
Lamina medialis processus pterygoideii

CROSS-SECTIONAL VIEW

Bones of the Skull

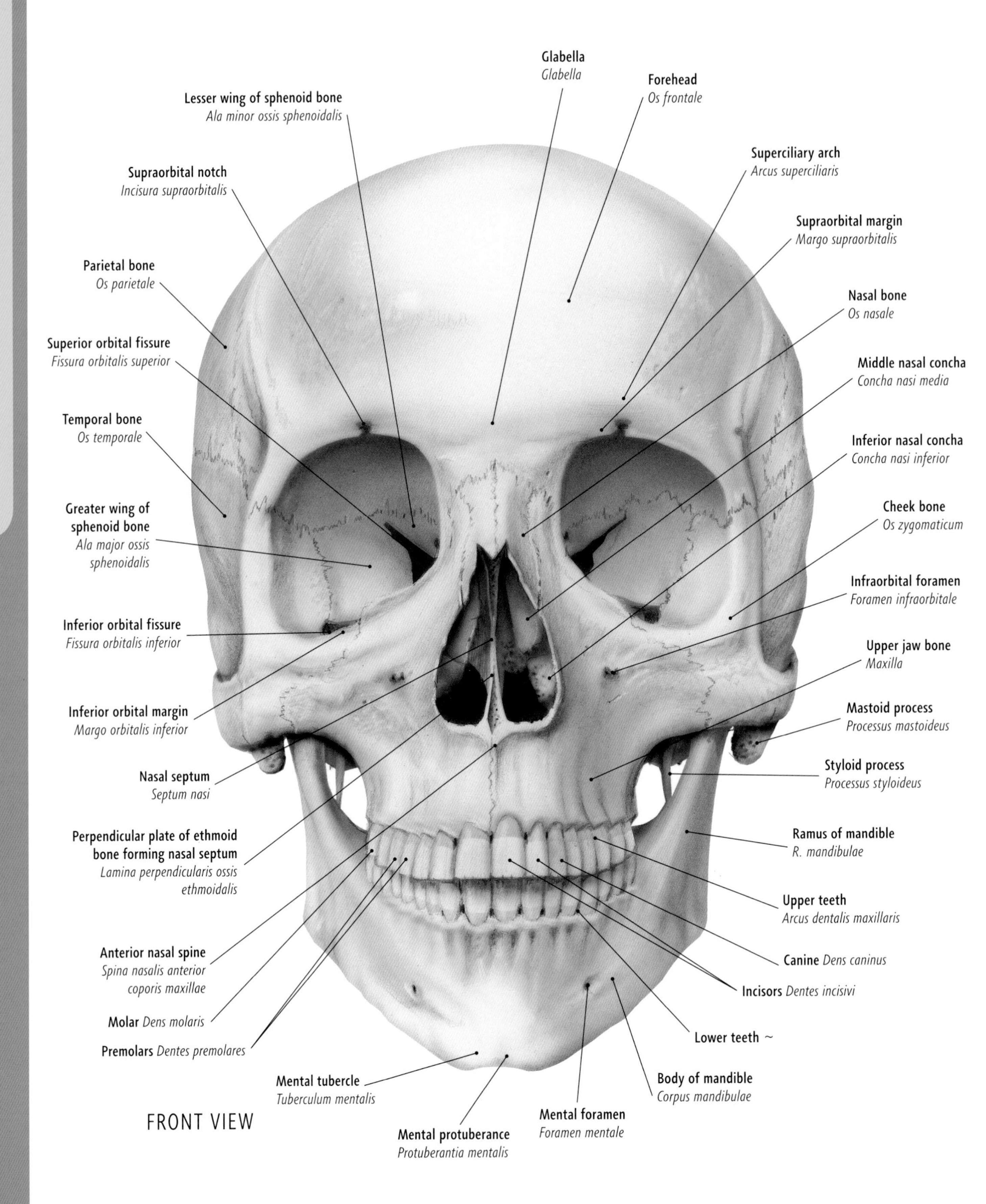
Glabella
Glabella
Forehead
Os frontale
Lesser wing of sphenoid bone
Ala minor ossis sphenoidalis
Superciliary arch
Arcus superciliaris
Supraorbital notch
Incisura supraorbitalis
Supraorbital margin
Margo supraorbitalis
Parietal bone
Os parietale
Nasal bone
Os nasale
Superior orbital fissure
Fissura orbitalis superior
Middle nasal concha
Concha nasi media
Temporal bone
Os temporale
Inferior nasal concha
Concha nasi inferior
Greater wing of sphenoid bone
Ala major ossis sphenoidalis
Cheek bone
Os zygomaticum
Infraorbital foramen
Foramen infraorbitale
Inferior orbital fissure
Fissura orbitalis inferior
Upper jaw bone
Maxilla
Inferior orbital margin
Margo orbitalis inferior
Mastoid process
Processus mastoideus
Nasal septum
Septum nasi
Styloid process
Processus styloideus
Perpendicular plate of ethmoid bone forming nasal septum
Lamina perpendicularis ossis ethmoidalis
Ramus of mandible
R. mandibulae
Upper teeth
Arcus dentalis maxillaris
Anterior nasal spine
Spina nasalis anterior coporis maxillae
Canine Dens caninus
Incisors Dentes incisivi
Molar Dens molaris
Lower teeth ~
Premolars Dentes premolares
Mental tubercle
Tuberculum mentalis
Body of mandible
Corpus mandibulae
Mental foramen
Foramen mentale
Mental protuberance
Protuberantia mentalis
FRONT VIEW

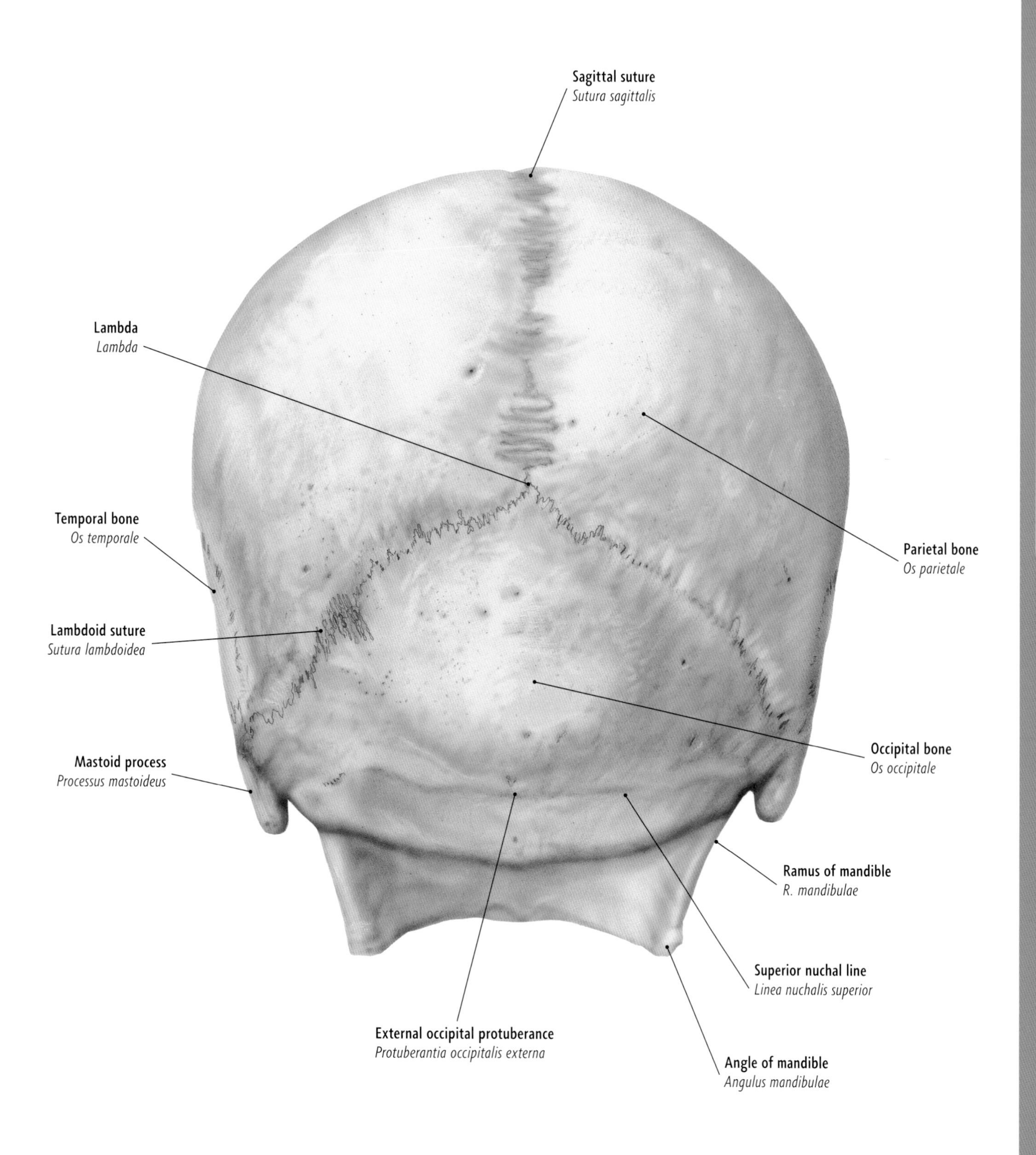
Sagittal suture
Sutura sagittalis
Lambda
Lambda
Temporal bone
Os temporale
Lambdoid suture
Sutura lambdoidea
Mastoid process
Processus mastoideus
Parietal bone
Os parietale
Occipital bone
Os occipitale
Ramus of mandible
R. mandibulae
Superior nuchal line
Linea nuchalis superior
External occipital protuberance
Protuberantia occipitalis externa
Angle of mandible
Angulus mandibulae

REAR VIEW

Bones of the Skull

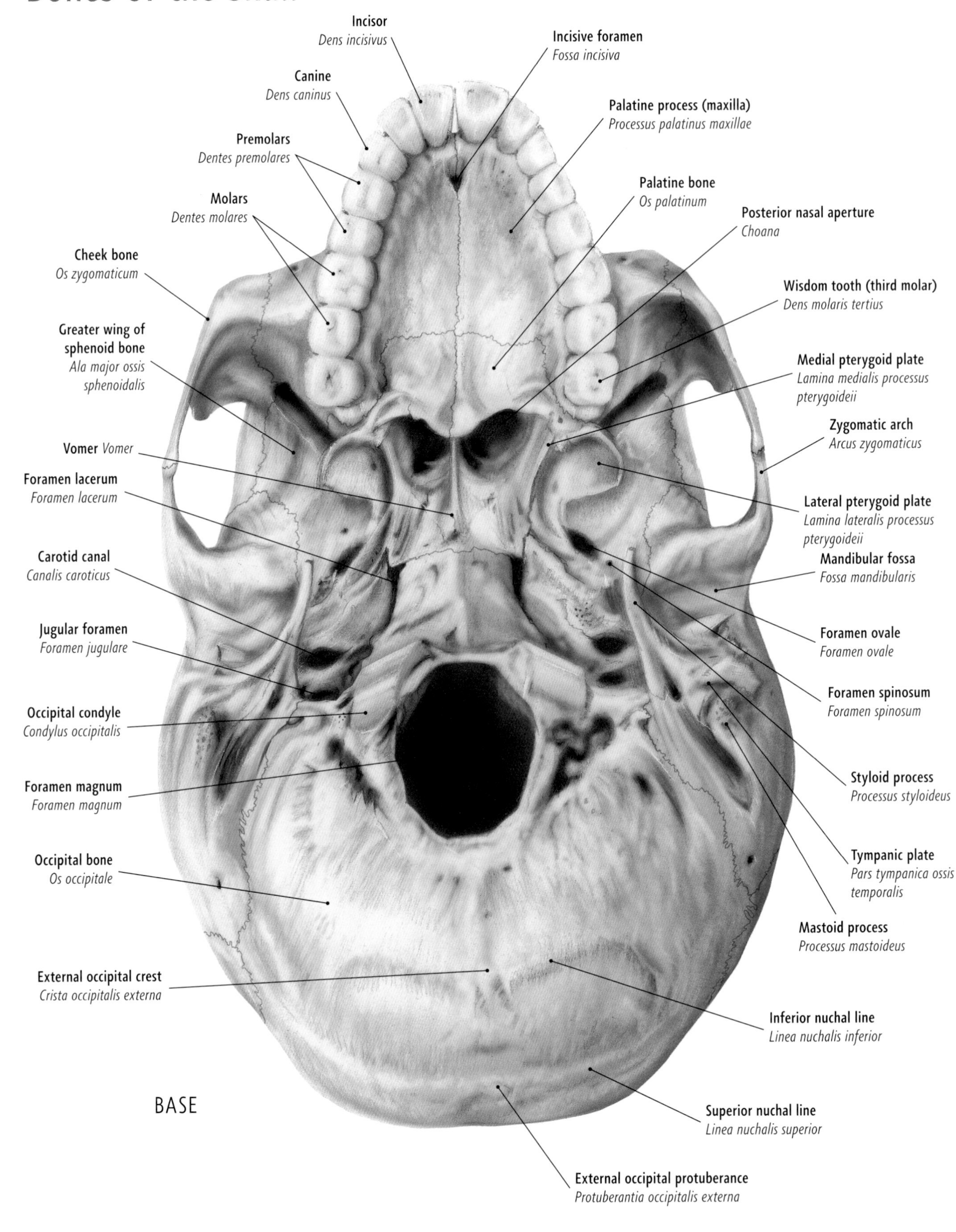

Incisor
Dens incisivus
Incisive foramen
Fossa incisiva
Canine
Dens caninus
Palatine process (maxilla)
Processus palatinus maxillae
Premolars
Dentes premolares
Palatine bone
Os palatinum
Molars
Dentes molares
Posterior nasal aperture
Choana
Cheek bone
Os zygomaticum
Wisdom tooth (third molar)
Dens molaris tertius
Greater wing of sphenoid bone
Ala major ossis sphenoidalis
Medial pterygoid plate
Lamina medialis processus pterygoideii
Zygomatic arch
Arcus zygomaticus
Vomer Vomer
Foramen lacerum
Foramen lacerum
Lateral pterygoid plate
Lamina lateralis processus pterygoideii
Mandibular fossa
Fossa mandibularis
Carotid canal
Canalis caroticus
Foramen ovale
Foramen ovale
Jugular foramen
Foramen jugulare
Foramen spinosum
Foramen spinosum
Occipital condyle
Condylus occipitalis
Styloid process
Processus styloideus
Foramen magnum
Foramen magnum
Tympanic plate
Pars tympanica ossis temporalis
Occipital bone
Os occipitale
Mastoid process
Processus mastoideus
External occipital crest
Crista occipitalis externa
Inferior nuchal line
Linea nuchalis inferior
BASE
Superior nuchal line
Linea nuchalis superior
External occipital protuberance
Protuberantia occipitalis externa

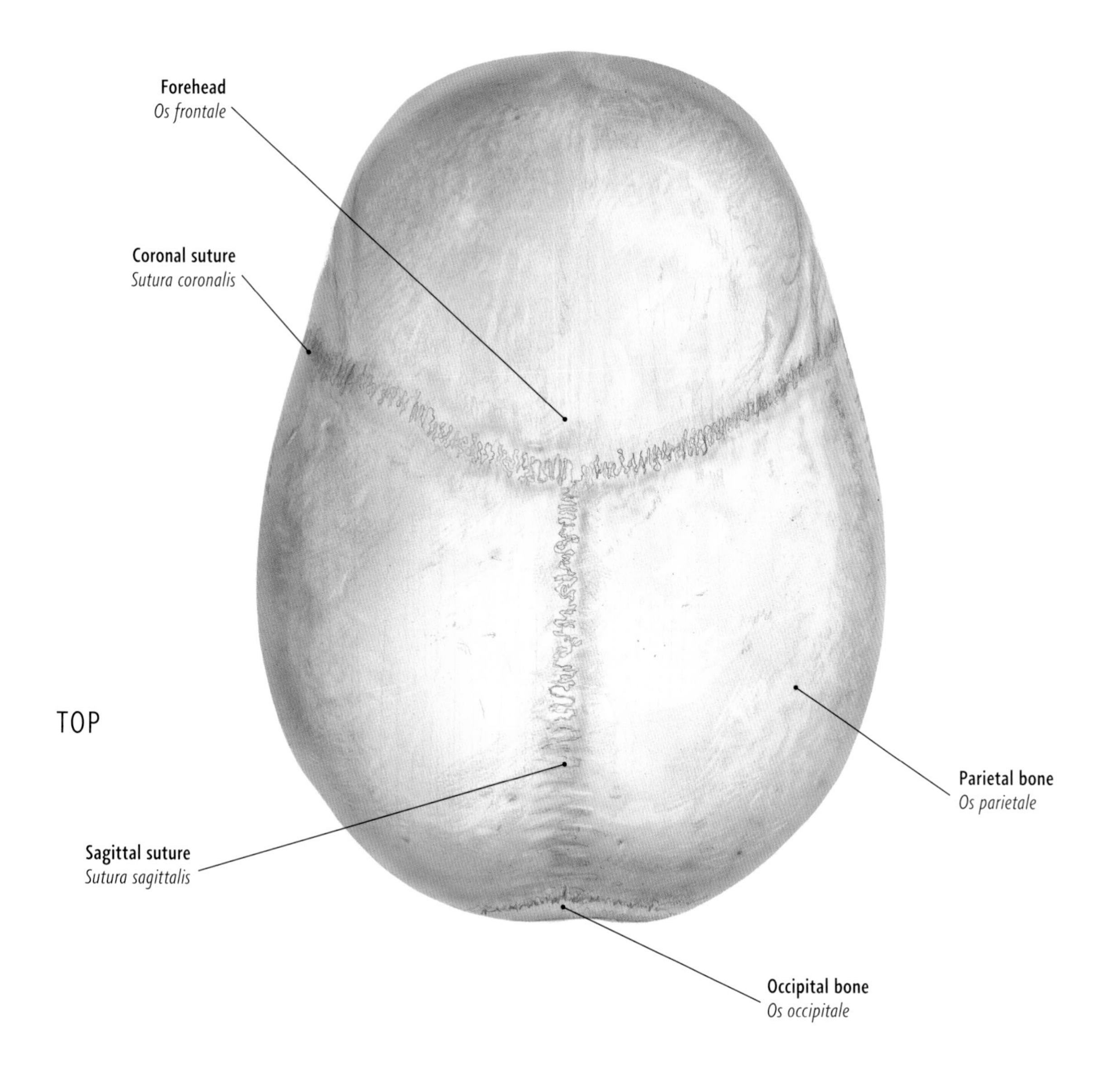
Forehead
Os frontale
Coronal suture
Sutura coronalis
TOP
Parietal bone
Os parietale
Sagittal suture
Sutura sagittalis
Occipital bone
Os occipitale

BONE FORMATION

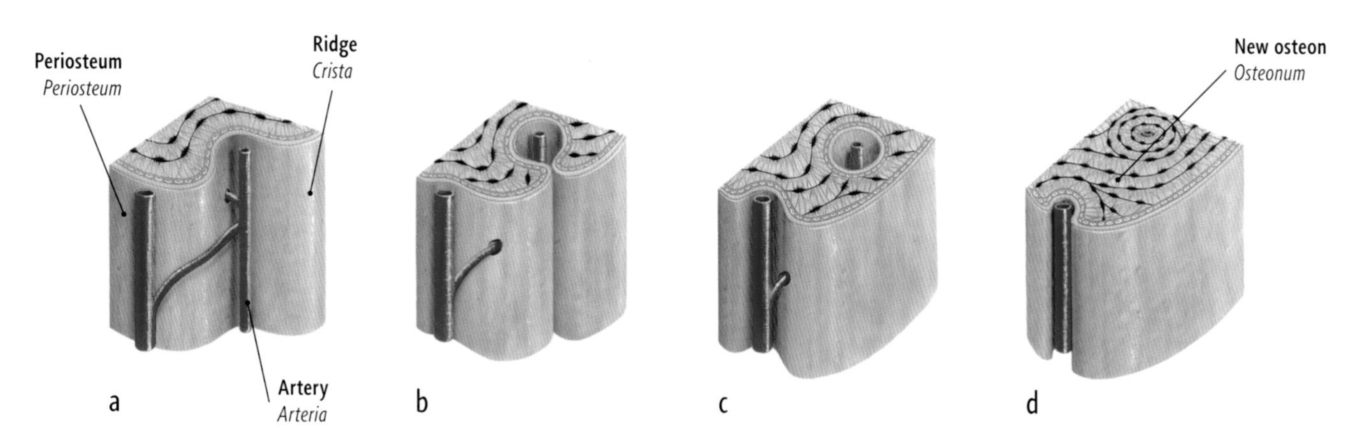
Periosteum
Periosteum
Ridge
Crista
New osteon
Osteonum
a
Artery
Arteria
b
c
d

Bone grows in width as new bone is laid down in ridges (a) either side of a blood vessel. The ridges grow together around the vessel (b). More bone is laid down, diminishing the space around the vessel (c) and eventually forming an osteon. The process continues with parallel blood vessels (d).

Bones of the Head and Face

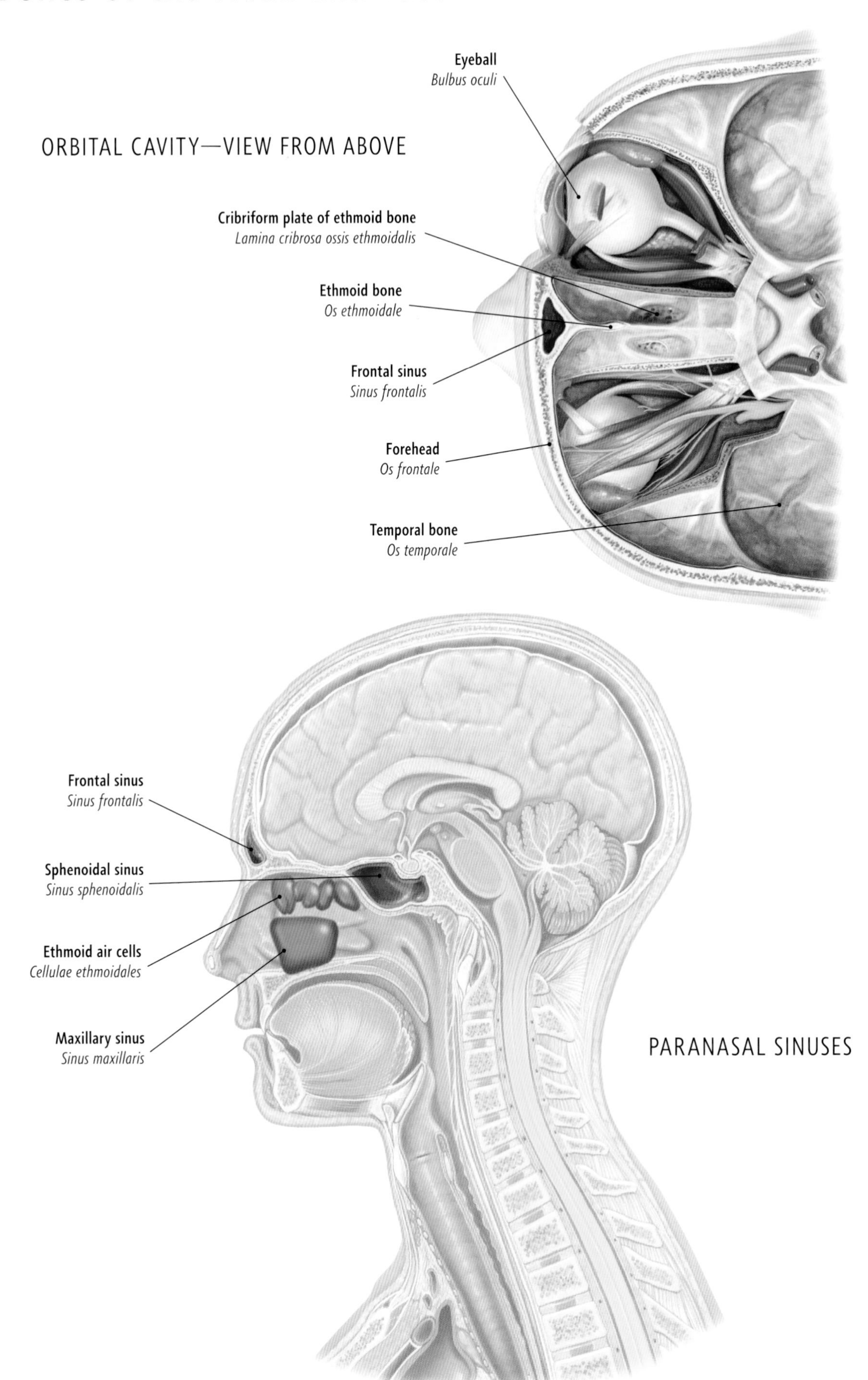
ORBITAL CAVITY—VIEW FROM ABOVE
Eyeball
Bulbus oculi
Cribriform plate of ethmoid bone
Lamina cribrosa ossis ethmoidalis
Ethmoid bone
Os ethmoidale
Frontal sinus
Sinus frontalis
Forehead
Os frontale
Temporal bone
Os temporale
Frontal sinus
Sinus frontalis
Sphenoidal sinus
Sinus sphenoidalis
Ethmoid air cells
Cellulae ethmoidales
Maxillary sinus
Sinus maxillaris
PARANASAL SINUSES

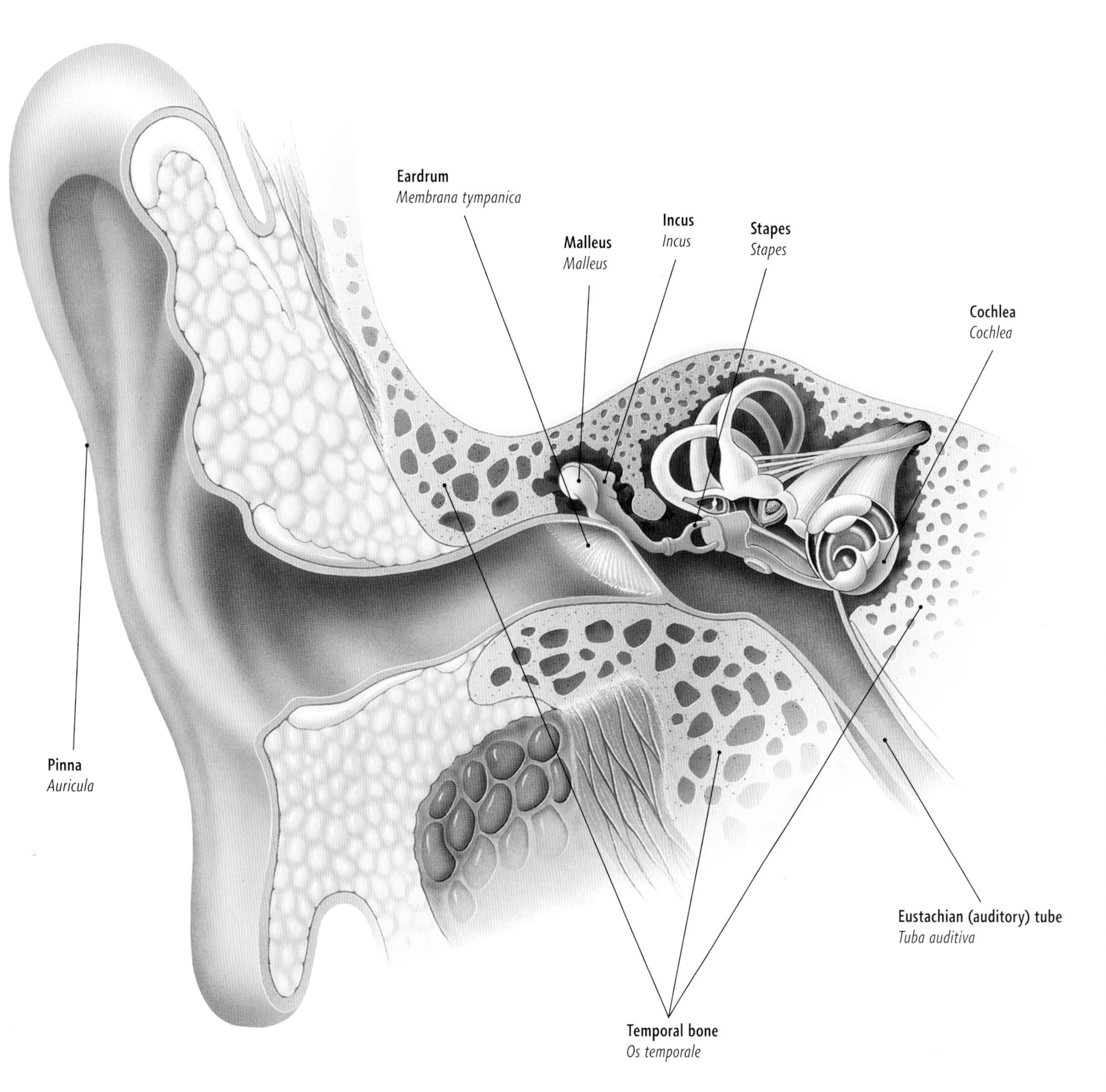
Eardrum
Membrana tympanica
Malleus
Malleus
Incus
Incus
Stapes
Stapes
Cochlea
Cochlea
Pinna
Auricula
Eustachian (auditory) tube
Tuba auditiva
Temporal bone
Os temporale

BONES OF THE EAR

Spine

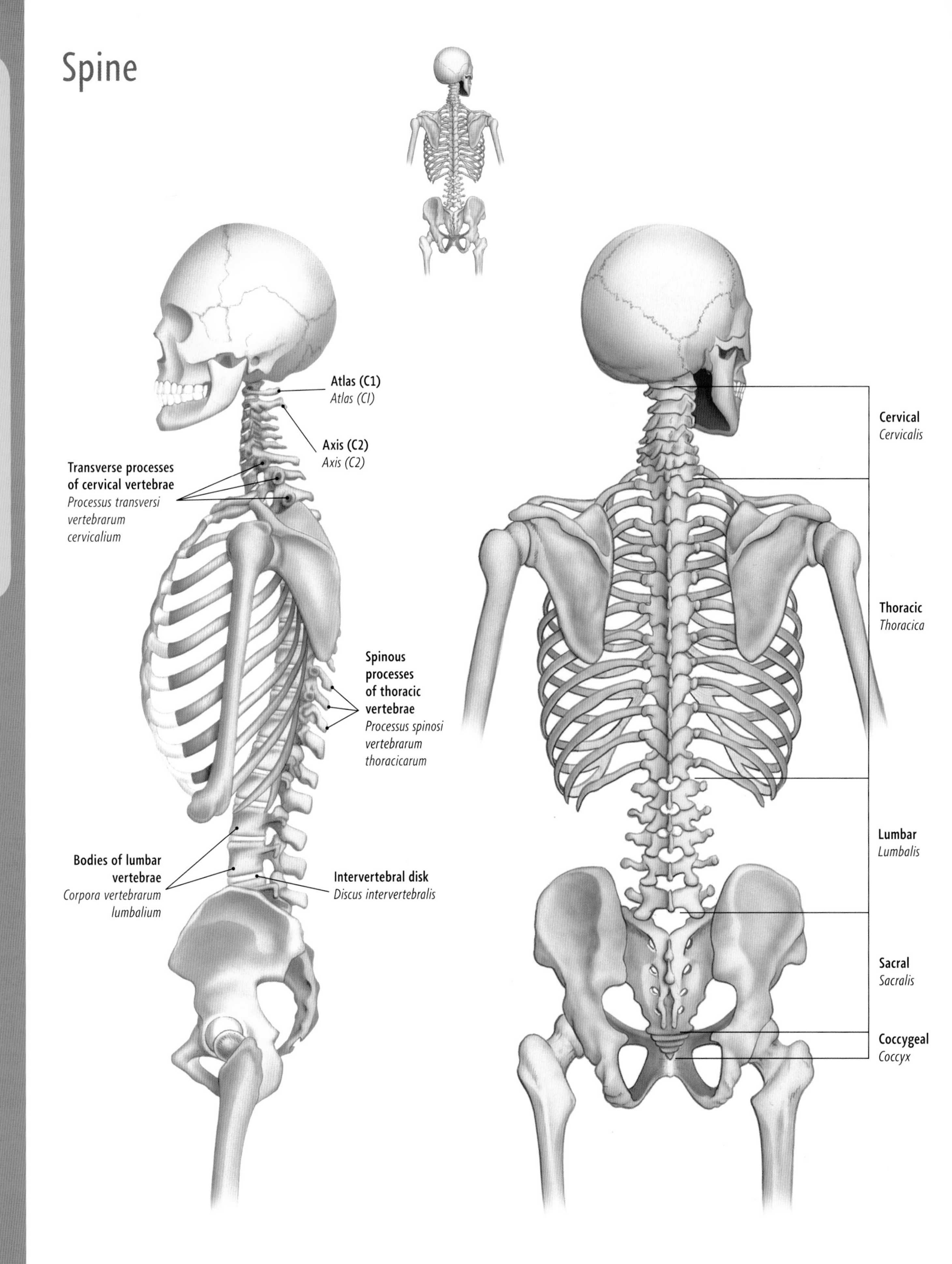

Atlas (C1)
Atlas (CI)
Axis (C2)
Axis (C2)
Transverse processes of cervical vertebrae
Processus transversi vertebrarum cervicalium
Spinous processes of thoracic vertebrae
Processus spinosi vertebrarum thoracicarum
Bodies of lumbar vertebrae
Corpora vertebrarum lumbalium
Intervertebral disk
Discus intervertebralis
Cervical
Cervicalis
Thoracic
Thoracica
Lumbar
Lumbalis
Sacral
Sacralis
Coccygeal
Coccyx

SPINE IN SITU—SIDE VIEW

SPINE IN SITU—REAR VIEW

[~] = no direct Latin equivalent

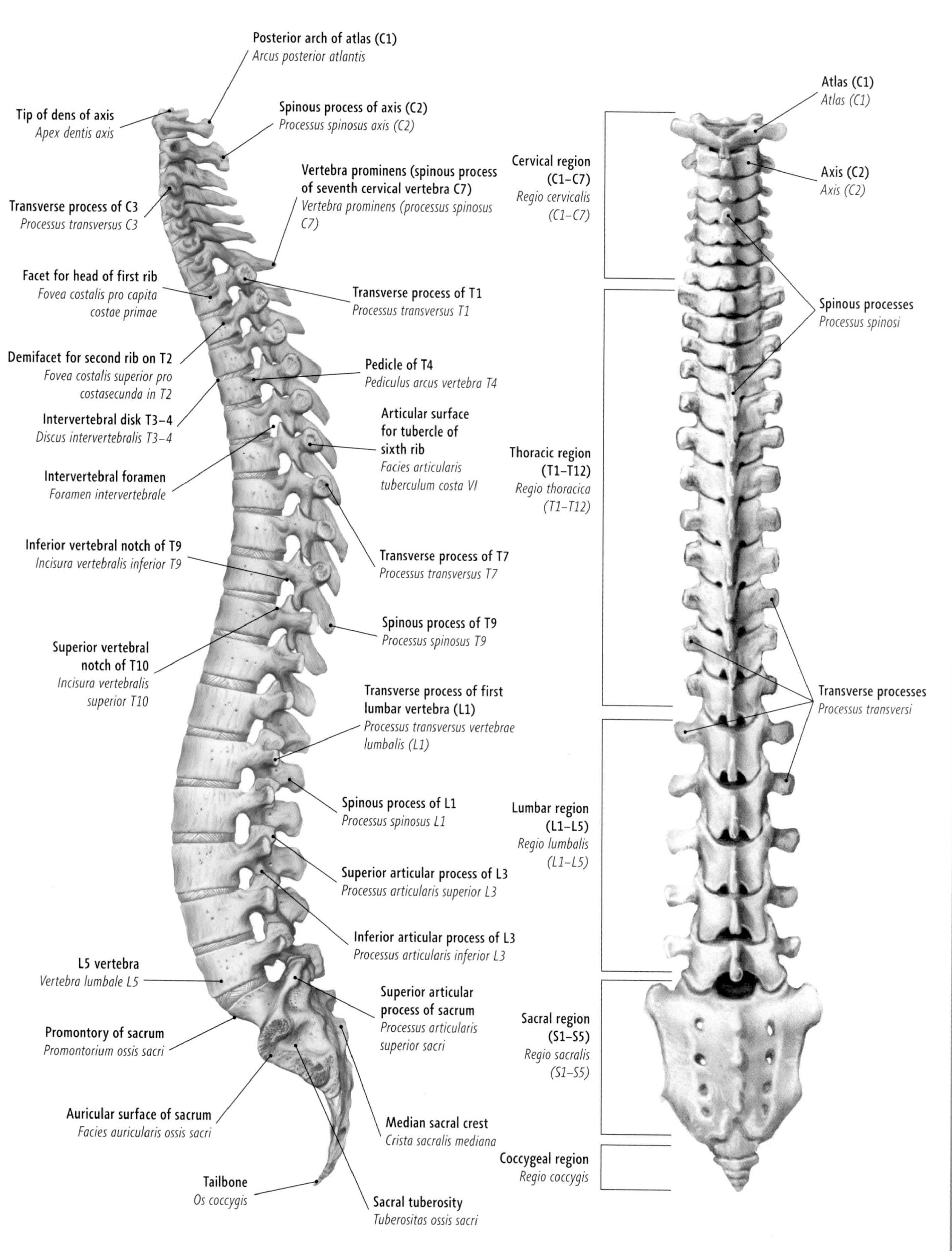
Posterior arch of atlas (C1)
Arcus posterior atlantis
Tip of dens of axis
Apex dentis axis
Spinous process of axis (C2)
Processus spinosus axis (C2)
Vertebra prominens (spinous process of seventh cervical vertebra C7)
Vertebra prominens (processus spinosus C7)
Transverse process of C3
Processus transversus C3
Facet for head of first rib
Fovea costalis pro capita costae primae
Transverse process of T1
Processus transversus T1
Demifacet for second rib on T2
Fovea costalis superior pro costasecunda in T2
Pedicle of T4
Pediculus arcus vertebra T4
Intervertebral disk T3–4
Discus intervertebralis T3–4
Articular surface for tubercle of sixth rib
Facies articularis tuberculum costa VI
Intervertebral foramen
Foramen intervertebrale
Inferior vertebral notch of T9
Incisura vertebralis inferior T9
Transverse process of T7
Processus transversus T7
Spinous process of T9
Processus spinosus T9
Superior vertebral notch of T10
Incisura vertebralis superior T10
Transverse process of first lumbar vertebra (L1)
Processus transversus vertebrae lumbalis (L1)
Spinous process of L1
Processus spinosus L1
Superior articular process of L3
Processus articularis superior L3
Inferior articular process of L3
Processus articularis inferior L3
L5 vertebra
Vertebra lumbale L5
Superior articular process of sacrum
Processus articularis superior sacri
Promontory of sacrum
Promontorium ossis sacri
Auricular surface of sacrum
Facies auricularis ossis sacri
Median sacral crest
Crista sacralis mediana
Tailbone
Os coccygis
Sacral tuberosity
Tuberositas ossis sacri
Atlas (C1)
Atlas (C1)
Axis (C2)
Axis (C2)
Cervical region (C1–C7)
Regio cervicalis (C1–C7)
Spinous processes
Processus spinosi
Thoracic region (T1–T12)
Regio thoracica (T1–T12)
Transverse processes
Processus transversi
Lumbar region (L1–L5)
Regio lumbalis (L1–L5)
Sacral region (S1–S5)
Regio sacralis (S1–S5)
Coccygeal region
Regio coccygis

[~] = no direct Latin equivalent

SPINE—SIDE VIEW

SPINE—REAR VIEW

Vertebrae

CERVICAL VERTEBRAE: AXIS AND ATLAS

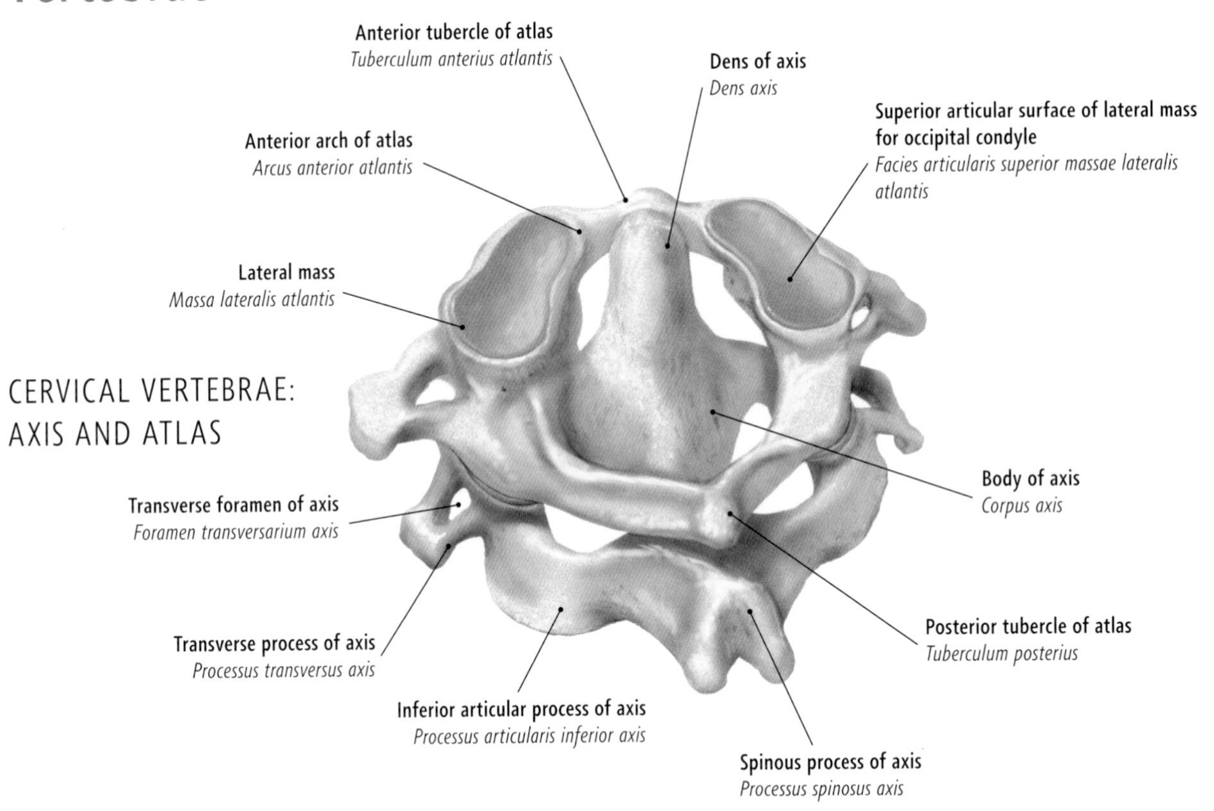
Anterior tubercle of atlas
Tuberculum anterius atlantis
Dens of axis
Dens axis
Superior articular surface of lateral mass for occipital condyle
Facies articularis superior massae lateralis atlantis
Anterior arch of atlas
Arcus anterior atlantis
Lateral mass
Massa lateralis atlantis
Body of axis
Corpus axis
Transverse foramen of axis
Foramen transversarium axis
Posterior tubercle of atlas
Tuberculum posterius
Transverse process of axis
Processus transversus axis
Inferior articular process of axis
Processus articularis inferior axis
Spinous process of axis
Processus spinosus axis

TYPICAL CERVICAL VERTEBRA

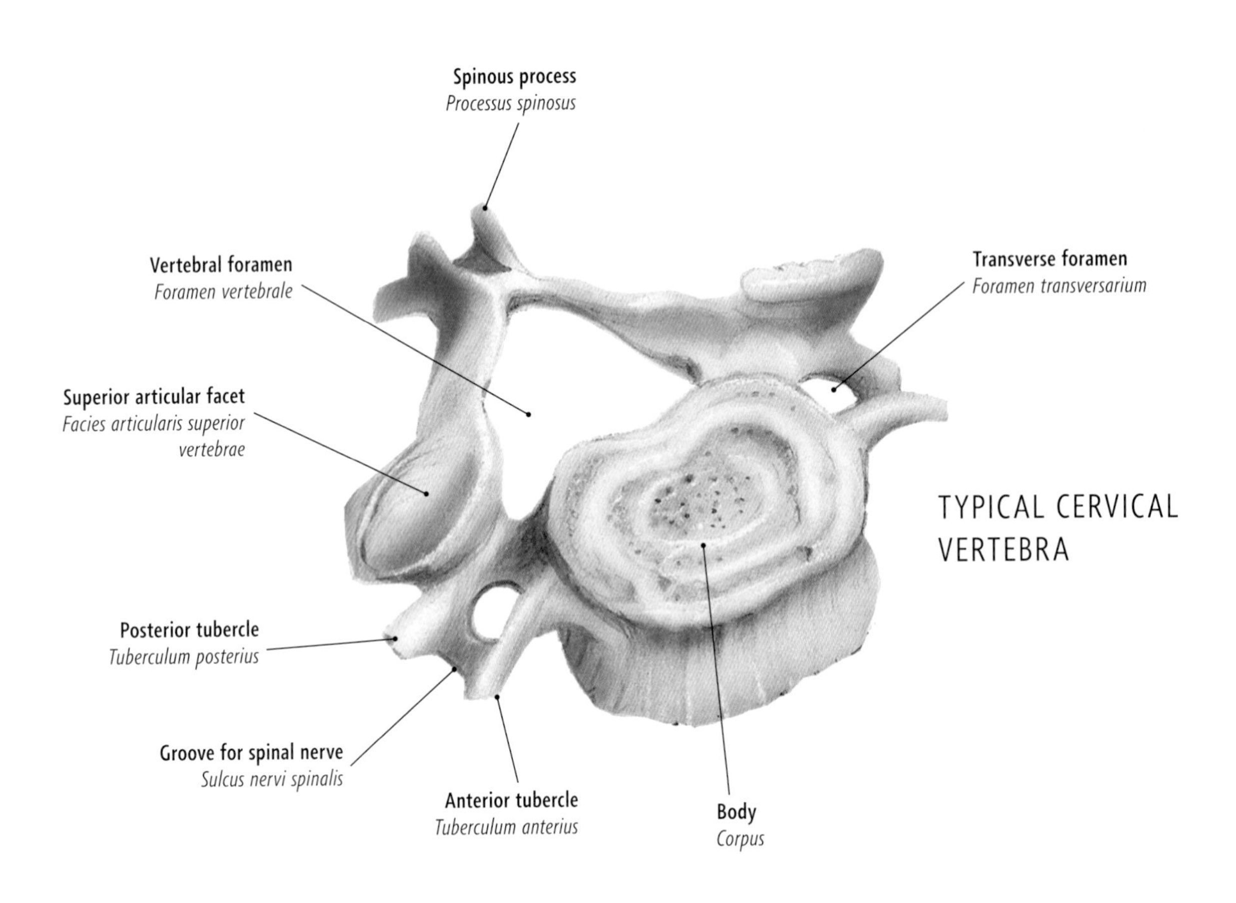
Spinous process
Processus spinosus
Vertebral foramen
Foramen vertebrale
Transverse foramen
Foramen transversarium
Superior articular facet
Facies articularis superior vertebrae
Posterior tubercle
Tuberculum posterius
Groove for spinal nerve
Sulcus nervi spinalis
Anterior tubercle
Tuberculum anterius
Body
Corpus

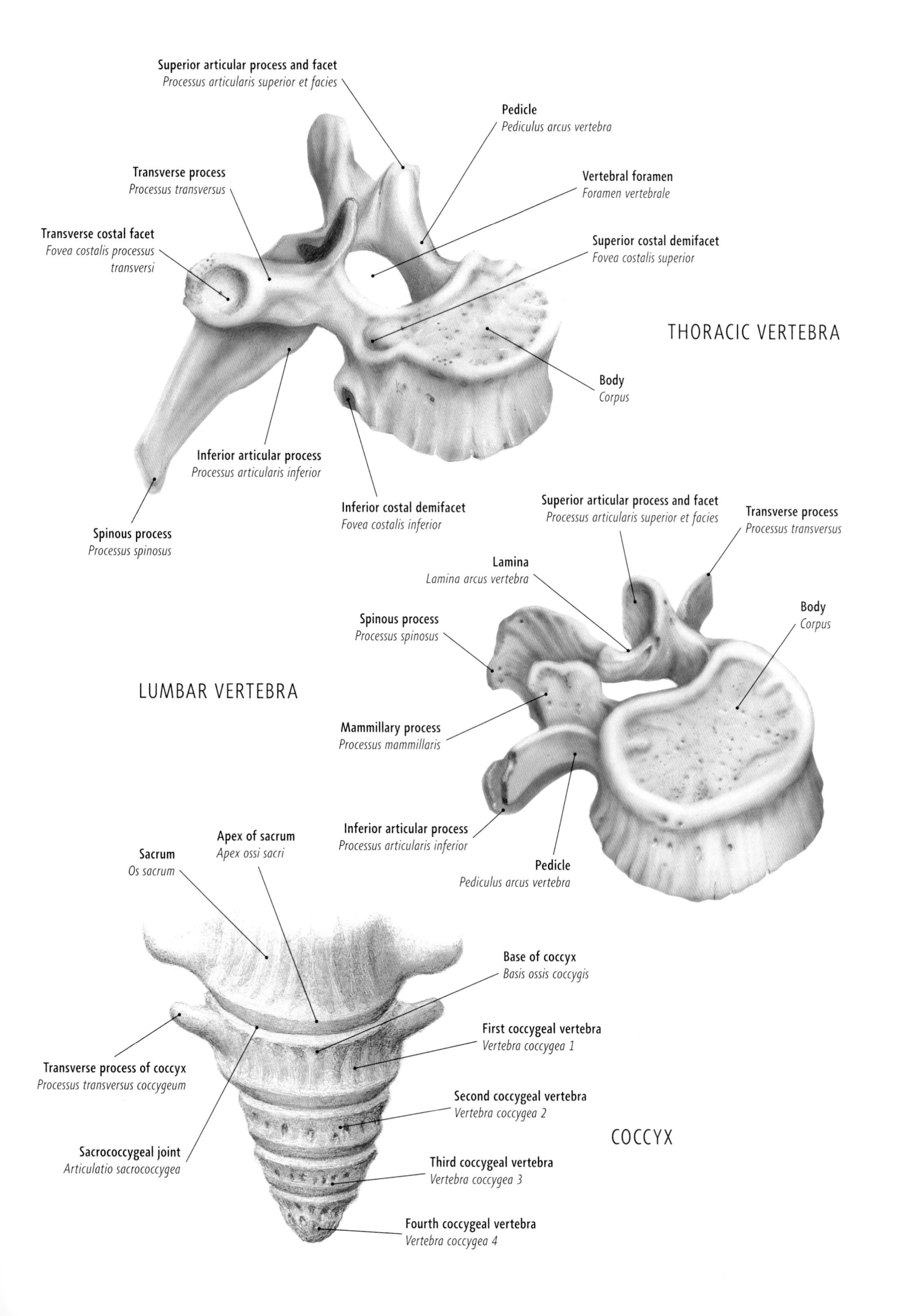
Superior articular process and facet
Processus articularis superior et facies
Pedicle
Pediculus arcus vertebra
Transverse process
Processus transversus
Vertebral foramen
Foramen vertebrale
Transverse costal facet
Fovea costalis processus transversi
Superior costal demifacet
Fovea costalis superior
THORACIC VERTEBRA
Body
Corpus
Inferior articular process
Processus articularis inferior
Inferior costal demifacet
Fovea costalis inferior
Spinous process
Processus spinosus
Superior articular process and facet
Processus articularis superior et facies
Transverse process
Processus transversus
Lamina
Lamina arcus vertebra
Body
Corpus
Spinous process
Processus spinosus
LUMBAR VERTEBRA
Mammillary process
Processus mammillaris
Inferior articular process
Processus articularis inferior
Pedicle
Pediculus arcus vertebra
Apex of sacrum
Apex ossi sacri
Sacrum
Os sacrum
Base of coccyx
Basis ossis coccygis
First coccygeal vertebra
Vertebra coccygea 1
Transverse process of coccyx
Processus transversus coccygeum
Second coccygeal vertebra
Vertebra coccygea 2
COCCYX
Sacrococcygeal joint
Articulatio sacrococcygea
Third coccygeal vertebra
Vertebra coccygea 3
Fourth coccygeal vertebra
Vertebra coccygea 4

Rib Cage and Clavicles

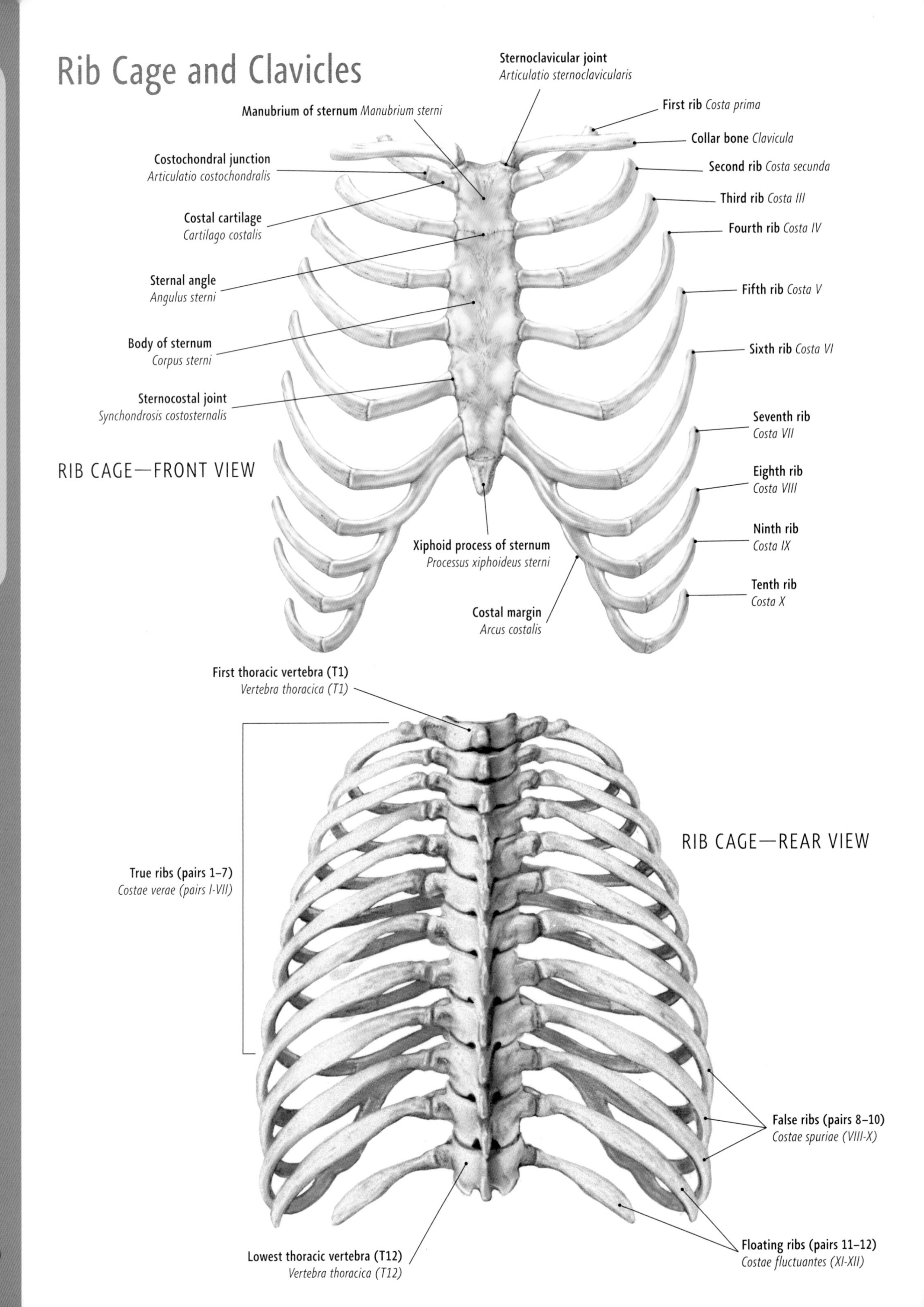
Sternoclavicular joint
Articulatio sternoclavicularis
Manubrium of sternum Manubrium sterni
First rib Costa prima
Collar bone Clavicula
Costochondral junction
Articulatio costochondralis
Second rib Costa secunda
Third rib Costa III
Costal cartilage
Cartilago costalis
Fourth rib Costa IV
Sternal angle
Angulus sterni
Fifth rib Costa V
Body of sternum
Corpus sterni
Sixth rib Costa VI
Sternocostal joint
Synchondrosis costosternalis
Seventh rib
Costa VII
RIB CAGE—FRONT VIEW
Eighth rib
Costa VIII
Ninth rib
Costa IX
Xiphoid process of sternum
Processus xiphoideus sterni
Tenth rib
Costa X
Costal margin
Arcus costalis
First thoracic vertebra (T1)
Vertebra thoracica (T1)
RIB CAGE—REAR VIEW
True ribs (pairs 1–7)
Costae verae (pairs I-VII)
False ribs (pairs 8–10)
Costae spuriae (VIII-X)
Lowest thoracic vertebra (T12)
Vertebra thoracica (T12)
Floating ribs (pairs 11–12)
Costae fluctuantes (XI-XII)

CLAVICLES

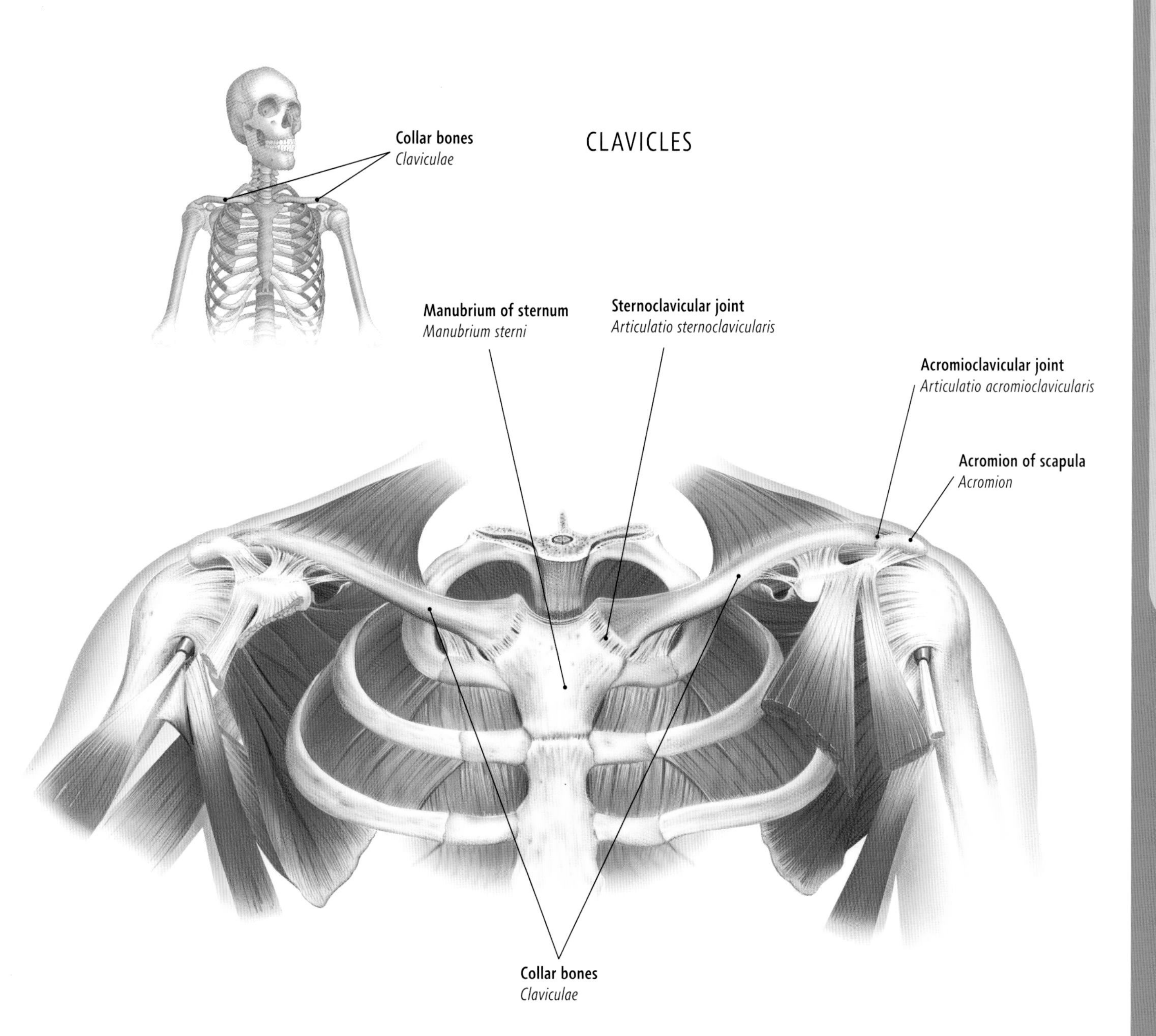
Collar bones
Claviculae
Manubrium of sternum
Manubrium sterni
Sternoclavicular joint
Articulatio sternoclavicularis
Acromioclavicular joint
Articulatio acromioclavicularis
Acromion of scapula
Acromion
Collar bones
Claviculae

Bones of the Arm

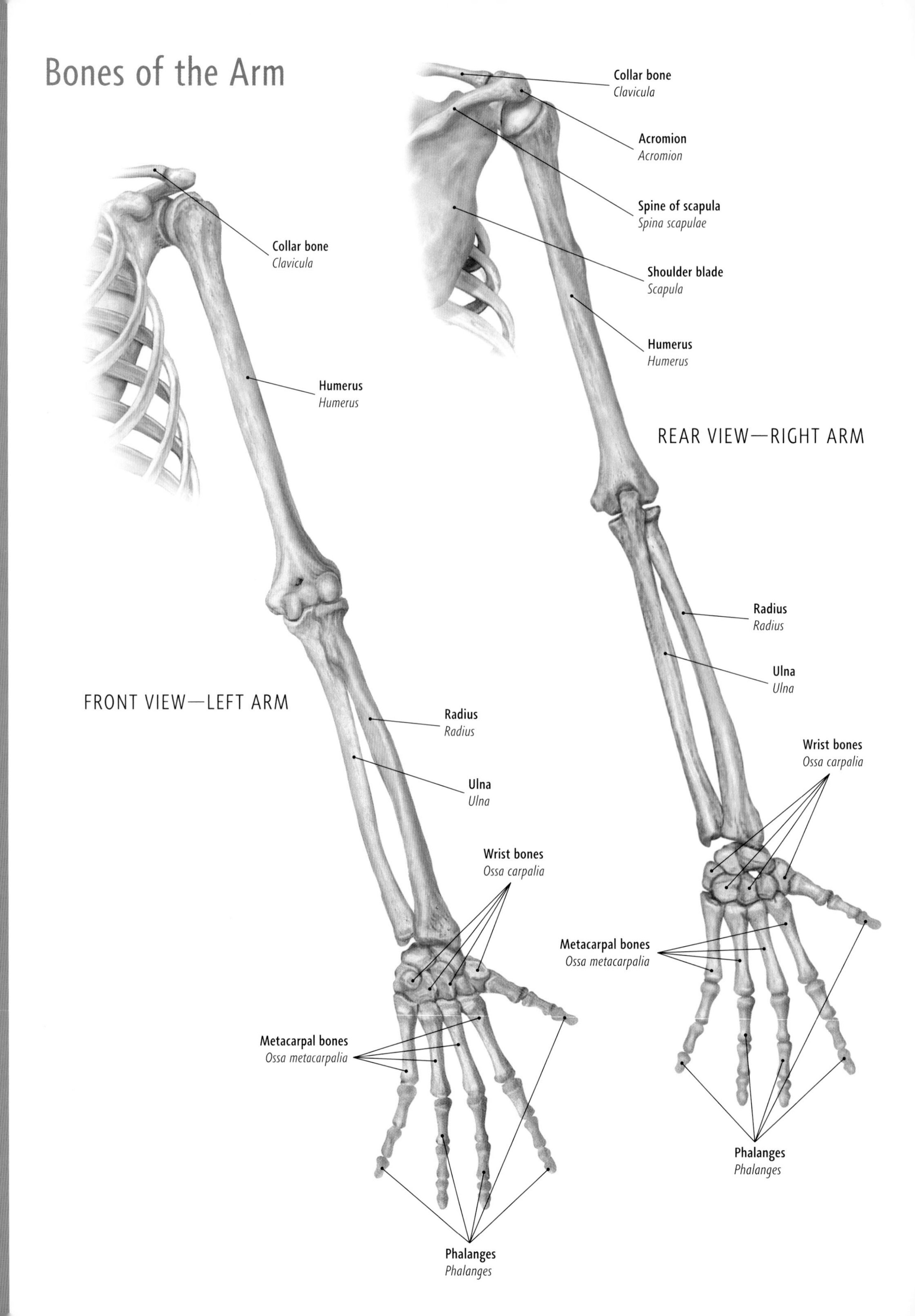

Collar bone
Clavicula
Acromion
Acromion
Spine of scapula
Spina scapulae
Shoulder blade
Scapula
Humerus
Humerus
REAR VIEW—RIGHT ARM
Radius
Radius
Ulna
Ulna
Wrist bones
Ossa carpalia
Metacarpal bones
Ossa metacarpalia
Phalanges
Phalanges
Collar bone
Clavicula
Humerus
Humerus
FRONT VIEW—LEFT ARM
Radius
Radius
Ulna
Ulna
Wrist bones
Ossa carpalia
Metacarpal bones
Ossa metacarpalia
Phalanges
Phalanges

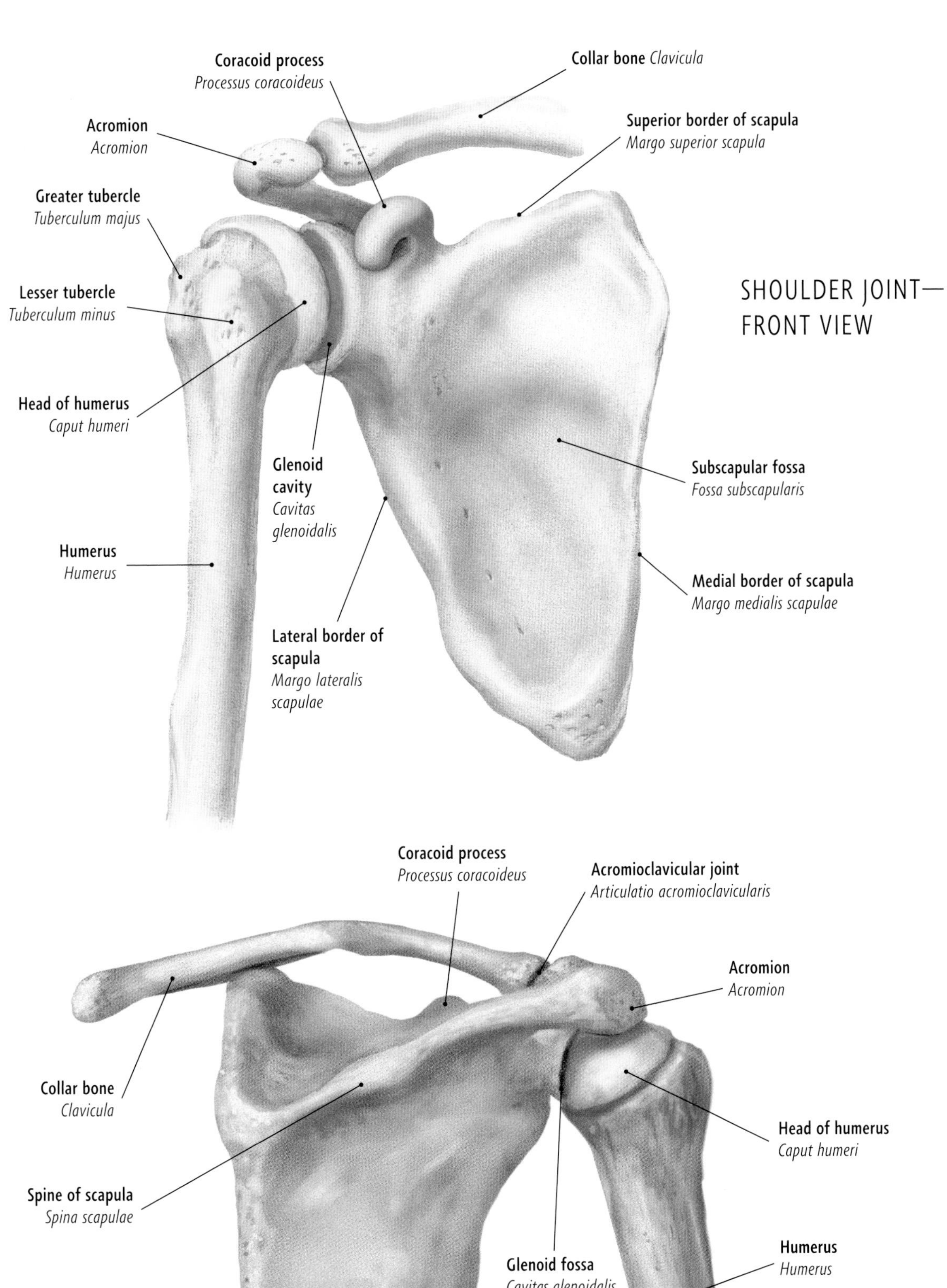
Coracoid process
Processus coracoideus
Collar bone Clavicula
Acromion
Acromion
Superior border of scapula
Margo superior scapula
Greater tubercle
Tuberculum majus
Lesser tubercle
Tuberculum minus
Head of humerus
Caput humeri
Glenoid cavity
Cavitas glenoidalis
Subscapular fossa
Fossa subscapularis
Humerus
Humerus
Medial border of scapula
Margo medialis scapulae
Lateral border of scapula
Margo lateralis scapulae

SHOULDER JOINT—FRONT VIEW

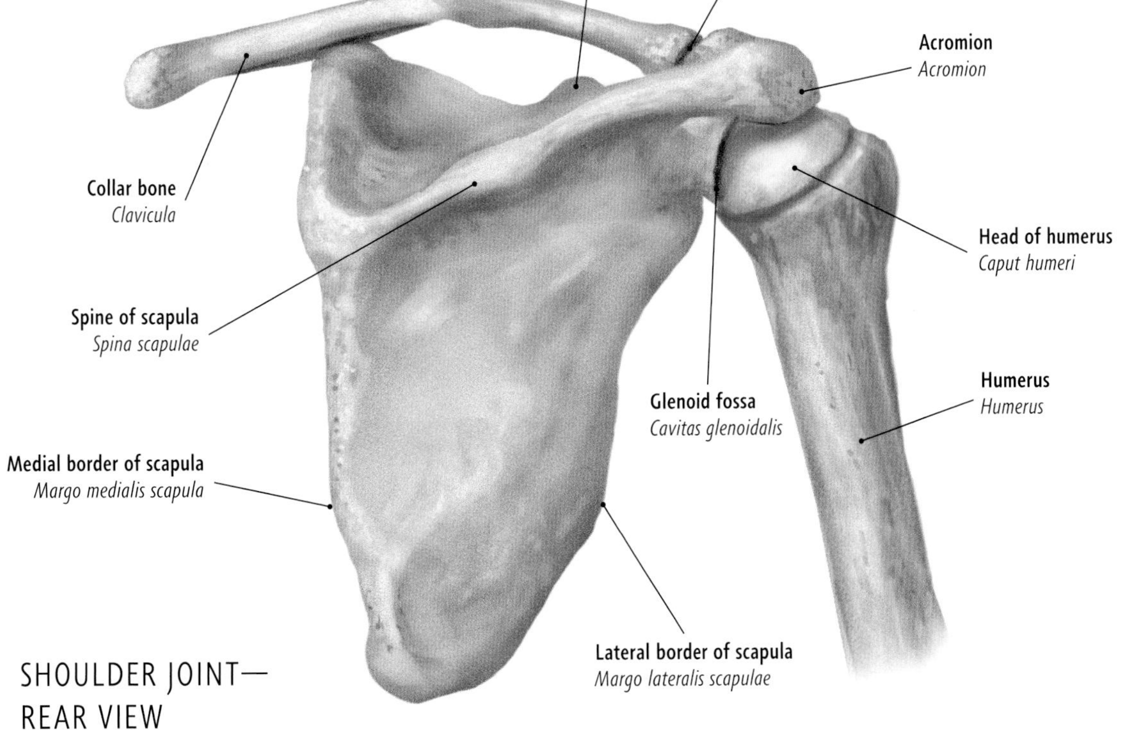
Coracoid process
Processus coracoideus
Acromioclavicular joint
Articulatio acromioclavicularis
Acromion
Acromion
Collar bone
Clavicula
Head of humerus
Caput humeri
Spine of scapula
Spina scapulae
Humerus
Humerus
Glenoid fossa
Cavitas glenoidalis
Medial border of scapula
Margo medialis scapula
Lateral border of scapula
Margo lateralis scapulae

SHOULDER JOINT—REAR VIEW

Bones of the Upper Arm

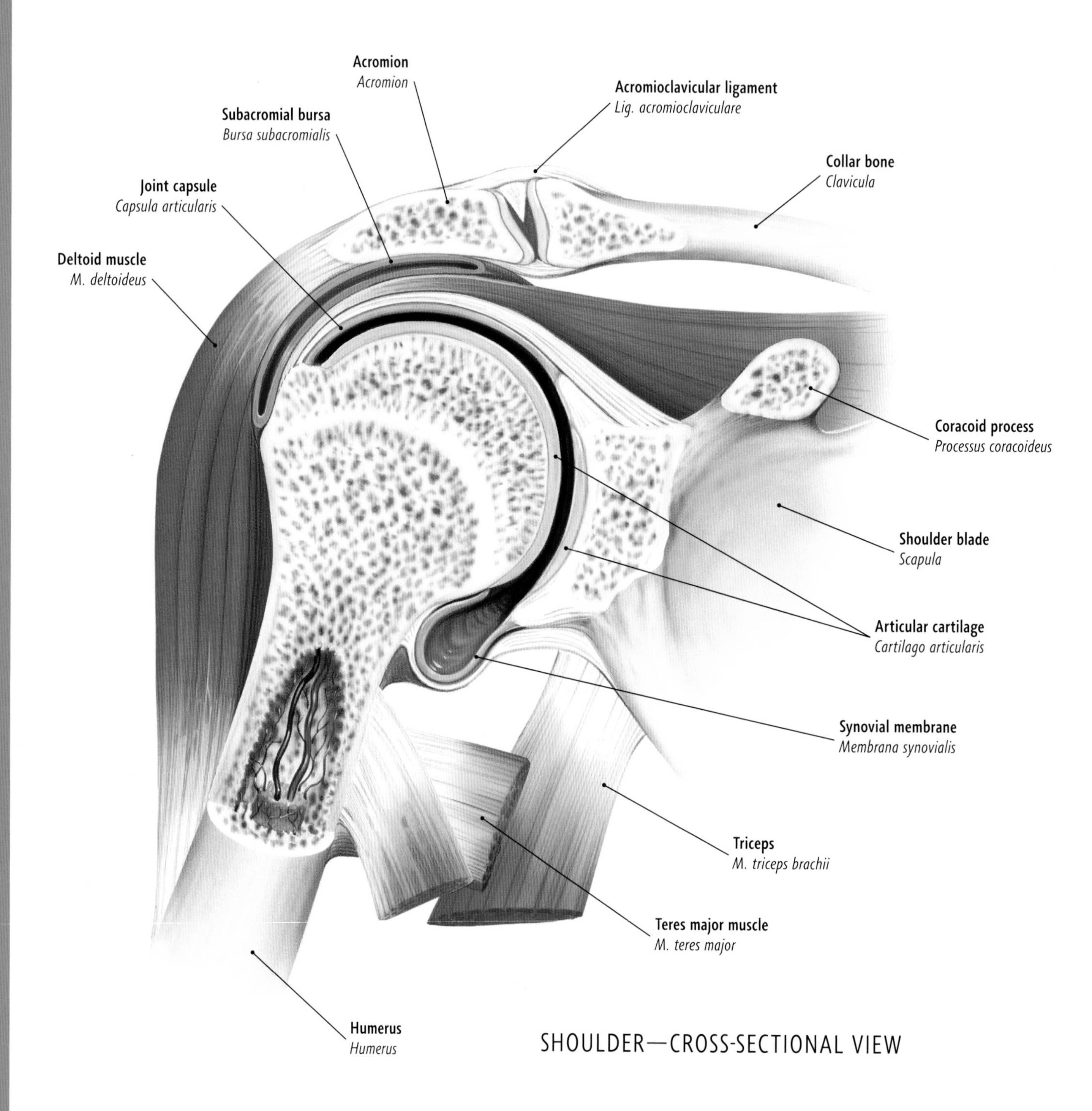
Acromion
Acromion
Acromioclavicular ligament
Lig. acromioclaviculare
Subacromial bursa
Bursa subacromialis
Collar bone
Clavicula
Joint capsule
Capsula articularis
Deltoid muscle
M. deltoideus
Coracoid process
Processus coracoideus
Shoulder blade
Scapula
Articular cartilage
Cartilago articularis
Synovial membrane
Membrana synovialis
Triceps
M. triceps brachii
Teres major muscle
M. teres major
Humerus
Humerus

SHOULDER—CROSS-SECTIONAL VIEW

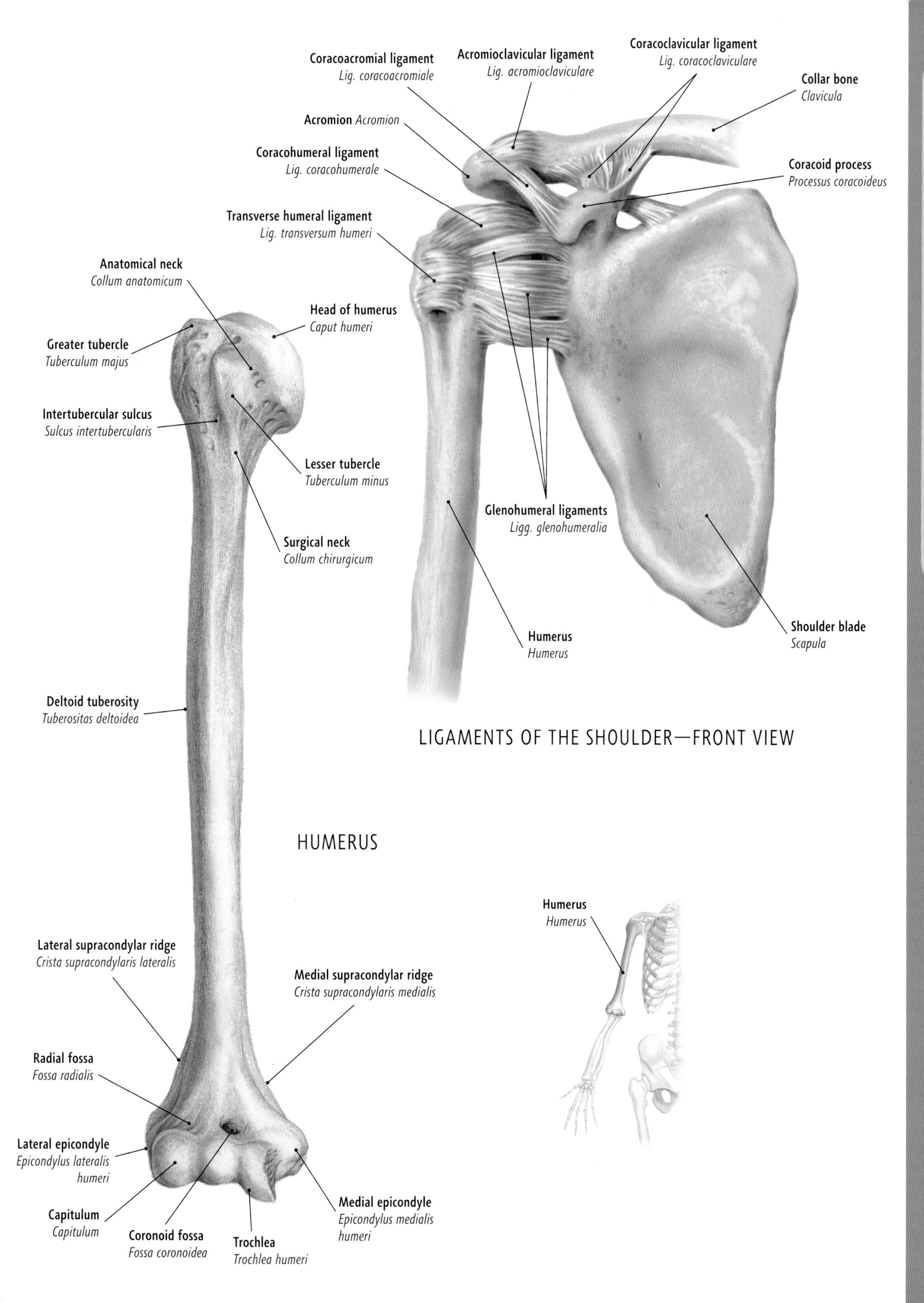
Coracoacromial ligament
Lig. coracoacromiale
Acromioclavicular ligament
Lig. acromioclaviculare
Coracoclavicular ligament
Lig. coracoclaviculare
Collar bone
Clavicula
Acromion Acromion
Coracohumeral ligament
Lig. coracohumerale
Coracoid process
Processus coracoideus
Transverse humeral ligament
Lig. transversum humeri
Glenohumeral ligaments
Ligg. glenohumeralia
Humerus
Humerus
Shoulder blade
Scapula
Anatomical neck
Collum anatomicum
Head of humerus
Caput humeri
Greater tubercle
Tuberculum majus
Intertubercular sulcus
Sulcus intertubercularis
Lesser tubercle
Tuberculum minus
Surgical neck
Collum chirurgicum
Deltoid tuberosity
Tuberositas deltoidea
Lateral supracondylar ridge
Crista supracondylaris lateralis
Medial supracondylar ridge
Crista supracondylaris medialis
Radial fossa
Fossa radialis
Lateral epicondyle
Epicondylus lateralis humeri
Capitulum
Capitulum
Coronoid fossa
Fossa coronoidea
Trochlea
Trochlea humeri
Medial epicondyle
Epicondylus medialis humeri
Humerus
Humerus

LIGAMENTS OF THE SHOULDER—FRONT VIEW

HUMERUS

Bones of the Lower Arm

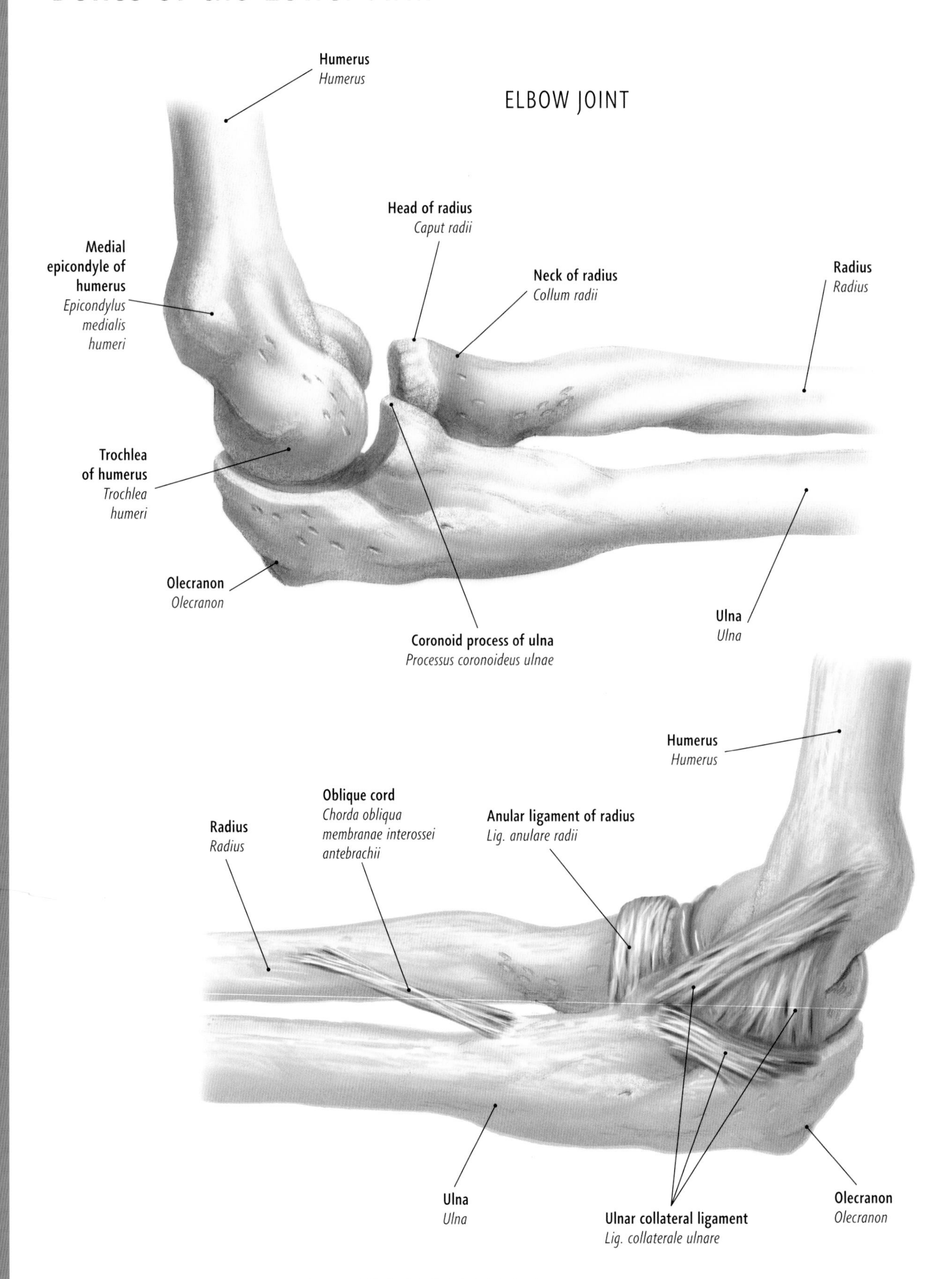
ELBOW JOINT
Humerus
Humerus
Head of radius
Caput radii
Medial epicondyle of humerus
Epicondylus medialis humeri
Neck of radius
Collum radii
Radius
Radius
Trochlea of humerus
Trochlea humeri
Olecranon
Olecranon
Coronoid process of ulna
Processus coronoideus ulnae
Ulna
Ulna
Humerus
Humerus
Oblique cord
Chorda obliqua membranae interossei antebrachii
Radius
Radius
Anular ligament of radius
Lig. anulare radii
Ulna
Ulna
Ulnar collateral ligament
Lig. collaterale ulnare
Olecranon
Olecranon
LIGAMENTS OF THE ELBOW

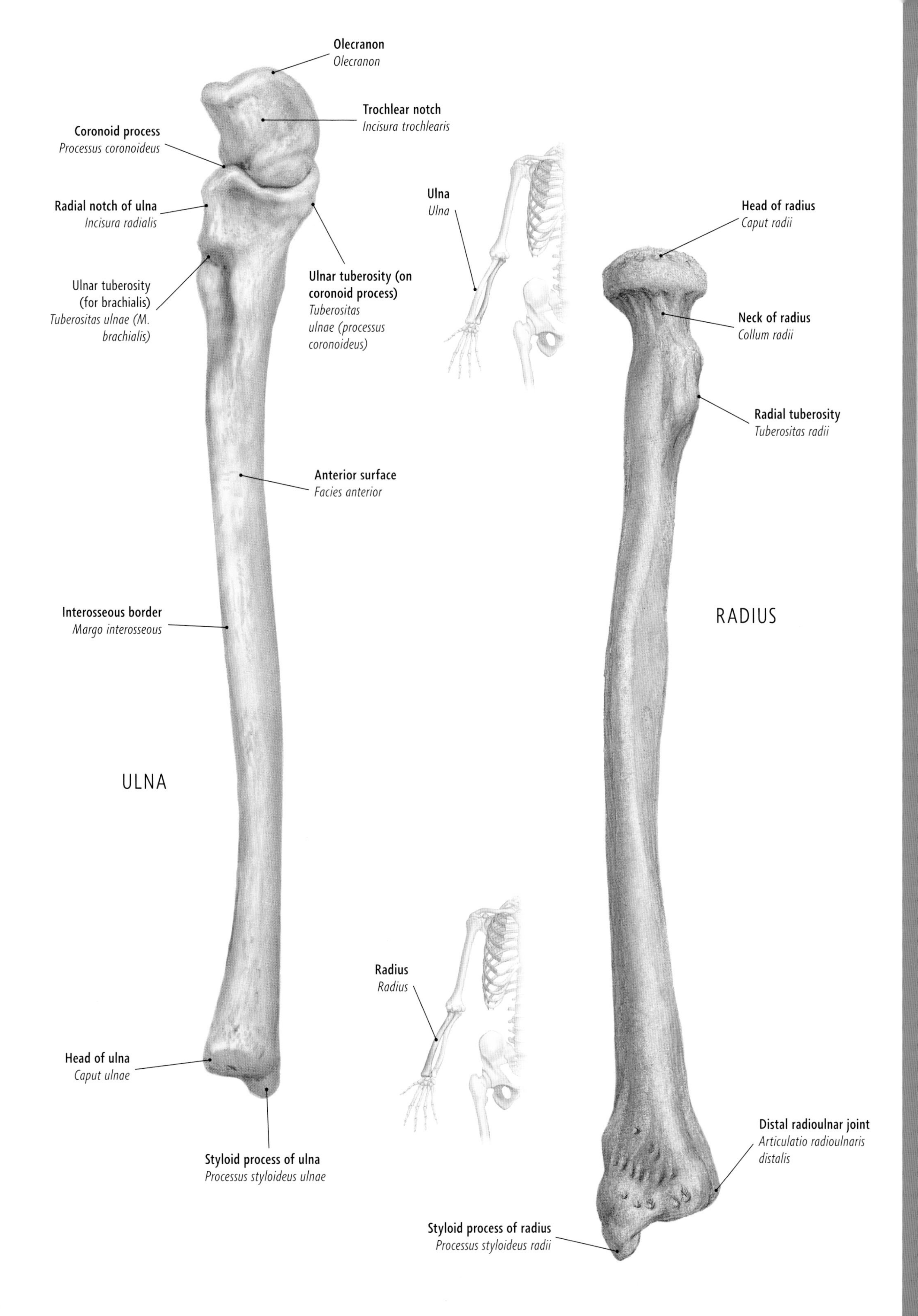
Olecranon
Olecranon
Trochlear notch
Incisura trochlearis
Coronoid process
Processus coronoideus
Radial notch of ulna
Incisura radialis
Ulnar tuberosity
(for brachialis)
Tuberositas ulnae (M. brachialis)
Ulnar tuberosity (on coronoid process)
Tuberositas ulnae (processus coronoideus)
Ulna
Ulna
Head of radius
Caput radii
Neck of radius
Collum radii
Radial tuberosity
Tuberositas radii
Anterior surface
Facies anterior
Interosseous border
Margo interosseous
RADIUS
ULNA
Radius
Radius
Head of ulna
Caput ulnae
Styloid process of ulna
Processus styloideus ulnae
Distal radioulnar joint
Articulatio radioulnaris distalis
Styloid process of radius
Processus styloideus radii

Bones of the Wrist and Hand

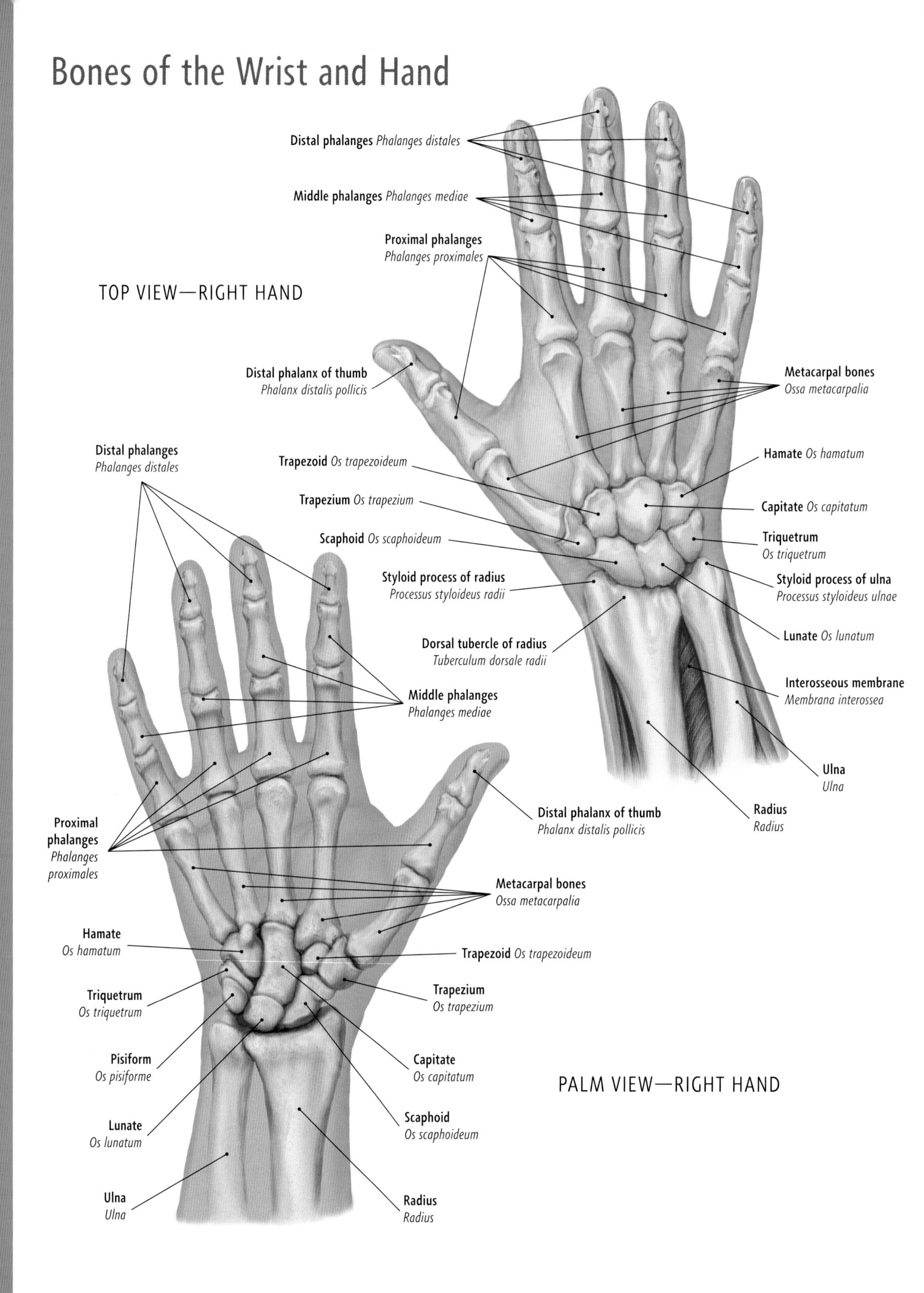
TOP VIEW—RIGHT HAND
Distal phalanges Phalanges distales
Middle phalanges Phalanges mediae
Proximal phalanges Phalanges proximales
Distal phalanx of thumb Phalanx distalis pollicis
Metacarpal bones Ossa metacarpalia
Trapezoid Os trapezoideum
Hamate Os hamatum
Trapezium Os trapezium
Capitate Os capitatum
Scaphoid Os scaphoideum
Triquetrum Os triquetrum
Styloid process of radius Processus styloideus radii
Styloid process of ulna Processus styloideus ulnae
Lunate Os lunatum
Dorsal tubercle of radius Tuberculum dorsale radii
Interosseous membrane Membrana interossea
Ulna Ulna
Radius Radius
PALM VIEW—RIGHT HAND
Distal phalanges Phalanges distales
Middle phalanges Phalanges mediae
Proximal phalanges Phalanges proximales
Distal phalanx of thumb Phalanx distalis pollicis
Metacarpal bones Ossa metacarpalia
Hamate Os hamatum
Trapezoid Os trapezoideum
Triquetrum Os triquetrum
Trapezium Os trapezium
Pisiform Os pisiforme
Capitate Os capitatum
Lunate Os lunatum
Scaphoid Os scaphoideum
Ulna Ulna
Radius Radius

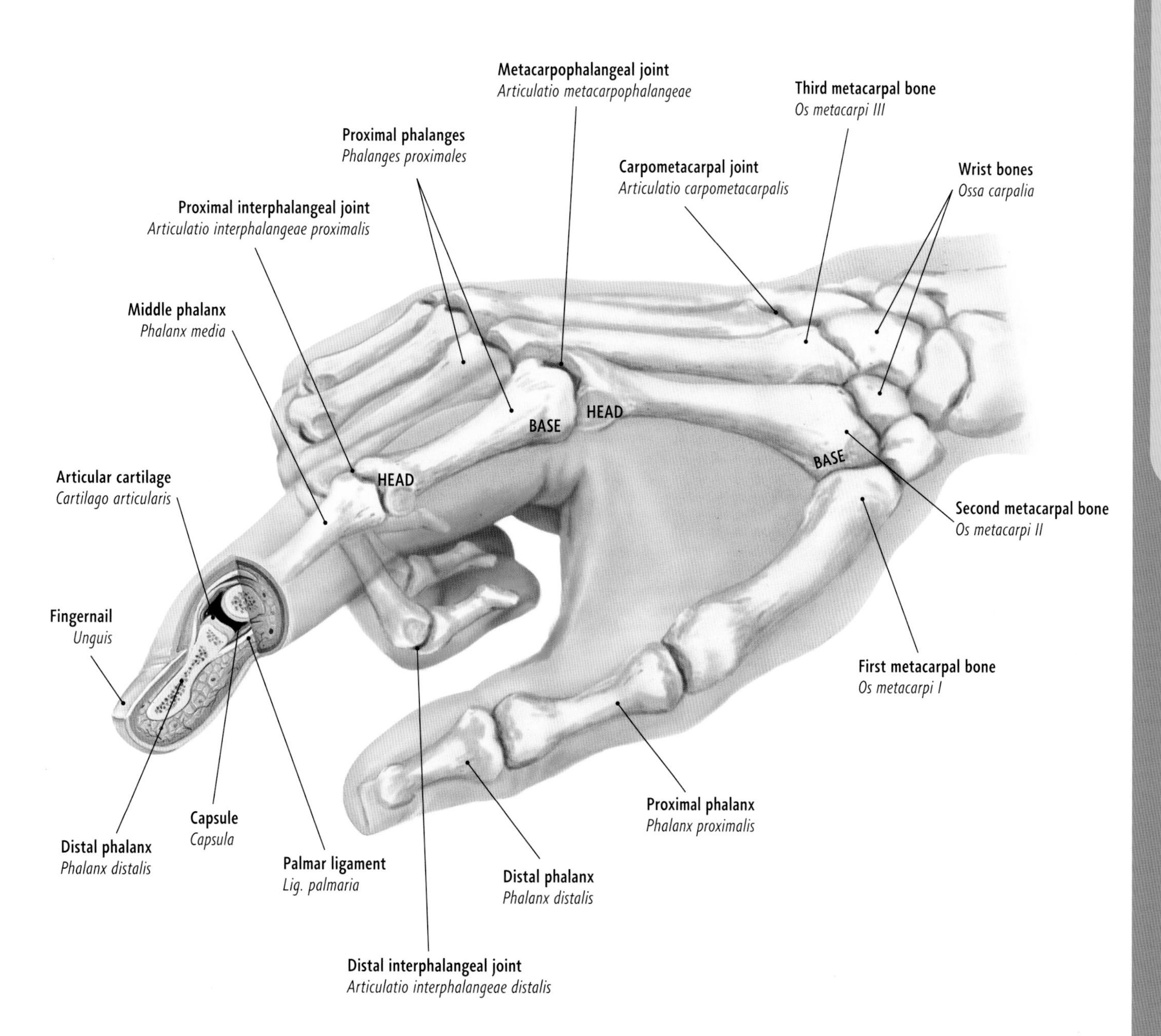
Metacarpophalangeal joint
Articulatio metacarpophalangeae
Third metacarpal bone
Os metacarpi III
Proximal phalanges
Phalanges proximales
Carpometacarpal joint
Articulatio carpometacarpalis
Wrist bones
Ossa carpalia
Proximal interphalangeal joint
Articulatio interphalangeae proximalis
Middle phalanx
Phalanx media
HEAD
BASE
HEAD
BASE
Articular cartilage
Cartilago articularis
Second metacarpal bone
Os metacarpi II
Fingernail
Unguis
First metacarpal bone
Os metacarpi I
Proximal phalanx
Phalanx proximalis
Capsule
Capsula
Distal phalanx
Phalanx distalis
Palmar ligament
Lig. palmaria
Distal phalanx
Phalanx distalis
Distal interphalangeal joint
Articulatio interphalangeae distalis

FINGER

Bones of the Pelvis

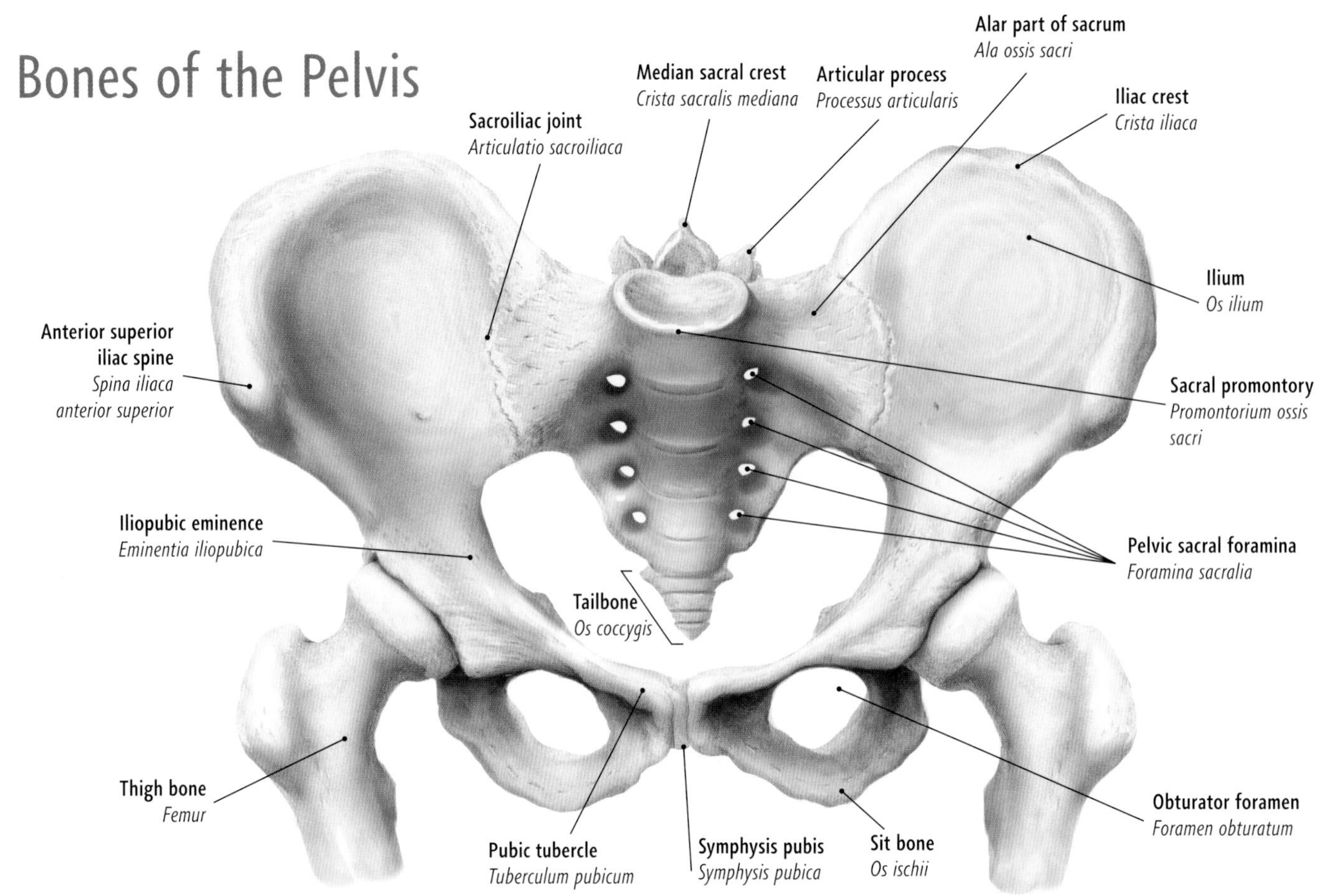

Alar part of sacrum
Ala ossis sacri
Median sacral crest
Crista sacralis mediana
Articular process
Processus articularis
Iliac crest
Crista iliaca
Sacroiliac joint
Articulatio sacroiliaca
Ilium
Os ilium
Anterior superior iliac spine
Spina iliaca anterior superior
Sacral promontory
Promontorium ossis sacri
Iliopubic eminence
Eminentia iliopubica
Pelvic sacral foramina
Foramina sacralia
Tailbone
Os coccygis
Thigh bone
Femur
Obturator foramen
Foramen obturatum
Pubic tubercle
Tuberculum pubicum
Symphysis pubis
Symphysis pubica
Sit bone
Os ischii

FEMALE PELVIS—FRONT VIEW

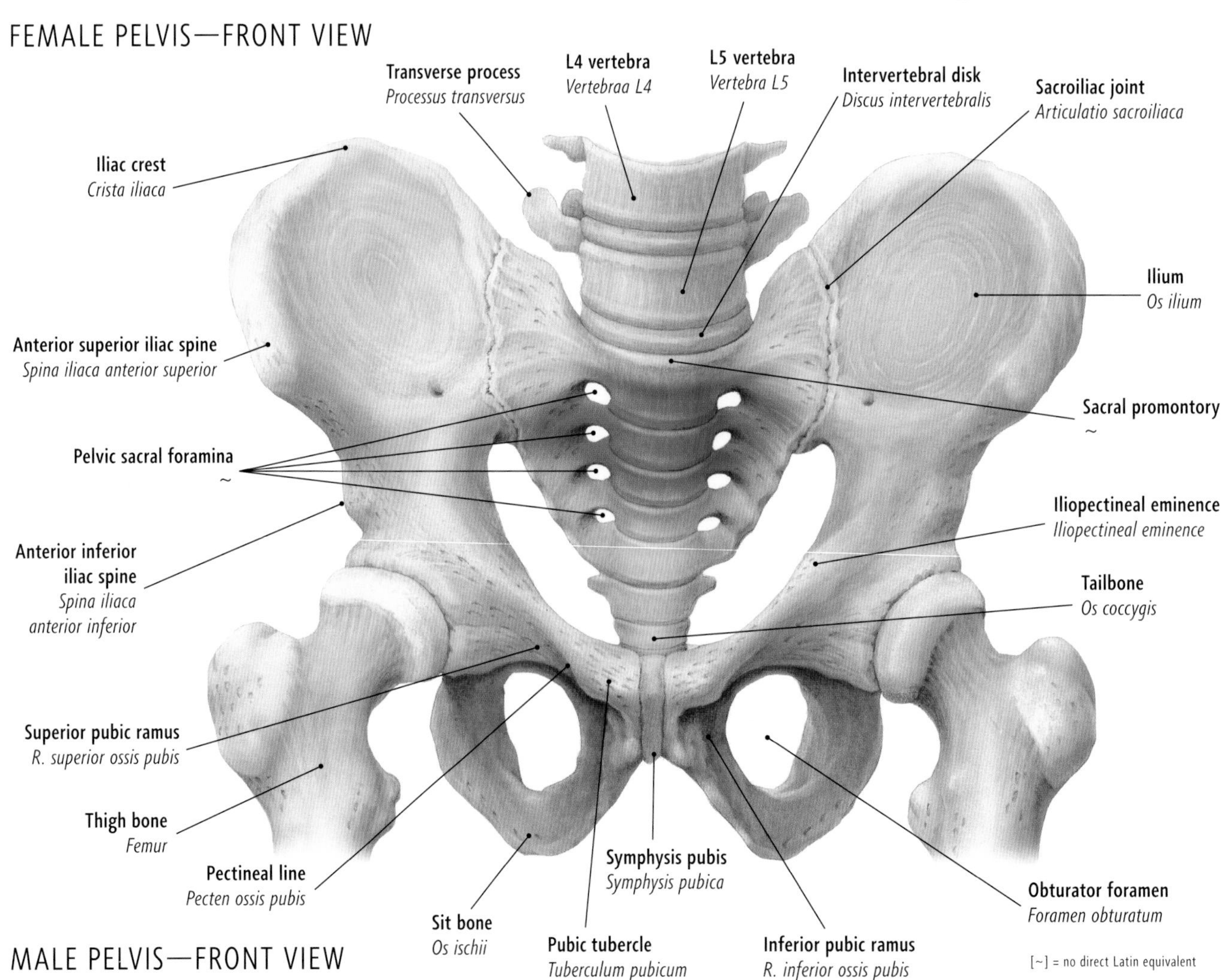

Transverse process
Processus transversus
L4 vertebra
Vertebraa L4
L5 vertebra
Vertebra L5
Intervertebral disk
Discus intervertebralis
Sacroiliac joint
Articulatio sacroiliaca
Iliac crest
Crista iliaca
Ilium
Os ilium
Anterior superior iliac spine
Spina iliaca anterior superior
Sacral promontory
~
Pelvic sacral foramina
~
Iliopectineal eminence
Iliopectineal eminence
Anterior inferior iliac spine
Spina iliaca anterior inferior
Tailbone
Os coccygis
Superior pubic ramus
R. superior ossis pubis
Thigh bone
Femur
Pectineal line
Pecten ossis pubis
Symphysis pubis
Symphysis pubica
Obturator foramen
Foramen obturatum
Sit bone
Os ischii
Pubic tubercle
Tuberculum pubicum
Inferior pubic ramus
R. inferior ossis pubis

MALE PELVIS—FRONT VIEW

[~] = no direct Latin equivalent

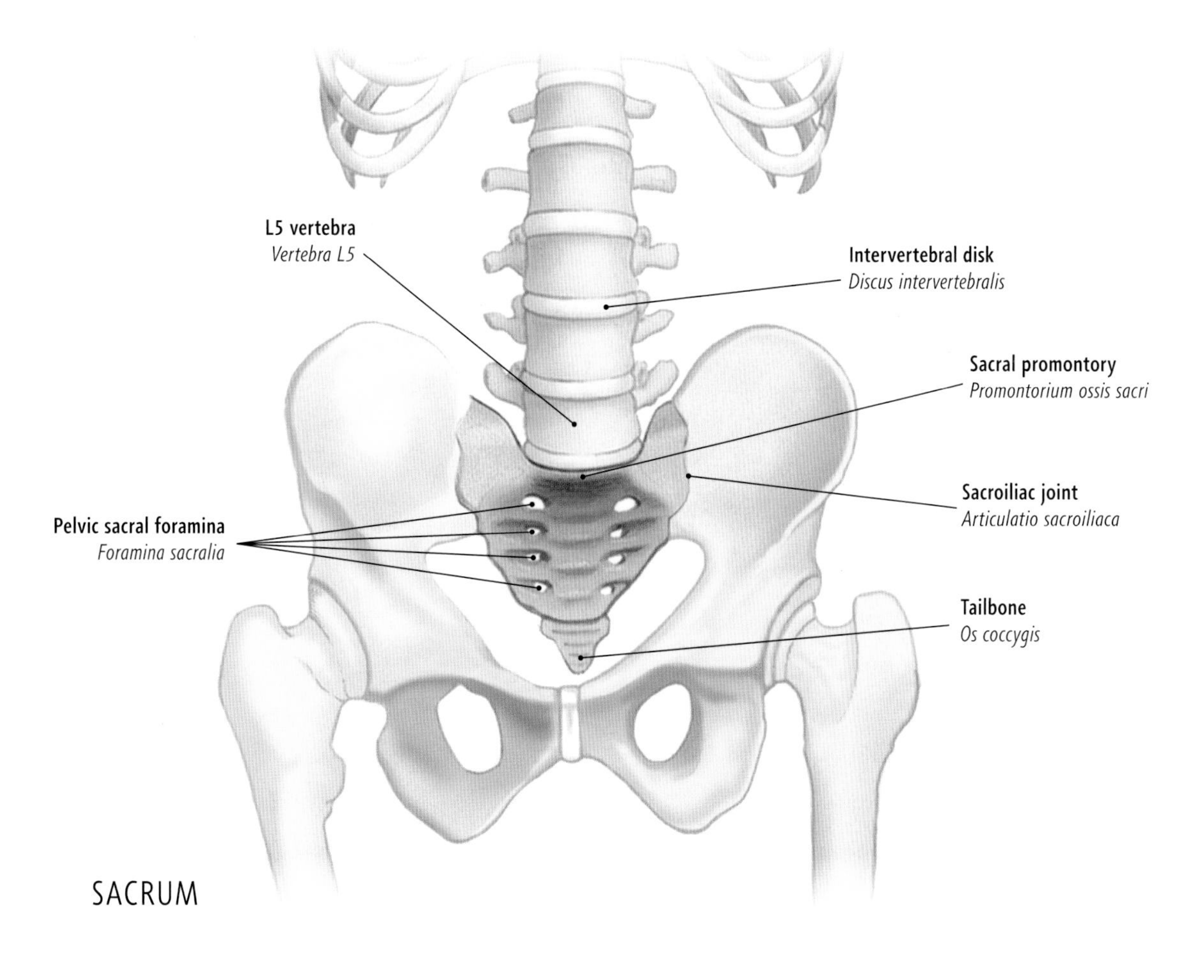
L5 vertebra
Vertebra L5
Intervertebral disk
Discus intervertebralis
Sacral promontory
Promontorium ossis sacri
Sacroiliac joint
Articulatio sacroiliaca
Pelvic sacral foramina
Foramina sacralia
Tailbone
Os coccygis

SACRUM

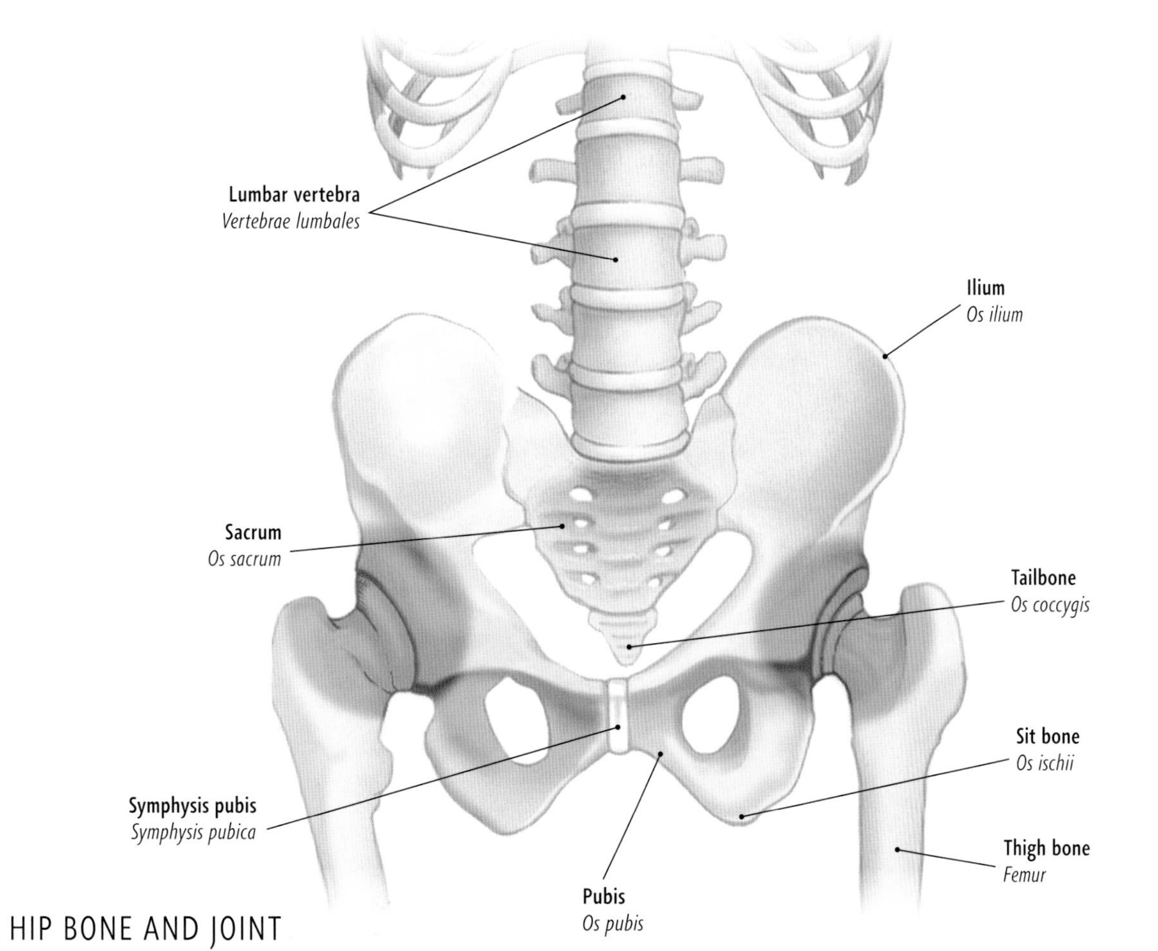
Lumbar vertebra
Vertebrae lumbales
Ilium
Os ilium
Sacrum
Os sacrum
Tailbone
Os coccygis
Sit bone
Os ischii
Symphysis pubis
Symphysis pubica
Thigh bone
Femur
Pubis
Os pubis

HIP BONE AND JOINT

Bones of the Leg

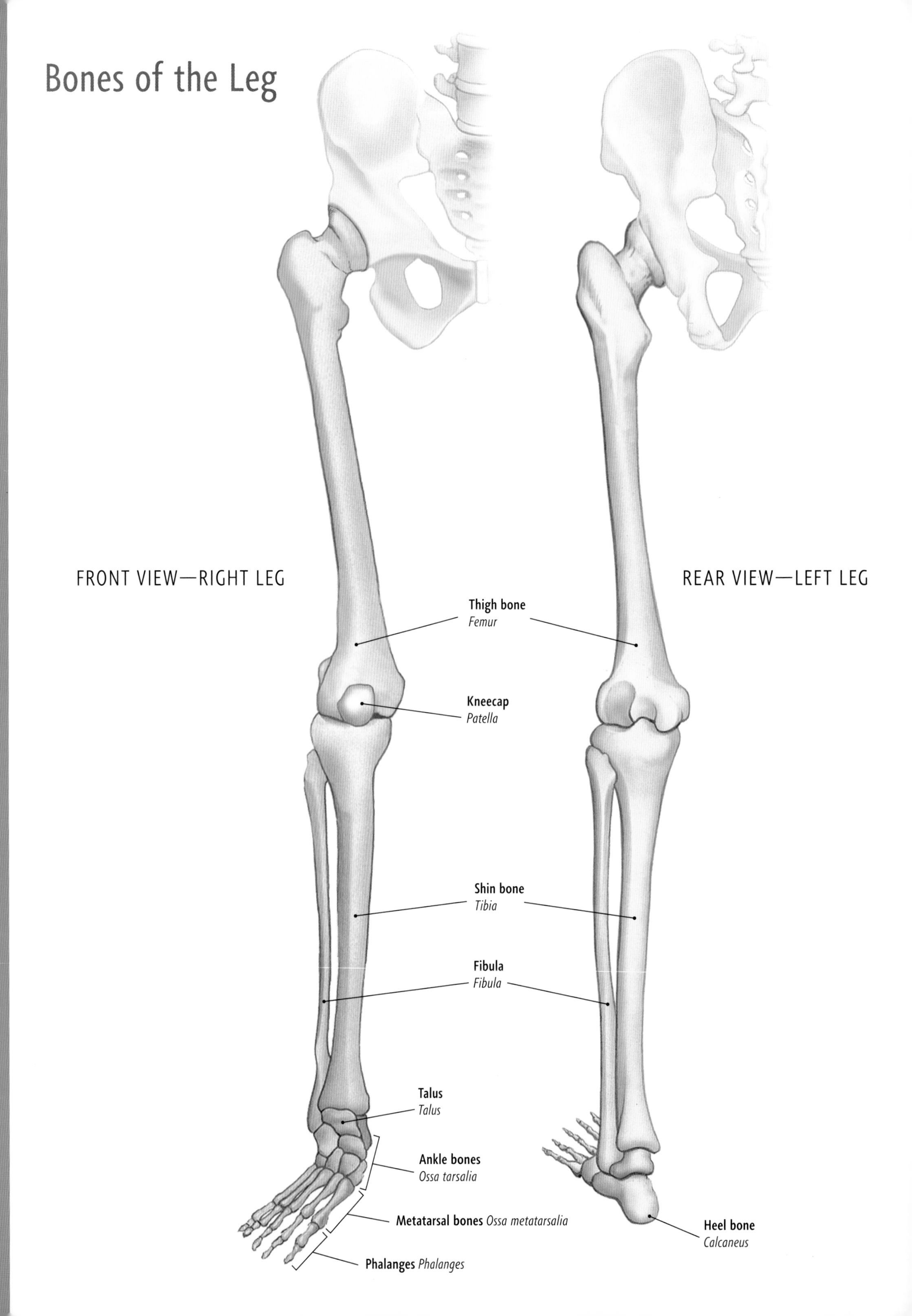
FRONT VIEW—RIGHT LEG
REAR VIEW—LEFT LEG
Thigh bone
Femur
Kneecap
Patella
Shin bone
Tibia
Fibula
Fibula
Talus
Talus
Ankle bones
Ossa tarsalia
Metatarsal bones Ossa metatarsalia
Phalanges Phalanges
Heel bone
Calcaneus

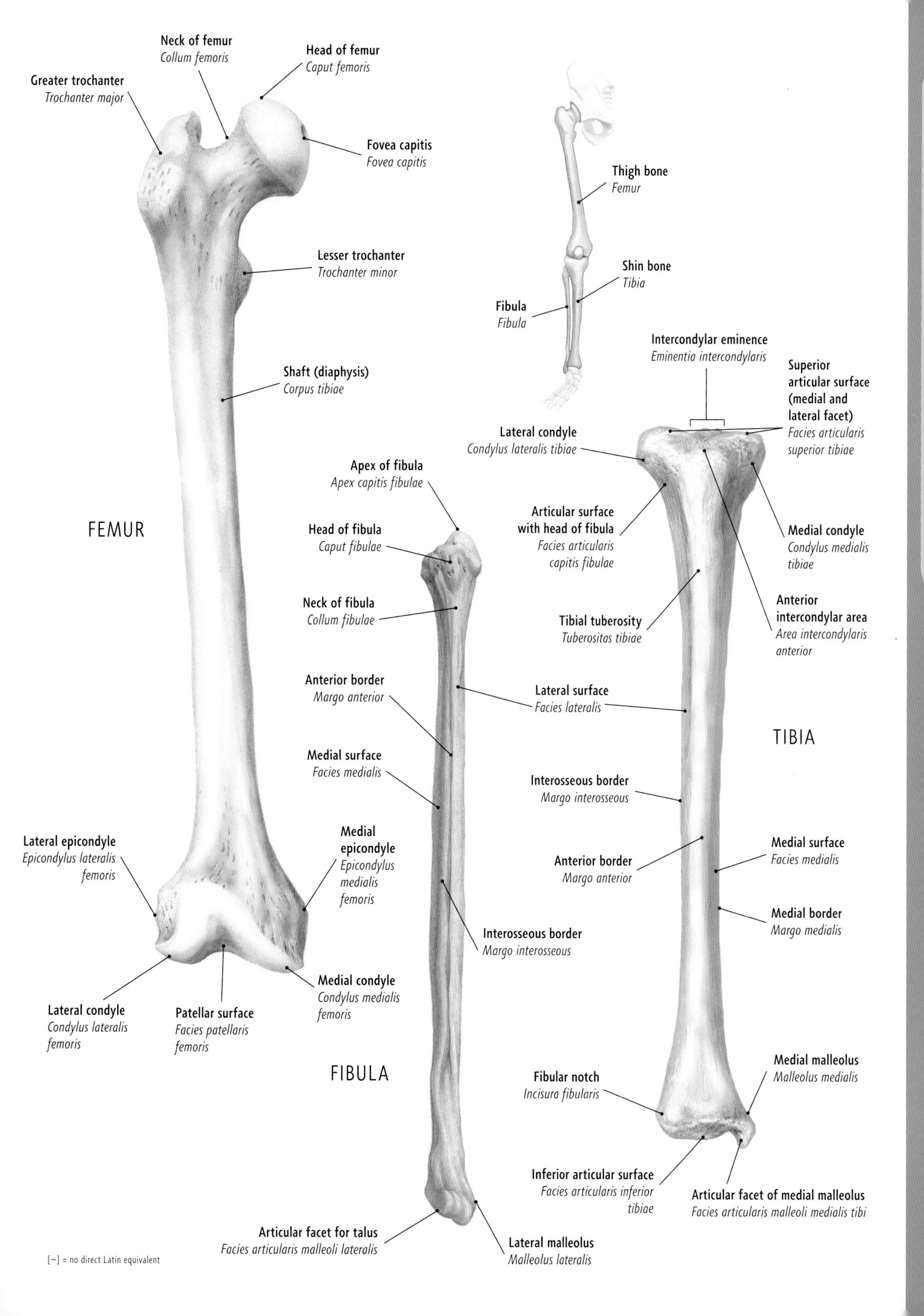
Neck of femur
Collum femoris
Head of femur
Caput femoris
Greater trochanter
Trochanter major
Fovea capitis
Fovea capitis
Thigh bone
Femur
Lesser trochanter
Trochanter minor
Shin bone
Tibia
Fibula
Fibula
Intercondylar eminence
Eminentia intercondylaris
Superior
articular surface
(medial and
lateral facet)
Facies articularis
superior tibiae
Shaft (diaphysis)
Corpus tibiae
Lateral condyle
Condylus lateralis tibiae
Apex of fibula
Apex capitis fibulae
FEMUR
Head of fibula
Caput fibulae
Articular surface
with head of fibula
Facies articularis
capitis fibulae
Medial condyle
Condylus medialis
tibiae
Neck of fibula
Collum fibulae
Tibial tuberosity
Tuberositas tibiae
Anterior
intercondylar area
Area intercondylaris
anterior
Anterior border
Margo anterior
Lateral surface
Facies lateralis
TIBIA
Medial surface
Facies medialis
Interosseous border
Margo interosseous
Lateral epicondyle
Epicondylus lateralis
femoris
Medial
epicondyle
Epicondylus
medialis
femoris
Anterior border
Margo anterior
Medial surface
Facies medialis
Medial border
Margo medialis
Interosseous border
Margo interosseous
Medial condyle
Condylus medialis
femoris
Lateral condyle
Condylus lateralis
femoris
Patellar surface
Facies patellaris
femoris
FIBULA
Medial malleolus
Malleolus medialis
Fibular notch
Incisura fibularis
Inferior articular surface
Facies articularis inferior
tibiae
Articular facet of medial malleolus
Facies articularis malleoli medialis tibi
Articular facet for talus
Facies articularis malleoli lateralis
Lateral malleolus
Malleolus lateralis

[~] = no direct Latin equivalent

Bones of the Knee

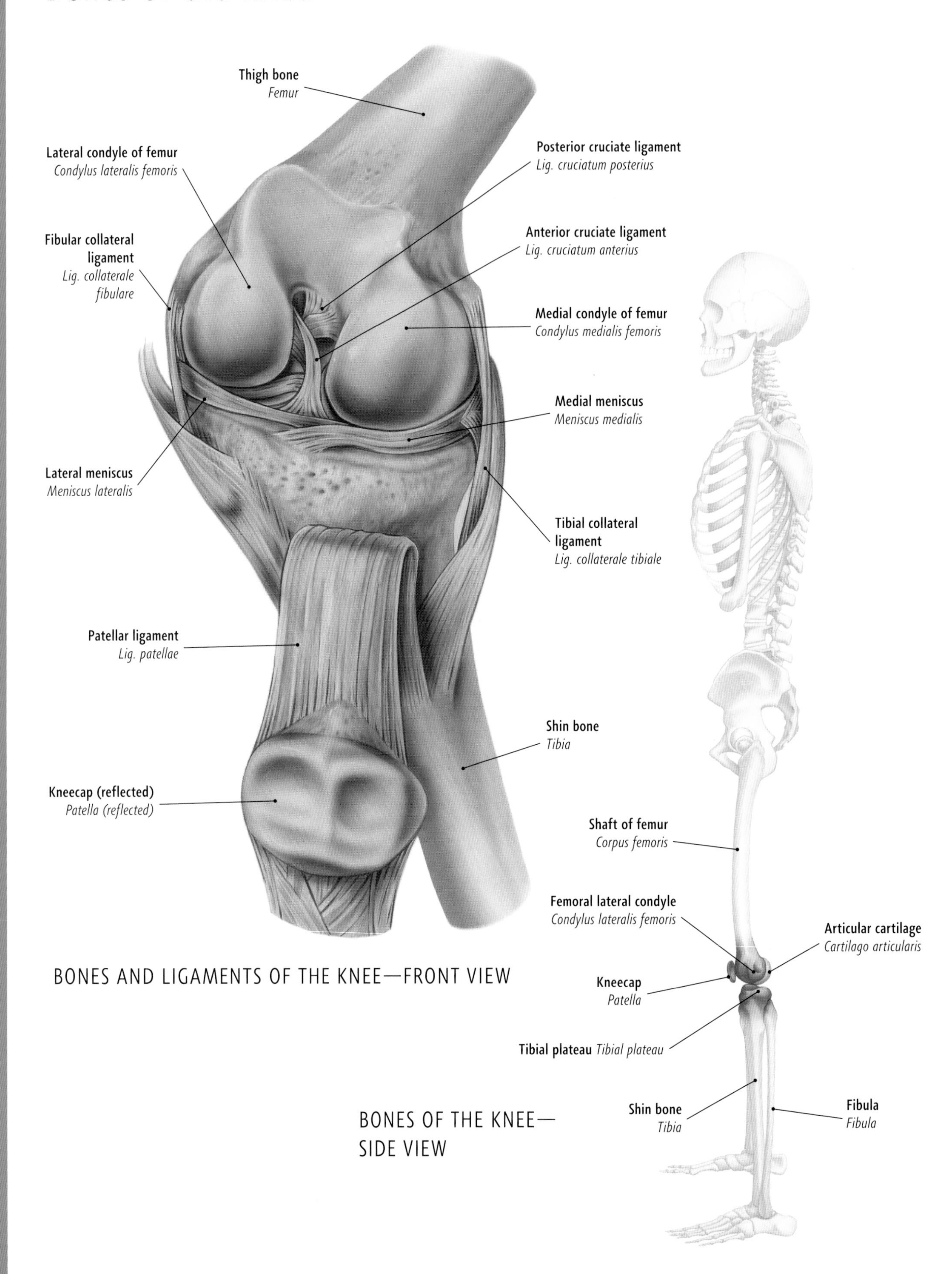
Thigh bone
Femur
Lateral condyle of femur
Condylus lateralis femoris
Posterior cruciate ligament
Lig. cruciatum posterius
Fibular collateral ligament
Lig. collaterale fibulare
Anterior cruciate ligament
Lig. cruciatum anterius
Medial condyle of femur
Condylus medialis femoris
Medial meniscus
Meniscus medialis
Lateral meniscus
Meniscus lateralis
Tibial collateral ligament
Lig. collaterale tibiale
Patellar ligament
Lig. patellae
Shin bone
Tibia
Kneecap (reflected)
Patella (reflected)
Shaft of femur
Corpus femoris
Femoral lateral condyle
Condylus lateralis femoris
Articular cartilage
Cartilago articularis
Kneecap
Patella
Tibial plateau Tibial plateau
Shin bone
Tibia
Fibula
Fibula

BONES AND LIGAMENTS OF THE KNEE—FRONT VIEW

BONES OF THE KNEE—SIDE VIEW

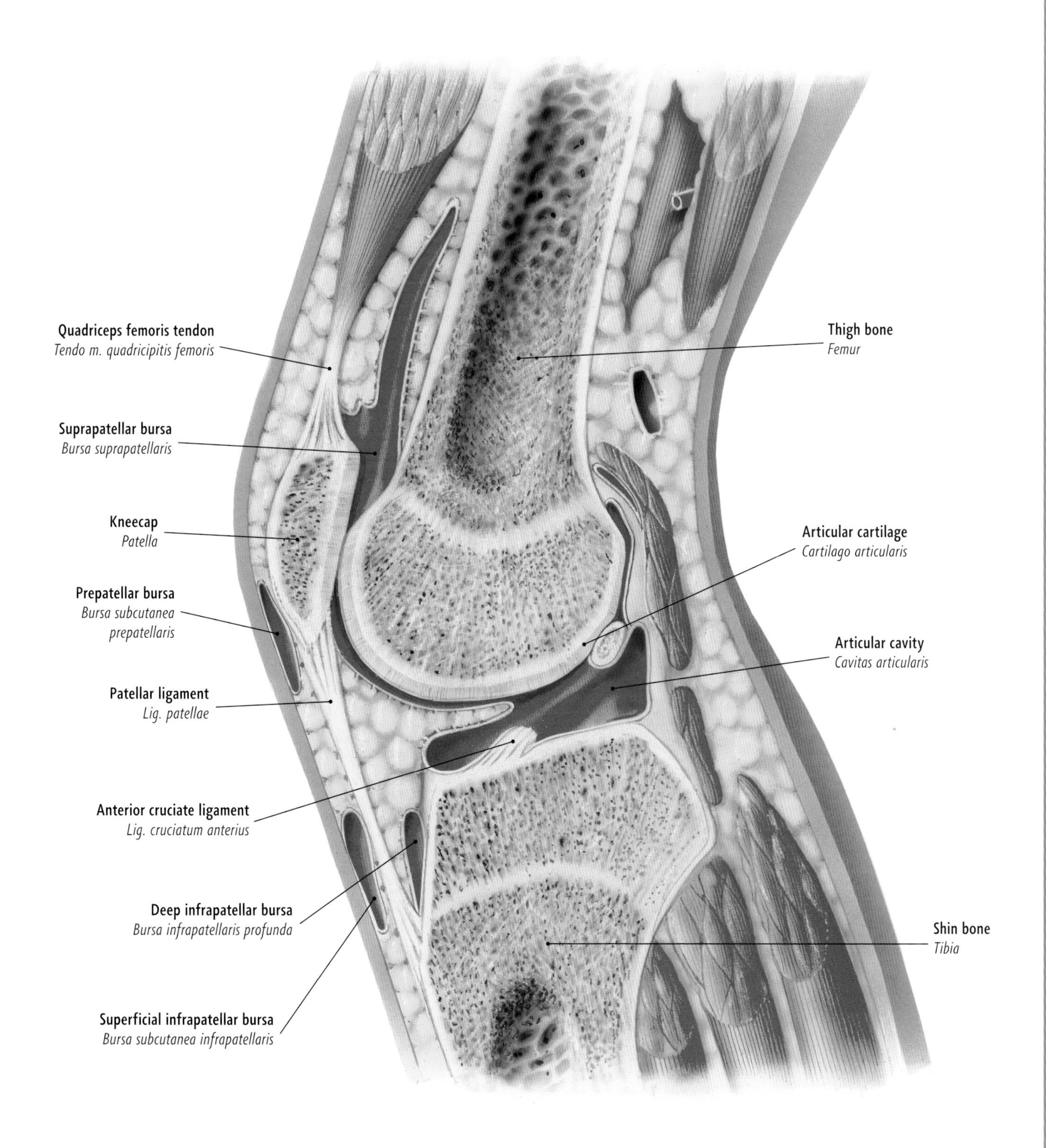
Quadriceps femoris tendon
Tendo m. quadricipitis femoris
Suprapatellar bursa
Bursa suprapatellaris
Kneecap
Patella
Prepatellar bursa
Bursa subcutanea prepatellaris
Patellar ligament
Lig. patellae
Anterior cruciate ligament
Lig. cruciatum anterius
Deep infrapatellar bursa
Bursa infrapatellaris profunda
Superficial infrapatellar bursa
Bursa subcutanea infrapatellaris
Thigh bone
Femur
Articular cartilage
Cartilago articularis
Articular cavity
Cavitas articularis
Shin bone
Tibia

KNEE JOINT—CROSS-SECTIONAL VIEW

Bones of the Ankle and Foot

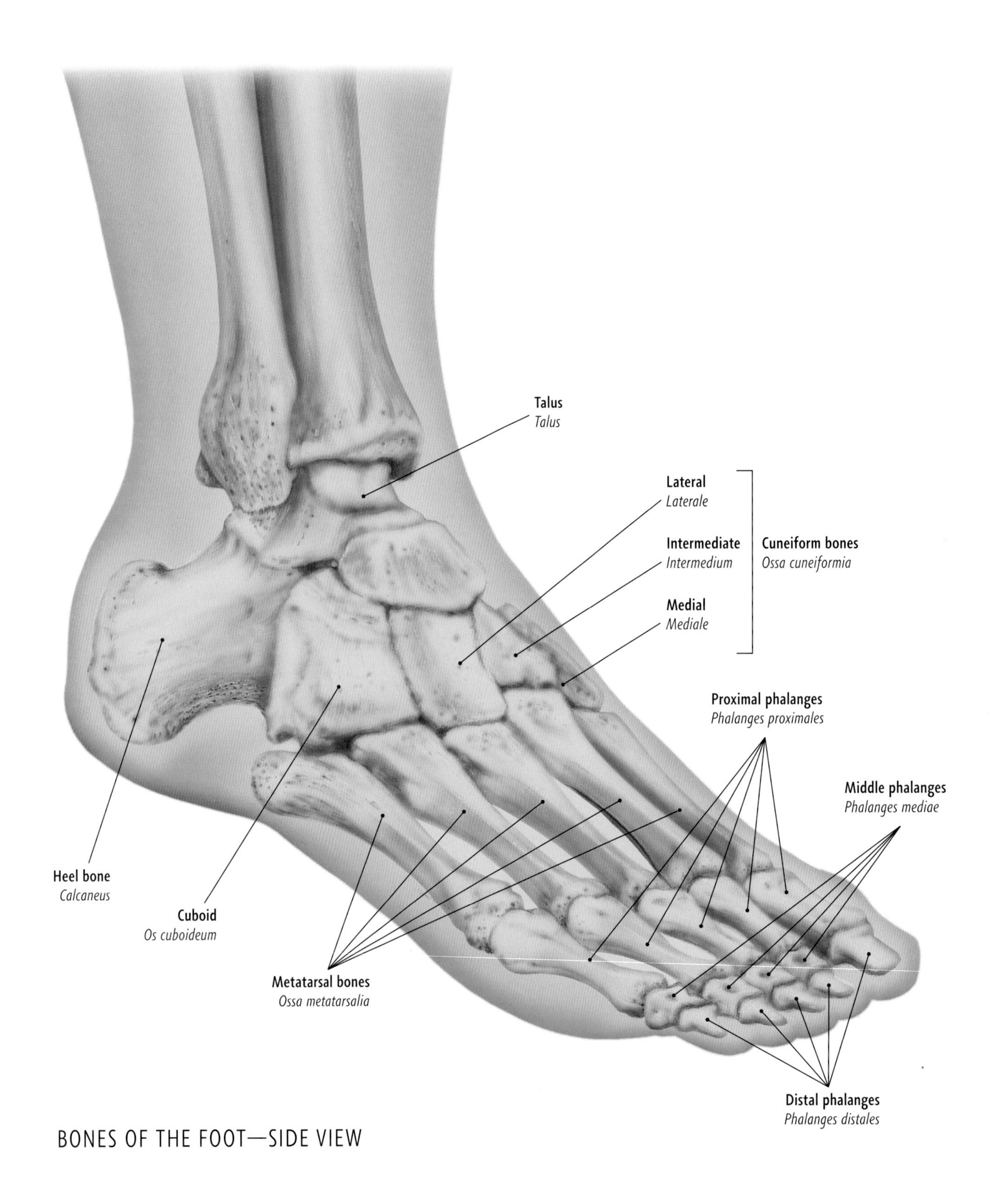

Talus
Talus
Lateral
Laterale
Intermediate
Intermedium
Medial
Mediale
Cuneiform bones
Ossa cuneiformia
Proximal phalanges
Phalanges proximales
Middle phalanges
Phalanges mediae
Heel bone
Calcaneus
Cuboid
Os cuboideum
Metatarsal bones
Ossa metatarsalia
Distal phalanges
Phalanges distales

BONES OF THE FOOT—SIDE VIEW

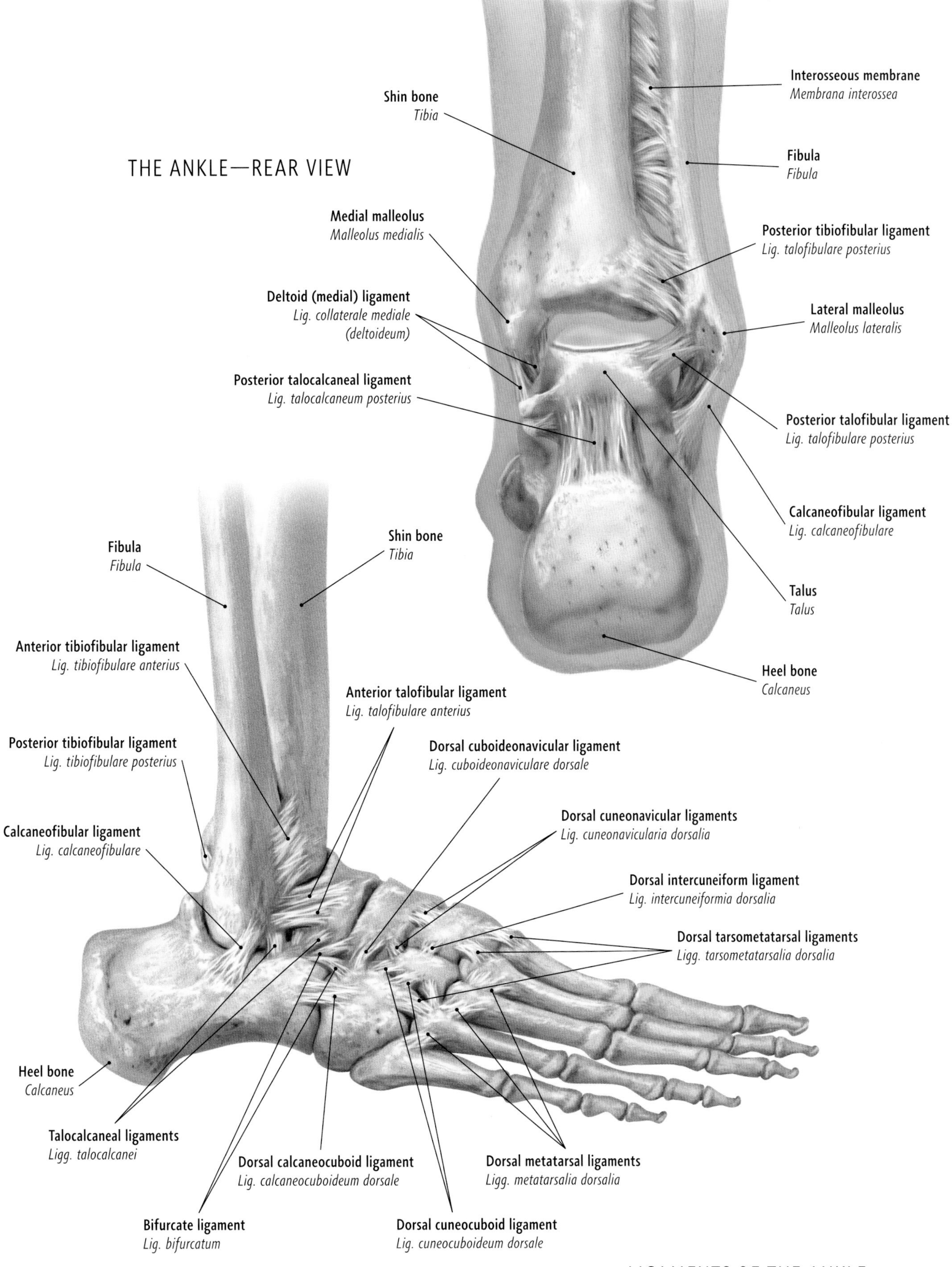
THE ANKLE—REAR VIEW
Shin bone
Tibia
Interosseous membrane
Membrana interossea
Fibula
Fibula
Medial malleolus
Malleolus medialis
Posterior tibiofibular ligament
Lig. talofibulare posterius
Deltoid (medial) ligament
Lig. collaterale mediale (deltoideum)
Lateral malleolus
Malleolus lateralis
Posterior talocalcaneal ligament
Lig. talocalcaneum posterius
Posterior talofibular ligament
Lig. talofibulare posterius
Calcaneofibular ligament
Lig. calcaneofibulare
Talus
Talus
Heel bone
Calcaneus
Fibula
Fibula
Shin bone
Tibia
Anterior tibiofibular ligament
Lig. tibiofibulare anterius
Anterior talofibular ligament
Lig. talofibulare anterius
Posterior tibiofibular ligament
Lig. tibiofibulare posterius
Dorsal cuboideonavicular ligament
Lig. cuboideonaviculare dorsale
Calcaneofibular ligament
Lig. calcaneofibulare
Dorsal cuneonavicular ligaments
Lig. cuneonavicularia dorsalia
Dorsal intercuneiform ligament
Lig. intercuneiformia dorsalia
Dorsal tarsometatarsal ligaments
Ligg. tarsometatarsalia dorsalia
Heel bone
Calcaneus
Talocalcaneal ligaments
Ligg. talocalcanei
Dorsal calcaneocuboid ligament
Lig. calcaneocuboideum dorsale
Dorsal metatarsal ligaments
Ligg. metatarsalia dorsalia
Bifurcate ligament
Lig. bifurcatum
Dorsal cuneocuboid ligament
Lig. cuneocuboideum dorsale
LIGAMENTS OF THE ANKLE AND FOOT—SIDE VIEW

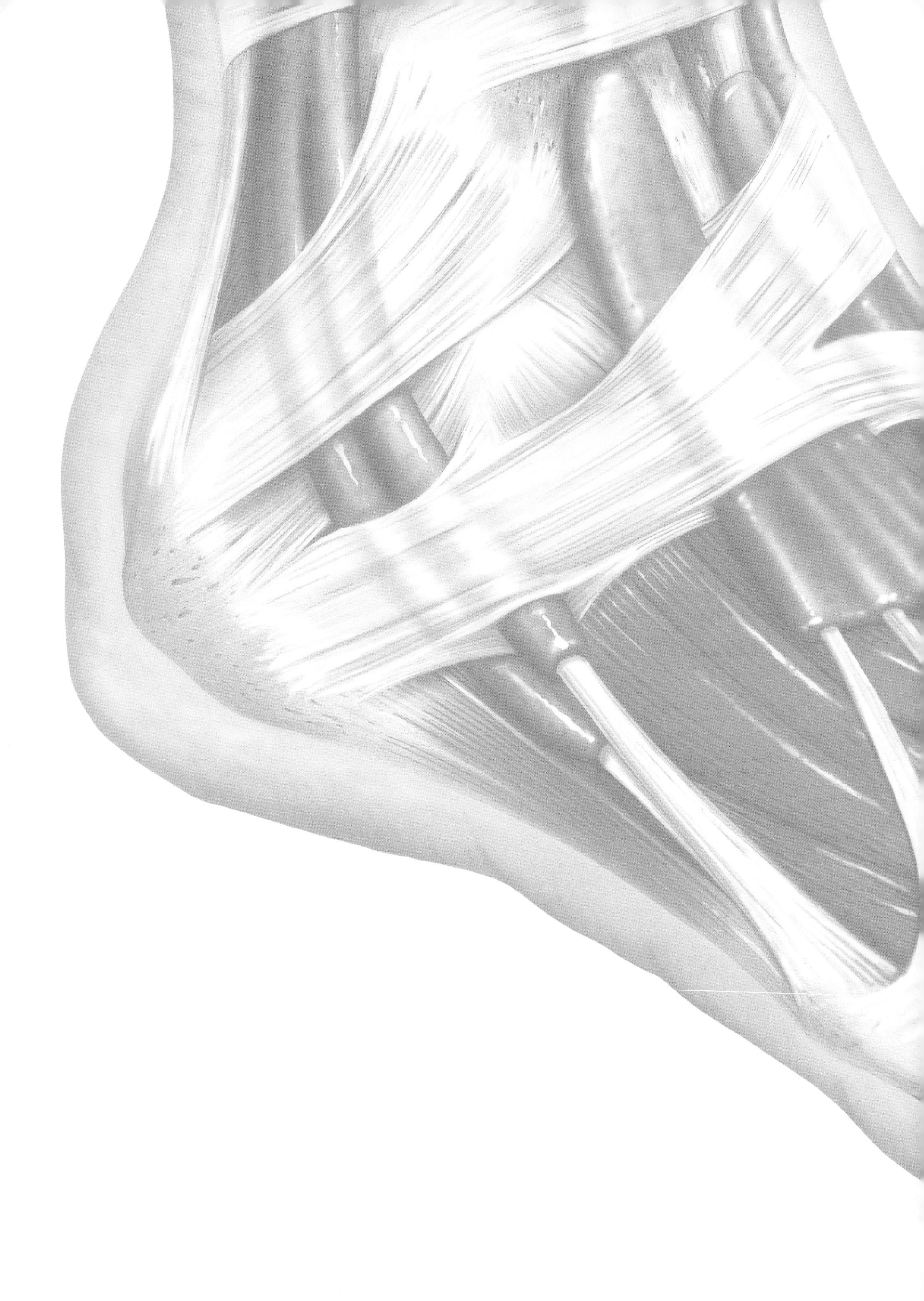

Body Movement

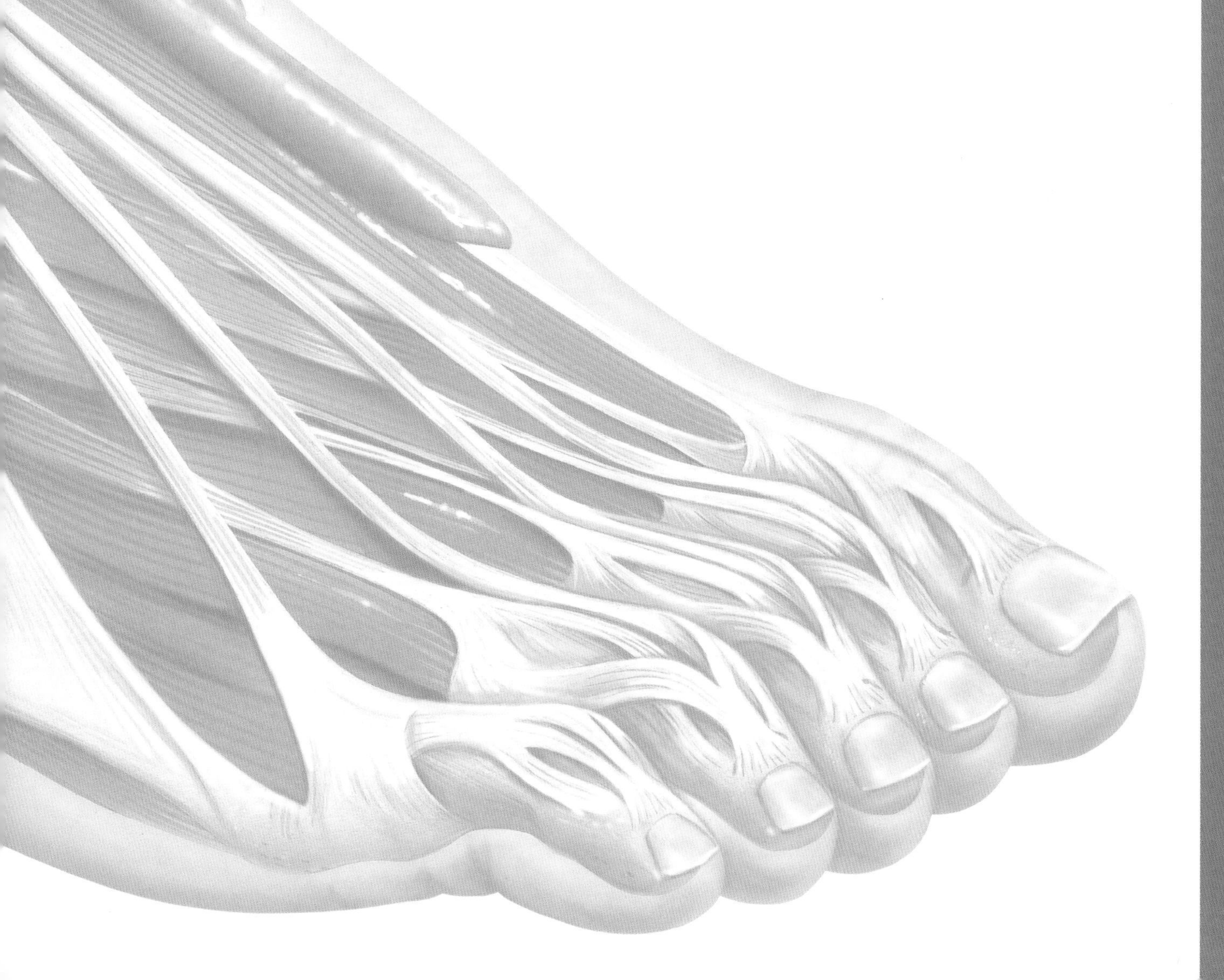

Movements of the Body

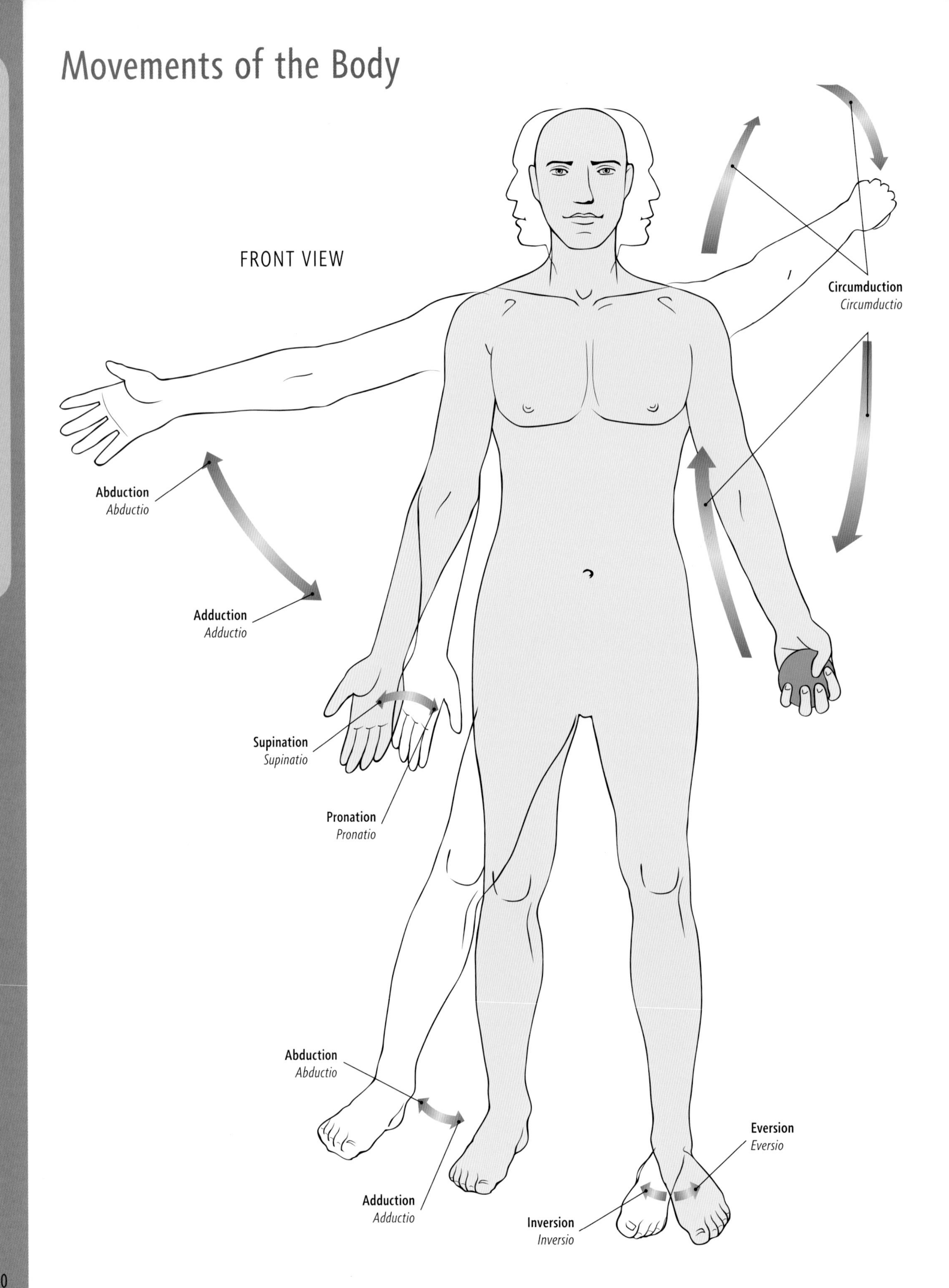
FRONT VIEW
Circumduction
Circumductio
Abduction
Abductio
Adduction
Adductio
Supination
Supinatio
Pronation
Pronatio
Abduction
Abductio
Adduction
Adductio
Eversion
Eversio
Inversion
Inversio

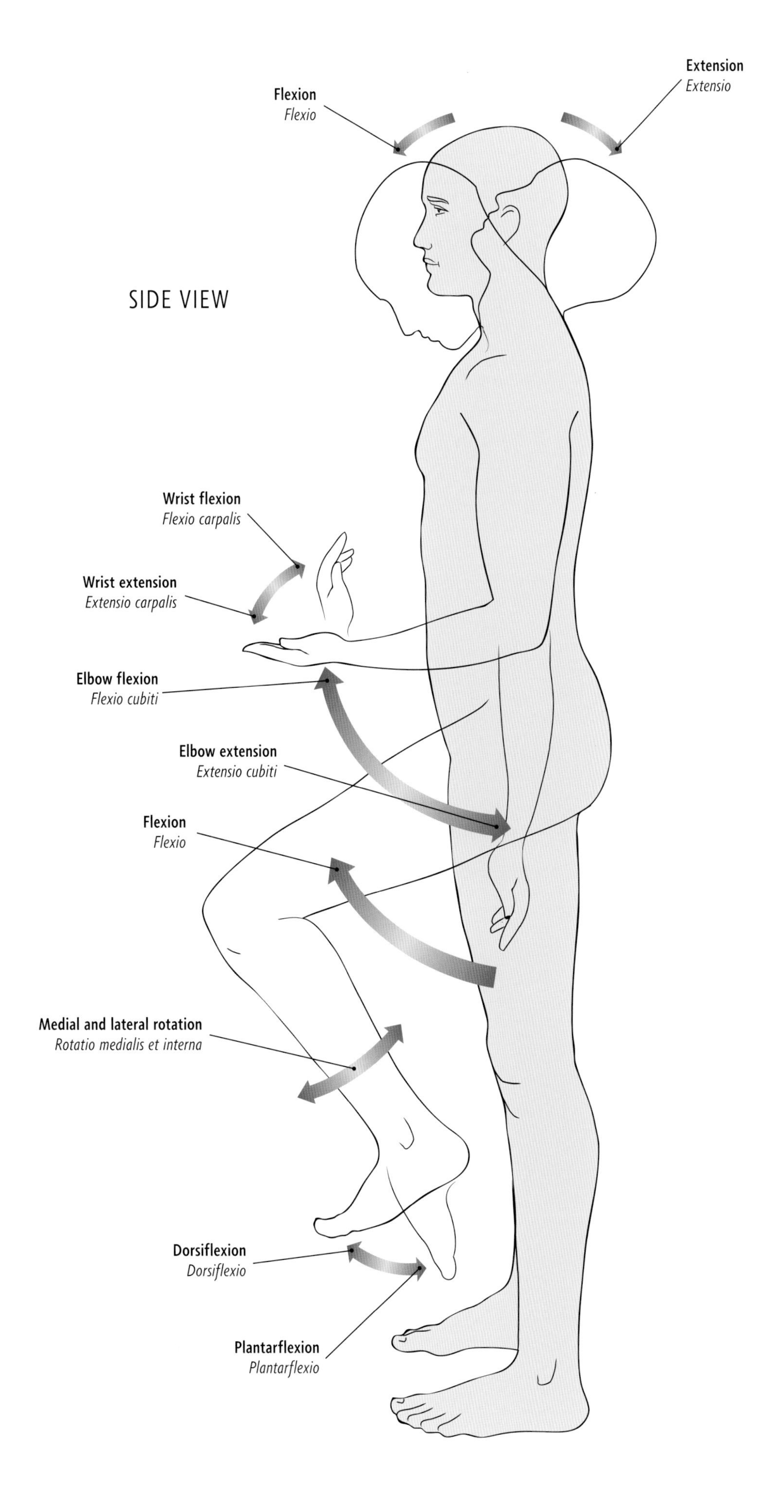
Flexion
Flexio
Extension
Extensio
SIDE VIEW
Wrist flexion
Flexio carpalis
Wrist extension
Extensio carpalis
Elbow flexion
Flexio cubiti
Elbow extension
Extensio cubiti
Flexion
Flexio
Medial and lateral rotation
Rotatio medialis et interna
Dorsiflexion
Dorsiflexio
Plantarflexion
Plantarflexio

Types of Joints

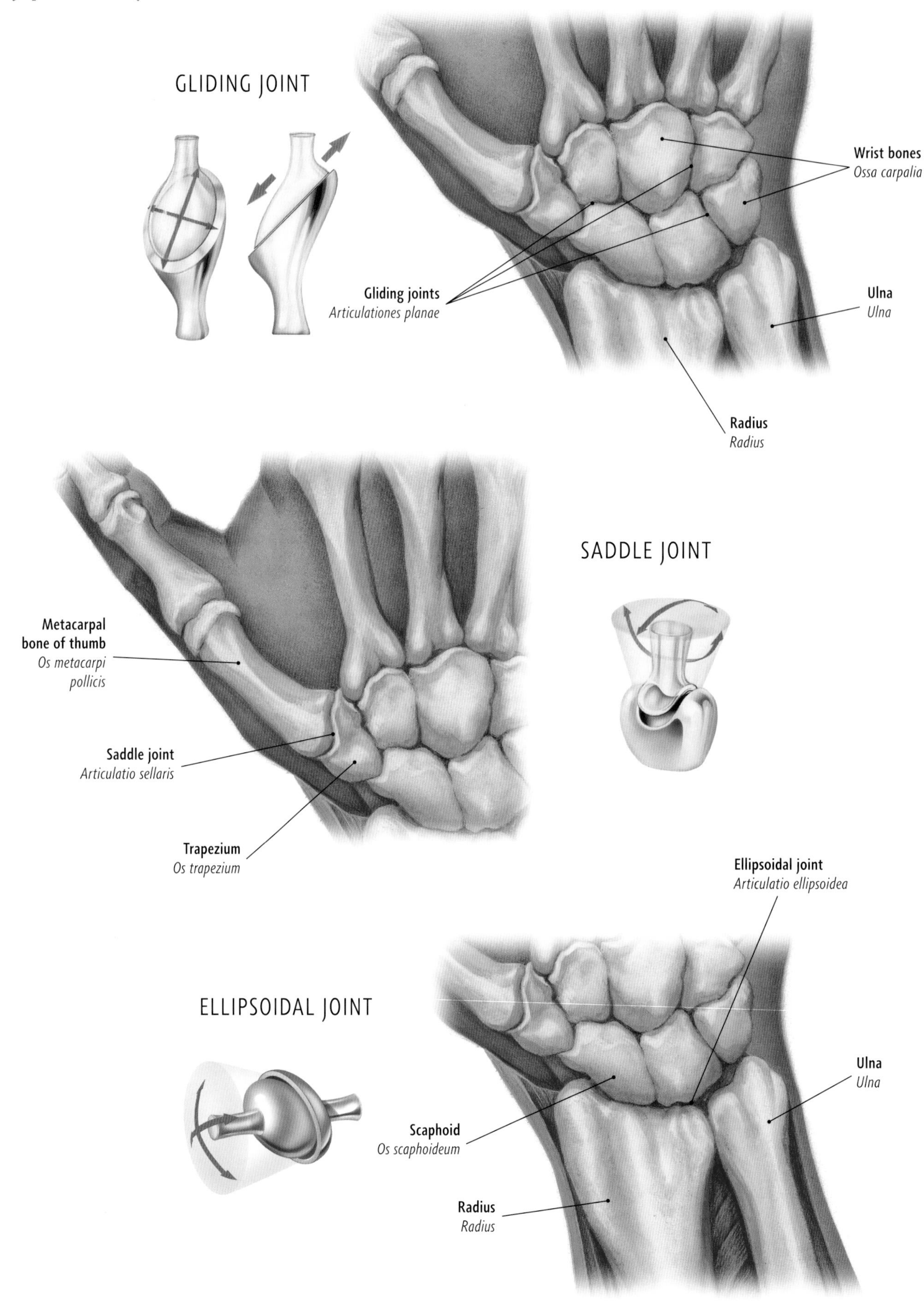
GLIDING JOINT
Wrist bones
Ossa carpalia
Gliding joints
Articulationes planae
Ulna
Ulna
Radius
Radius
SADDLE JOINT
Metacarpal
bone of thumb
Os metacarpi
pollicis
Saddle joint
Articulatio sellaris
Trapezium
Os trapezium
Ellipsoidal joint
Articulatio ellipsoidea
ELLIPSOIDAL JOINT
Ulna
Ulna
Scaphoid
Os scaphoideum
Radius
Radius

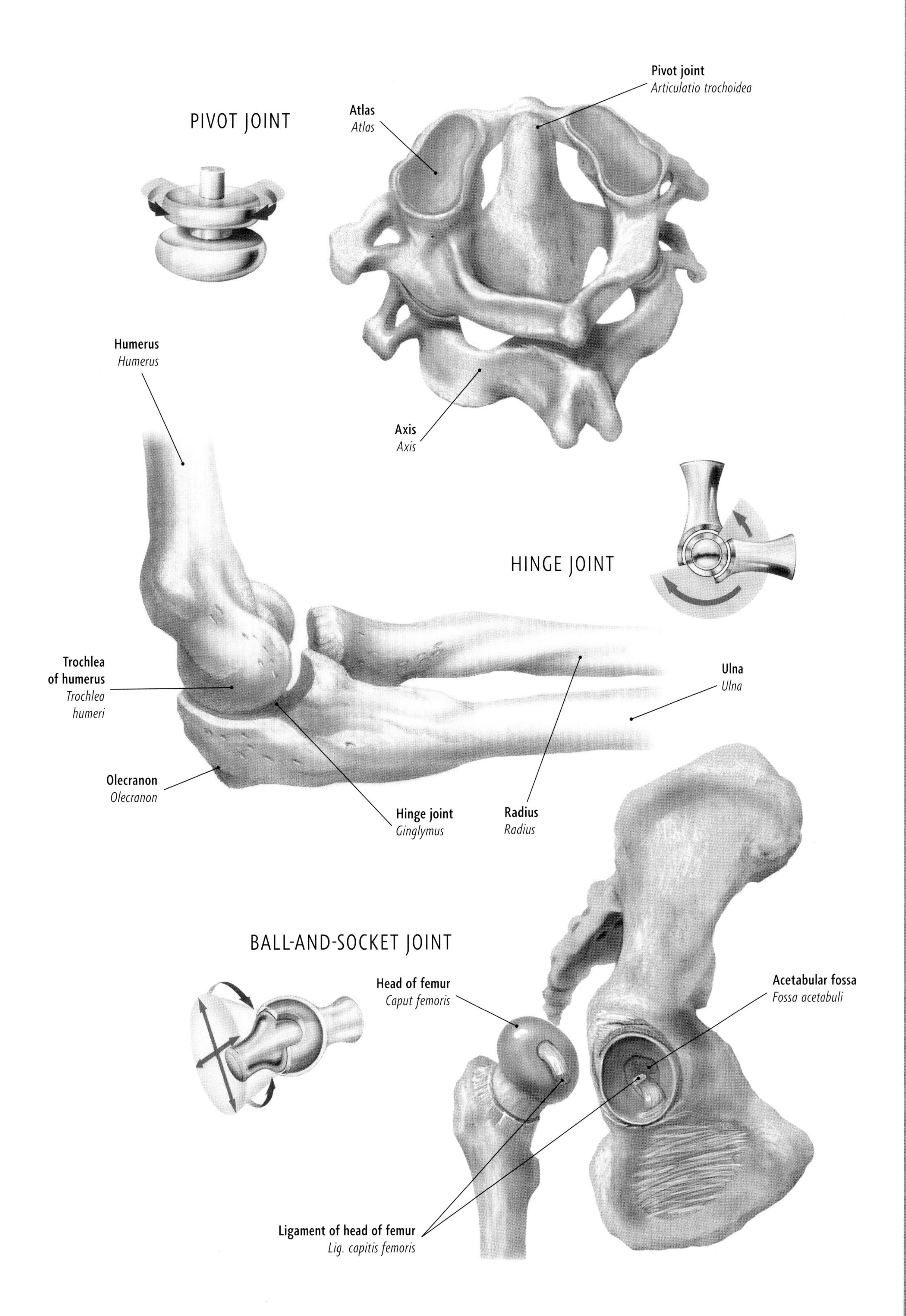
PIVOT JOINT
Atlas
Atlas
Pivot joint
Articulatio trochoidea
Humerus
Humerus
Axis
Axis
HINGE JOINT
Trochlea of humerus
Trochlea humeri
Ulna
Ulna
Olecranon
Olecranon
Hinge joint
Ginglymus
Radius
Radius
BALL-AND-SOCKET JOINT
Head of femur
Caput femoris
Acetabular fossa
Fossa acetabuli
Ligament of head of femur
Lig. capitis femoris

Joints

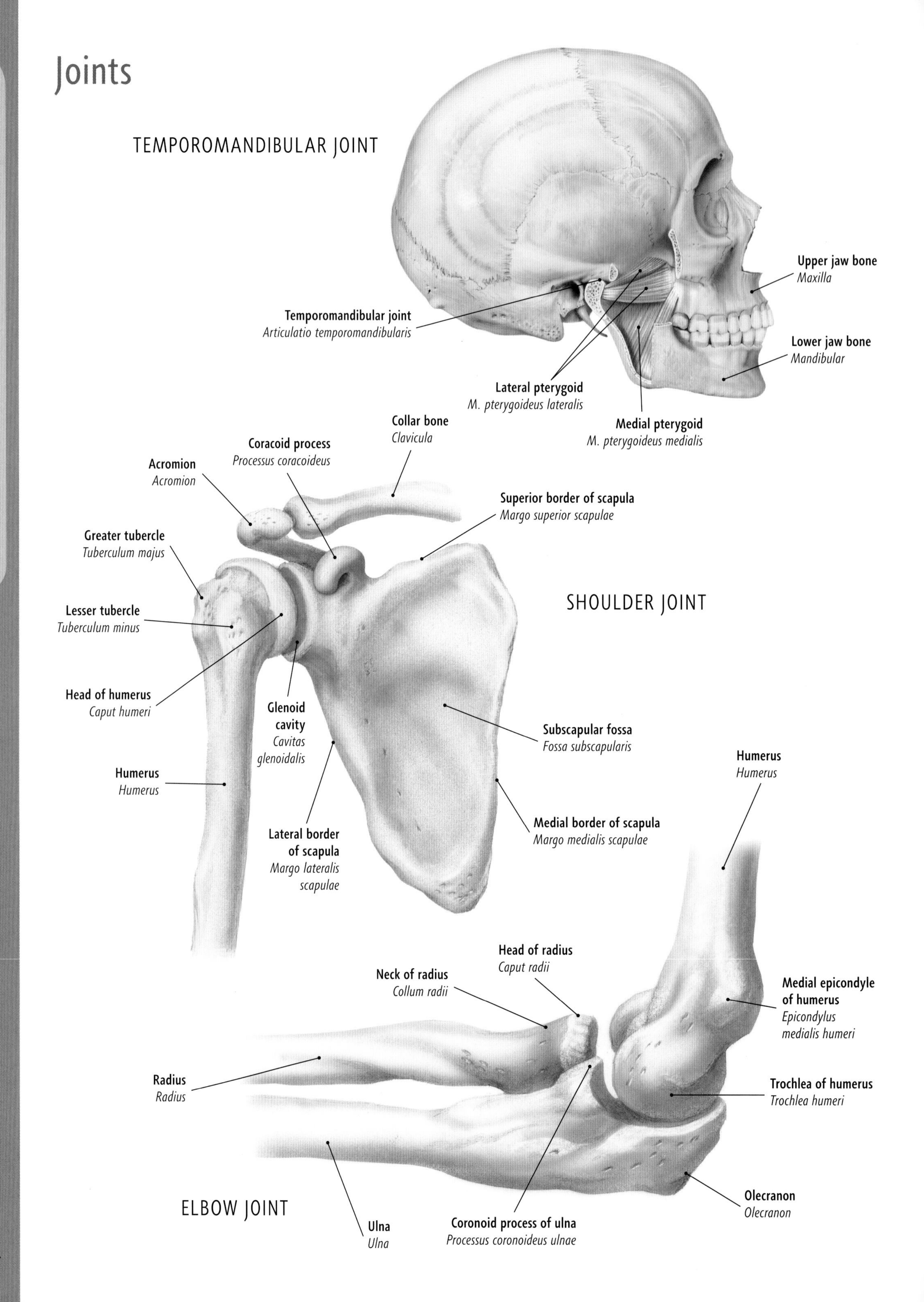
TEMPOROMANDIBULAR JOINT
Upper jaw bone
Maxilla
Temporomandibular joint
Articulatio temporomandibularis
Lower jaw bone
Mandibular
Lateral pterygoid
M. pterygoideus lateralis
Medial pterygoid
M. pterygoideus medialis
Collar bone
Clavicula
Coracoid process
Processus coracoideus
Acromion
Acromion
Superior border of scapula
Margo superior scapulae
Greater tubercle
Tuberculum majus
SHOULDER JOINT
Lesser tubercle
Tuberculum minus
Head of humerus
Caput humeri
Glenoid cavity
Cavitas glenoidalis
Subscapular fossa
Fossa subscapularis
Humerus
Humerus
Humerus
Humerus
Medial border of scapula
Margo medialis scapulae
Lateral border of scapula
Margo lateralis scapulae
Head of radius
Caput radii
Neck of radius
Collum radii
Medial epicondyle of humerus
Epicondylus medialis humeri
Radius
Radius
Trochlea of humerus
Trochlea humeri
Olecranon
Olecranon
ELBOW JOINT
Ulna
Ulna
Coronoid process of ulna
Processus coronoideus ulnae

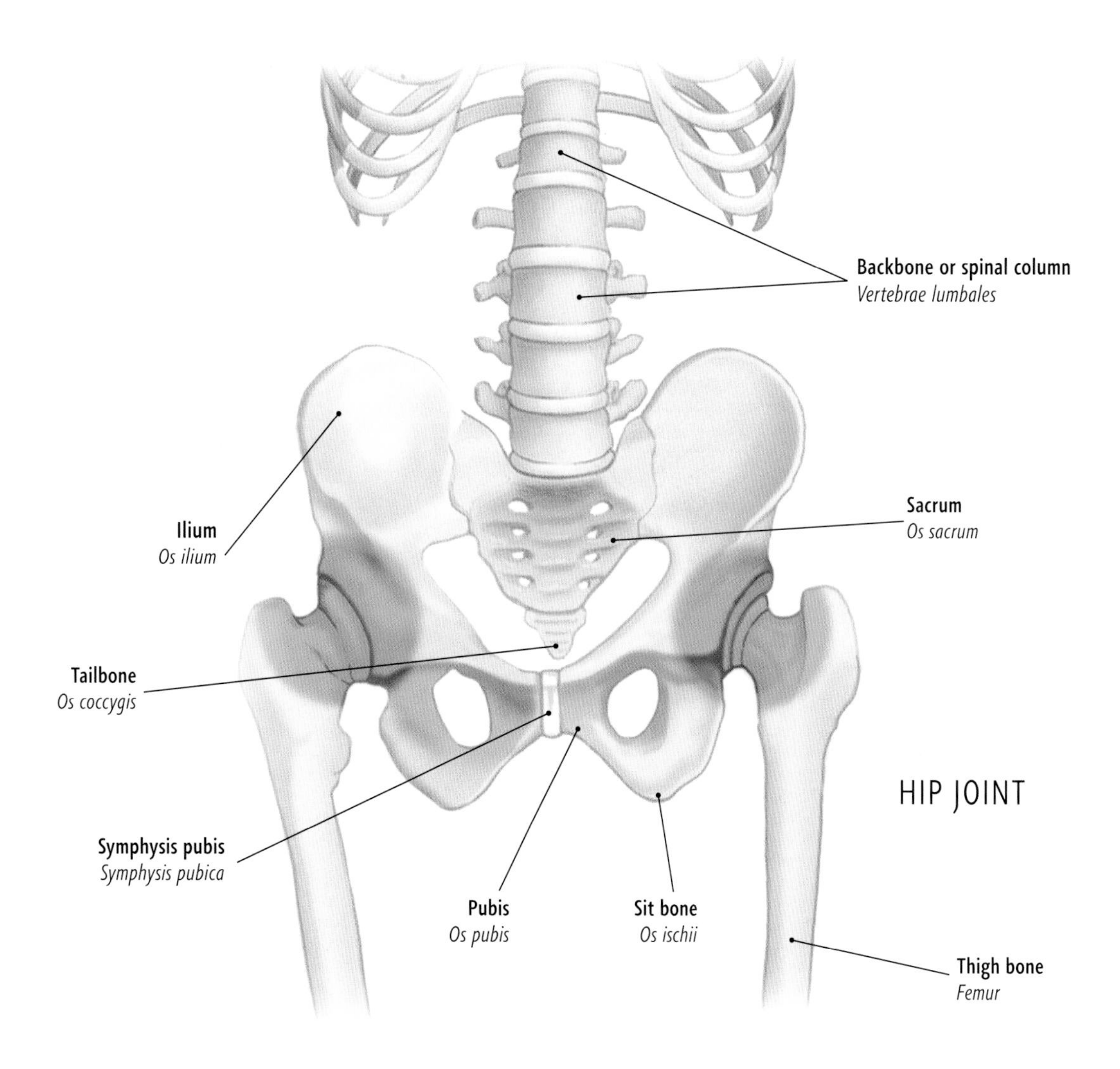
Backbone or spinal column
Vertebrae lumbales
Ilium
Os ilium
Sacrum
Os sacrum
Tailbone
Os coccygis
HIP JOINT
Symphysis pubis
Symphysis pubica
Pubis
Os pubis
Sit bone
Os ischii
Thigh bone
Femur

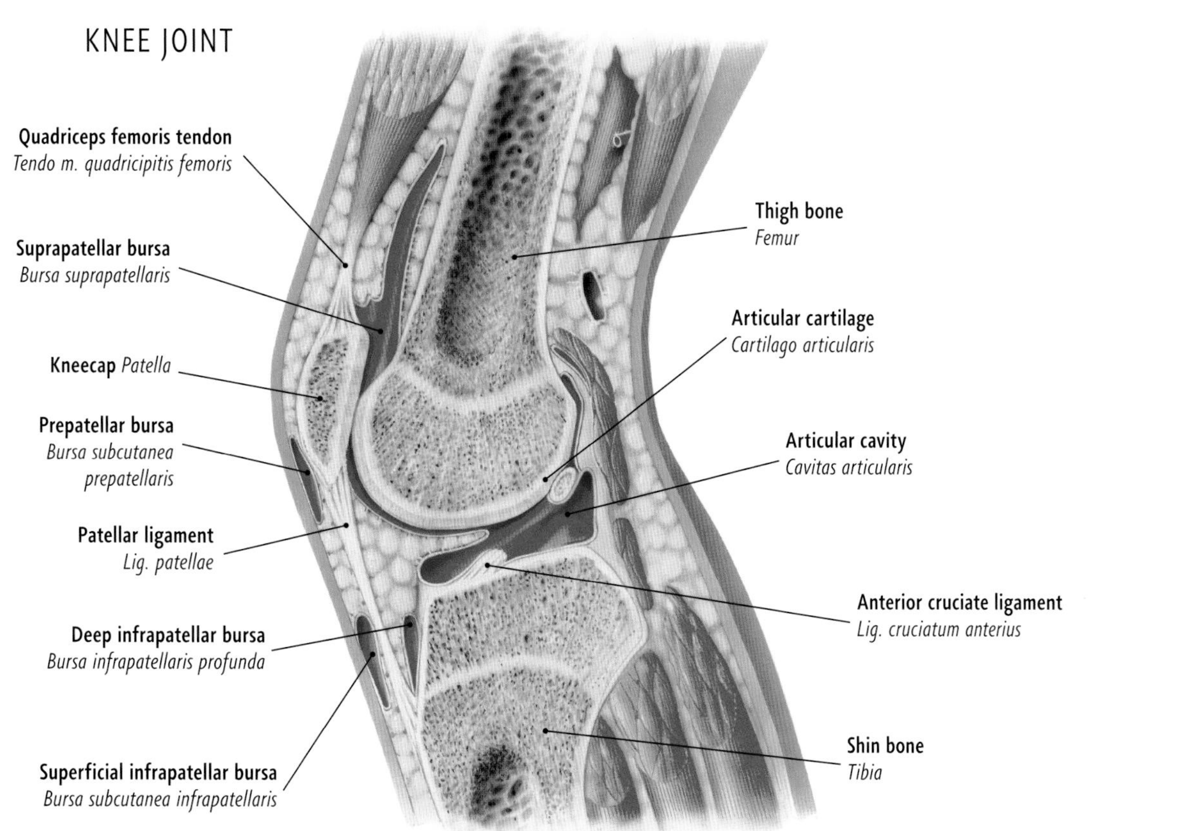
KNEE JOINT
Quadriceps femoris tendon
Tendo m. quadricipitis femoris
Thigh bone
Femur
Suprapatellar bursa
Bursa suprapatellaris
Articular cartilage
Cartilago articularis
Kneecap Patella
Prepatellar bursa
Bursa subcutanea prepatellaris
Articular cavity
Cavitas articularis
Patellar ligament
Lig. patellae
Anterior cruciate ligament
Lig. cruciatum anterius
Deep infrapatellar bursa
Bursa infrapatellaris profunda
Shin bone
Tibia
Superficial infrapatellar bursa
Bursa subcutanea infrapatellaris

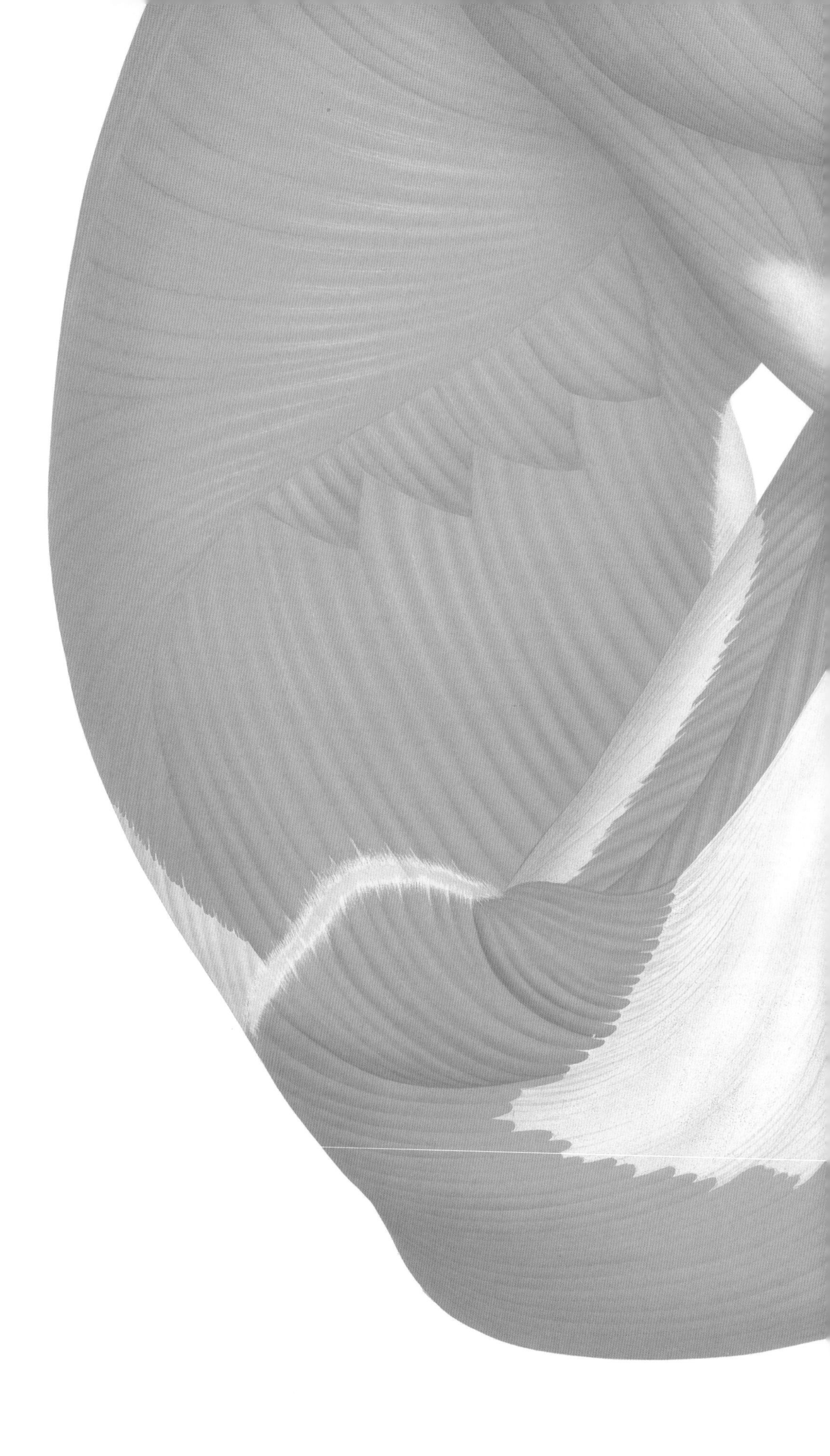

The Muscular System

Muscles of the Body

Frontalis *M. occipitofrontalis*

Temporalis *M. temporalis*

Occipitalis *M. occipitalis*

Levator labii superioris *M. levator labii superioris*

Zygomaticus major *M. zygomaticus major*

Orbicularis oculi *M. orbicularis oculi*

Masseter *M. masseter*

Orbicularis oris *M. orbicularis oris*

Depressor anguli oris *M. depressor anguli oris*

Trapezius *M. trapezius*

Trapezius *M. trapezius*

Pectoralis major *M. pectoralis major*

Sternohyoid *M. sternohyoideus*

Deltoid *M. deltoideus*

Sternocleidomastoid *M. sternocleidomastoideus*

Abdominals *Mm. recti abdomines*

External abdominal oblique *M. obliquus externus abdominis*

Serratus anterior *M. serratus anterior*

Tendon of biceps brachii *Tendo m. bicipitis brachii*

Biceps *M. biceps brachii*

Brachialis *M. brachialis*

Triceps *M. triceps brachii*

Bicipital aponeurosis *Aponeurosis musculi bicipitis brachii*

Brachioradialis *M. brachioradialis*

Tendon of palmaris longus *Tendo m. palmaris longi*

Tendon of flexor carpi radialis *Tendo m. flexoris carpi radialis*

Tendon of flexor carpi ulnaris *Tendo m. flexoris carpi ulnaris*

Flexor digitorum superficialis *M. flexor digitorum superficialis*

Tensor fascia lata *M. tensor fasciae latae*

Abductor pollicis brevis *M. abductor pollicis brevis*

Sartorius *M. sartorius*

Lumbricals *Mm. lumbricales*

Adductor magnus *M. adductor magnus*

Iliopsoas *M. iliopsoas*

Gracilis *M. gracilis*

Pectineus *M. pectineus*

Vastus lateralis *M. vastus lateralis*

Adductor longus *M. adductor longus*

Rectus femoris *M. rectus femoris*

Fibularis (peroneus) longus *M. fibularis (peroneus) longus*

Vastus medialis *M. vastus medialis*

Gastrocnemius *M. gastrocnemius*

Tibialis anterior *M. tibialis anterior*

Extensor digitorum longus *M. extensor digitorum longus*

Extensor hallucis longus *M. extensor hallucis longus*

Soleus *M. soleus*

Superior extensor retinaculum *Retinaculum musculorum extensorum superius pedis*

FRONT VIEW

Inferior extensor retinaculum *Retinaculum musculorum extensorum inferius pedis*

Shin bone *Tibia*

Tendon of extensor digitorum longus *Tendo m. extensor digitorum longus*

Tendon of extensor hallucis longus *Tendo m. extensoris hallucis longi*

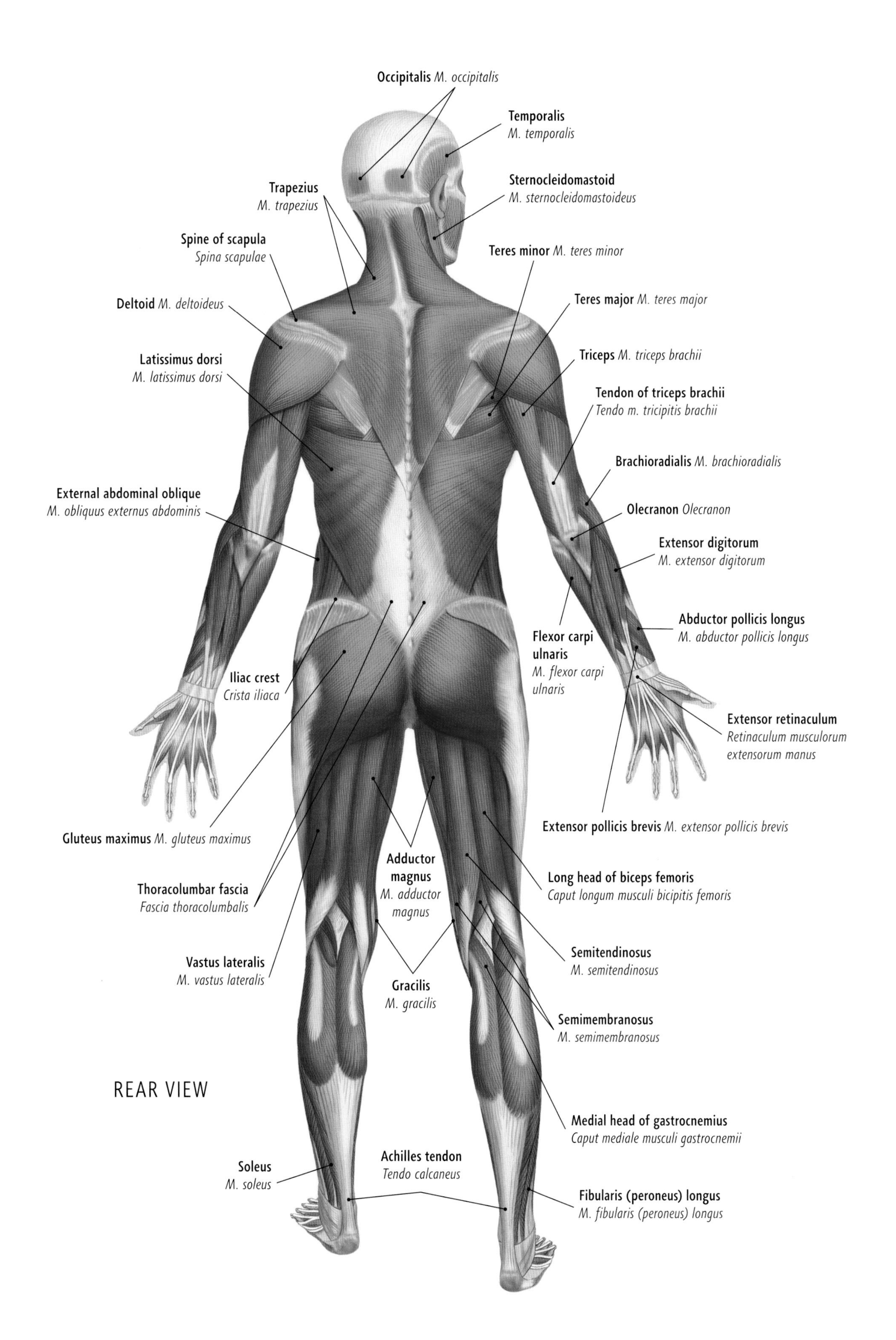
Occipitalis M. occipitalis
Temporalis
M. temporalis
Trapezius
M. trapezius
Sternocleidomastoid
M. sternocleidomastoideus
Spine of scapula
Spina scapulae
Teres minor M. teres minor
Deltoid M. deltoideus
Teres major M. teres major
Latissimus dorsi
M. latissimus dorsi
Triceps M. triceps brachii
Tendon of triceps brachii
Tendo m. tricipitis brachii
Brachioradialis M. brachioradialis
External abdominal oblique
M. obliquus externus abdominis
Olecranon Olecranon
Extensor digitorum
M. extensor digitorum
Abductor pollicis longus
M. abductor pollicis longus
Flexor carpi
ulnaris
M. flexor carpi
ulnaris
Iliac crest
Crista iliaca
Extensor retinaculum
Retinaculum musculorum
extensorum manus
Extensor pollicis brevis M. extensor pollicis brevis
Gluteus maximus M. gluteus maximus
Adductor
magnus
M. adductor
magnus
Long head of biceps femoris
Caput longum musculi bicipitis femoris
Thoracolumbar fascia
Fascia thoracolumbalis
Semitendinosus
M. semitendinosus
Vastus lateralis
M. vastus lateralis
Gracilis
M. gracilis
Semimembranosus
M. semimembranosus
REAR VIEW
Medial head of gastrocnemius
Caput mediale musculi gastrocnemii
Achilles tendon
Tendo calcaneus
Soleus
M. soleus
Fibularis (peroneus) longus
M. fibularis (peroneus) longus

Muscles of the Body

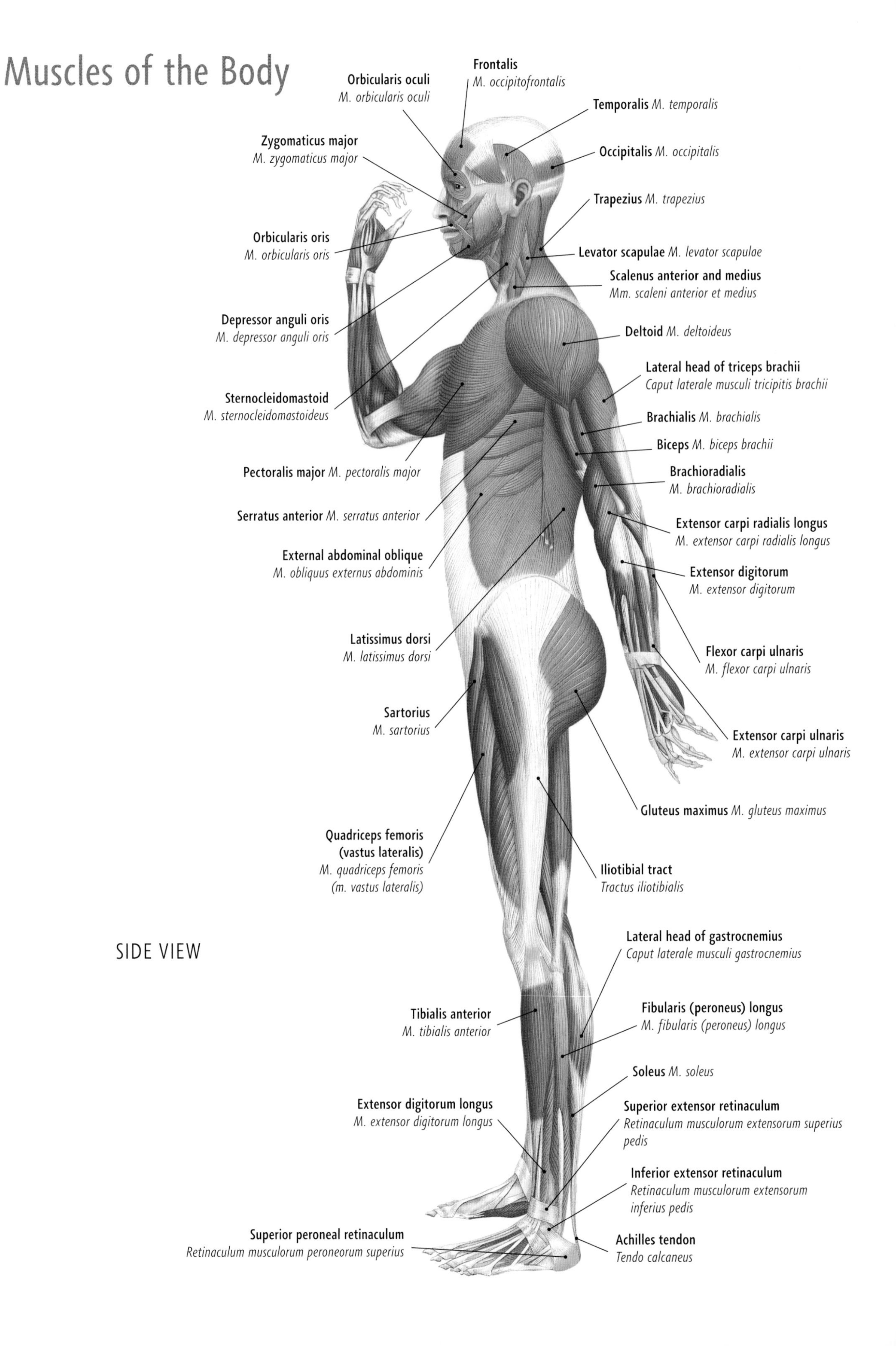
Frontalis
M. occipitofrontalis
Orbicularis oculi
M. orbicularis oculi
Temporalis M. temporalis
Zygomaticus major
M. zygomaticus major
Occipitalis M. occipitalis
Trapezius M. trapezius
Orbicularis oris
M. orbicularis oris
Levator scapulae M. levator scapulae
Scalenus anterior and medius
Mm. scaleni anterior et medius
Depressor anguli oris
M. depressor anguli oris
Deltoid M. deltoideus
Lateral head of triceps brachii
Caput laterale musculi tricipitis brachii
Sternocleidomastoid
M. sternocleidomastoideus
Brachialis M. brachialis
Biceps M. biceps brachii
Pectoralis major M. pectoralis major
Brachioradialis
M. brachioradialis
Serratus anterior M. serratus anterior
Extensor carpi radialis longus
M. extensor carpi radialis longus
External abdominal oblique
M. obliquus externus abdominis
Extensor digitorum
M. extensor digitorum
Latissimus dorsi
M. latissimus dorsi
Flexor carpi ulnaris
M. flexor carpi ulnaris
Sartorius
M. sartorius
Extensor carpi ulnaris
M. extensor carpi ulnaris
Gluteus maximus M. gluteus maximus
Quadriceps femoris
(vastus lateralis)
M. quadriceps femoris
(m. vastus lateralis)
Iliotibial tract
Tractus iliotibialis
SIDE VIEW
Lateral head of gastrocnemius
Caput laterale musculi gastrocnemius
Tibialis anterior
M. tibialis anterior
Fibularis (peroneus) longus
M. fibularis (peroneus) longus
Soleus M. soleus
Extensor digitorum longus
M. extensor digitorum longus
Superior extensor retinaculum
Retinaculum musculorum extensorum superius pedis
Inferior extensor retinaculum
Retinaculum musculorum extensorum inferius pedis
Superior peroneal retinaculum
Retinaculum musculorum peroneorum superius
Achilles tendon
Tendo calcaneus

Muscle Types

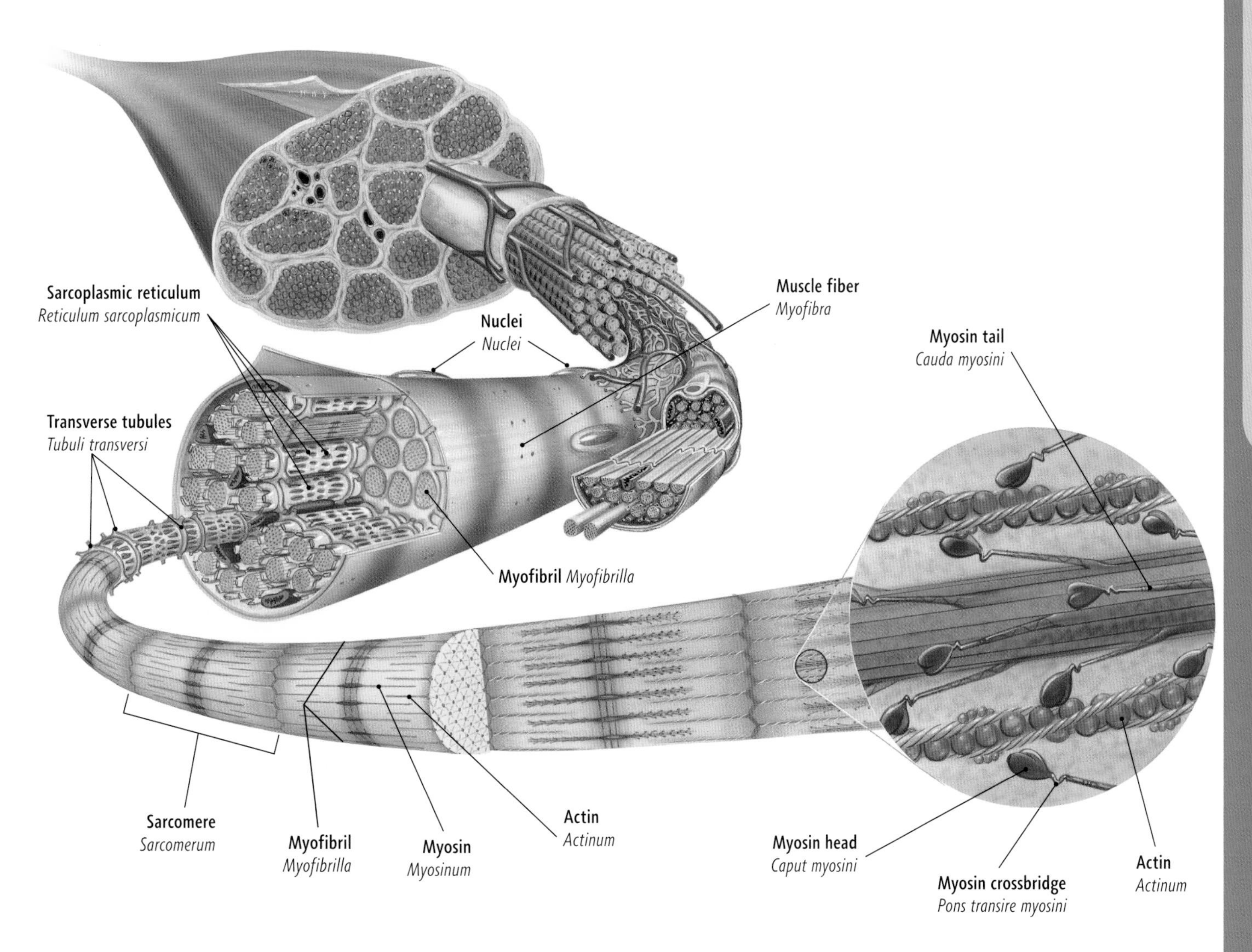
Sarcoplasmic reticulum
Reticulum sarcoplasmicum
Nuclei
Nuclei
Muscle fiber
Myofibra
Myosin tail
Cauda myosini
Transverse tubules
Tubuli transversi
Myofibril Myofibrilla
Sarcomere
Sarcomerum
Myofibril
Myofibrilla
Myosin
Myosinum
Actin
Actinum
Myosin head
Caput myosini
Myosin crossbridge
Pons transire myosini
Actin
Actinum

MUSCLE FIBER—MICROSTRUCTURE

Muscle fibers are elongated cells containing fine threads made of myofibrils. Myofibrils consist of contractile protein myofilaments, arranged in regular arrays. Contraction of the muscle is produced by the interaction of myosin heads with actin.

[~] = no direct Latin equivalent

MUSCLE SHAPES

Muscles are classified based on their general shape—some muscles have mainly parallel fibers, and others have oblique fibers. The shape and arrangement of muscle fibers reflects the function of the muscle (for example, muscle fibers that support organs are criss-crossed).

Unipennate
M. unipennatus

Bipennate
M. bipennatus

Multipennate
M. multipennatus

Spiral
M. spiralis

Spiral
M. spiralis

Radial
M. radialis

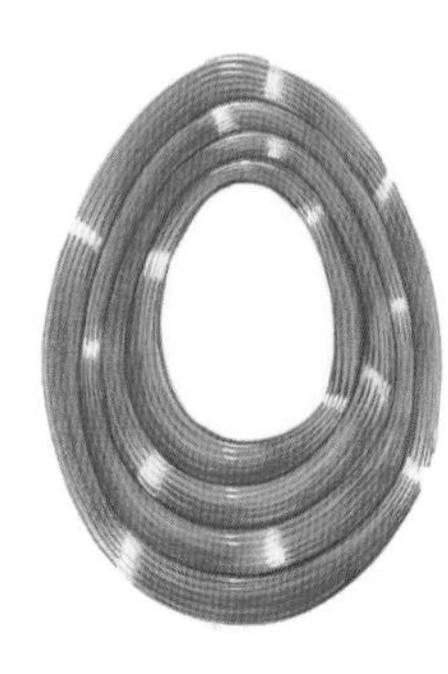

Circular
M. orbicularis

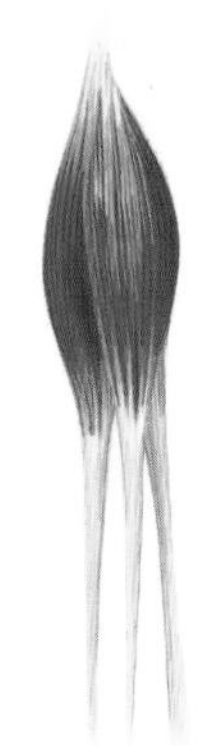

Multicaudal
M. multicaudalis

Quadrilateral
M. quadratus

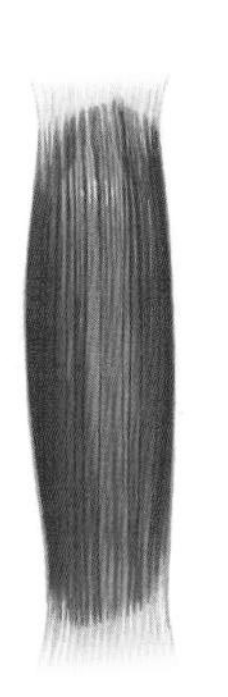

Strap
M. habenalis

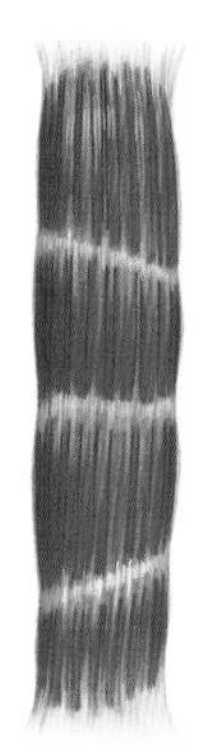

Strap (with tendinous intersections)
M. habenalis (cum intersectiones tendineae)

Cruciate
M. cruciatus

Triangular
M. triangularis

Fusiform
M. fusiformis

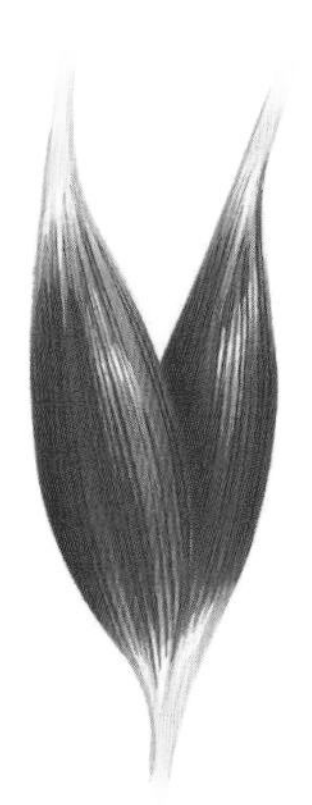

Bicipital
M. biceps

Tricipital
M. triceps

Quadricipital
M. quadriceps

Digastric
M. biventer

Muscle Action

EXERCISING MUSCLES

Muscles are arranged in antagonistic pairs to allow opposing movements such as flexion of a joint in one direction and extension in another. For example, biceps brachii controls flexion of the elbow, while triceps brachii controls elbow extension.

MOVEMENT AND SUPPORT

The muscles of the shoulder joint, known as the rotator cuff muscles, provide important support for the joint, while allowing a wide range of movement.

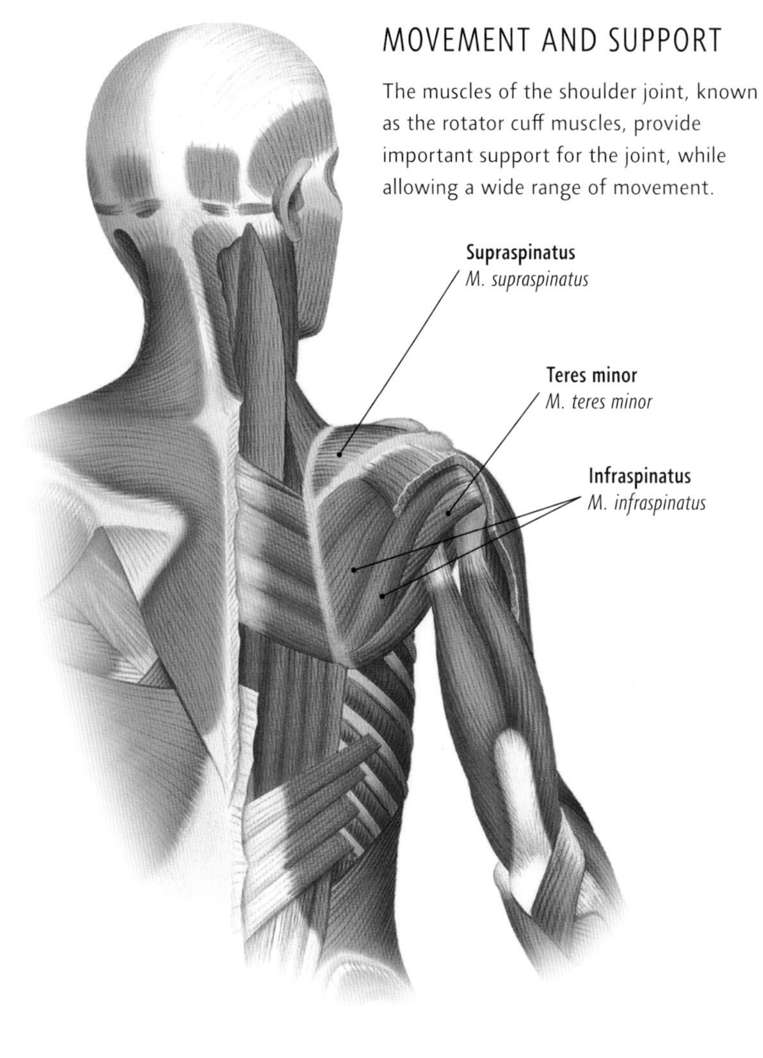

Supraspinatus
M. supraspinatus
Teres minor
M. teres minor
Infraspinatus
M. infraspinatus

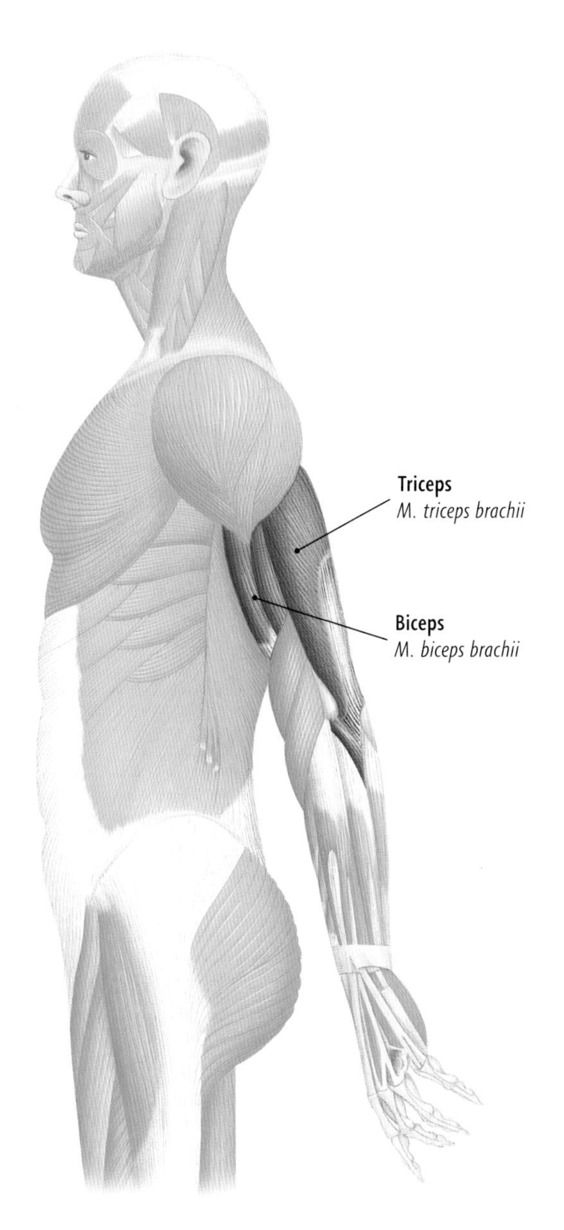

Triceps
M. triceps brachii
Biceps
M. biceps brachii

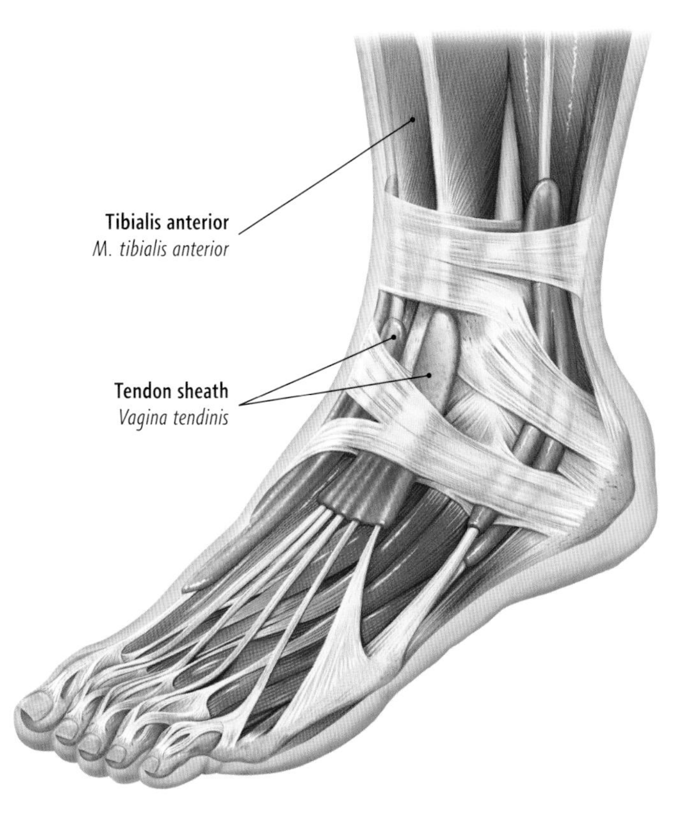

Tibialis anterior
M. tibialis anterior
Tendon sheath
Vagina tendinis

MUSCLE CONTRACTION

When a muscle contracts, it shortens and pulls the muscle attachment. The muscle's action depends on its position in relation to the joint it works on. The tibialis anterior, for example, crosses in front of the ankle and moves the foot upward (dorsiflexion).

Muscles of the Head

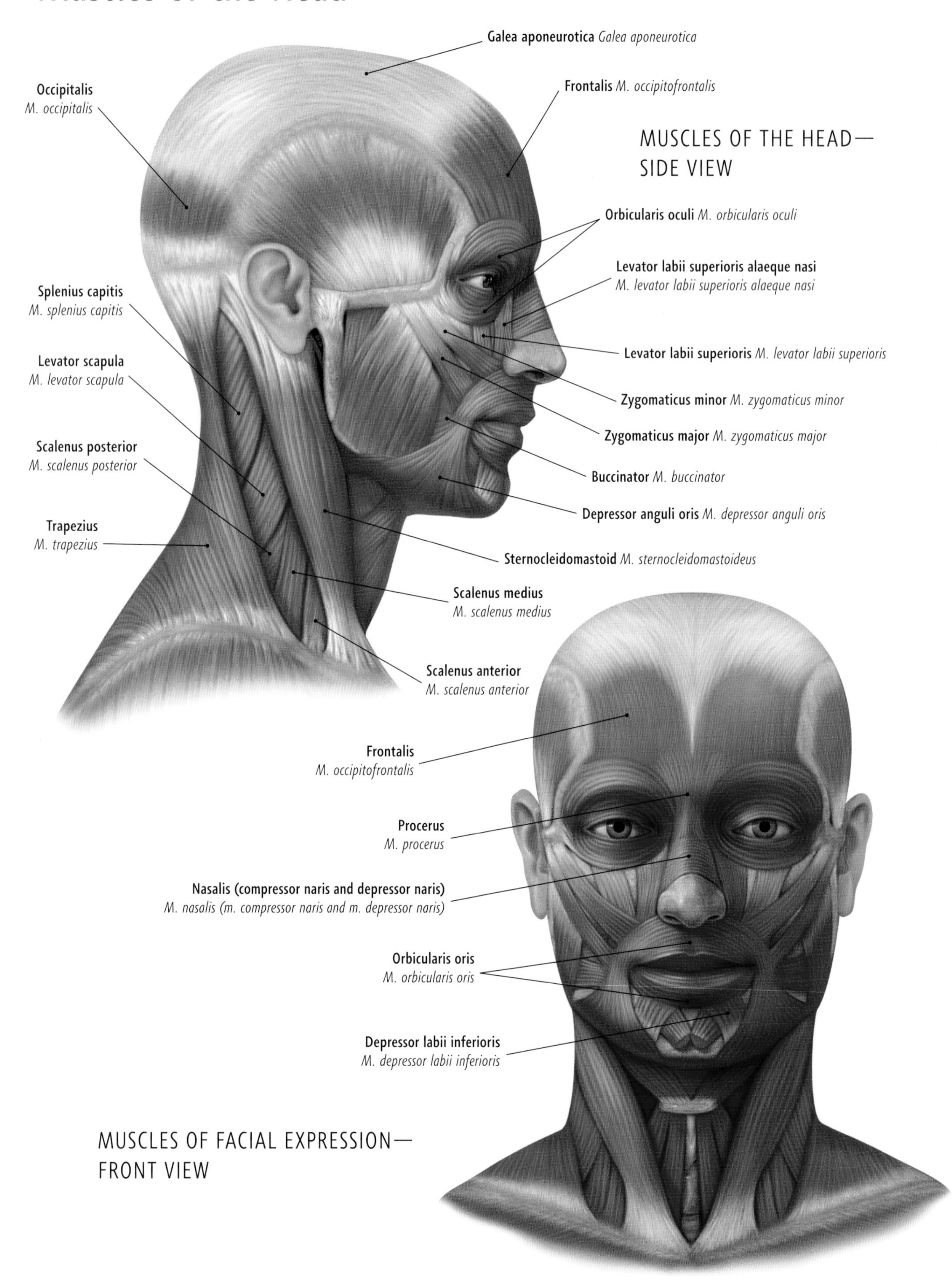
Galea aponeurotica Galea aponeurotica
Occipitalis
M. occipitalis
Frontalis M. occipitofrontalis
MUSCLES OF THE HEAD—
SIDE VIEW
Orbicularis oculi M. orbicularis oculi
Levator labii superioris alaeque nasi
M. levator labii superioris alaeque nasi
Splenius capitis
M. splenius capitis
Levator labii superioris M. levator labii superioris
Levator scapula
M. levator scapula
Zygomaticus minor M. zygomaticus minor
Zygomaticus major M. zygomaticus major
Scalenus posterior
M. scalenus posterior
Buccinator M. buccinator
Depressor anguli oris M. depressor anguli oris
Trapezius
M. trapezius
Sternocleidomastoid M. sternocleidomastoideus
Scalenus medius
M. scalenus medius
Scalenus anterior
M. scalenus anterior
Frontalis
M. occipitofrontalis
Procerus
M. procerus
Nasalis (compressor naris and depressor naris)
M. nasalis (m. compressor naris and m. depressor naris)
Orbicularis oris
M. orbicularis oris
Depressor labii inferioris
M. depressor labii inferioris
MUSCLES OF FACIAL EXPRESSION—
FRONT VIEW

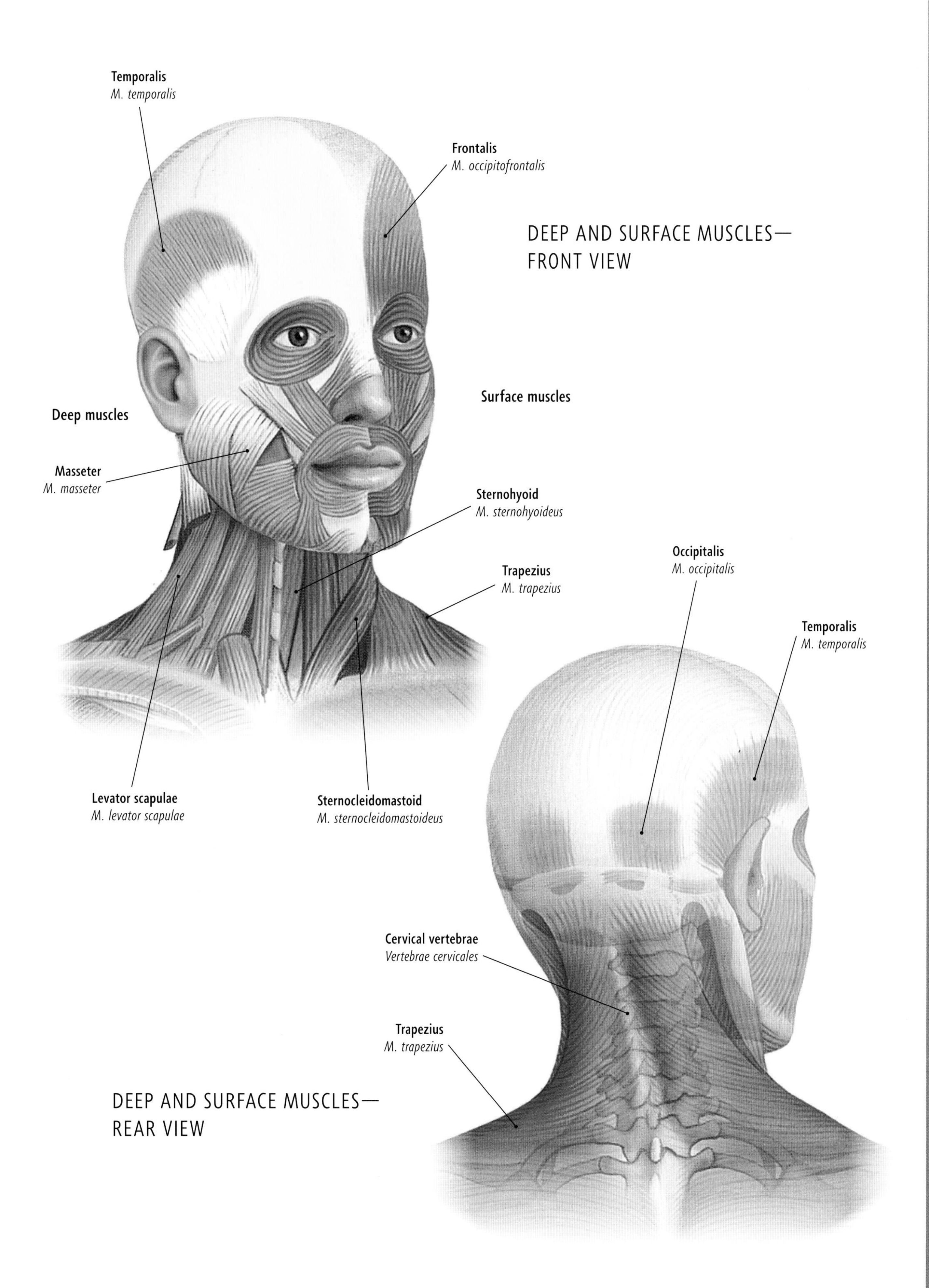
Temporalis
M. temporalis
Frontalis
M. occipitofrontalis
DEEP AND SURFACE MUSCLES—
FRONT VIEW
Surface muscles
Deep muscles
Masseter
M. masseter
Sternohyoid
M. sternohyoideus
Trapezius
M. trapezius
Levator scapulae
M. levator scapulae
Sternocleidomastoid
M. sternocleidomastoideus
Occipitalis
M. occipitalis
Temporalis
M. temporalis
Cervical vertebrae
Vertebrae cervicales
Trapezius
M. trapezius
DEEP AND SURFACE MUSCLES—
REAR VIEW

Muscles of the Eye and Neck

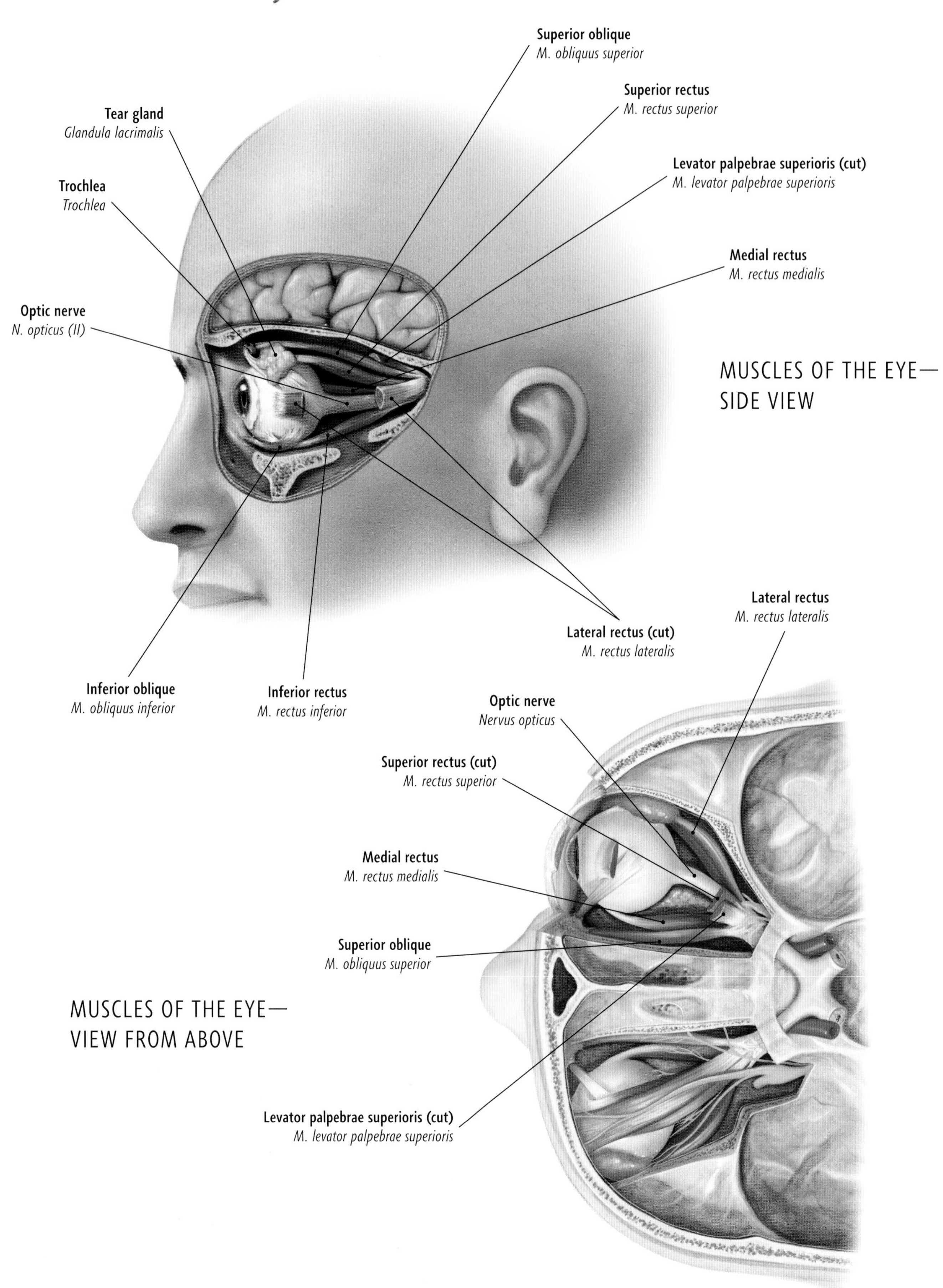
Superior oblique
M. obliquus superior
Superior rectus
M. rectus superior
Tear gland
Glandula lacrimalis
Levator palpebrae superioris (cut)
M. levator palpebrae superioris
Trochlea
Trochlea
Medial rectus
M. rectus medialis
Optic nerve
N. opticus (II)
MUSCLES OF THE EYE—
SIDE VIEW
Lateral rectus
M. rectus lateralis
Lateral rectus (cut)
M. rectus lateralis
Inferior oblique
M. obliquus inferior
Inferior rectus
M. rectus inferior
Optic nerve
Nervus opticus
Superior rectus (cut)
M. rectus superior
Medial rectus
M. rectus medialis
Superior oblique
M. obliquus superior
MUSCLES OF THE EYE—
VIEW FROM ABOVE
Levator palpebrae superioris (cut)
M. levator palpebrae superioris

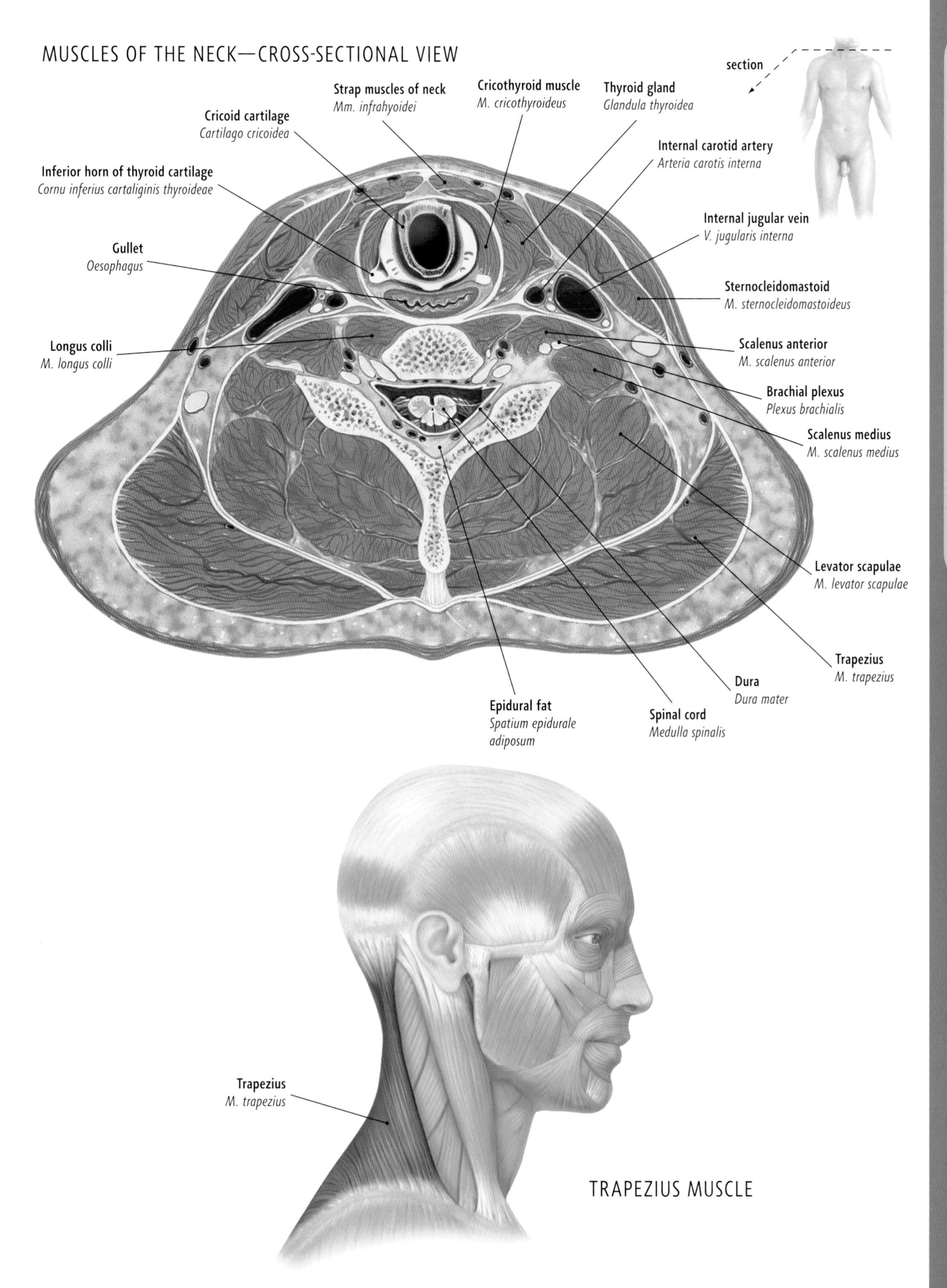
MUSCLES OF THE NECK—CROSS-SECTIONAL VIEW
section
Strap muscles of neck
Mm. infrahyoidei
Cricothyroid muscle
M. cricothyroideus
Thyroid gland
Glandula thyroidea
Cricoid cartilage
Cartilago cricoidea
Internal carotid artery
Arteria carotis interna
Inferior horn of thyroid cartilage
Cornu inferius cartaliginis thyroideae
Internal jugular vein
V. jugularis interna
Gullet
Oesophagus
Sternocleidomastoid
M. sternocleidomastoideus
Longus colli
M. longus colli
Scalenus anterior
M. scalenus anterior
Brachial plexus
Plexus brachialis
Scalenus medius
M. scalenus medius
Levator scapulae
M. levator scapulae
Trapezius
M. trapezius
Dura
Dura mater
Epidural fat
Spatium epidurale adiposum
Spinal cord
Medulla spinalis
Trapezius
M. trapezius
TRAPEZIUS MUSCLE

Muscles of the Jaw and Throat

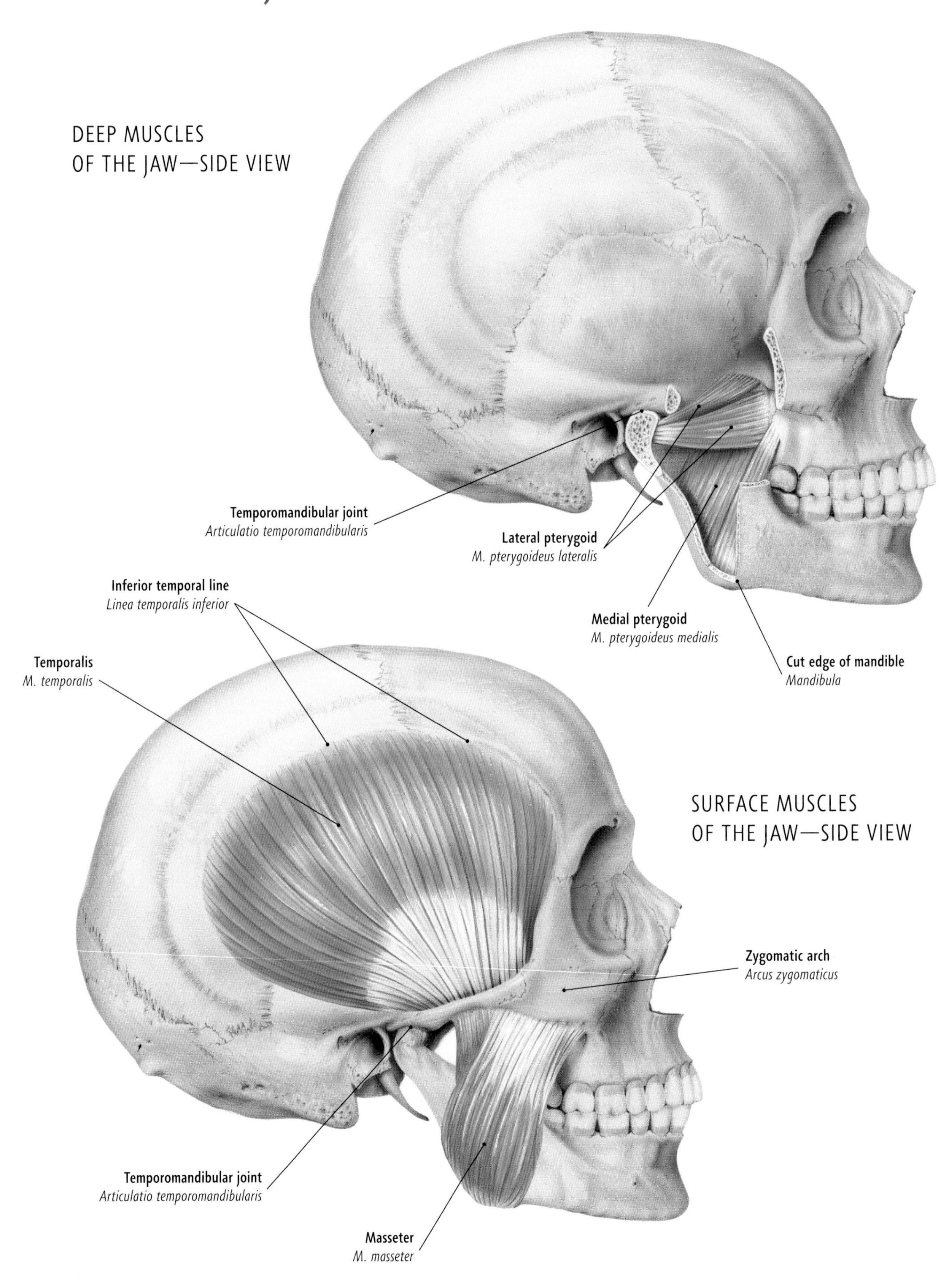
DEEP MUSCLES
OF THE JAW—SIDE VIEW
Temporomandibular joint
Articulatio temporomandibularis
Lateral pterygoid
M. pterygoideus lateralis
Medial pterygoid
M. pterygoideus medialis
Cut edge of mandible
Mandibula
Inferior temporal line
Linea temporalis inferior
Temporalis
M. temporalis
SURFACE MUSCLES
OF THE JAW—SIDE VIEW
Zygomatic arch
Arcus zygomaticus
Temporomandibular joint
Articulatio temporomandibularis
Masseter
M. masseter

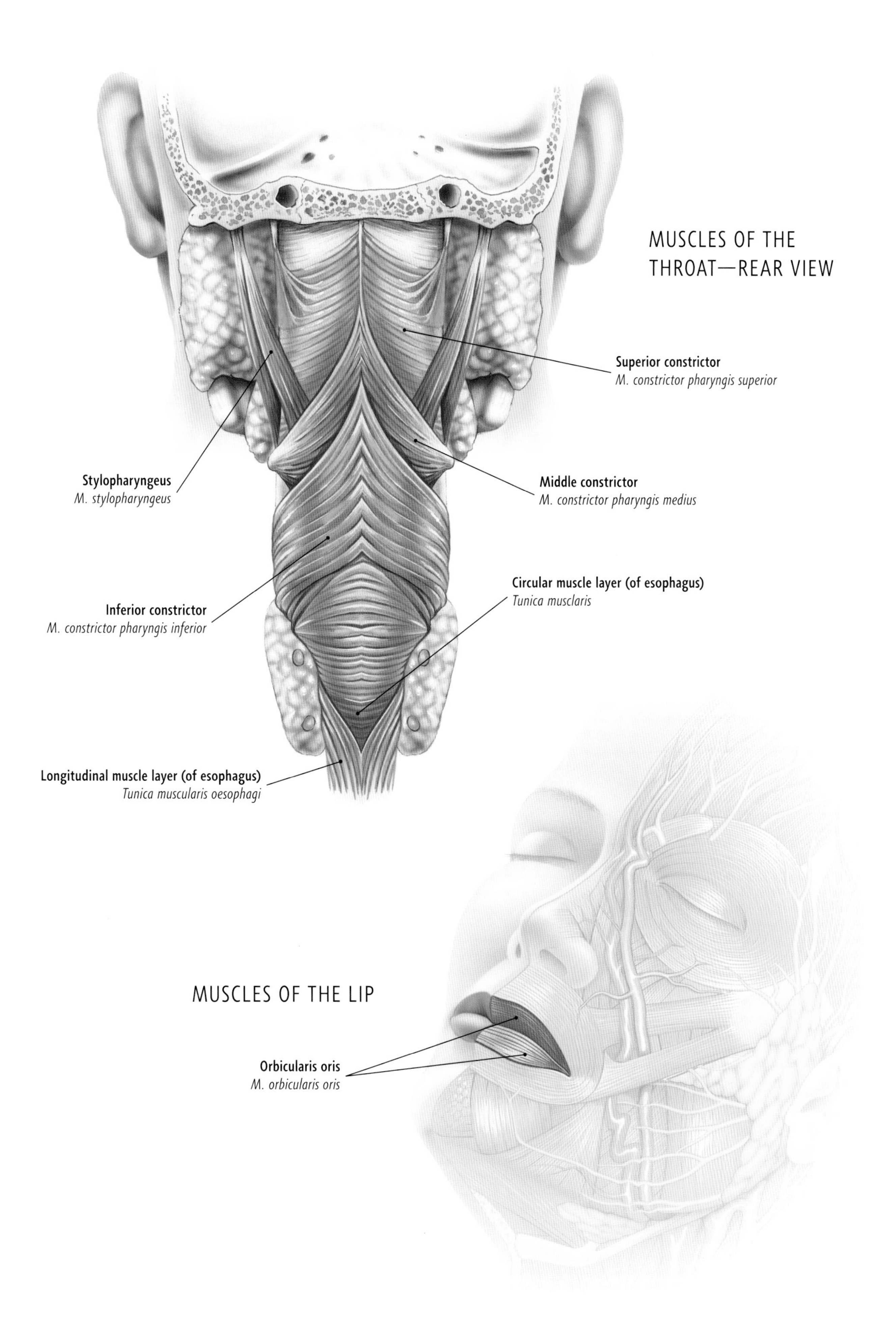
MUSCLES OF THE
THROAT—REAR VIEW
Superior constrictor
M. constrictor pharyngis superior
Stylopharyngeus
M. stylopharyngeus
Middle constrictor
M. constrictor pharyngis medius
Circular muscle layer (of esophagus)
Tunica musclaris
Inferior constrictor
M. constrictor pharyngis inferior
Longitudinal muscle layer (of esophagus)
Tunica muscularis oesophagi
MUSCLES OF THE LIP
Orbicularis oris
M. orbicularis oris

Muscles of the Back and Abdomen

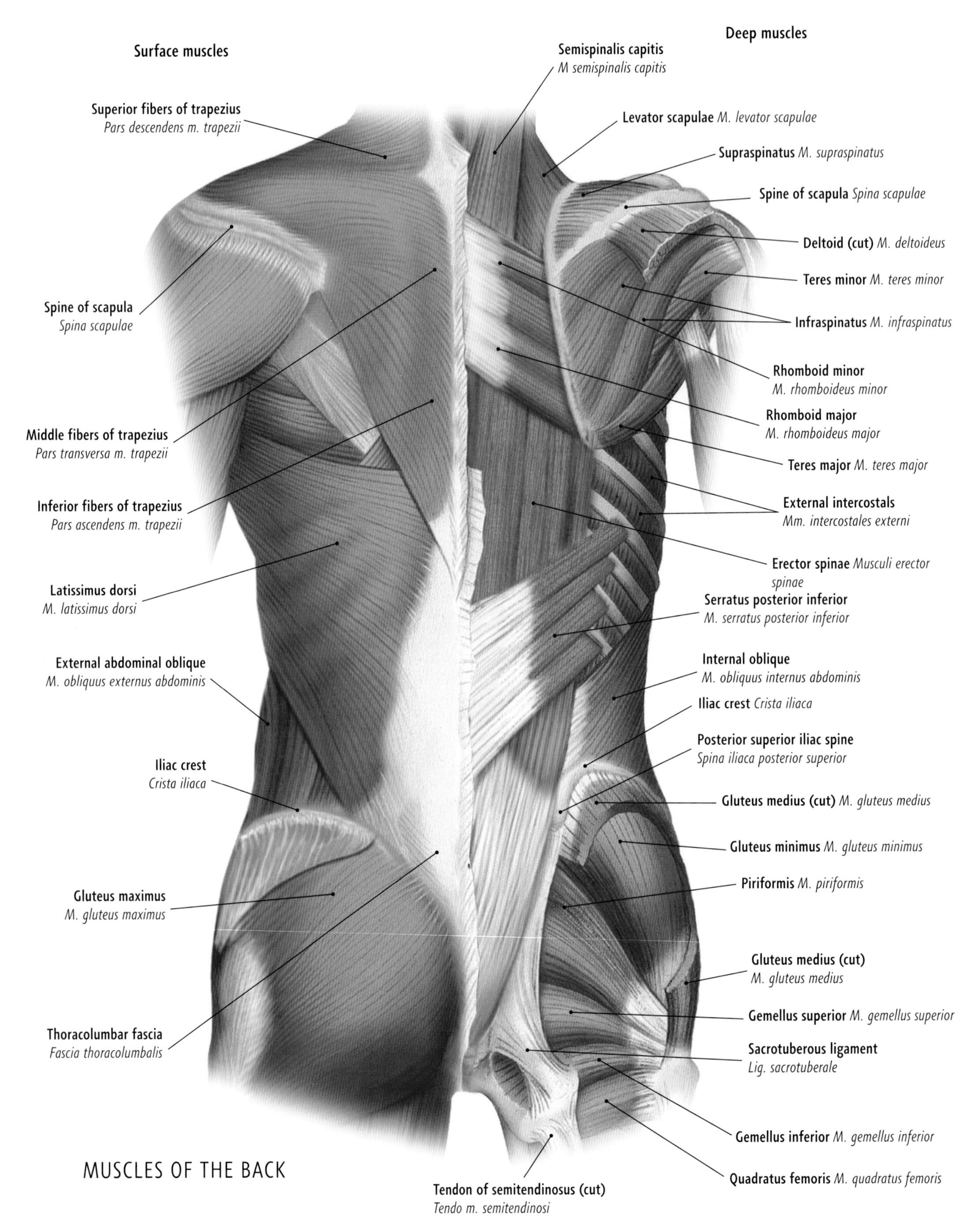
Surface muscles
Deep muscles
Superior fibers of trapezius
Pars descendens m. trapezii
Spine of scapula
Spina scapulae
Middle fibers of trapezius
Pars transversa m. trapezii
Inferior fibers of trapezius
Pars ascendens m. trapezii
Latissimus dorsi
M. latissimus dorsi
External abdominal oblique
M. obliquus externus abdominis
Iliac crest
Crista iliaca
Gluteus maximus
M. gluteus maximus
Thoracolumbar fascia
Fascia thoracolumbalis
Semispinalis capitis
M semispinalis capitis
Levator scapulae M. levator scapulae
Supraspinatus M. supraspinatus
Spine of scapula Spina scapulae
Deltoid (cut) M. deltoideus
Teres minor M. teres minor
Infraspinatus M. infraspinatus
Rhomboid minor
M. rhomboideus minor
Rhomboid major
M. rhomboideus major
Teres major M. teres major
External intercostals
Mm. intercostales externi
Erector spinae Musculi erector spinae
Serratus posterior inferior
M. serratus posterior inferior
Internal oblique
M. obliquus internus abdominis
Iliac crest Crista iliaca
Posterior superior iliac spine
Spina iliaca posterior superior
Gluteus medius (cut) M. gluteus medius
Gluteus minimus M. gluteus minimus
Piriformis M. piriformis
Gluteus medius (cut)
M. gluteus medius
Gemellus superior M. gemellus superior
Sacrotuberous ligament
Lig. sacrotuberale
Gemellus inferior M. gemellus inferior
Quadratus femoris M. quadratus femoris
Tendon of semitendinosus (cut)
Tendo m. semitendinosi

MUSCLES OF THE BACK

[~] = no direct Latin equivalent

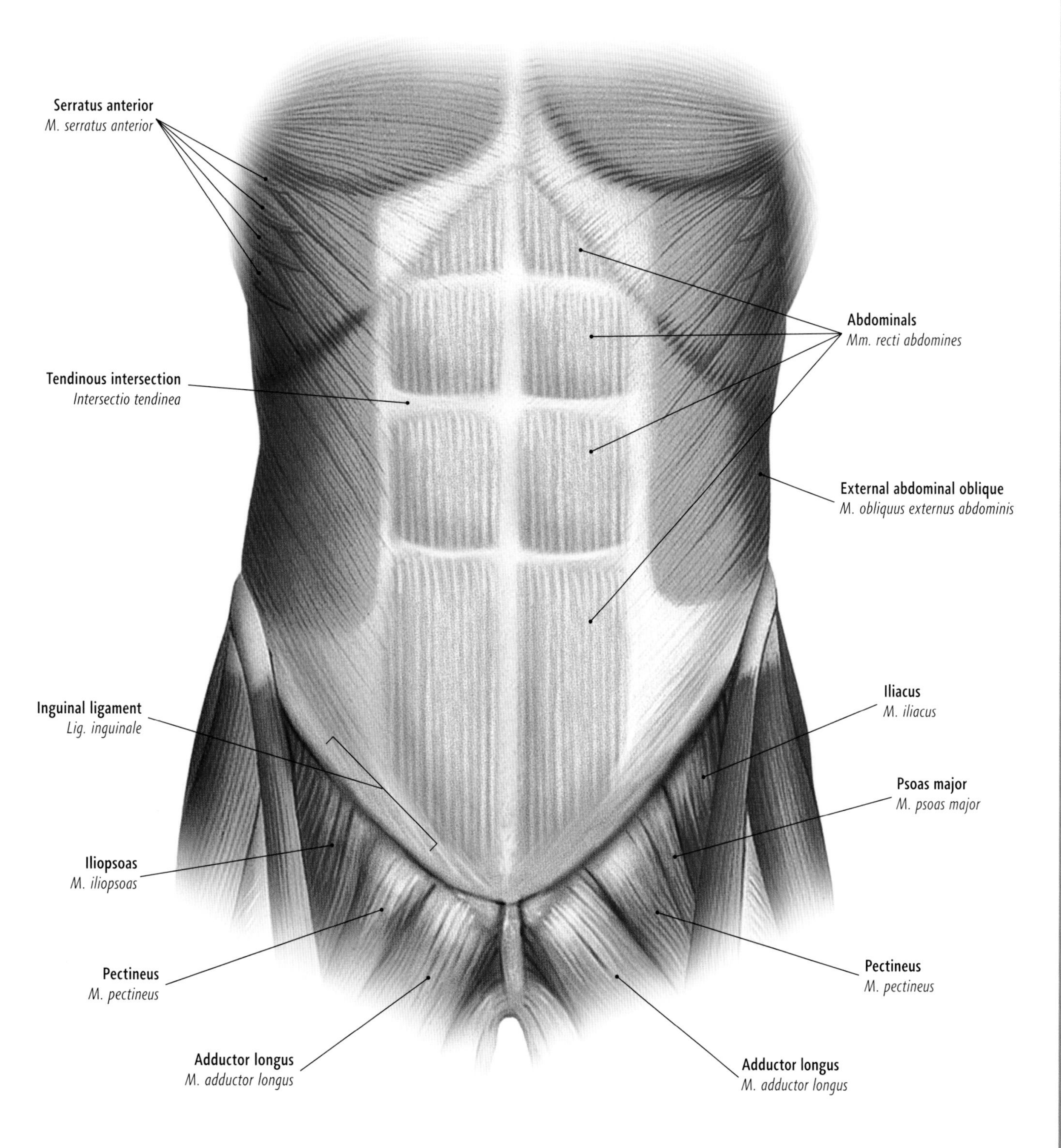
Serratus anterior
M. serratus anterior
Abdominals
Mm. recti abdomines
Tendinous intersection
Intersectio tendinea
External abdominal oblique
M. obliquus externus abdominis
Iliacus
M. iliacus
Inguinal ligament
Lig. inguinale
Psoas major
M. psoas major
Iliopsoas
M. iliopsoas
Pectineus
M. pectineus
Pectineus
M. pectineus
Adductor longus
M. adductor longus
Adductor longus
M. adductor longus

MUSCLES OF THE ABDOMEN

Muscles of Breathing and the Pelvis

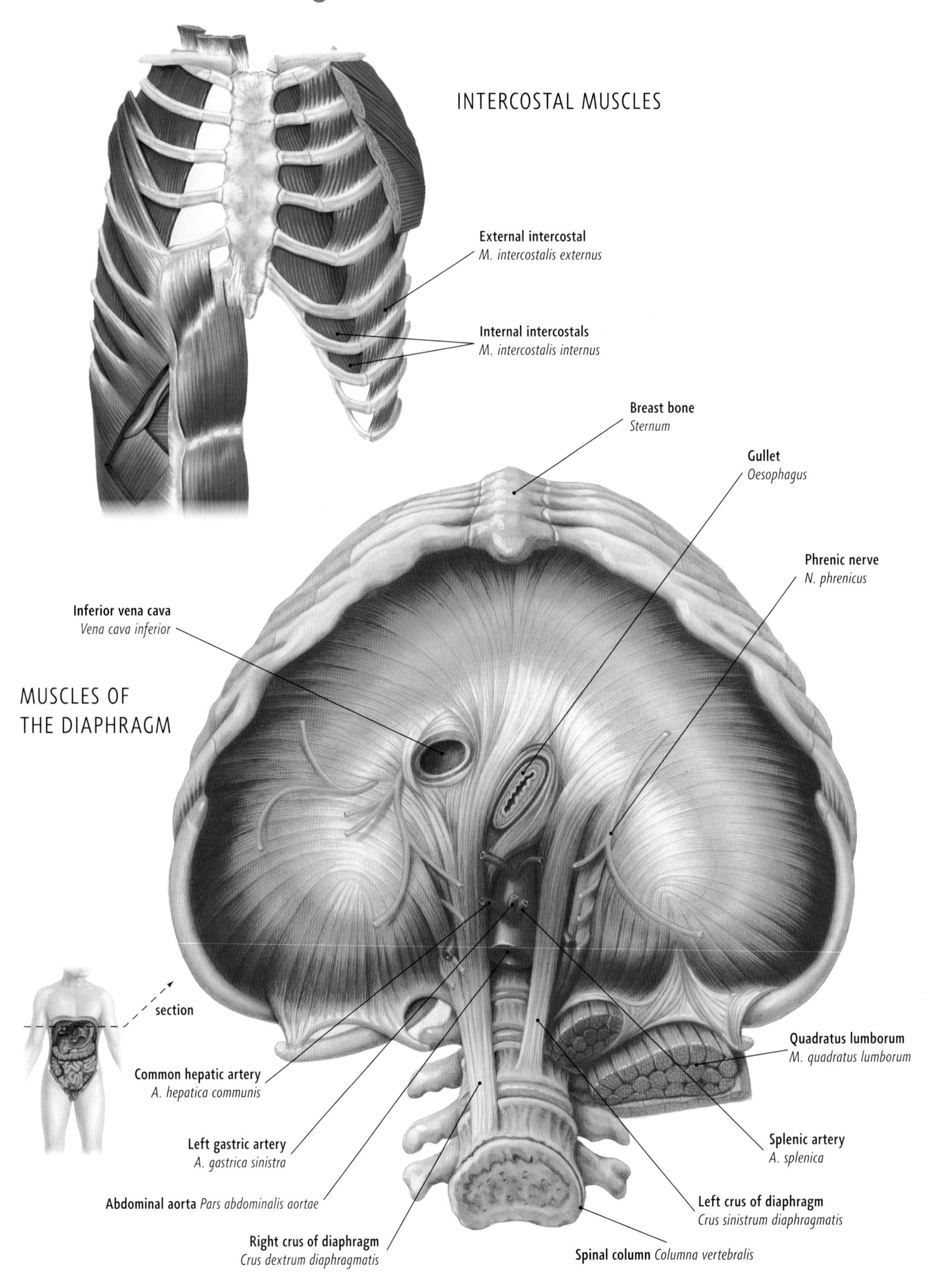
INTERCOSTAL MUSCLES
External intercostal
M. intercostalis externus
Internal intercostals
M. intercostalis internus
Breast bone
Sternum
Gullet
Oesophagus
Phrenic nerve
N. phrenicus
Inferior vena cava
Vena cava inferior
MUSCLES OF
THE DIAPHRAGM
section
Quadratus lumborum
M. quadratus lumborum
Common hepatic artery
A. hepatica communis
Left gastric artery
A. gastrica sinistra
Splenic artery
A. splenica
Abdominal aorta Pars abdominalis aortae
Left crus of diaphragm
Crus sinistrum diaphragmatis
Right crus of diaphragm
Crus dextrum diaphragmatis
Spinal column Columna vertebralis

PELVIC FLOOR MUSCLES

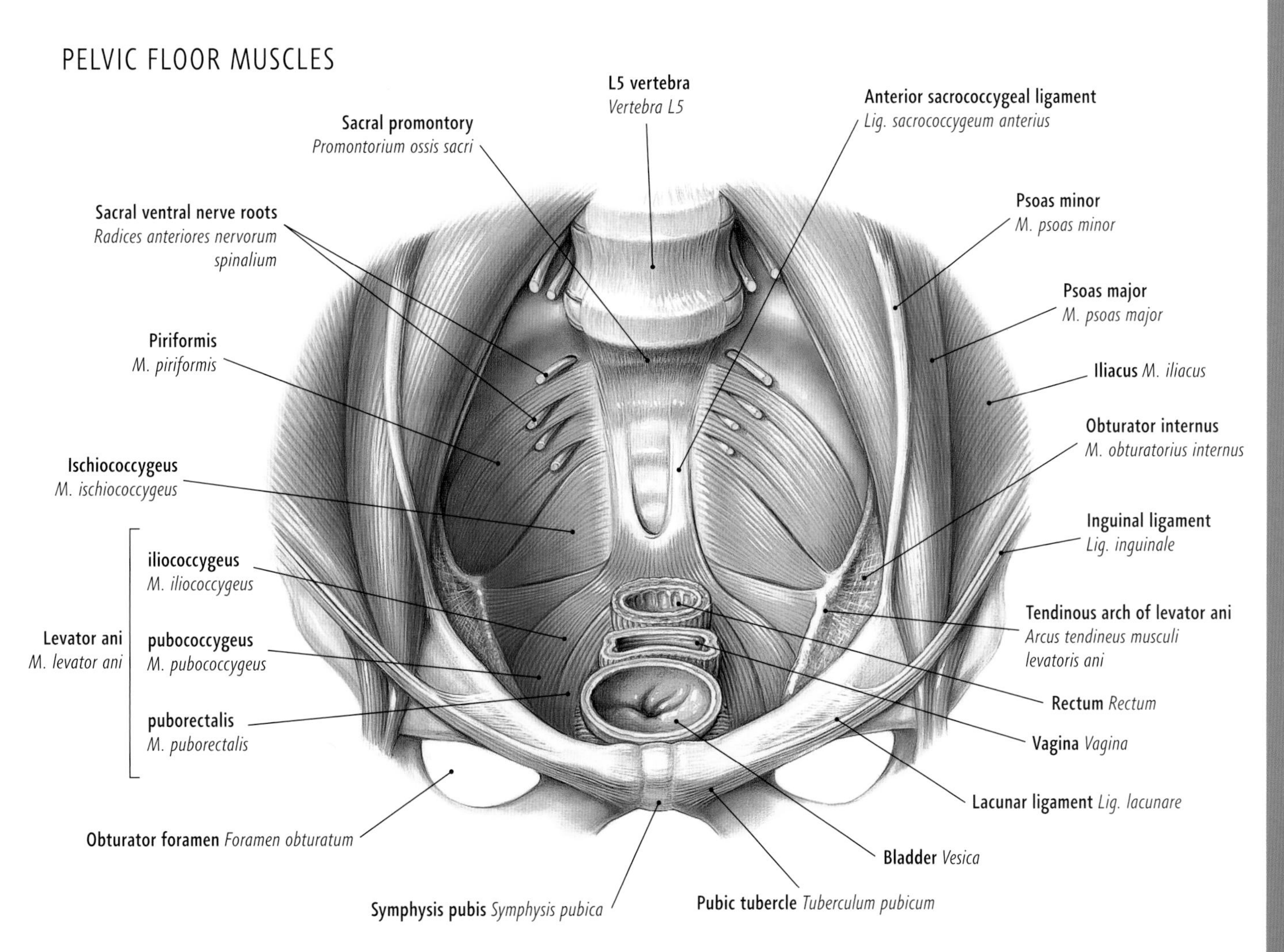

L5 vertebra
Vertebra L5
Sacral promontory
Promontorium ossis sacri
Anterior sacrococcygeal ligament
Lig. sacrococcygeum anterius
Sacral ventral nerve roots
Radices anteriores nervorum spinalium
Psoas minor
M. psoas minor
Psoas major
M. psoas major
Piriformis
M. piriformis
Iliacus M. iliacus
Obturator internus
M. obturatorius internus
Ischiococcygeus
M. ischiococcygeus
Inguinal ligament
Lig. inguinale
Levator ani
M. levator ani
iliococcygeus
M. iliococcygeus
pubococcygeus
M. pubococcygeus
puborectalis
M. puborectalis
Tendinous arch of levator ani
Arcus tendineus musculi levatoris ani
Rectum Rectum
Vagina Vagina
Lacunar ligament Lig. lacunare
Obturator foramen Foramen obturatum
Bladder Vesica
Symphysis pubis Symphysis pubica
Pubic tubercle Tuberculum pubicum

MUSCLES OF THE GROIN

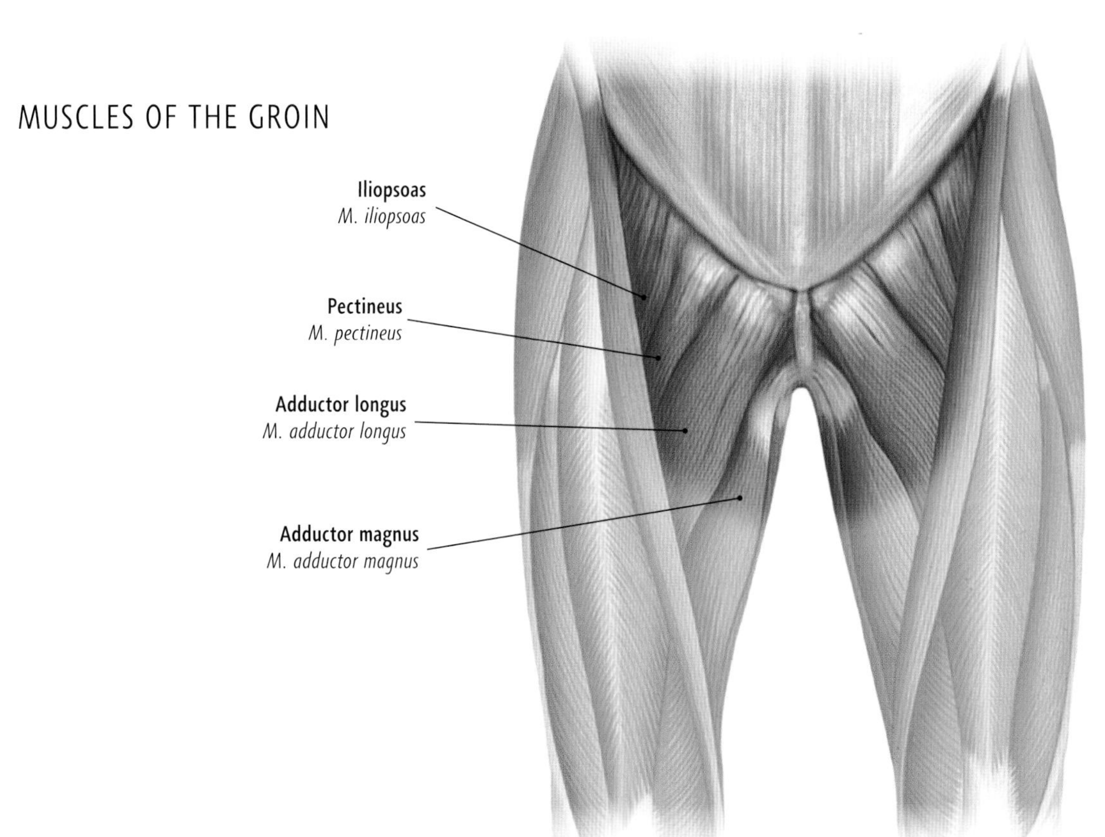

Iliopsoas
M. iliopsoas
Pectineus
M. pectineus
Adductor longus
M. adductor longus
Adductor magnus
M. adductor magnus

[~] = no direct Latin equivalent

Muscles of the Arm

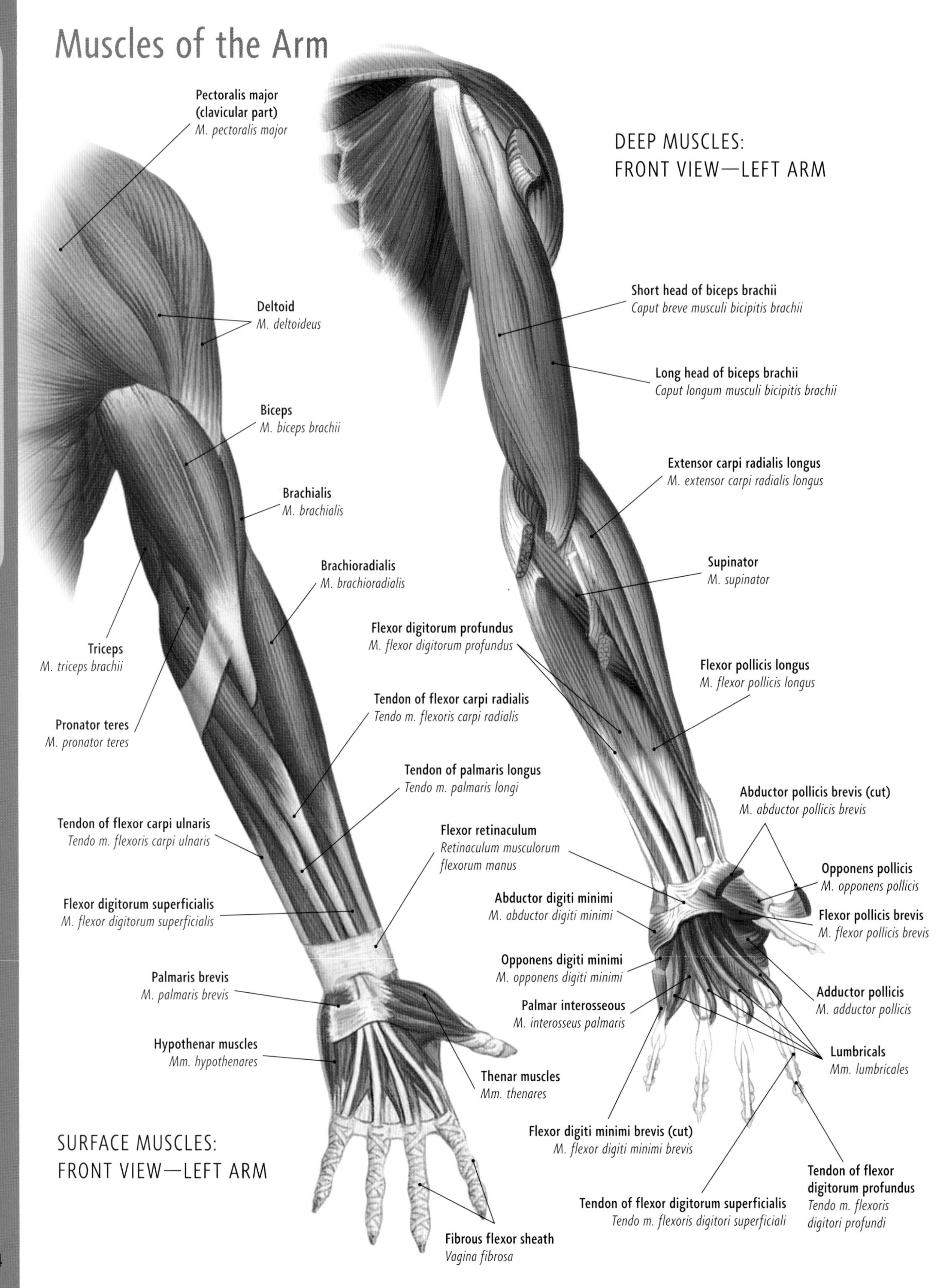
Pectoralis major
(clavicular part)
M. pectoralis major
Deltoid
M. deltoideus
Biceps
M. biceps brachii
Brachialis
M. brachialis
Brachioradialis
M. brachioradialis
Triceps
M. triceps brachii
Pronator teres
M. pronator teres
Tendon of flexor carpi radialis
Tendo m. flexoris carpi radialis
Tendon of palmaris longus
Tendo m. palmaris longi
Tendon of flexor carpi ulnaris
Tendo m. flexoris carpi ulnaris
Flexor retinaculum
Retinaculum musculorum
flexorum manus
Flexor digitorum superficialis
M. flexor digitorum superficialis
Palmaris brevis
M. palmaris brevis
Hypothenar muscles
Mm. hypothenares
Thenar muscles
Mm. thenares
SURFACE MUSCLES:
FRONT VIEW—LEFT ARM
Fibrous flexor sheath
Vagina fibrosa
DEEP MUSCLES:
FRONT VIEW—LEFT ARM
Short head of biceps brachii
Caput breve musculi bicipitis brachii
Long head of biceps brachii
Caput longum musculi bicipitis brachii
Extensor carpi radialis longus
M. extensor carpi radialis longus
Supinator
M. supinator
Flexor digitorum profundus
M. flexor digitorum profundus
Flexor pollicis longus
M. flexor pollicis longus
Abductor pollicis brevis (cut)
M. abductor pollicis brevis
Opponens pollicis
M. opponens pollicis
Abductor digiti minimi
M. abductor digiti minimi
Flexor pollicis brevis
M. flexor pollicis brevis
Opponens digiti minimi
M. opponens digiti minimi
Adductor pollicis
M. adductor pollicis
Palmar interosseous
M. interosseus palmaris
Lumbricals
Mm. lumbricales
Flexor digiti minimi brevis (cut)
M. flexor digiti minimi brevis
Tendon of flexor
digitorum profundus
Tendo m. flexoris
digitori profundi
Tendon of flexor digitorum superficialis
Tendo m. flexoris digitori superficiali

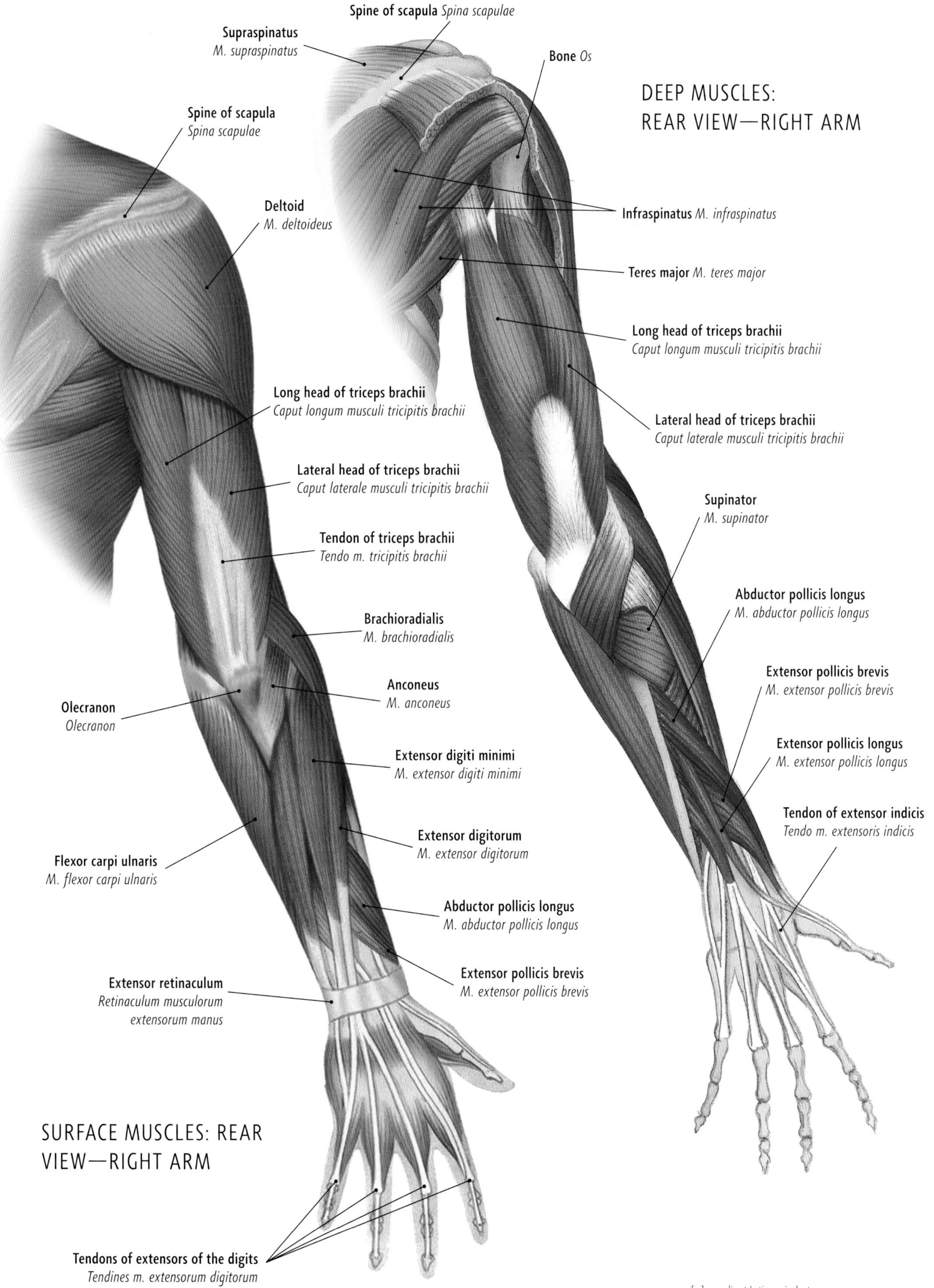
Spine of scapula Spina scapulae
Supraspinatus
M. supraspinatus
Bone Os
DEEP MUSCLES:
REAR VIEW—RIGHT ARM
Spine of scapula
Spina scapulae
Deltoid
M. deltoideus
Infraspinatus M. infraspinatus
Teres major M. teres major
Long head of triceps brachii
Caput longum musculi tricipitis brachii
Long head of triceps brachii
Caput longum musculi tricipitis brachii
Lateral head of triceps brachii
Caput laterale musculi tricipitis brachii
Lateral head of triceps brachii
Caput laterale musculi tricipitis brachii
Supinator
M. supinator
Tendon of triceps brachii
Tendo m. tricipitis brachii
Abductor pollicis longus
M. abductor pollicis longus
Brachioradialis
M. brachioradialis
Extensor pollicis brevis
M. extensor pollicis brevis
Anconeus
M. anconeus
Olecranon
Olecranon
Extensor pollicis longus
M. extensor pollicis longus
Extensor digiti minimi
M. extensor digiti minimi
Tendon of extensor indicis
Tendo m. extensoris indicis
Extensor digitorum
M. extensor digitorum
Flexor carpi ulnaris
M. flexor carpi ulnaris
Abductor pollicis longus
M. abductor pollicis longus
Extensor pollicis brevis
M. extensor pollicis brevis
Extensor retinaculum
Retinaculum musculorum
extensorum manus
SURFACE MUSCLES: REAR
VIEW—RIGHT ARM
Tendons of extensors of the digits
Tendines m. extensorum digitorum

[~] = no direct Latin equivalent

Muscles of the Arm

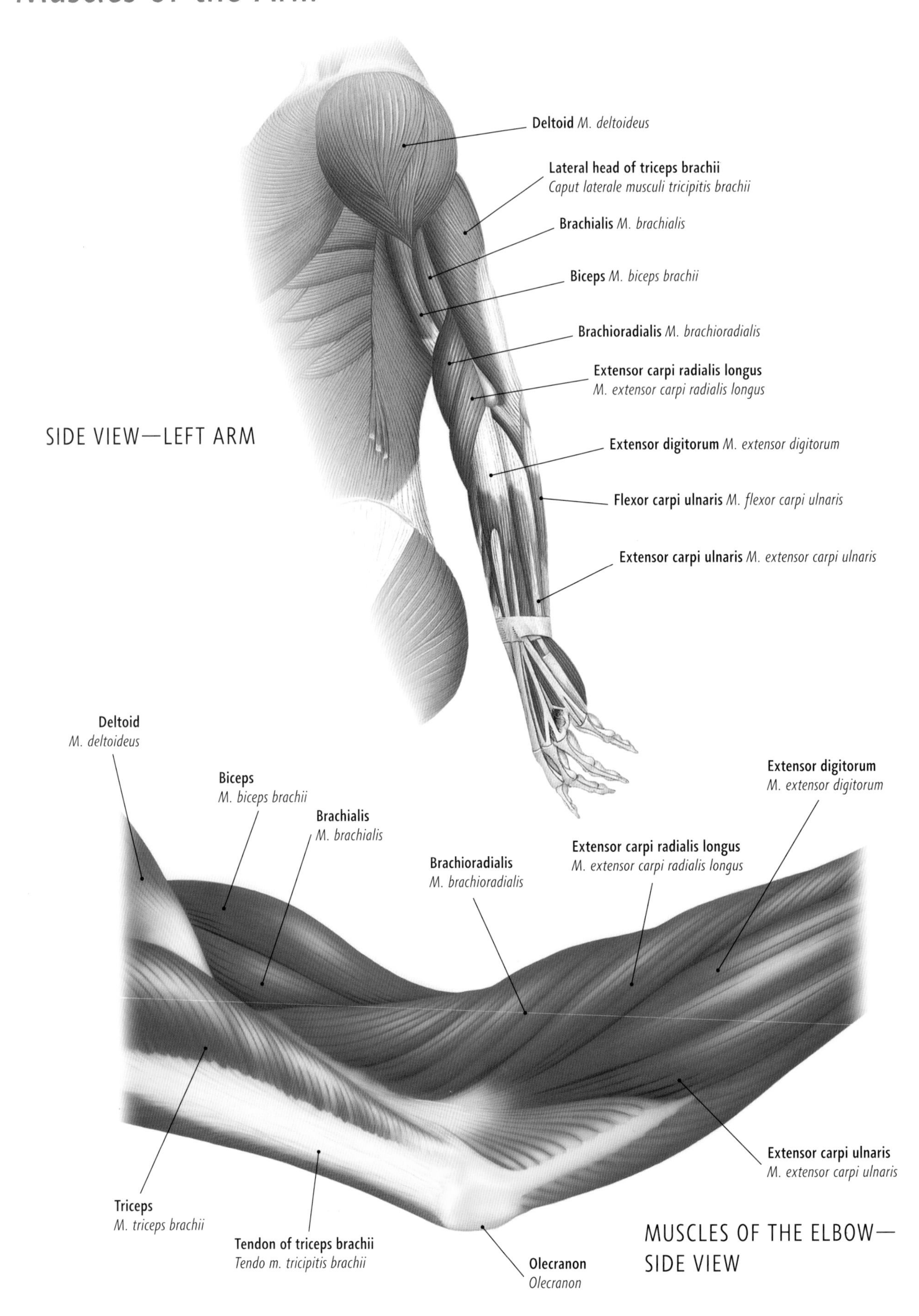

Deltoid M. deltoideus
Lateral head of triceps brachii
Caput laterale musculi tricipitis brachii
Brachialis M. brachialis
Biceps M. biceps brachii
Brachioradialis M. brachioradialis
Extensor carpi radialis longus
M. extensor carpi radialis longus
Extensor digitorum M. extensor digitorum
Flexor carpi ulnaris M. flexor carpi ulnaris
Extensor carpi ulnaris M. extensor carpi ulnaris
SIDE VIEW—LEFT ARM
Deltoid
M. deltoideus
Biceps
M. biceps brachii
Brachialis
M. brachialis
Brachioradialis
M. brachioradialis
Extensor carpi radialis longus
M. extensor carpi radialis longus
Extensor digitorum
M. extensor digitorum
Extensor carpi ulnaris
M. extensor carpi ulnaris
Triceps
M. triceps brachii
Tendon of triceps brachii
Tendo m. tricipitis brachii
Olecranon
Olecranon
MUSCLES OF THE ELBOW—
SIDE VIEW

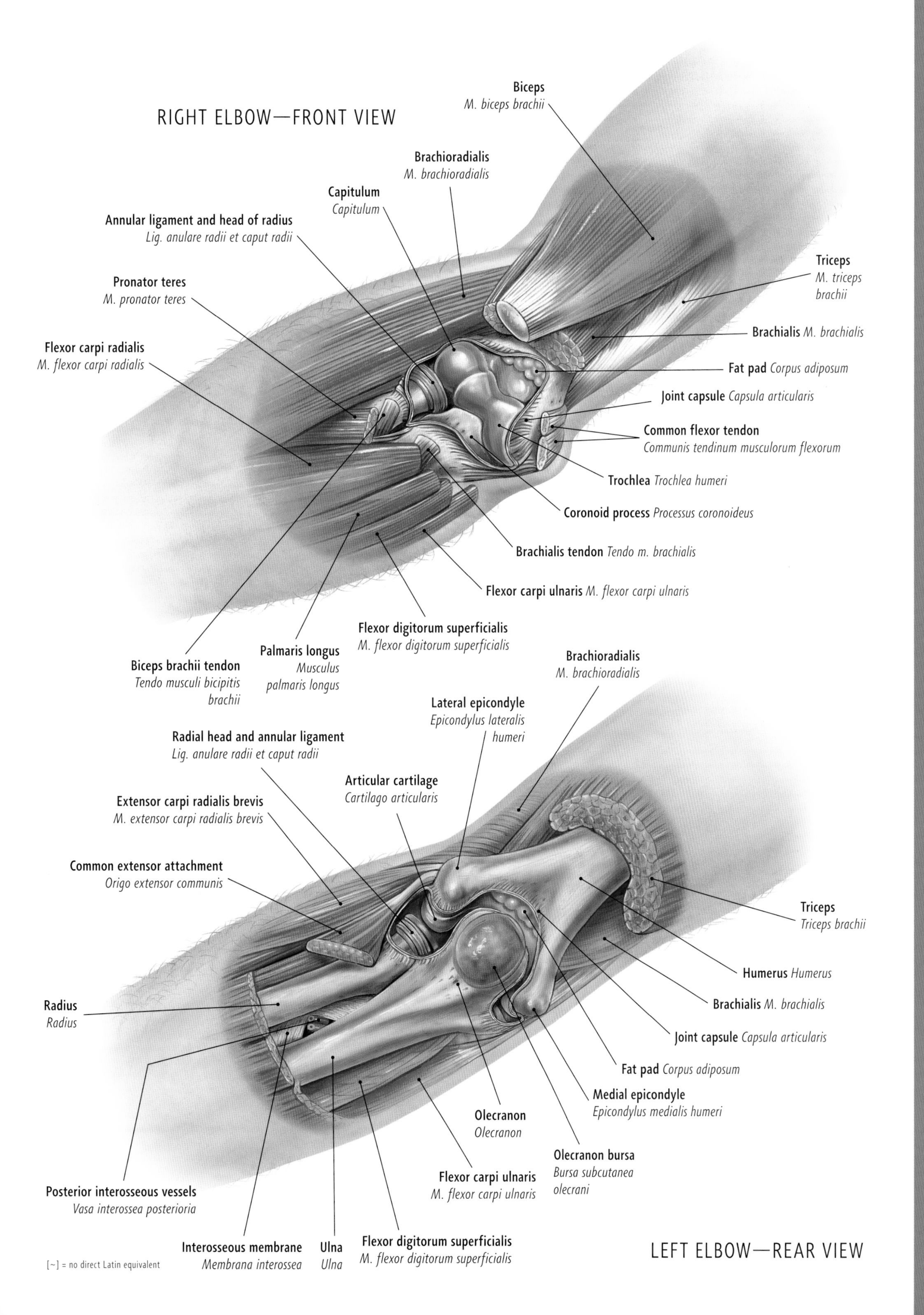
RIGHT ELBOW—FRONT VIEW
Biceps
M. biceps brachii
Brachioradialis
M. brachioradialis
Capitulum
Capitulum
Annular ligament and head of radius
Lig. anulare radii et caput radii
Pronator teres
M. pronator teres
Flexor carpi radialis
M. flexor carpi radialis
Triceps
M. triceps brachii
Brachialis M. brachialis
Fat pad Corpus adiposum
Joint capsule Capsula articularis
Common flexor tendon
Communis tendinum musculorum flexorum
Trochlea Trochlea humeri
Coronoid process Processus coronoideus
Brachialis tendon Tendo m. brachialis
Flexor carpi ulnaris M. flexor carpi ulnaris
Flexor digitorum superficialis
M. flexor digitorum superficialis
Biceps brachii tendon
Tendo musculi bicipitis brachii
Palmaris longus
Musculus palmaris longus
Brachioradialis
M. brachioradialis
Lateral epicondyle
Epicondylus lateralis humeri
Radial head and annular ligament
Lig. anulare radii et caput radii
Articular cartilage
Cartilago articularis
Extensor carpi radialis brevis
M. extensor carpi radialis brevis
Common extensor attachment
Origo extensor communis
Triceps
Triceps brachii
Humerus Humerus
Brachialis M. brachialis
Radius
Radius
Joint capsule Capsula articularis
Fat pad Corpus adiposum
Medial epicondyle
Epicondylus medialis humeri
Olecranon
Olecranon
Olecranon bursa
Bursa subcutanea olecrani
Flexor carpi ulnaris
M. flexor carpi ulnaris
Posterior interosseous vessels
Vasa interossea posterioria
Interosseous membrane
Membrana interossea
Ulna
Ulna
Flexor digitorum superficialis
M. flexor digitorum superficialis
LEFT ELBOW—REAR VIEW

[~] = no direct Latin equivalent

Muscles of the Arm and Hand

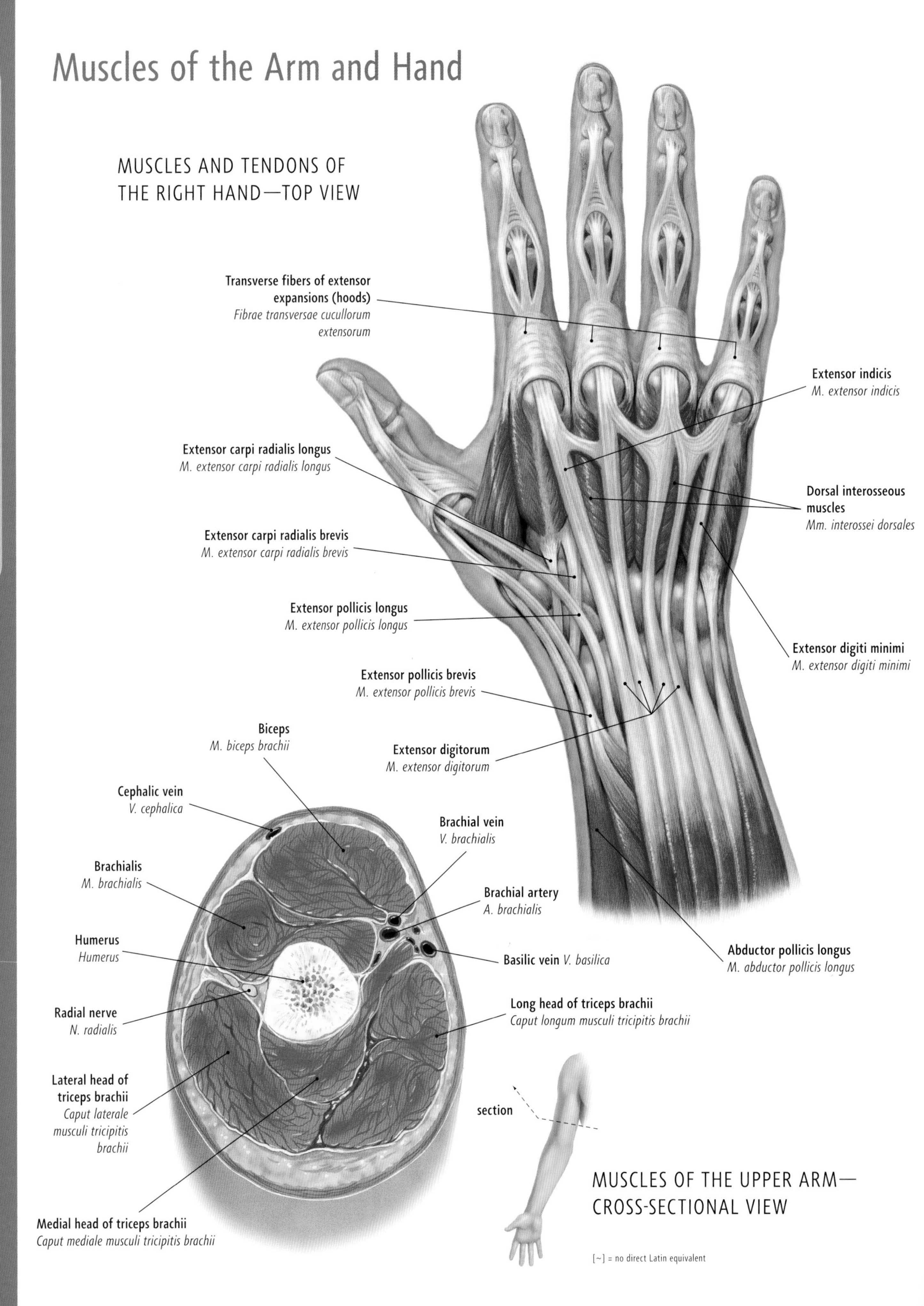
MUSCLES AND TENDONS OF THE RIGHT HAND—TOP VIEW
Transverse fibers of extensor expansions (hoods)
Fibrae transversae cucullorum extensorum
Extensor indicis
M. extensor indicis
Extensor carpi radialis longus
M. extensor carpi radialis longus
Dorsal interosseous muscles
Mm. interossei dorsales
Extensor carpi radialis brevis
M. extensor carpi radialis brevis
Extensor pollicis longus
M. extensor pollicis longus
Extensor digiti minimi
M. extensor digiti minimi
Extensor pollicis brevis
M. extensor pollicis brevis
Biceps
M. biceps brachii
Extensor digitorum
M. extensor digitorum
Cephalic vein
V. cephalica
Brachial vein
V. brachialis
Brachialis
M. brachialis
Brachial artery
A. brachialis
Humerus
Humerus
Basilic vein V. basilica
Abductor pollicis longus
M. abductor pollicis longus
Radial nerve
N. radialis
Long head of triceps brachii
Caput longum musculi tricipitis brachii
Lateral head of triceps brachii
Caput laterale musculi tricipitis brachii
section
Medial head of triceps brachii
Caput mediale musculi tricipitis brachii
MUSCLES OF THE UPPER ARM—CROSS-SECTIONAL VIEW
[~] = no direct Latin equivalent

ROTATOR CUFF MUSCLES—REAR VIEW

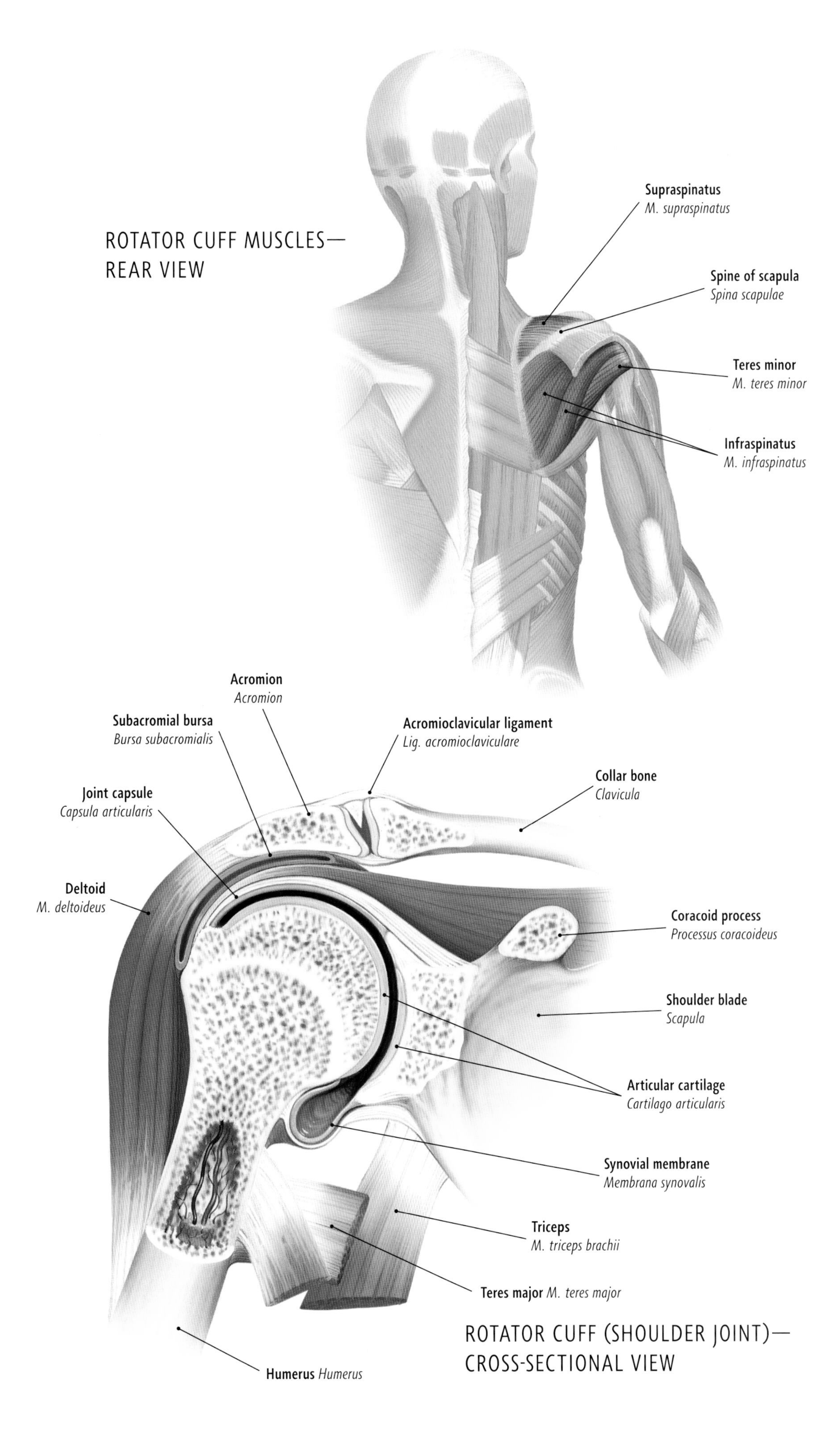
Supraspinatus
M. supraspinatus
Spine of scapula
Spina scapulae
Teres minor
M. teres minor
Infraspinatus
M. infraspinatus
Acromion
Acromion
Subacromial bursa
Bursa subacromialis
Acromioclavicular ligament
Lig. acromioclaviculare
Collar bone
Clavicula
Joint capsule
Capsula articularis
Deltoid
M. deltoideus
Coracoid process
Processus coracoideus
Shoulder blade
Scapula
Articular cartilage
Cartilago articularis
Synovial membrane
Membrana synovalis
Triceps
M. triceps brachii
Teres major M. teres major
Humerus Humerus

ROTATOR CUFF (SHOULDER JOINT)—CROSS-SECTIONAL VIEW

Muscles of the Leg

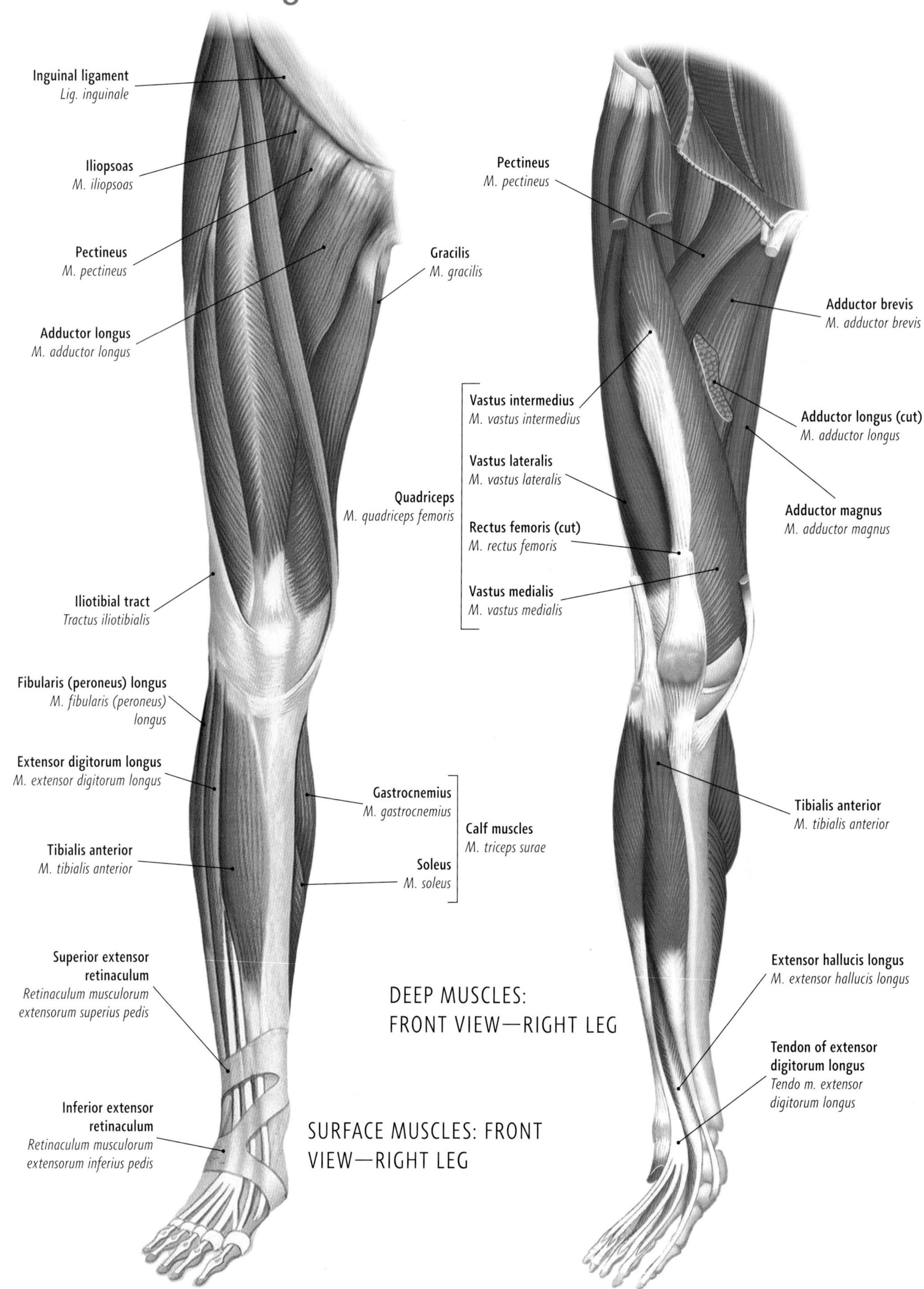
Inguinal ligament
Lig. inguinale
Iliopsoas
M. iliopsoas
Pectineus
M. pectineus
Adductor longus
M. adductor longus
Gracilis
M. gracilis
Iliotibial tract
Tractus iliotibialis
Fibularis (peroneus) longus
M. fibularis (peroneus) longus
Extensor digitorum longus
M. extensor digitorum longus
Tibialis anterior
M. tibialis anterior
Gastrocnemius
M. gastrocnemius
Calf muscles
M. triceps surae
Soleus
M. soleus
Superior extensor retinaculum
Retinaculum musculorum extensorum superius pedis
Inferior extensor retinaculum
Retinaculum musculorum extensorum inferius pedis
SURFACE MUSCLES: FRONT VIEW—RIGHT LEG
Pectineus
M. pectineus
Adductor brevis
M. adductor brevis
Quadriceps
M. quadriceps femoris
Vastus intermedius
M. vastus intermedius
Vastus lateralis
M. vastus lateralis
Rectus femoris (cut)
M. rectus femoris
Vastus medialis
M. vastus medialis
Adductor longus (cut)
M. adductor longus
Adductor magnus
M. adductor magnus
Tibialis anterior
M. tibialis anterior
Extensor hallucis longus
M. extensor hallucis longus
Tendon of extensor digitorum longus
Tendo m. extensor digitorum longus
DEEP MUSCLES: FRONT VIEW—RIGHT LEG

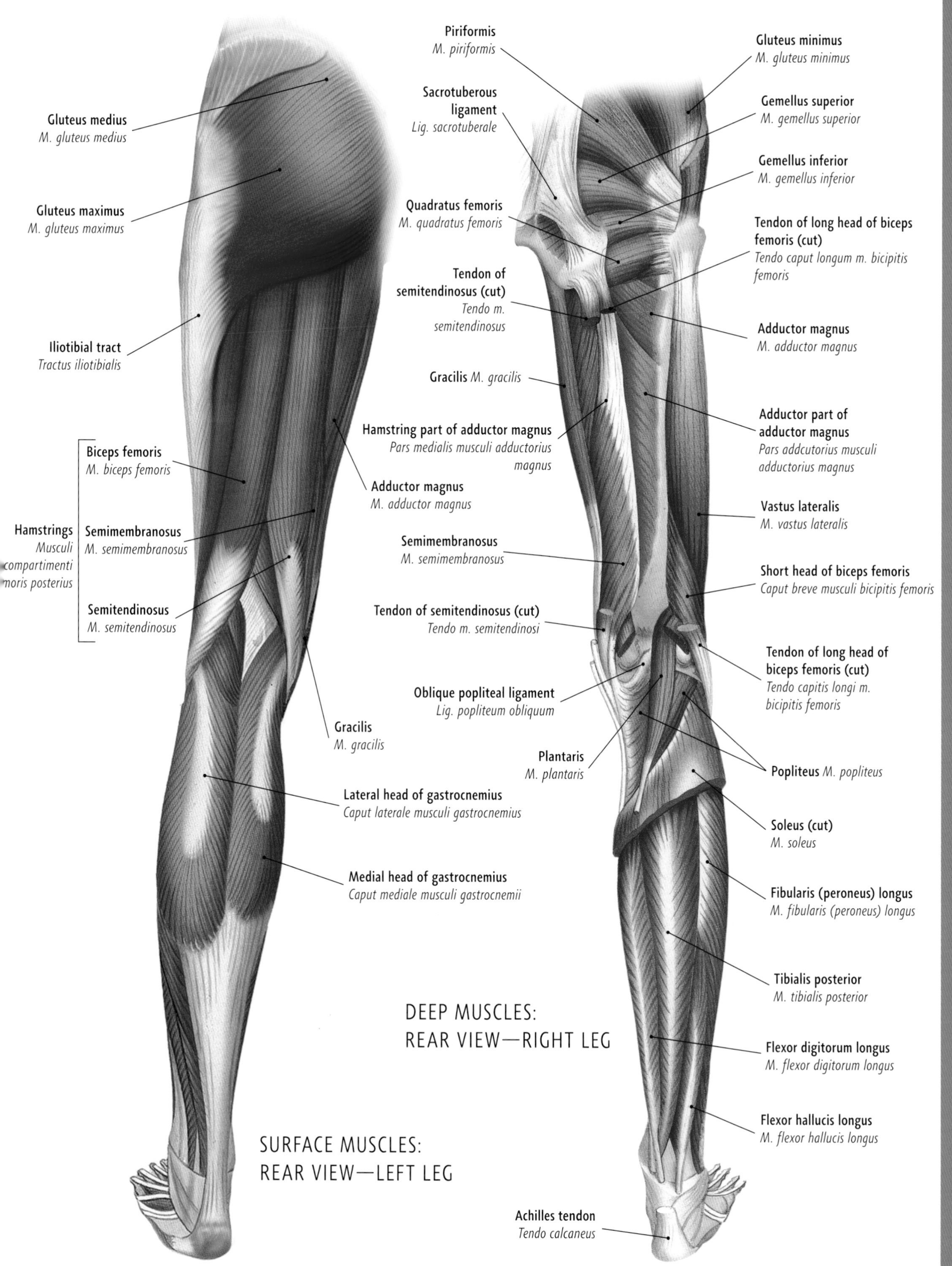
Gluteus medius
M. gluteus medius
Gluteus maximus
M. gluteus maximus
Iliotibial tract
Tractus iliotibialis
Biceps femoris
M. biceps femoris
Hamstrings
Musculi
compartimenti
femoris posterius
Semimembranosus
M. semimembranosus
Semitendinosus
M. semitendinosus
Adductor magnus
M. adductor magnus
Gracilis
M. gracilis
Lateral head of gastrocnemius
Caput laterale musculi gastrocnemius
Medial head of gastrocnemius
Caput mediale musculi gastrocnemii
SURFACE MUSCLES:
REAR VIEW—LEFT LEG
Piriformis
M. piriformis
Sacrotuberous
ligament
Lig. sacrotuberale
Quadratus femoris
M. quadratus femoris
Tendon of
semitendinosus (cut)
Tendo m.
semitendinosus
Gracilis M. gracilis
Hamstring part of adductor magnus
Pars medialis musculi adductorius
magnus
Semimembranosus
M. semimembranosus
Tendon of semitendinosus (cut)
Tendo m. semitendinosi
Oblique popliteal ligament
Lig. popliteum obliquum
Plantaris
M. plantaris
DEEP MUSCLES:
REAR VIEW—RIGHT LEG
Achilles tendon
Tendo calcaneus
Gluteus minimus
M. gluteus minimus
Gemellus superior
M. gemellus superior
Gemellus inferior
M. gemellus inferior
Tendon of long head of biceps
femoris (cut)
Tendo caput longum m. bicipitis
femoris
Adductor magnus
M. adductor magnus
Adductor part of
adductor magnus
Pars addcutorius musculi
adductorius magnus
Vastus lateralis
M. vastus lateralis
Short head of biceps femoris
Caput breve musculi bicipitis femoris
Tendon of long head of
biceps femoris (cut)
Tendo capitis longi m.
bicipitis femoris
Popliteus M. popliteus
Soleus (cut)
M. soleus
Fibularis (peroneus) longus
M. fibularis (peroneus) longus
Tibialis posterior
M. tibialis posterior
Flexor digitorum longus
M. flexor digitorum longus
Flexor hallucis longus
M. flexor hallucis longus

[~] = no direct Latin equivalent

Muscles of the Leg and Foot

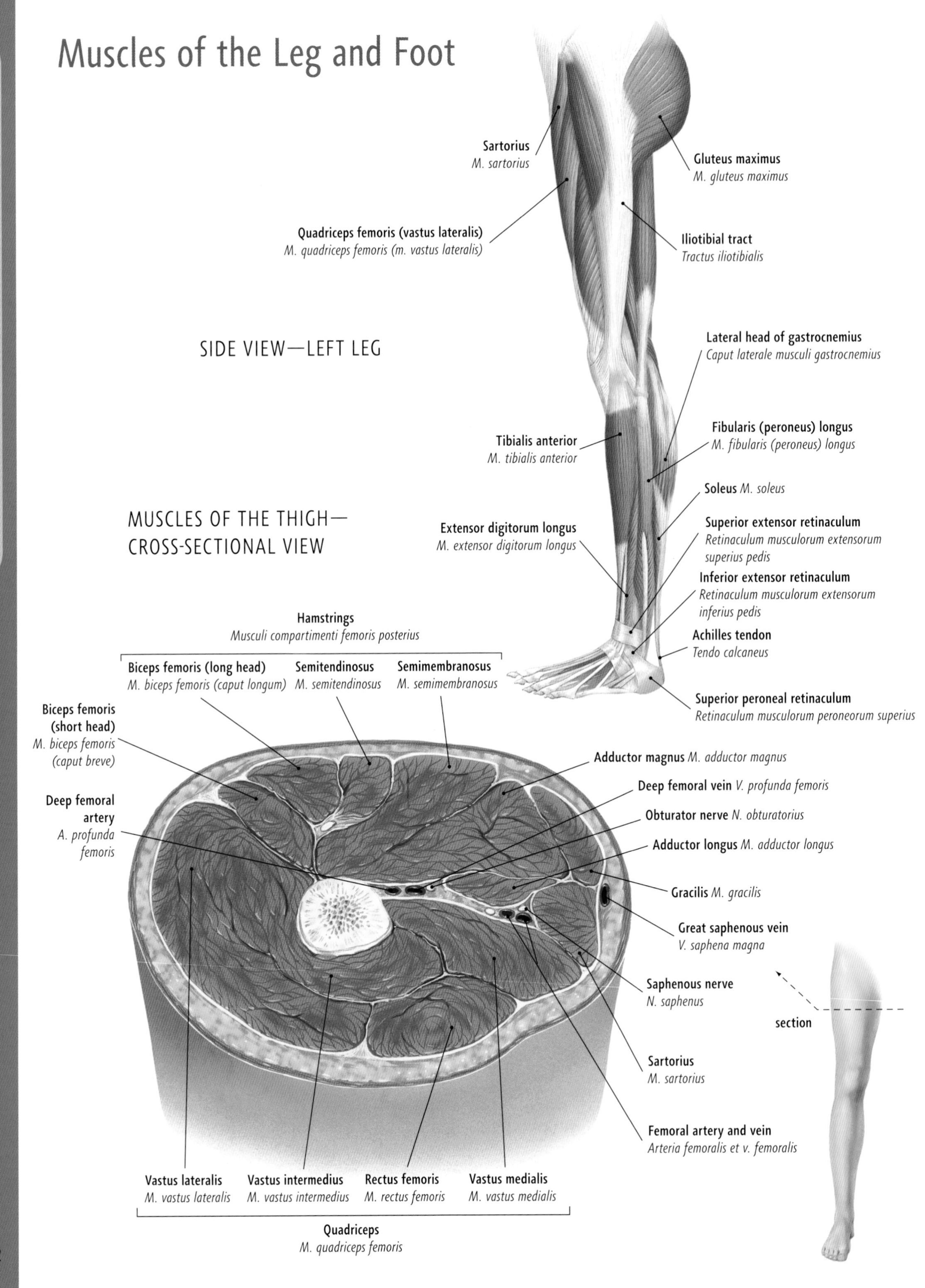
SIDE VIEW—LEFT LEG
Sartorius
M. sartorius
Gluteus maximus
M. gluteus maximus
Quadriceps femoris (vastus lateralis)
M. quadriceps femoris (m. vastus lateralis)
Iliotibial tract
Tractus iliotibialis
Lateral head of gastrocnemius
Caput laterale musculi gastrocnemius
Tibialis anterior
M. tibialis anterior
Fibularis (peroneus) longus
M. fibularis (peroneus) longus
Soleus M. soleus
Extensor digitorum longus
M. extensor digitorum longus
Superior extensor retinaculum
Retinaculum musculorum extensorum superius pedis
Inferior extensor retinaculum
Retinaculum musculorum extensorum inferius pedis
Achilles tendon
Tendo calcaneus
Superior peroneal retinaculum
Retinaculum musculorum peroneorum superius
MUSCLES OF THE THIGH—CROSS-SECTIONAL VIEW
Hamstrings
Musculi compartimenti femoris posterius
Biceps femoris (long head)
M. biceps femoris (caput longum)
Semitendinosus
M. semitendinosus
Semimembranosus
M. semimembranosus
Biceps femoris (short head)
M. biceps femoris (caput breve)
Deep femoral artery
A. profunda femoris
Adductor magnus M. adductor magnus
Deep femoral vein V. profunda femoris
Obturator nerve N. obturatorius
Adductor longus M. adductor longus
Gracilis M. gracilis
Great saphenous vein
V. saphena magna
Saphenous nerve
N. saphenus
section
Sartorius
M. sartorius
Femoral artery and vein
Arteria femoralis et v. femoralis
Vastus lateralis
M. vastus lateralis
Vastus intermedius
M. vastus intermedius
Rectus femoris
M. rectus femoris
Vastus medialis
M. vastus medialis
Quadriceps
M. quadriceps femoris

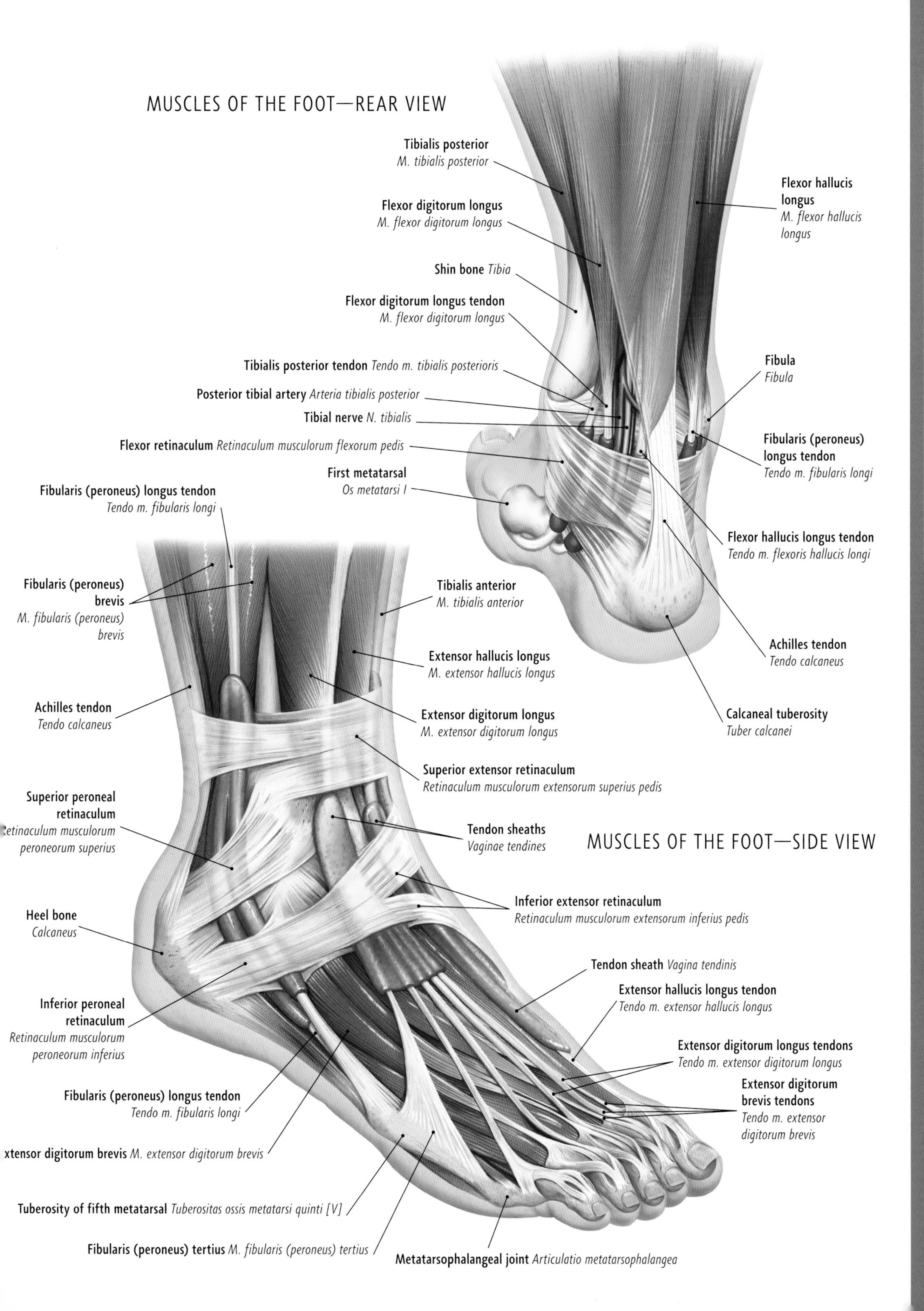
MUSCLES OF THE FOOT—REAR VIEW
Tibialis posterior
M. tibialis posterior
Flexor digitorum longus
M. flexor digitorum longus
Shin bone Tibia
Flexor digitorum longus tendon
M. flexor digitorum longus
Tibialis posterior tendon Tendo m. tibialis posterioris
Posterior tibial artery Arteria tibialis posterior
Tibial nerve N. tibialis
Flexor retinaculum Retinaculum musculorum flexorum pedis
First metatarsal
Os metatarsi I
Flexor hallucis longus
M. flexor hallucis longus
Fibula
Fibula
Fibularis (peroneus) longus tendon
Tendo m. fibularis longi
Flexor hallucis longus tendon
Tendo m. flexoris hallucis longi
Achilles tendon
Tendo calcaneus
Calcaneal tuberosity
Tuber calcanei
MUSCLES OF THE FOOT—SIDE VIEW
Fibularis (peroneus) longus tendon
Tendo m. fibularis longi
Fibularis (peroneus) brevis
M. fibularis (peroneus) brevis
Tibialis anterior
M. tibialis anterior
Extensor hallucis longus
M. extensor hallucis longus
Achilles tendon
Tendo calcaneus
Extensor digitorum longus
M. extensor digitorum longus
Superior extensor retinaculum
Retinaculum musculorum extensorum superius pedis
Superior peroneal retinaculum
etinaculum musculorum peroneorum superius
Tendon sheaths
Vaginae tendines
Inferior extensor retinaculum
Retinaculum musculorum extensorum inferius pedis
Heel bone
Calcaneus
Tendon sheath Vagina tendinis
Extensor hallucis longus tendon
Tendo m. extensor hallucis longus
Inferior peroneal retinaculum
Retinaculum musculorum peroneorum inferius
Extensor digitorum longus tendons
Tendo m. extensor digitorum longus
Extensor digitorum brevis tendons
Tendo m. extensor digitorum brevis
Fibularis (peroneus) longus tendon
Tendo m. fibularis longi
xtensor digitorum brevis M. extensor digitorum brevis
Tuberosity of fifth metatarsal Tuberositas ossis metatarsi quinti [V]
Fibularis (peroneus) tertius M. fibularis (peroneus) tertius
Metatarsophalangeal joint Articulatio metatarsophalangea

Muscles of the Foot

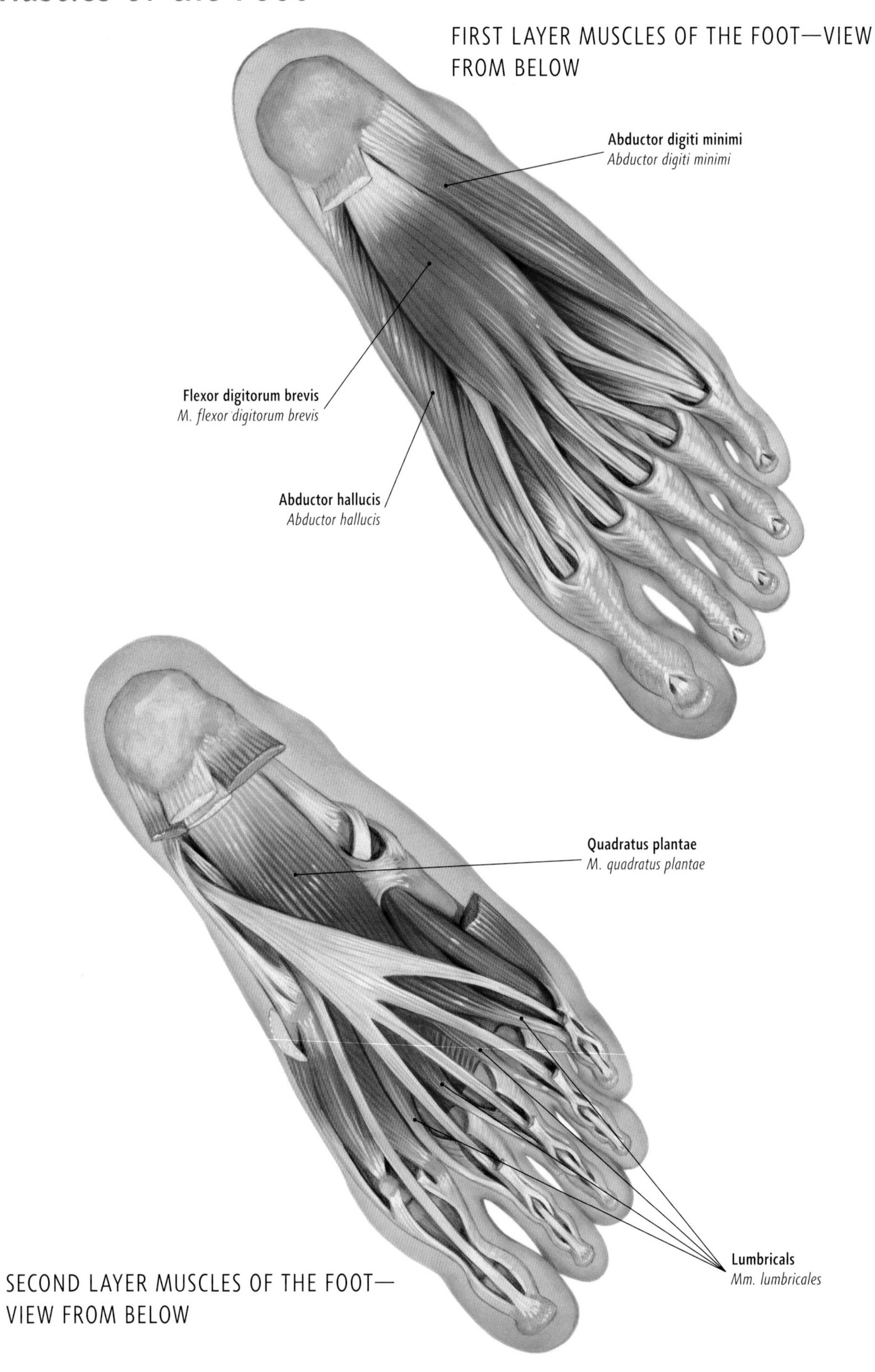
FIRST LAYER MUSCLES OF THE FOOT—VIEW FROM BELOW
Abductor digiti minimi
Abductor digiti minimi
Flexor digitorum brevis
M. flexor digitorum brevis
Abductor hallucis
Abductor hallucis
Quadratus plantae
M. quadratus plantae
Lumbricals
Mm. lumbricales
SECOND LAYER MUSCLES OF THE FOOT—VIEW FROM BELOW

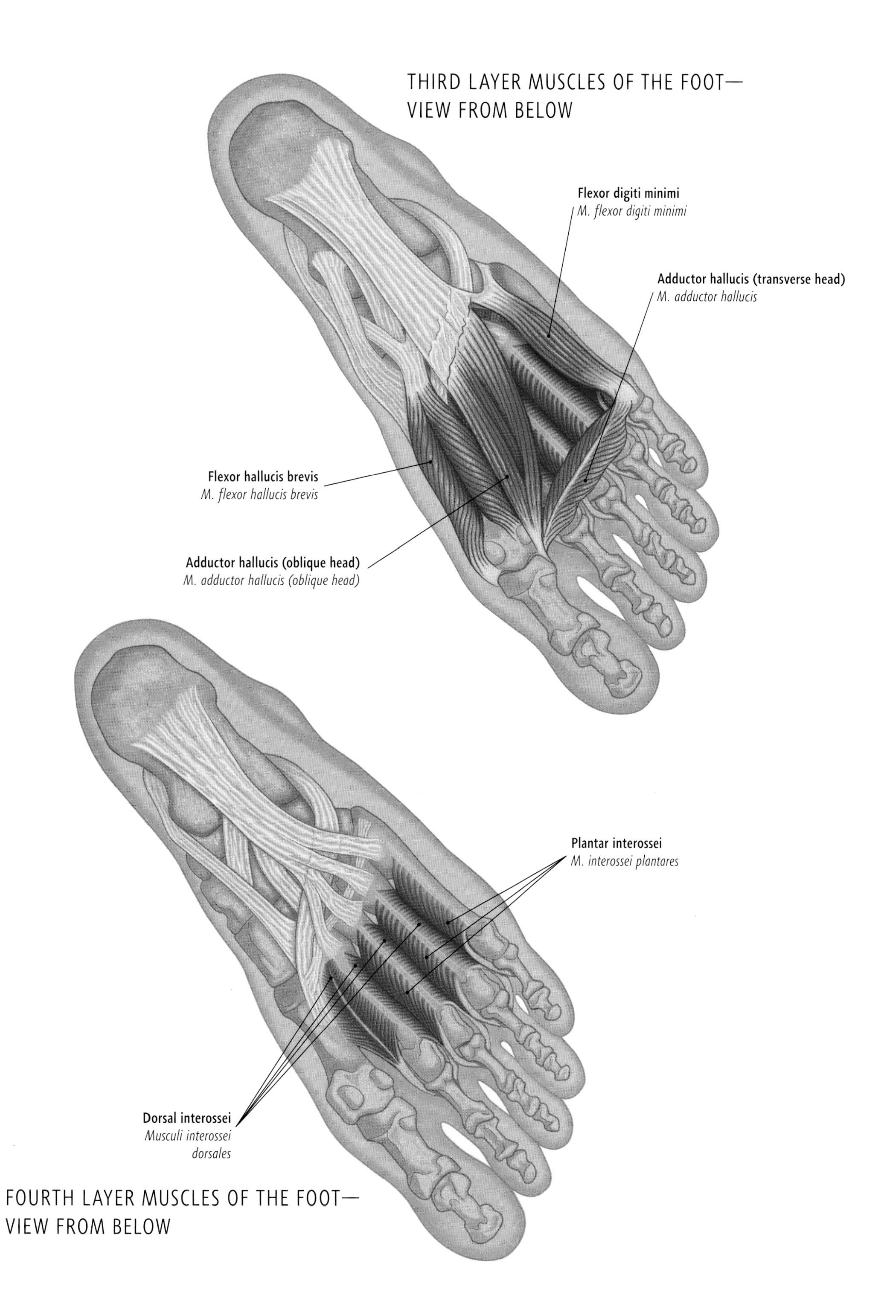
THIRD LAYER MUSCLES OF THE FOOT—
VIEW FROM BELOW
Flexor digiti minimi
M. flexor digiti minimi
Adductor hallucis (transverse head)
M. adductor hallucis
Flexor hallucis brevis
M. flexor hallucis brevis
Adductor hallucis (oblique head)
M. adductor hallucis (oblique head)
Plantar interossei
M. interossei plantares
Dorsal interossei
Musculi interossei
dorsales
FOURTH LAYER MUSCLES OF THE FOOT—
VIEW FROM BELOW

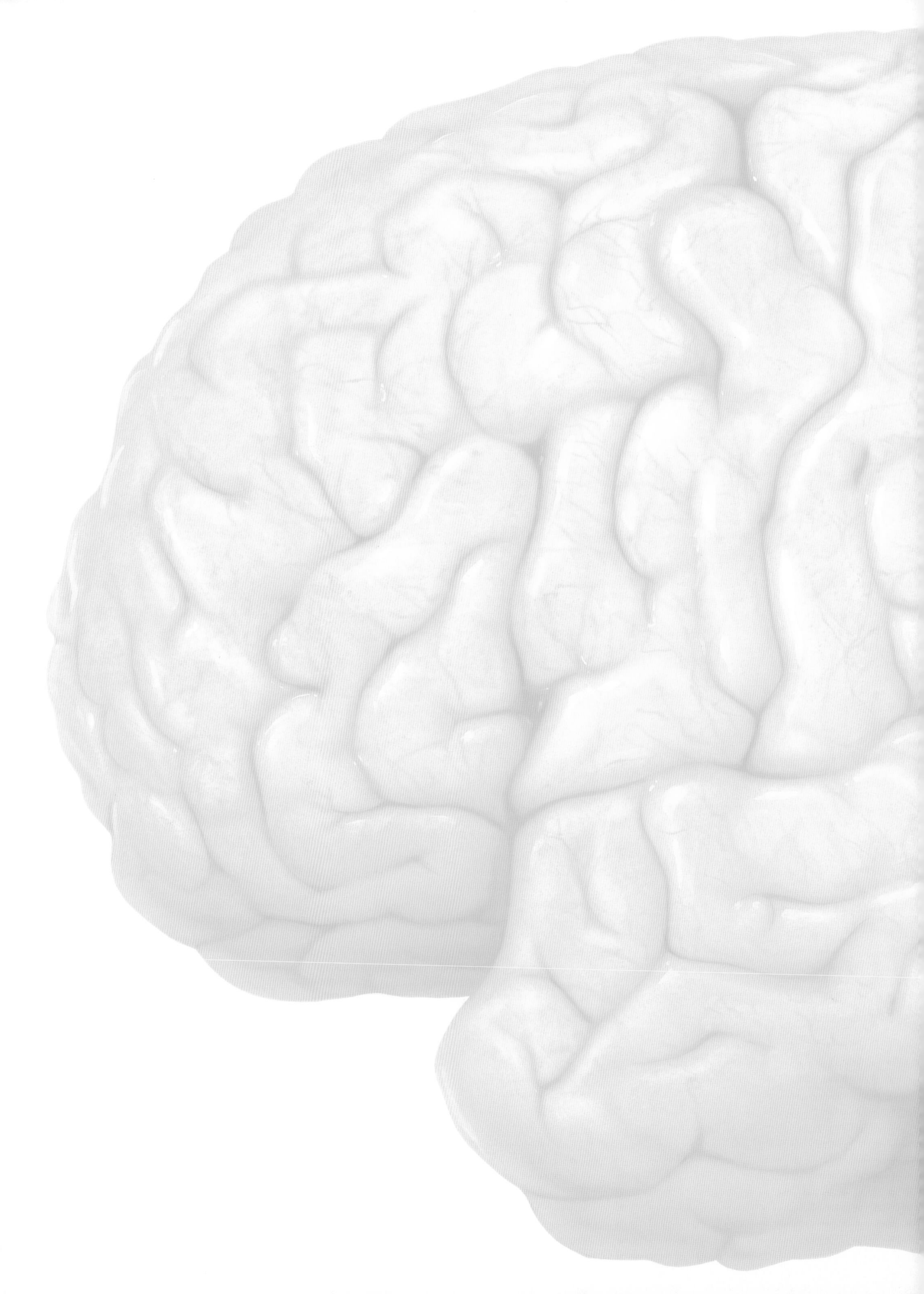

The Nervous System

The Nervous System

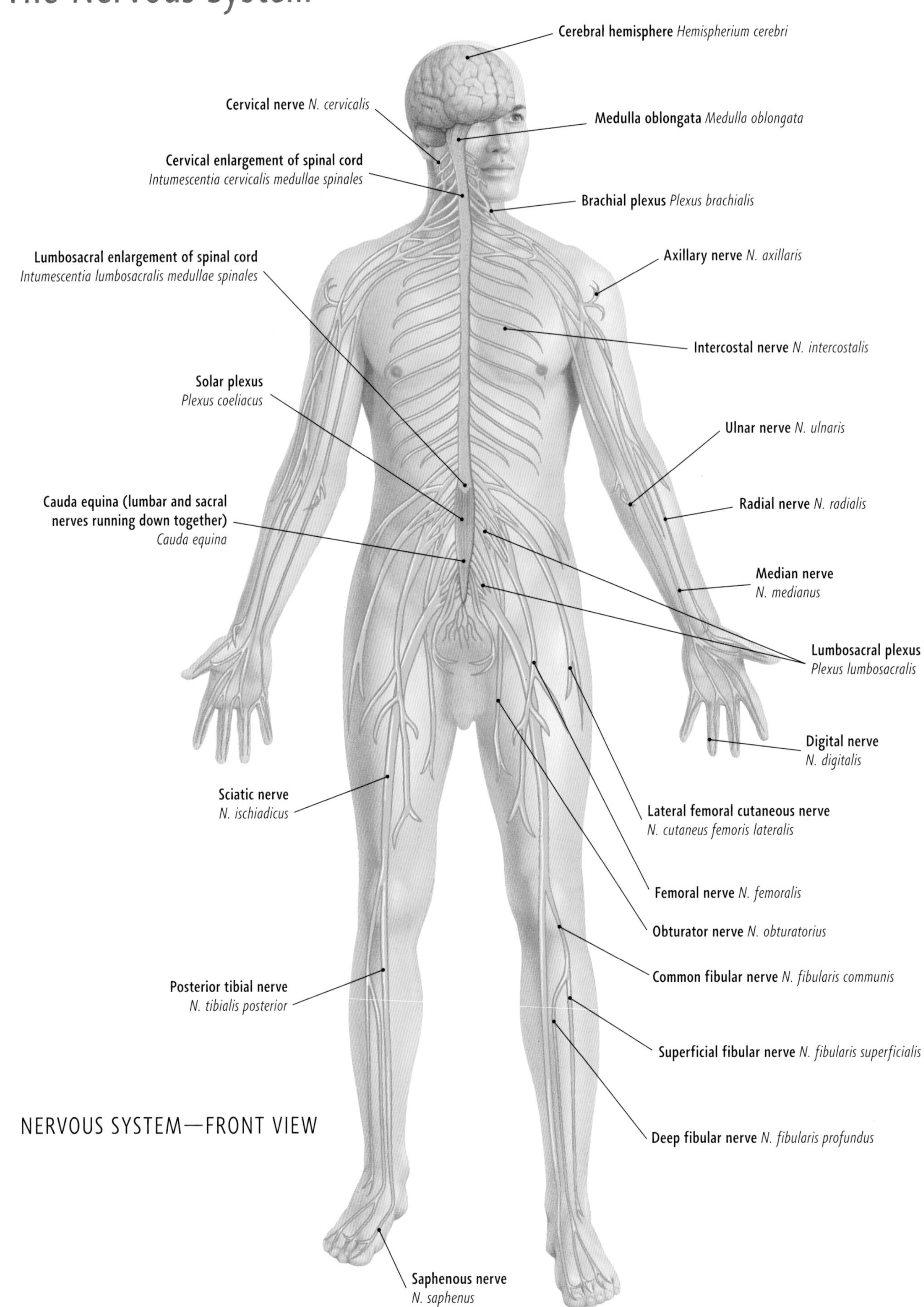
Cerebral hemisphere *Hemispherium cerebri*
Cervical nerve *N. cervicalis*
Medulla oblongata *Medulla oblongata*
Cervical enlargement of spinal cord *Intumescentia cervicalis medullae spinales*
Brachial plexus *Plexus brachialis*
Lumbosacral enlargement of spinal cord *Intumescentia lumbosacralis medullae spinales*
Axillary nerve *N. axillaris*
Intercostal nerve *N. intercostalis*
Solar plexus *Plexus coeliacus*
Ulnar nerve *N. ulnaris*
Cauda equina (lumbar and sacral nerves running down together) *Cauda equina*
Radial nerve *N. radialis*
Median nerve *N. medianus*
Lumbosacral plexus *Plexus lumbosacralis*
Digital nerve *N. digitalis*
Sciatic nerve *N. ischiadicus*
Lateral femoral cutaneous nerve *N. cutaneus femoris lateralis*
Femoral nerve *N. femoralis*
Obturator nerve *N. obturatorius*
Common fibular nerve *N. fibularis communis*
Posterior tibial nerve *N. tibialis posterior*
Superficial fibular nerve *N. fibularis superficialis*
Deep fibular nerve *N. fibularis profundus*
Saphenous nerve *N. saphenus*

NERVOUS SYSTEM—FRONT VIEW

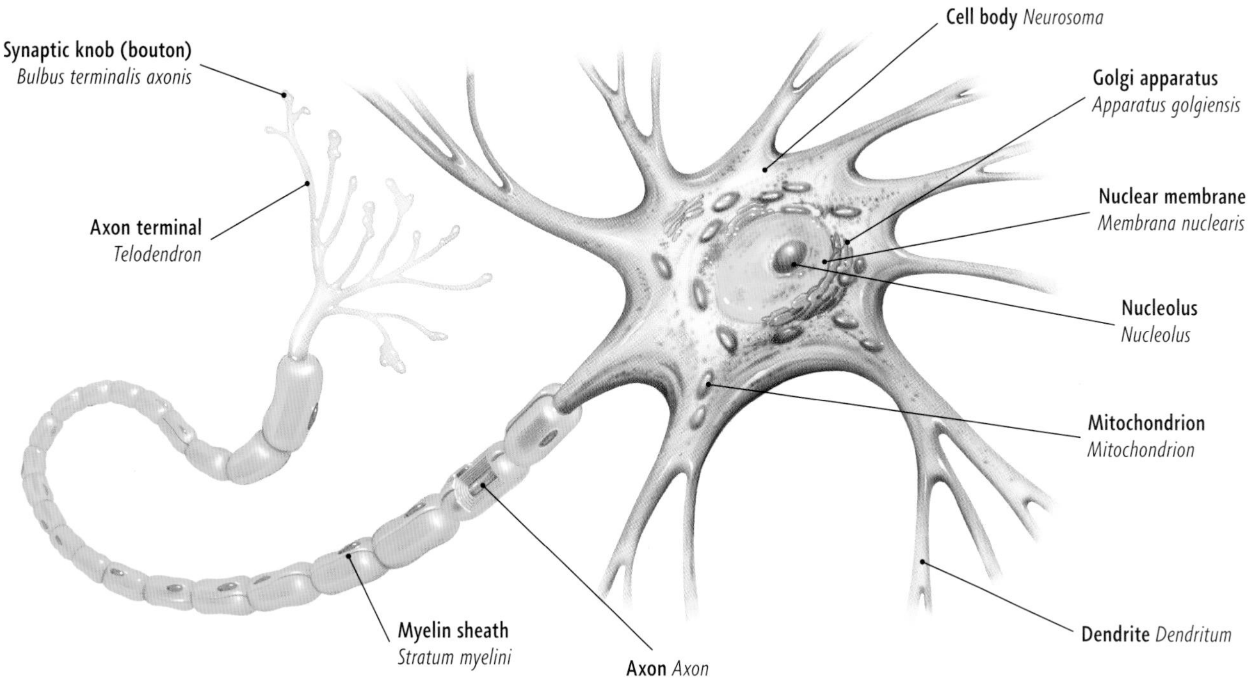
Synaptic knob (bouton)
Bulbus terminalis axonis
Axon terminal
Telodendron
Cell body *Neurosoma*
Golgi apparatus
Apparatus golgiensis
Nuclear membrane
Membrana nuclearis
Nucleolus
Nucleolus
Mitochondrion
Mitochondrion
Myelin sheath
Stratum myelini
Axon *Axon*
Dendrite *Dendritum*

NEURON

Neurons are specialized cells that conduct nerve impulses. Each neuron has three main parts: a cell body, several branching dendrites that carry impulses to the cell body, and an axon that conveys impulses away from the cell body.

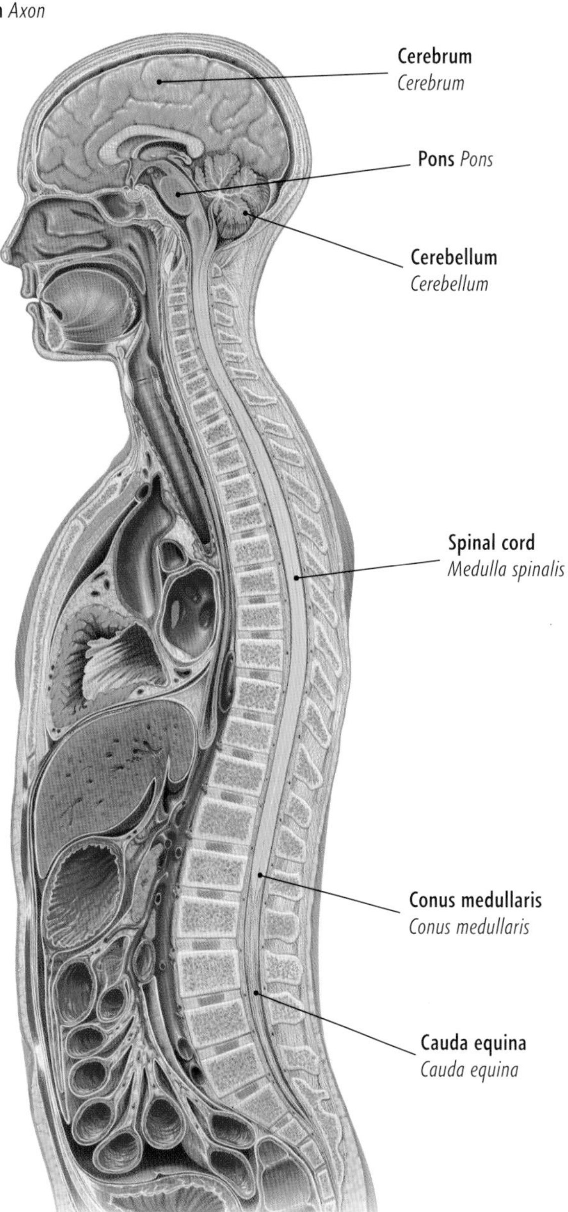
Cerebrum
Cerebrum
Pons *Pons*
Cerebellum
Cerebellum
Spinal cord
Medulla spinalis
Conus medullaris
Conus medullaris
Cauda equina
Cauda equina

CENTRAL NERVOUS SYSTEM—CROSS-SECTIONAL VIEW

The central nervous system (CNS) comprises the brain and spinal cord. Nerves connect the CNS to peripheral regions such as the muscles, skin, and internal organs. Most nerve cells of the autonomic nervous system (which controls involuntary body functions) are found peripherally, but some are located in the CNS.

[~] = no direct Latin equivalent

The Brain

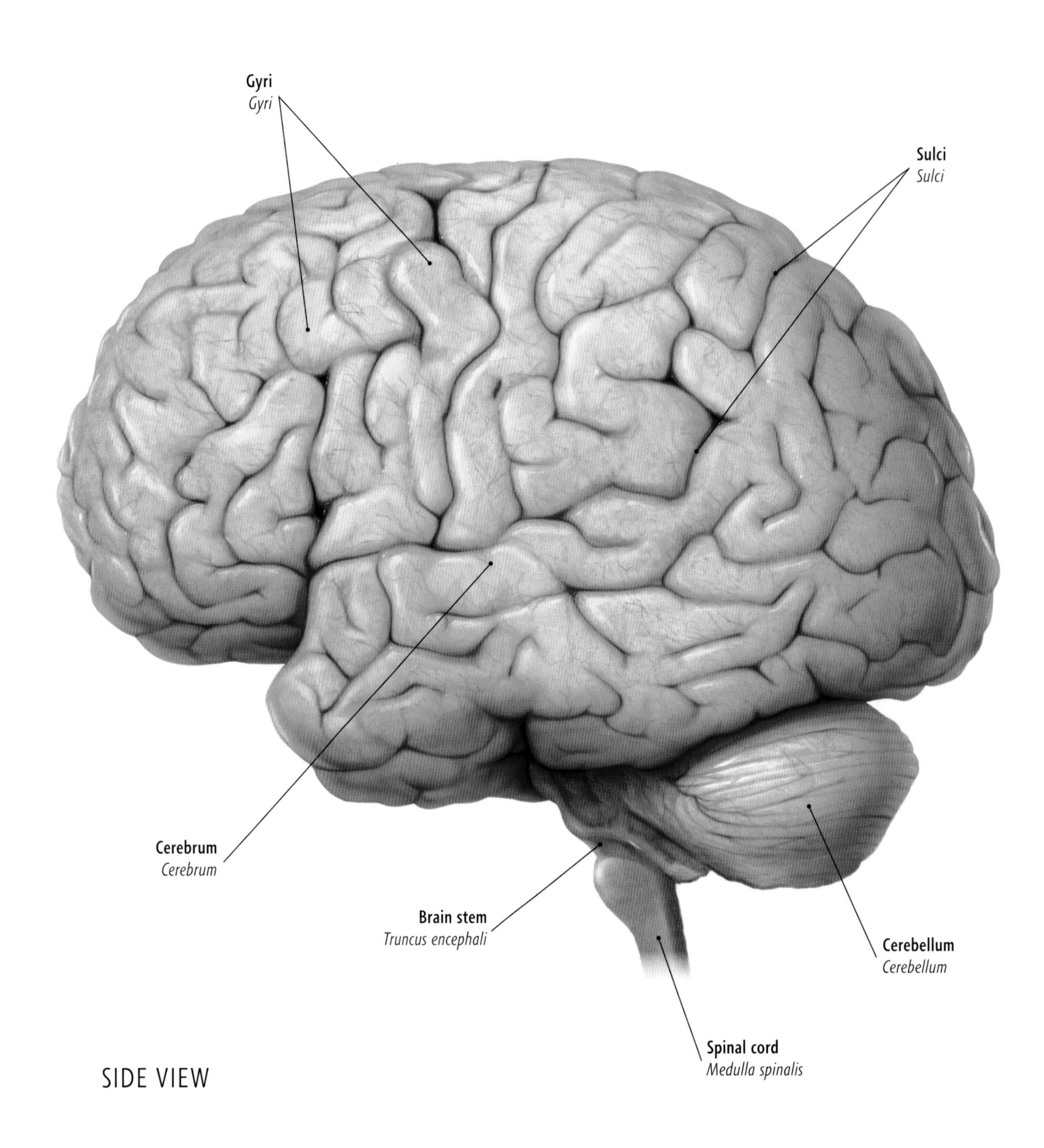
Gyri
Gyri
Sulci
Sulci
Cerebrum
Cerebrum
Brain stem
Truncus encephali
Cerebellum
Cerebellum
Spinal cord
Medulla spinalis

SIDE VIEW

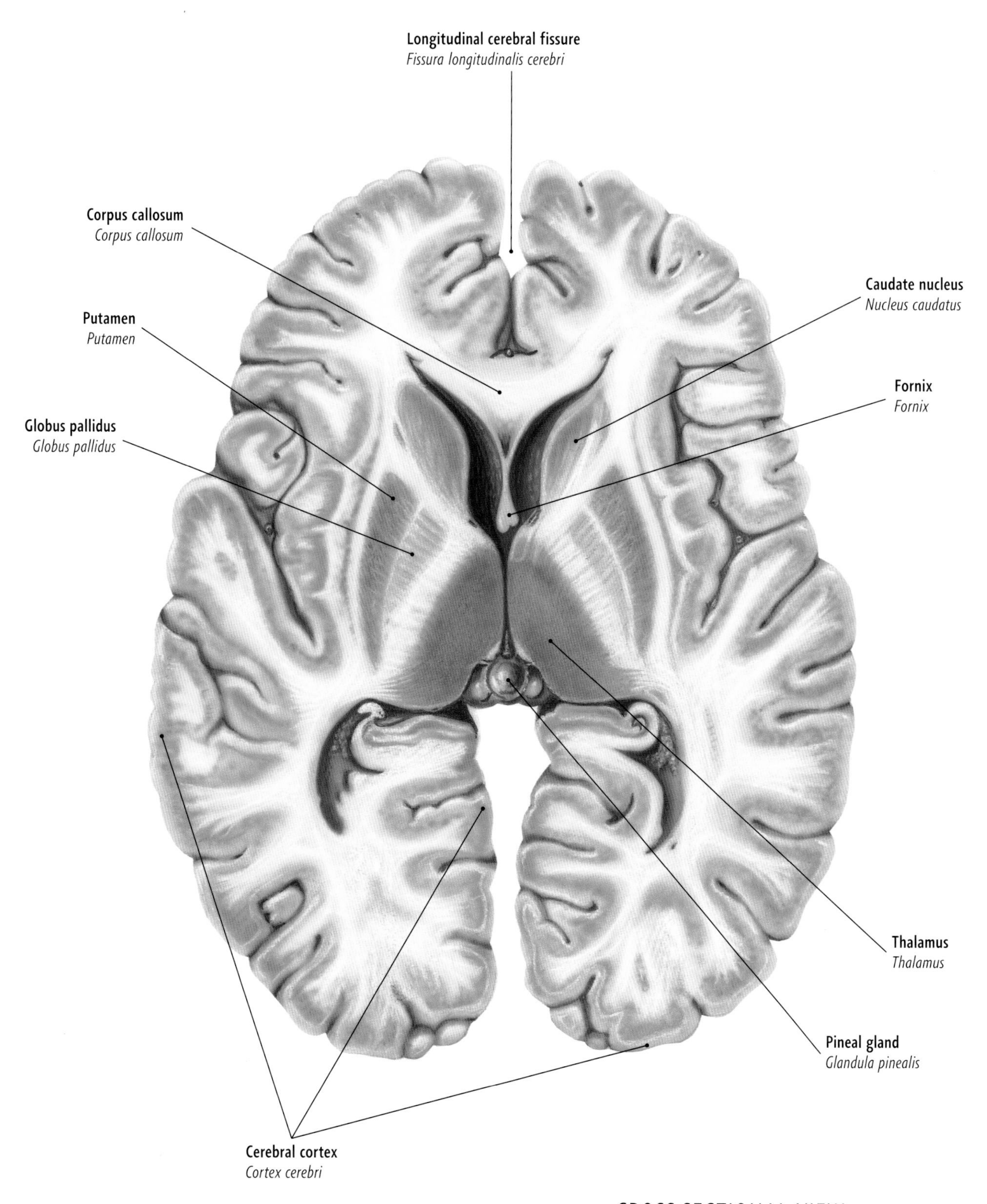

Longitudinal cerebral fissure
Fissura longitudinalis cerebri
Corpus callosum
Corpus callosum
Putamen
Putamen
Globus pallidus
Globus pallidus
Caudate nucleus
Nucleus caudatus
Fornix
Fornix
Thalamus
Thalamus
Pineal gland
Glandula pinealis
Cerebral cortex
Cortex cerebri

CROSS-SECTIONAL VIEW

The Brain

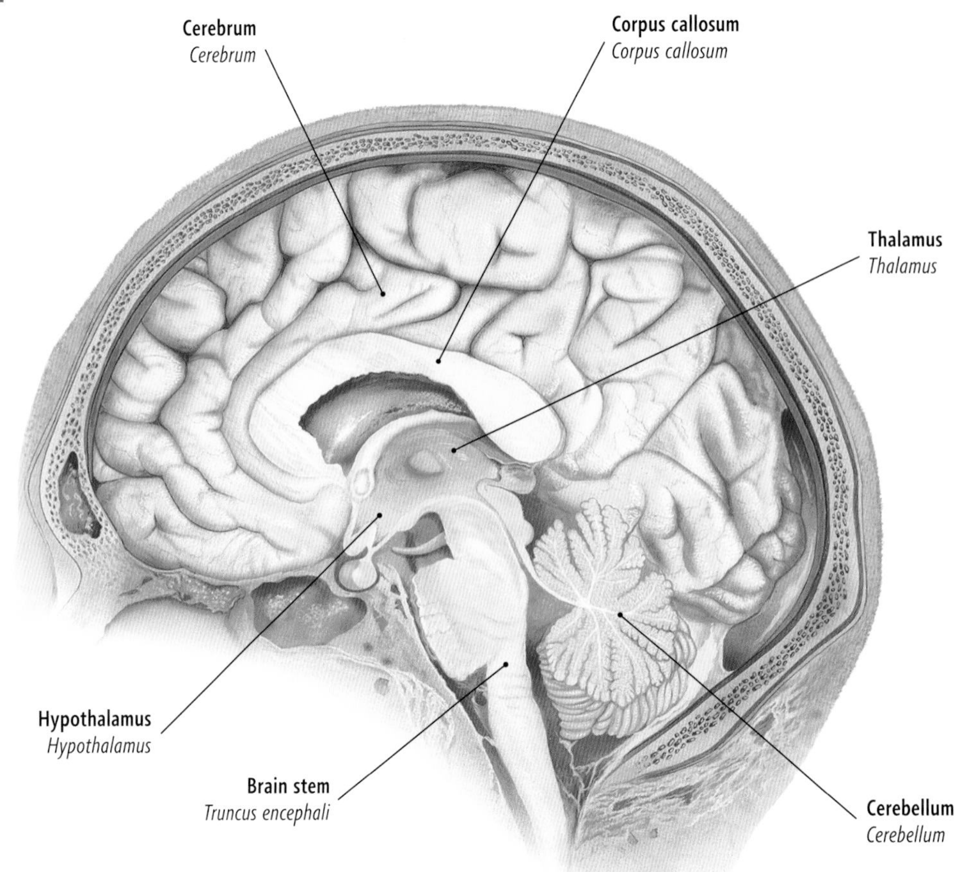

Cerebrum
Cerebrum
Corpus callosum
Corpus callosum
Thalamus
Thalamus
Hypothalamus
Hypothalamus
Brain stem
Truncus encephali
Cerebellum
Cerebellum

BRAIN—CROSS-SECTIONAL VIEW

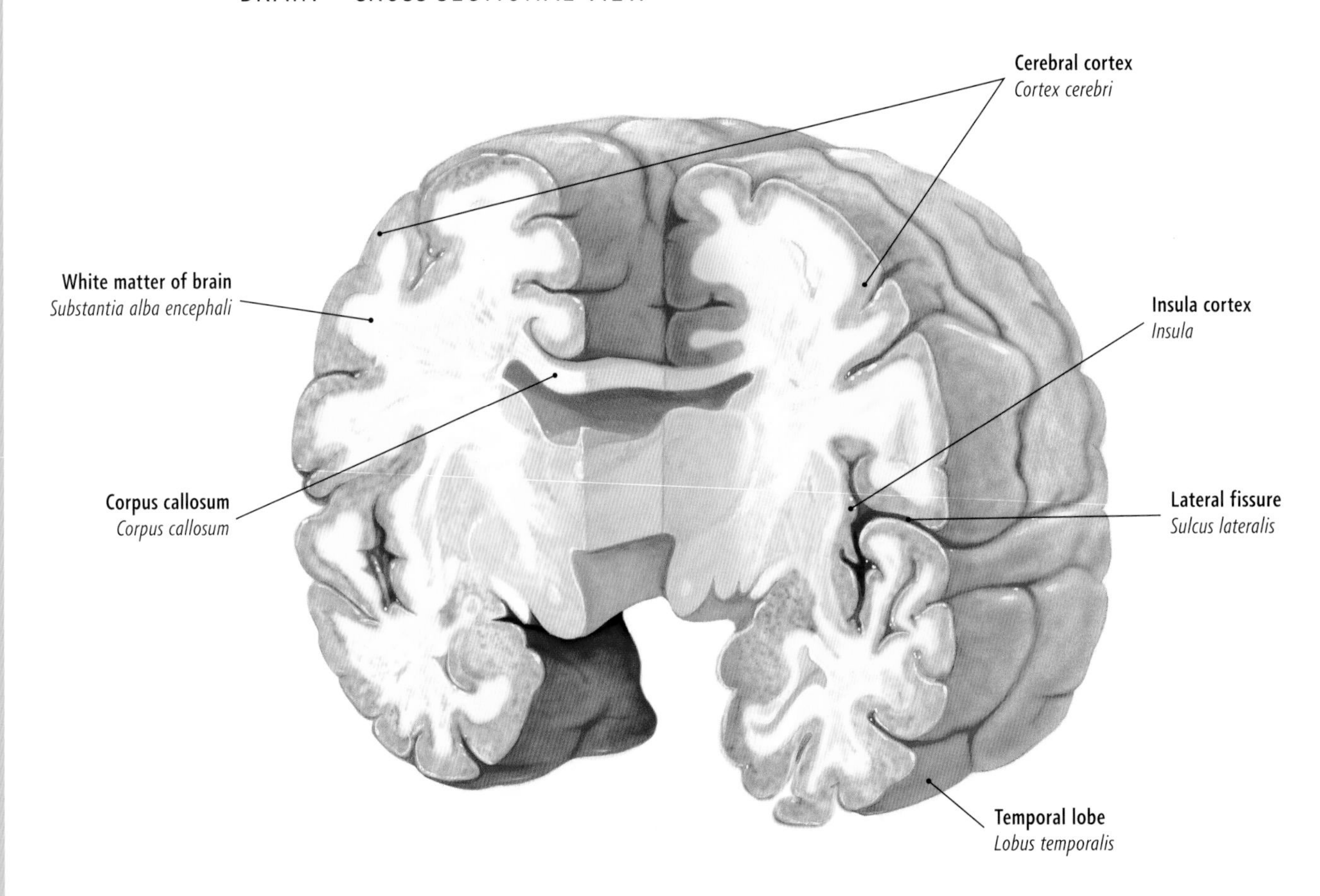

Cerebral cortex
Cortex cerebri
White matter of brain
Substantia alba encephali
Insula cortex
Insula
Corpus callosum
Corpus callosum
Lateral fissure
Sulcus lateralis
Temporal lobe
Lobus temporalis

CEREBRAL CORTEX—CROSS-SECTIONAL VIEW

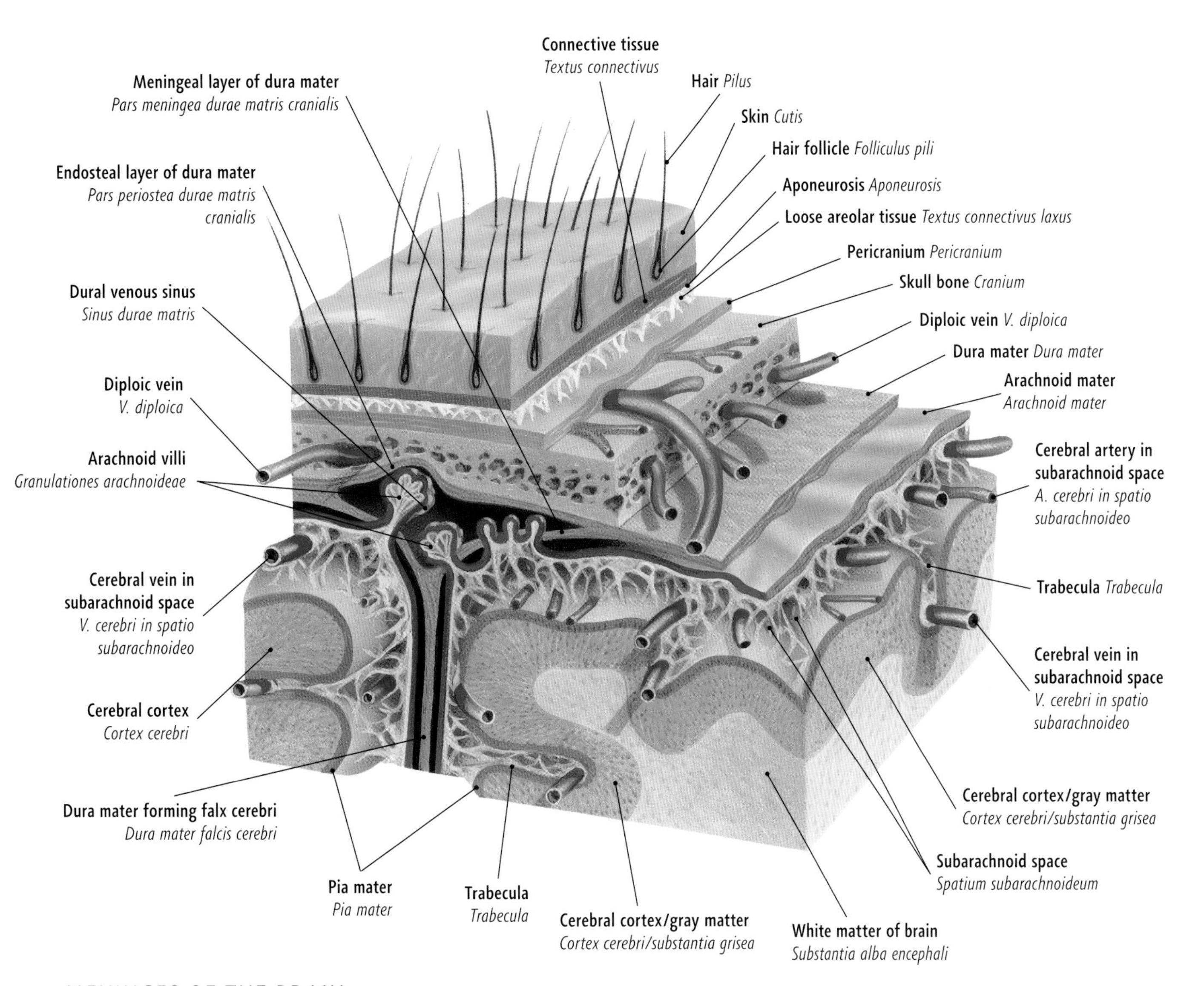
Connective tissue
Textus connectivus
Meningeal layer of dura mater
Pars meningea durae matris cranialis
Hair Pilus
Skin Cutis
Hair follicle Folliculus pili
Endosteal layer of dura mater
Pars periostea durae matris cranialis
Aponeurosis Aponeurosis
Loose areolar tissue Textus connectivus laxus
Pericranium Pericranium
Dural venous sinus
Sinus durae matris
Skull bone Cranium
Diploic vein V. diploica
Dura mater Dura mater
Diploic vein
V. diploica
Arachnoid mater
Arachnoid mater
Arachnoid villi
Granulationes arachnoideae
Cerebral artery in subarachnoid space
A. cerebri in spatio subarachnoideo
Trabecula Trabecula
Cerebral vein in subarachnoid space
V. cerebri in spatio subarachnoideo
Cerebral vein in subarachnoid space
V. cerebri in spatio subarachnoideo
Cerebral cortex
Cortex cerebri
Cerebral cortex/gray matter
Cortex cerebri/substantia grisea
Dura mater forming falx cerebri
Dura mater falcis cerebri
Subarachnoid space
Spatium subarachnoideum
Pia mater
Pia mater
Trabecula
Trabecula
Cerebral cortex/gray matter
Cortex cerebri/substantia grisea
White matter of brain
Substantia alba encephali

MENINGES OF THE BRAIN

There are three layers of meninges that cover and protect the brain and spinal cord: the fibrous outside layer (dura mater), the middle layer of collagen and elastin (arachnoid mater), and the inner pia mater. The subarachnoid space contains many blood vessels.

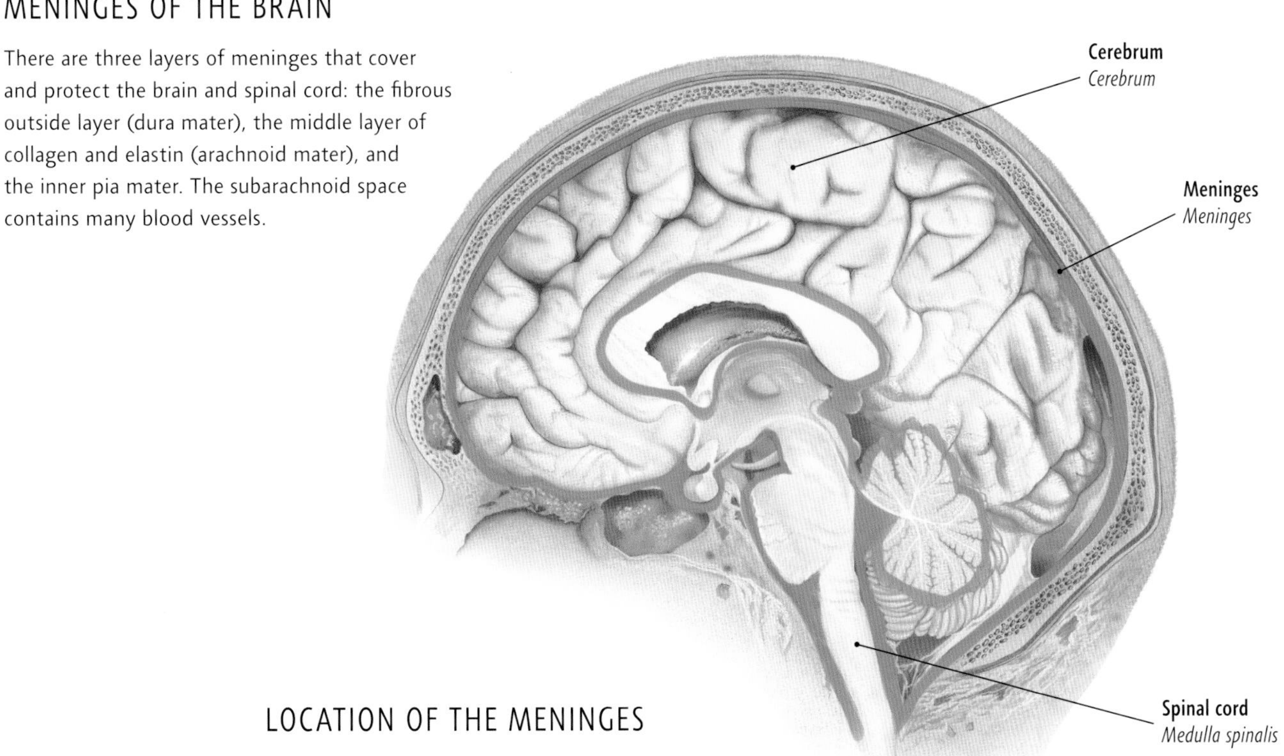
Cerebrum
Cerebrum
Meninges
Meninges
Spinal cord
Medulla spinalis

LOCATION OF THE MENINGES

[~] = no direct Latin equivalent

Brain Functions

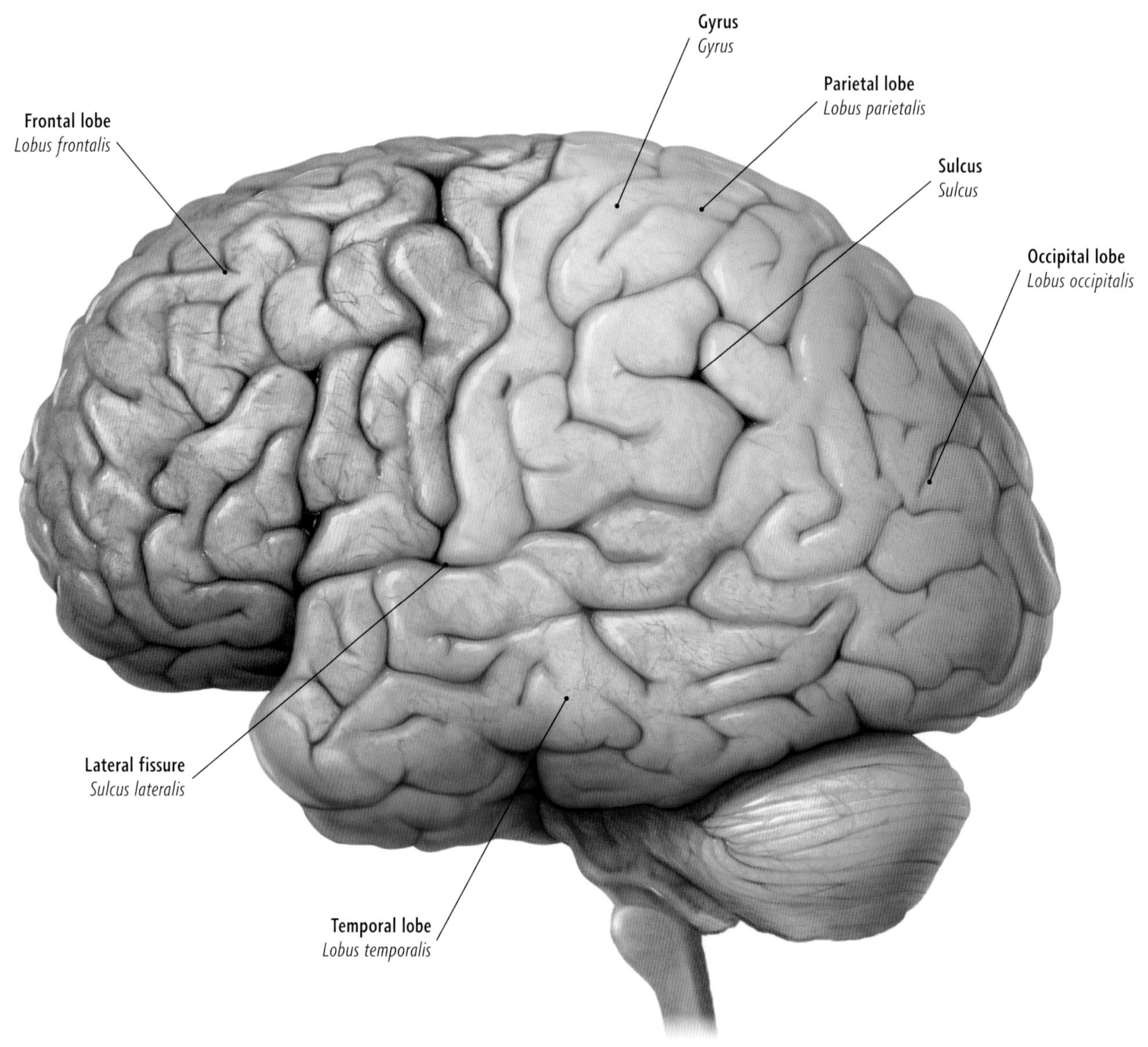
Gyrus
Gyrus
Parietal lobe
Lobus parietalis
Frontal lobe
Lobus frontalis
Sulcus
Sulcus
Occipital lobe
Lobus occipitalis
Lateral fissure
Sulcus lateralis
Temporal lobe
Lobus temporalis

LOBES OF THE BRAIN

A ridge on the surface of the cerebral cortex is called a gyrus. A groove is called a sulcus if shallow, or a fissure if deep. Fissures and sulci divide the cortex into separate areas called lobes.

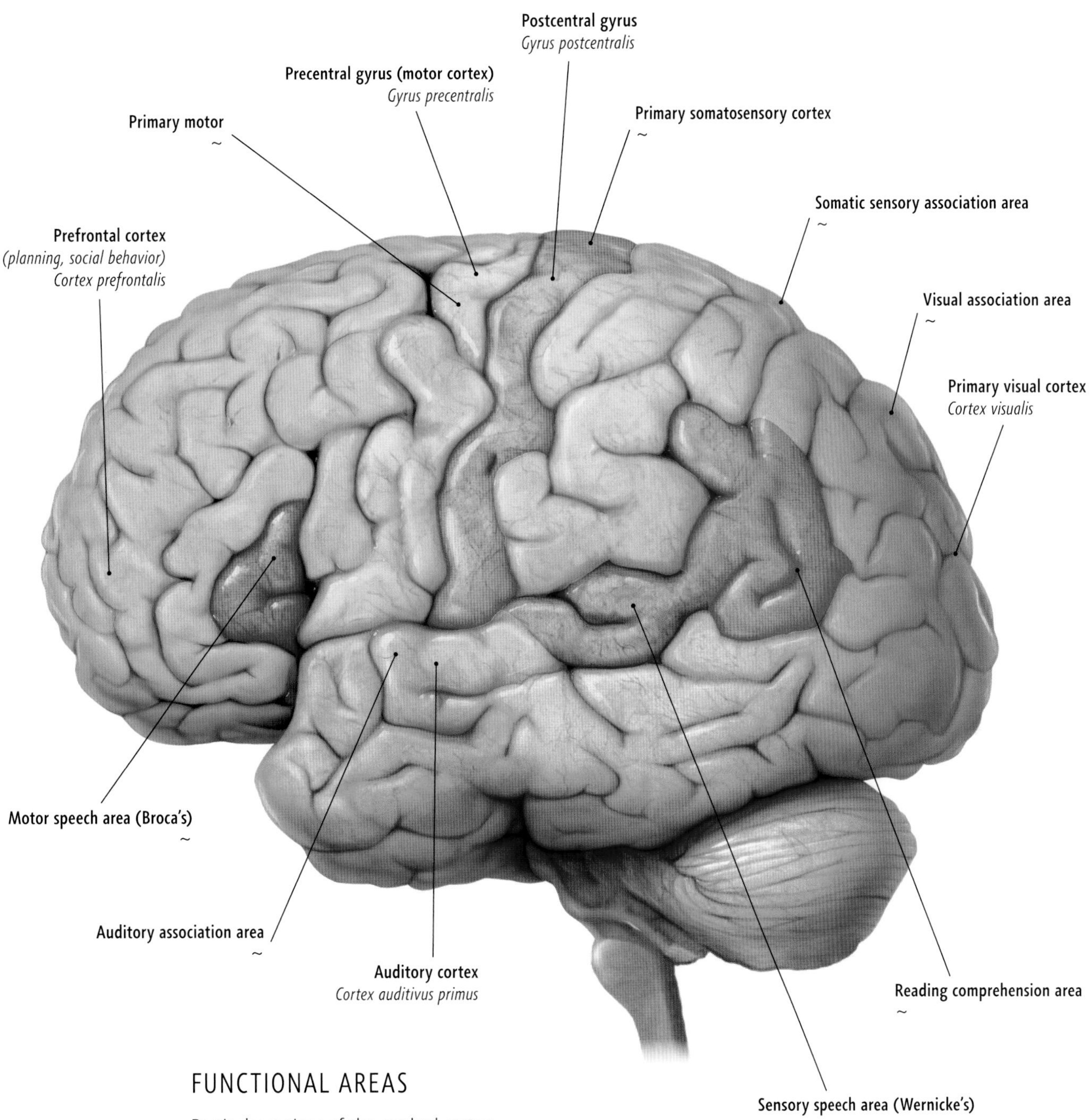
Postcentral gyrus
Gyrus postcentralis
Precentral gyrus (motor cortex)
Gyrus precentralis
Primary motor
~
Primary somatosensory cortex
~
Somatic sensory association area
~
Prefrontal cortex
(planning, social behavior)
Cortex prefrontalis
Visual association area
~
Primary visual cortex
Cortex visualis
Motor speech area (Broca's)
~
Auditory association area
~
Auditory cortex
Cortex auditivus primus
Reading comprehension area
~
Sensory speech area (Wernicke's)
~

FUNCTIONAL AREAS

Particular regions of the cerebral cortex are associated with certain functions. For example, the precentral gyrus is associated with the voluntary control of skeletal muscles, while the postcentral gyrus is associated with sensations from the skin, muscles, and joints (somatosensory function).

[~] = no direct Latin equivalent

Brain Functions

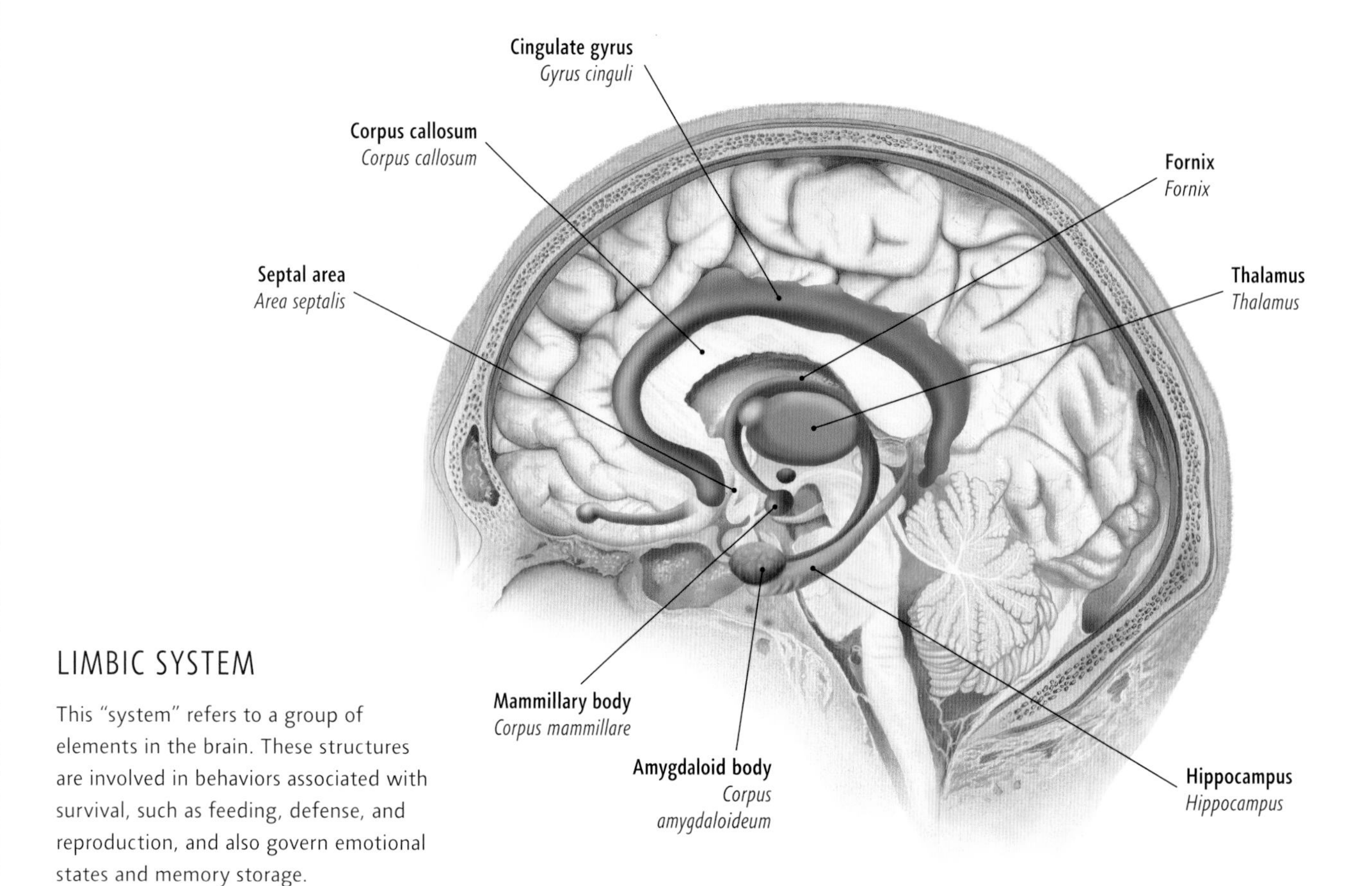
Cingulate gyrus
Gyrus cinguli
Corpus callosum
Corpus callosum
Fornix
Fornix
Septal area
Area septalis
Thalamus
Thalamus
Mammillary body
Corpus mammillare
Amygdaloid body
Corpus amygdaloideum
Hippocampus
Hippocampus

LIMBIC SYSTEM

This "system" refers to a group of elements in the brain. These structures are involved in behaviors associated with survival, such as feeding, defense, and reproduction, and also govern emotional states and memory storage.

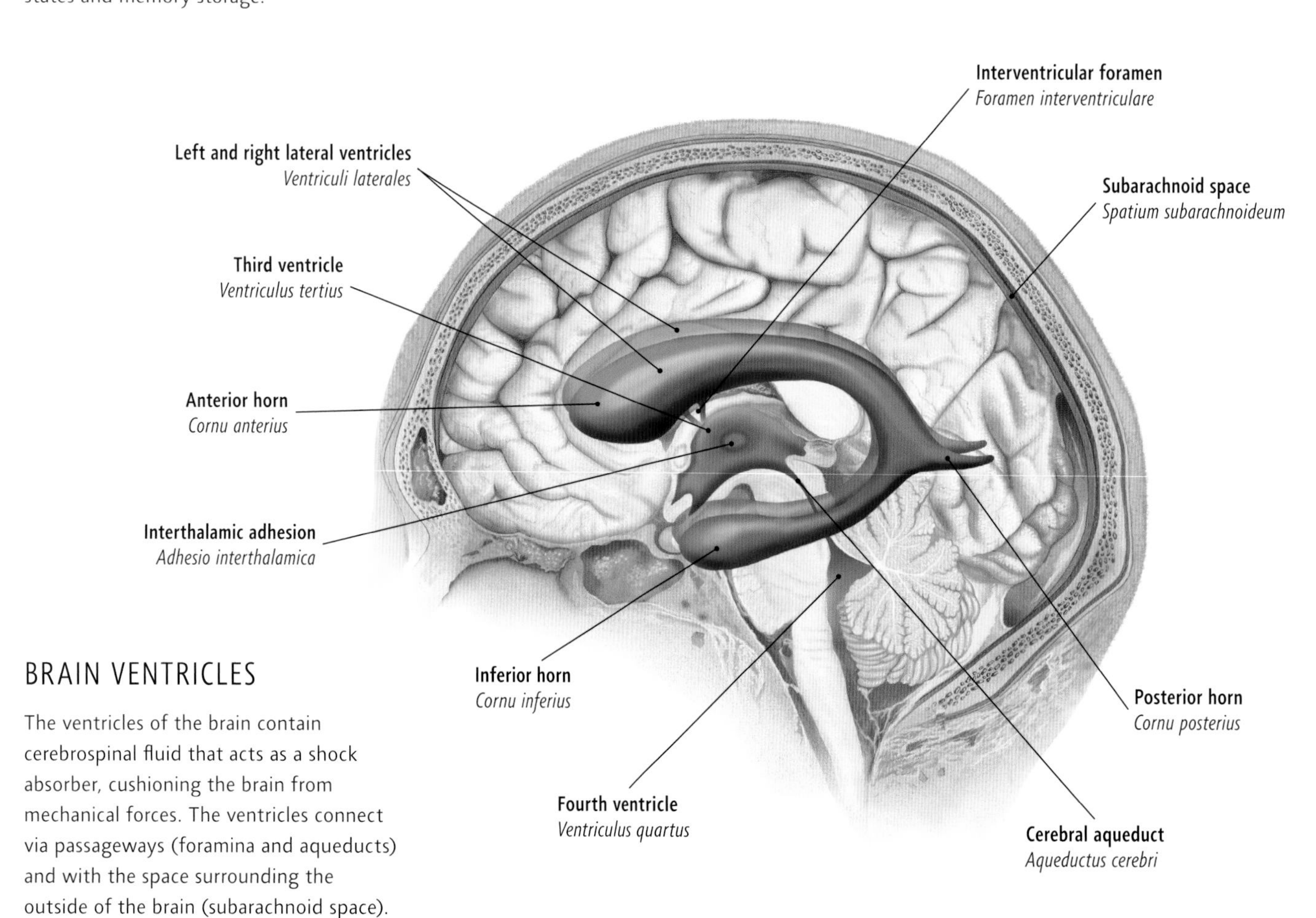
Interventricular foramen
Foramen interventriculare
Left and right lateral ventricles
Ventriculi laterales
Subarachnoid space
Spatium subarachnoideum
Third ventricle
Ventriculus tertius
Anterior horn
Cornu anterius
Interthalamic adhesion
Adhesio interthalamica
Inferior horn
Cornu inferius
Posterior horn
Cornu posterius
Fourth ventricle
Ventriculus quartus
Cerebral aqueduct
Aqueductus cerebri

BRAIN VENTRICLES

The ventricles of the brain contain cerebrospinal fluid that acts as a shock absorber, cushioning the brain from mechanical forces. The ventricles connect via passageways (foramina and aqueducts) and with the space surrounding the outside of the brain (subarachnoid space).

ORGANIZATION OF THE MOTOR AND SENSORY AREAS OF THE CEREBRAL CORTEX

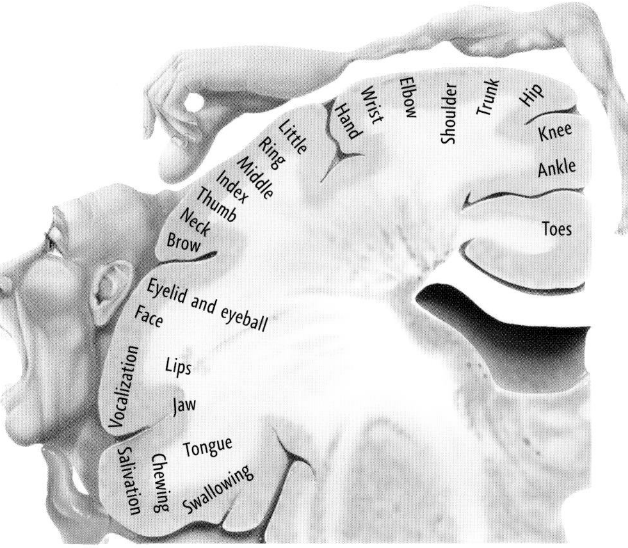
Motor activity
Hip
Trunk
Shoulder
Elbow
Wrist
Hand
Little
Ring
Middle
Index
Thumb
Neck
Brow
Knee
Ankle
Toes
Eyelid and eyeball
Face
Lips
Vocalization
Jaw
Tongue
Salivation
Chewing
Swallowing

Sensory activity
Leg
Hip
Trunk
Neck
Head
Shoulder
Arm
Elbow
Forearm
Wrist
Hand
Little
Ring
Middle
Index
Thumb
Eye
Nose
Face
Upper lip
Lips
Lower lip
Teeth, gums, jaw
Tongue
Pharynx
Intra-abdominal
Foot
Toes
Genitalia

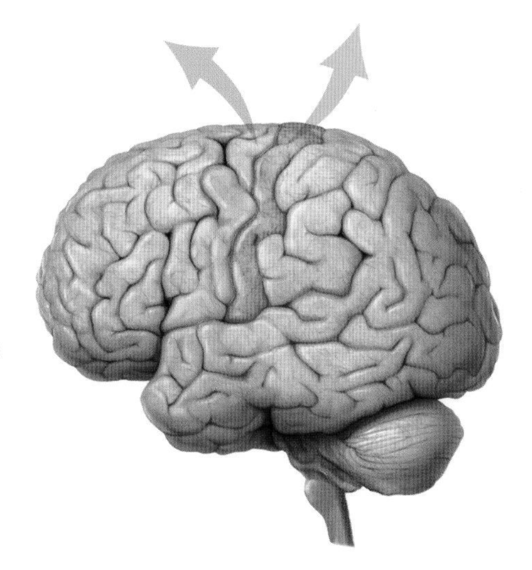

The size of the body part reflects the proportion of the precentral gyrus involved in motor activity in that specific area of the body (top left), and the proportion of the postcentral gyrus involved in sensory activity in that specific area of the body (top right).

Brain Functions

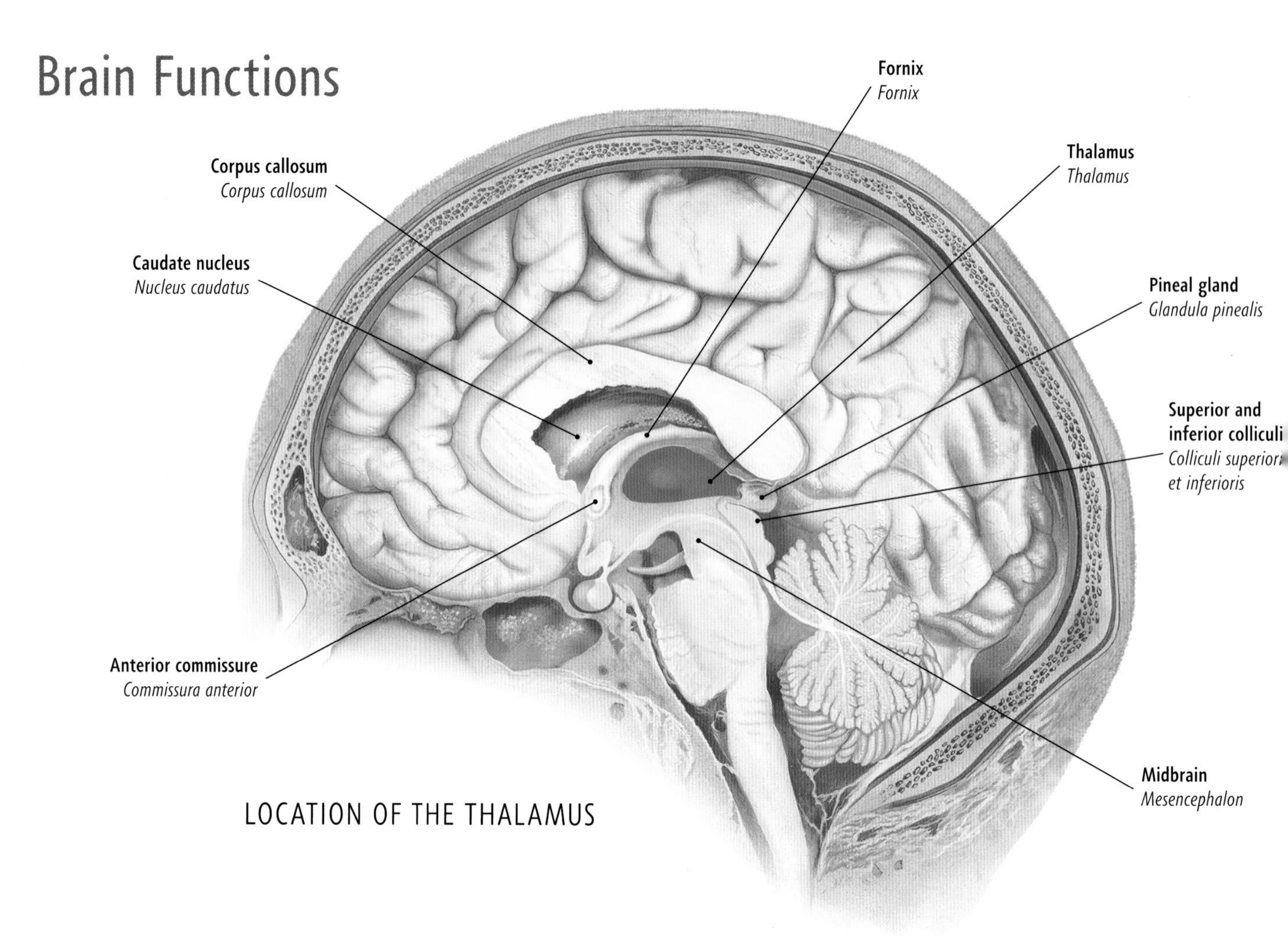
Fornix
Fornix
Corpus callosum
Corpus callosum
Thalamus
Thalamus
Caudate nucleus
Nucleus caudatus
Pineal gland
Glandula pinealis
Superior and inferior colliculi
Colliculi superior et inferioris
Anterior commissure
Commissura anterior
Midbrain
Mesencephalon

LOCATION OF THE THALAMUS

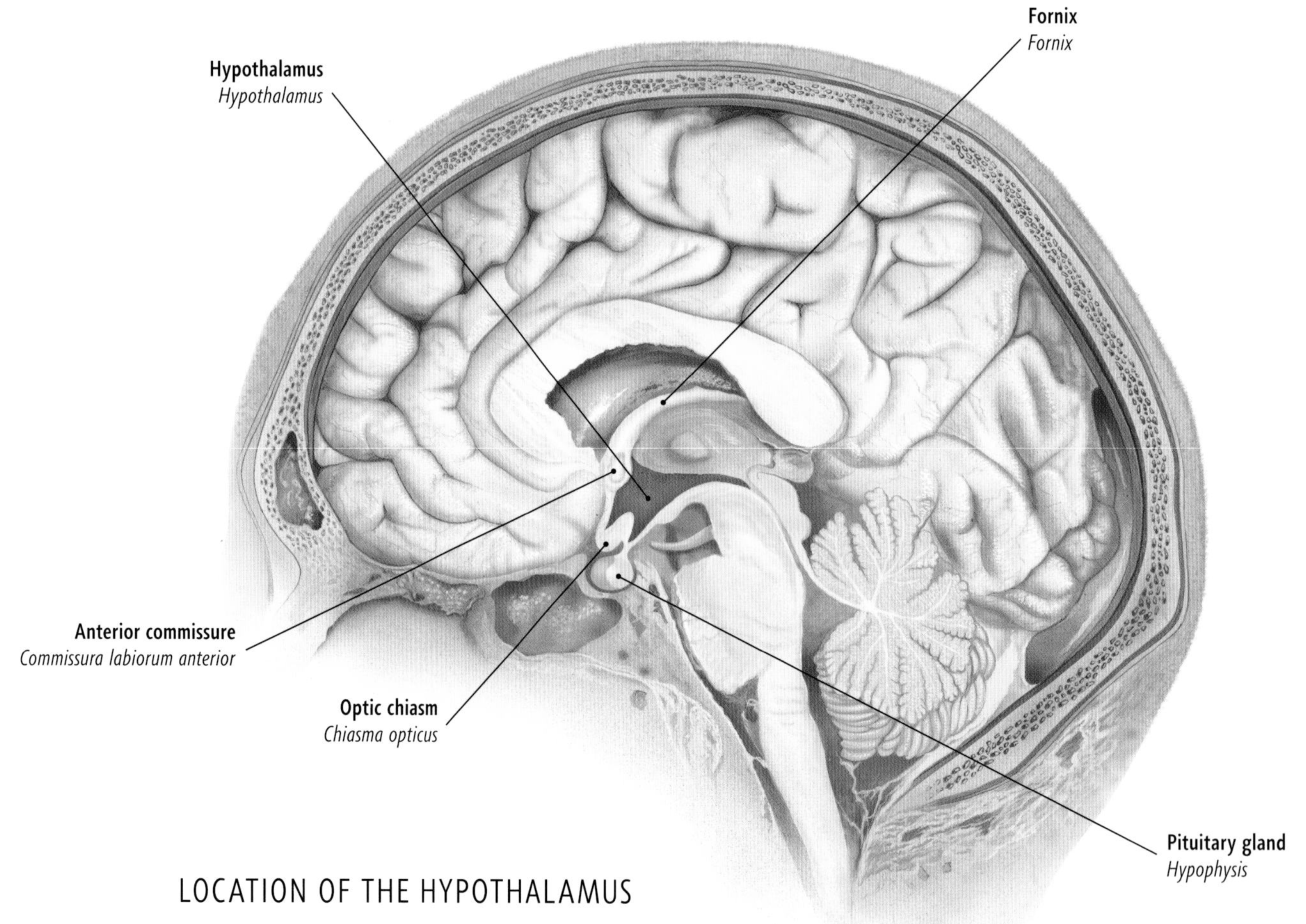
Fornix
Fornix
Hypothalamus
Hypothalamus
Anterior commissure
Commissura labiorum anterior
Optic chiasm
Chiasma opticus
Pituitary gland
Hypophysis

LOCATION OF THE HYPOTHALAMUS

Cranial Nerves

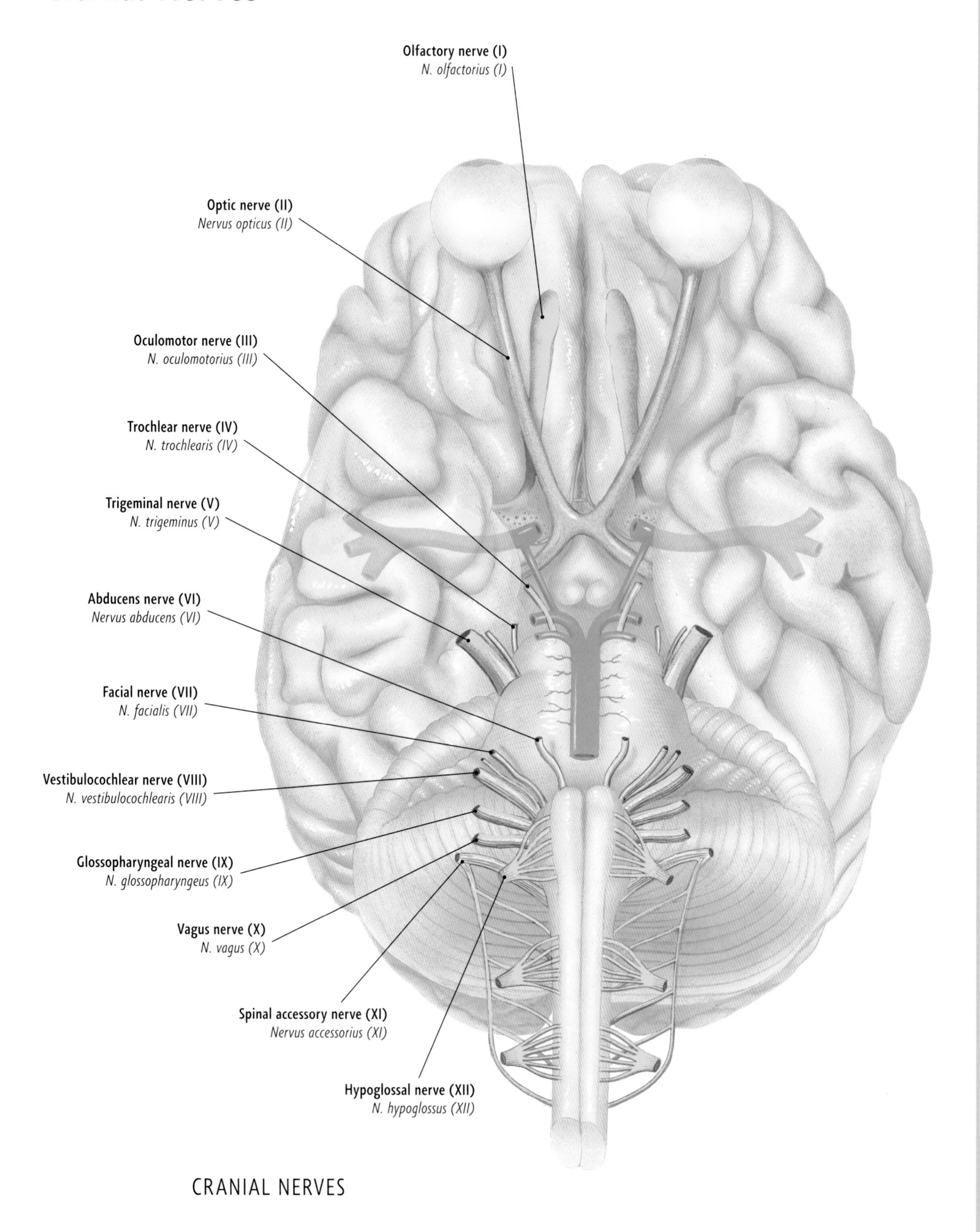

Olfactory nerve (I)
N. olfactorius (I)
Optic nerve (II)
Nervus opticus (II)
Oculomotor nerve (III)
N. oculomotorius (III)
Trochlear nerve (IV)
N. trochlearis (IV)
Trigeminal nerve (V)
N. trigeminus (V)
Abducens nerve (VI)
Nervus abducens (VI)
Facial nerve (VII)
N. facialis (VII)
Vestibulocochlear nerve (VIII)
N. vestibulocochlearis (VIII)
Glossopharyngeal nerve (IX)
N. glossopharyngeus (IX)
Vagus nerve (X)
N. vagus (X)
Spinal accessory nerve (XI)
Nervus accessorius (XI)
Hypoglossal nerve (XII)
N. hypoglossus (XII)

CRANIAL NERVES

Brain Stem

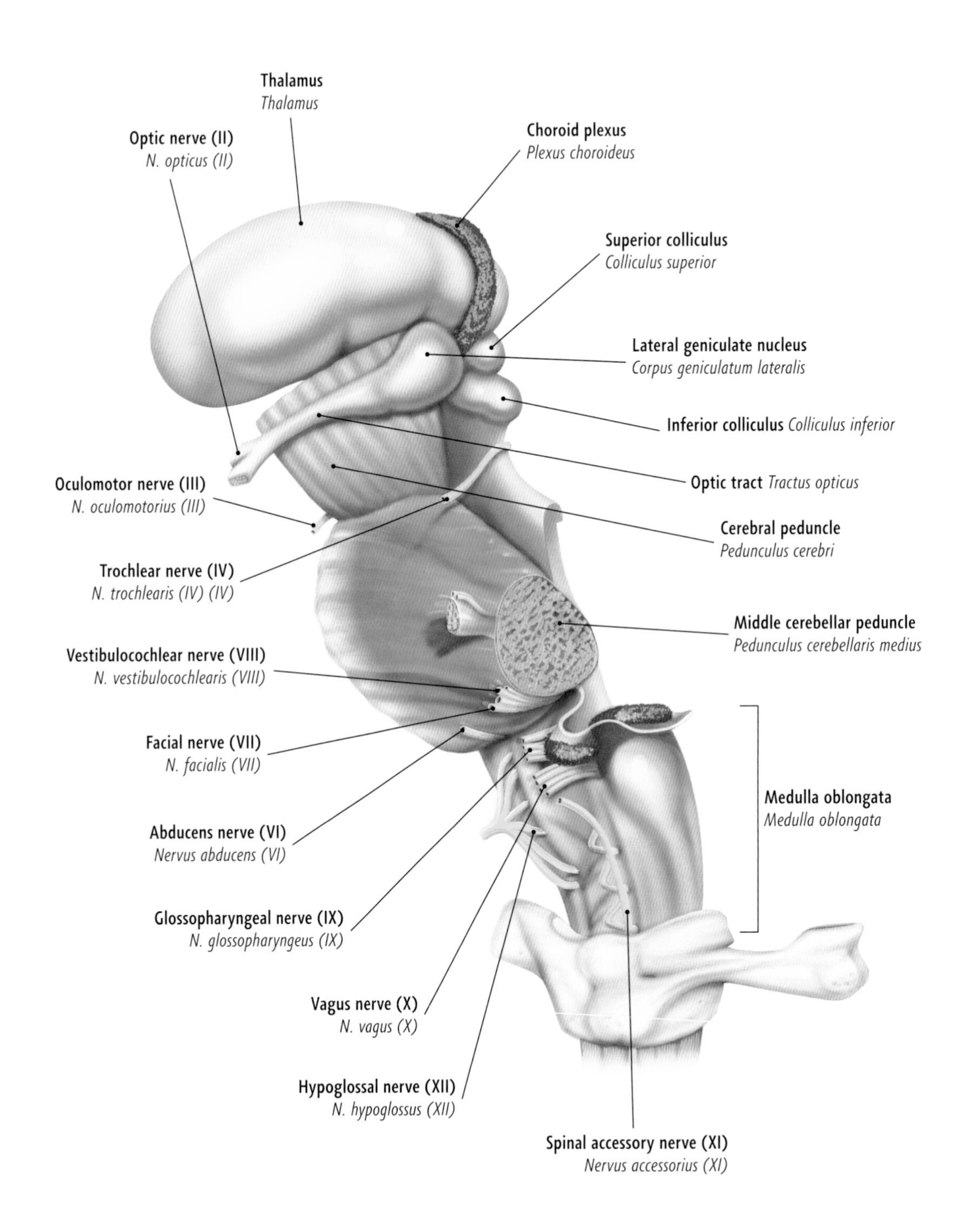
Thalamus
Thalamus
Optic nerve (II)
N. opticus (II)
Choroid plexus
Plexus choroideus
Superior colliculus
Colliculus superior
Lateral geniculate nucleus
Corpus geniculatum lateralis
Inferior colliculus Colliculus inferior
Optic tract Tractus opticus
Oculomotor nerve (III)
N. oculomotorius (III)
Cerebral peduncle
Pedunculus cerebri
Trochlear nerve (IV)
N. trochlearis (IV) (IV)
Middle cerebellar peduncle
Pedunculus cerebellaris medius
Vestibulocochlear nerve (VIII)
N. vestibulocochlearis (VIII)
Facial nerve (VII)
N. facialis (VII)
Medulla oblongata
Medulla oblongata
Abducens nerve (VI)
Nervus abducens (VI)
Glossopharyngeal nerve (IX)
N. glossopharyngeus (IX)
Vagus nerve (X)
N. vagus (X)
Hypoglossal nerve (XII)
N. hypoglossus (XII)
Spinal accessory nerve (XI)
Nervus accessorius (XI)

BRAIN STEM—SIDE VIEW

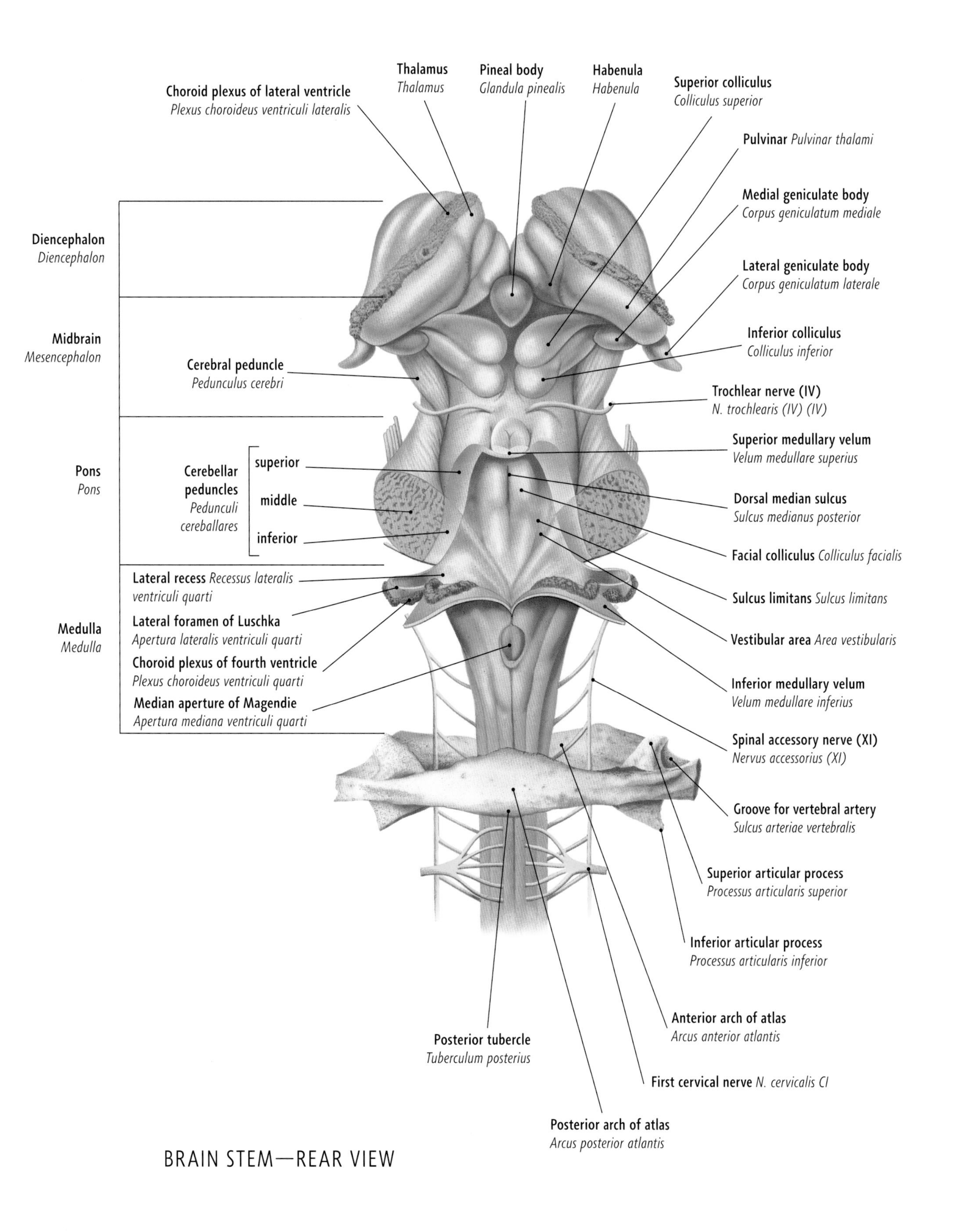
Thalamus
Thalamus
Pineal body
Glandula pinealis
Habenula
Habenula
Choroid plexus of lateral ventricle
Plexus choroideus ventriculi lateralis
Superior colliculus
Colliculus superior
Pulvinar Pulvinar thalami
Medial geniculate body
Corpus geniculatum mediale
Diencephalon
Diencephalon
Lateral geniculate body
Corpus geniculatum laterale
Midbrain
Mesencephalon
Inferior colliculus
Colliculus inferior
Cerebral peduncle
Pedunculus cerebri
Trochlear nerve (IV)
N. trochlearis (IV) (IV)
Superior medullary velum
Velum medullare superius
Pons
Pons
Cerebellar peduncles
Pedunculi cereballares
superior
middle
inferior
Dorsal median sulcus
Sulcus medianus posterior
Facial colliculus Colliculus facialis
Lateral recess Recessus lateralis ventriculi quarti
Sulcus limitans Sulcus limitans
Lateral foramen of Luschka
Apertura lateralis ventriculi quarti
Medulla
Medulla
Vestibular area Area vestibularis
Choroid plexus of fourth ventricle
Plexus choroideus ventriculi quarti
Inferior medullary velum
Velum medullare inferius
Median aperture of Magendie
Apertura mediana ventriculi quarti
Spinal accessory nerve (XI)
Nervus accessorius (XI)
Groove for vertebral artery
Sulcus arteriae vertebralis
Superior articular process
Processus articularis superior
Inferior articular process
Processus articularis inferior
Anterior arch of atlas
Arcus anterior atlantis
Posterior tubercle
Tuberculum posterius
First cervical nerve N. cervicalis CI
Posterior arch of atlas
Arcus posterior atlantis

BRAIN STEM—REAR VIEW

Spinal Cord

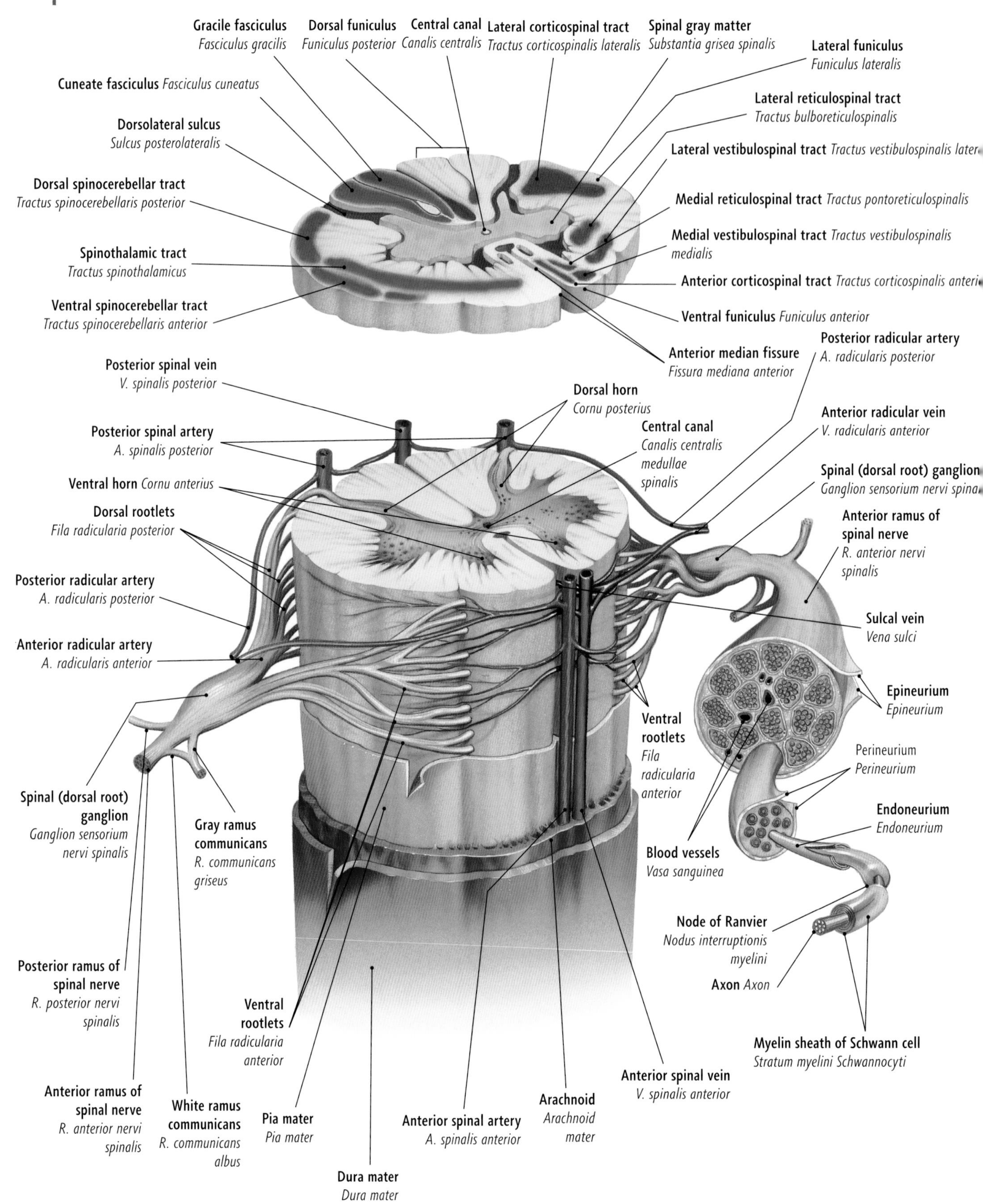

Gracile fasciculus
Fasciculus gracilis
Dorsal funiculus
Funiculus posterior
Central canal
Canalis centralis
Lateral corticospinal tract
Tractus corticospinalis lateralis
Spinal gray matter
Substantia grisea spinalis
Lateral funiculus
Funiculus lateralis
Cuneate fasciculus Fasciculus cuneatus
Lateral reticulospinal tract
Tractus bulboreticulospinalis
Dorsolateral sulcus
Sulcus posterolateralis
Lateral vestibulospinal tract Tractus vestibulospinalis later
Dorsal spinocerebellar tract
Tractus spinocerebellaris posterior
Medial reticulospinal tract Tractus pontoreticulospinalis
Medial vestibulospinal tract Tractus vestibulospinalis medialis
Spinothalamic tract
Tractus spinothalamicus
Anterior corticospinal tract Tractus corticospinalis anteri
Ventral spinocerebellar tract
Tractus spinocerebellaris anterior
Ventral funiculus Funiculus anterior
Anterior median fissure
Fissura mediana anterior
Posterior radicular artery
A. radicularis posterior
Posterior spinal vein
V. spinalis posterior
Dorsal horn
Cornu posterius
Central canal
Canalis centralis medullae spinalis
Anterior radicular vein
V. radicularis anterior
Posterior spinal artery
A. spinalis posterior
Spinal (dorsal root) ganglion
Ganglion sensorium nervi spina
Ventral horn Cornu anterius
Dorsal rootlets
Fila radicularia posterior
Anterior ramus of spinal nerve
R. anterior nervi spinalis
Posterior radicular artery
A. radicularis posterior
Sulcal vein
Vena sulci
Anterior radicular artery
A. radicularis anterior
Epineurium
Epineurium
Ventral rootlets
Fila radicularia anterior
Perineurium
Perineurium
Spinal (dorsal root) ganglion
Ganglion sensorium nervi spinalis
Gray ramus communicans
R. communicans griseus
Endoneurium
Endoneurium
Blood vessels
Vasa sanguinea
Node of Ranvier
Nodus interruptionis myelini
Axon Axon
Posterior ramus of spinal nerve
R. posterior nervi spinalis
Ventral rootlets
Fila radicularia anterior
Myelin sheath of Schwann cell
Stratum myelini Schwannocyti
Anterior spinal vein
V. spinalis anterior
Anterior ramus of spinal nerve
R. anterior nervi spinalis
White ramus communicans
R. communicans albus
Pia mater
Pia mater
Anterior spinal artery
A. spinalis anterior
Arachnoid
Arachnoid mater
Dura mater
Dura mater

[~] = no direct Latin equivalent

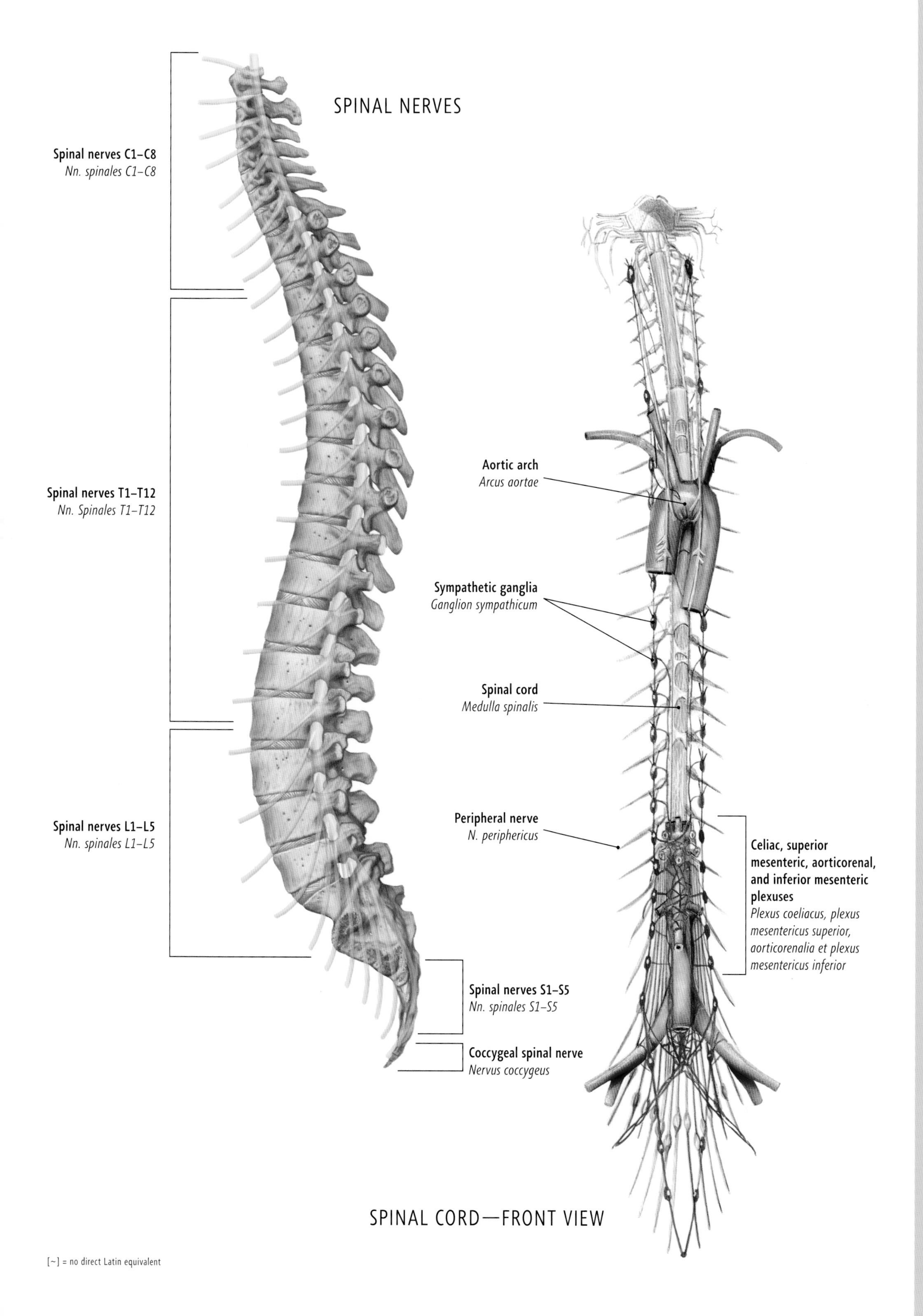
SPINAL NERVES
Spinal nerves C1–C8
Nn. spinales C1–C8
Spinal nerves T1–T12
Nn. Spinales T1–T12
Spinal nerves L1–L5
Nn. spinales L1–L5
Spinal nerves S1–S5
Nn. spinales S1–S5
Coccygeal spinal nerve
Nervus coccygeus
Aortic arch
Arcus aortae
Sympathetic ganglia
Ganglion sympathicum
Spinal cord
Medulla spinalis
Peripheral nerve
N. periphericus
Celiac, superior mesenteric, aorticorenal, and inferior mesenteric plexuses
Plexus coeliacus, plexus mesentericus superior, aorticorenalia et plexus mesentericus inferior
SPINAL CORD—FRONT VIEW

[~] = no direct Latin equivalent

Dermatomes

Spinal nerves are numbered and correspond closely to the spinal vertebrae. Each pair of nerves supplies a specific dermatome (skin area) of the body. However, there is usually some overlap between adjacent dermatomes. The skin of the face is supplied by branches of the trigeminal nerve (cranial nerves V1, V2, and V3).

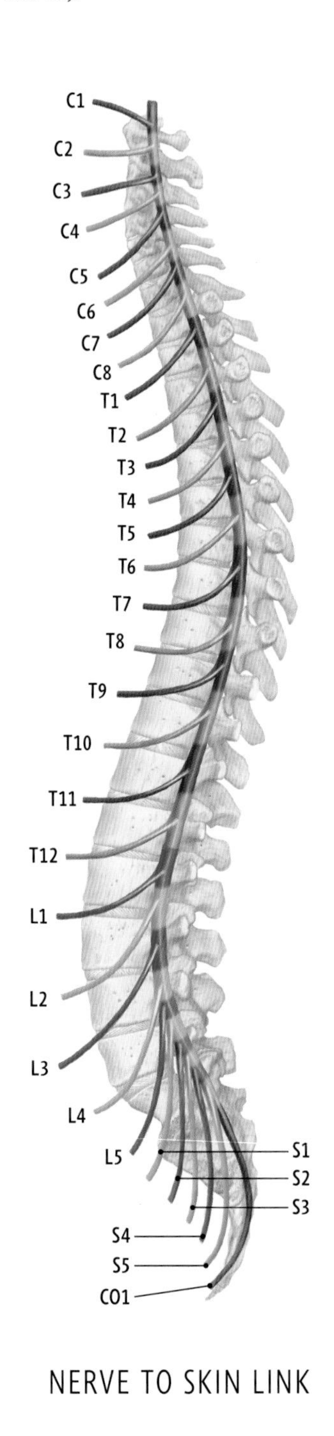
C1
C2
C3
C4
C5
C6
C7
C8
T1
T2
T3
T4
T5
T6
T7
T8
T9
T10
T11
T12
L1
L2
L3
L4
L5
S1
S2
S3
S4
S5
CO1

NERVE TO SKIN LINK

DERMATOMES—SIDE VIEW

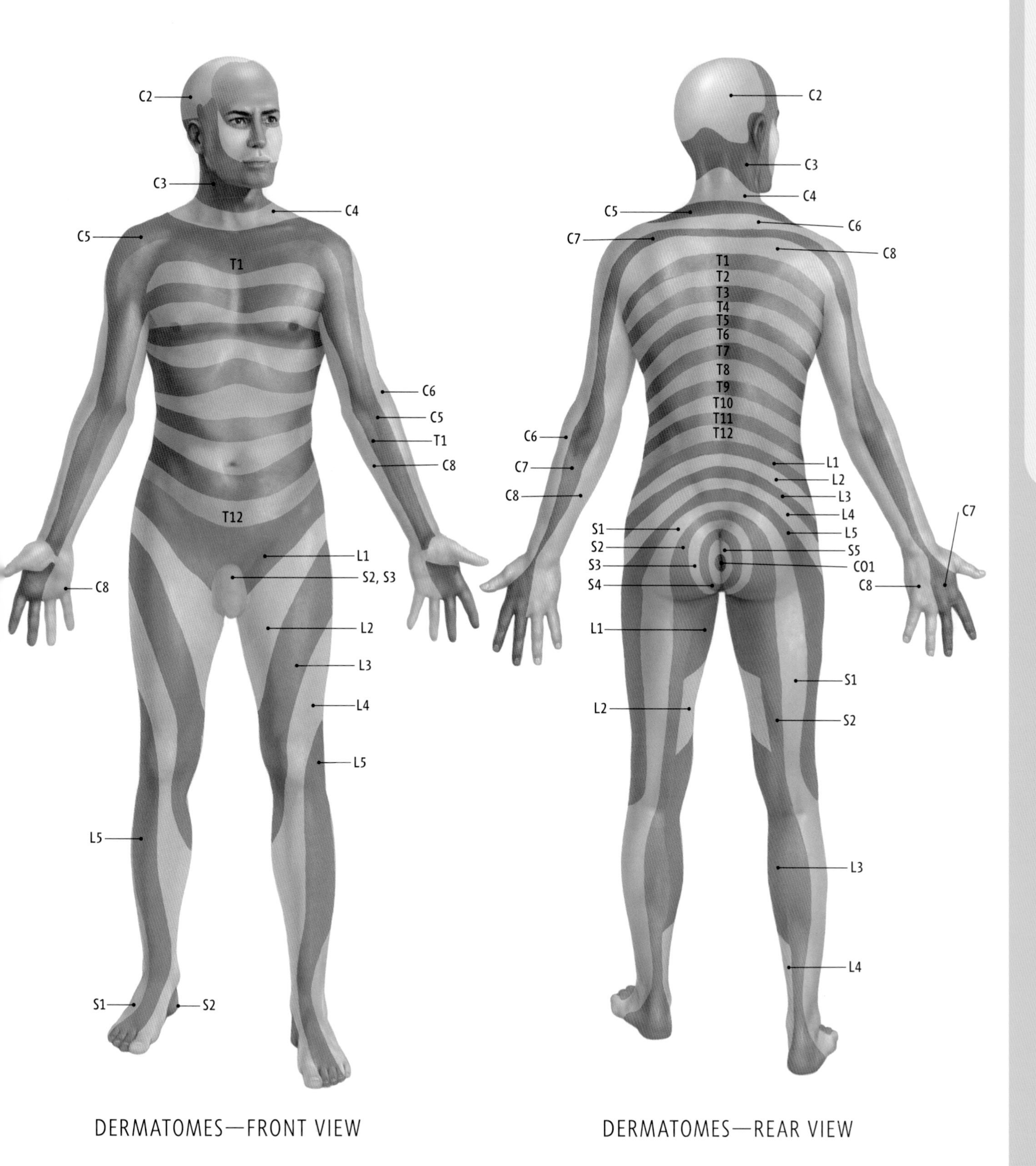
C2
C3
C4
C5
T1
C6
C5
T1
C8
T12
C8
L1
S2, S3
L2
L3
L4
L5
L5
S1
S2
C2
C3
C4
C5
C6
C7
C8
T1
T2
T3
T4
T5
T6
T7
T8
T9
T10
T11
T12
C6
C7
C8
L1
L2
L3
L4
L5
S1
S2
S5
S3
CO1
S4
C7
C8
L1
S1
L2
S2
L3
L4

DERMATOMES—FRONT VIEW

DERMATOMES—REAR VIEW

Nerves of the Head, Arms, and Legs

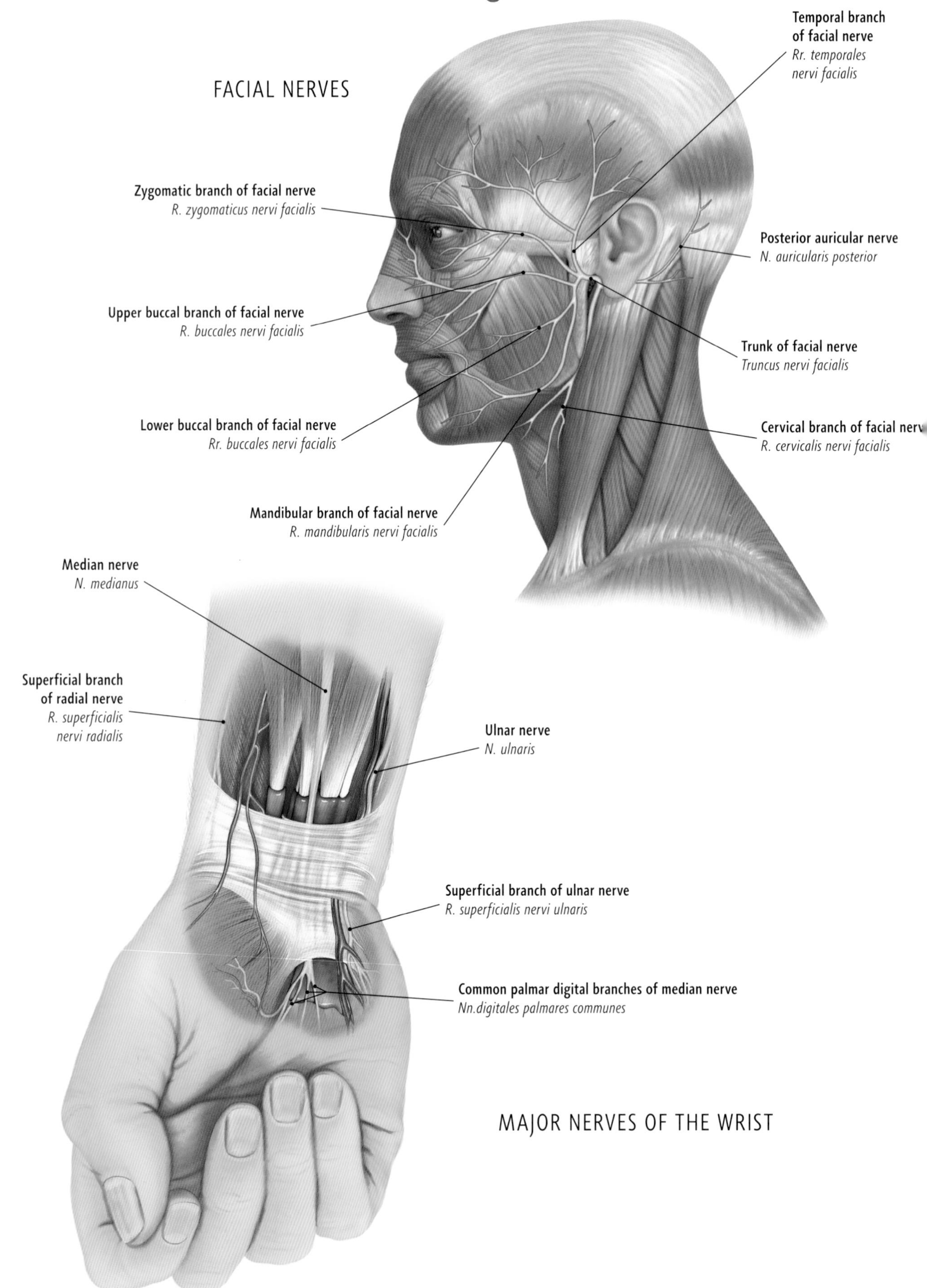
FACIAL NERVES
Temporal branch of facial nerve
Rr. temporales nervi facialis
Zygomatic branch of facial nerve
R. zygomaticus nervi facialis
Posterior auricular nerve
N. auricularis posterior
Upper buccal branch of facial nerve
R. buccales nervi facialis
Trunk of facial nerve
Truncus nervi facialis
Lower buccal branch of facial nerve
Rr. buccales nervi facialis
Cervical branch of facial nerv
R. cervicalis nervi facialis
Mandibular branch of facial nerve
R. mandibularis nervi facialis
Median nerve
N. medianus
Superficial branch of radial nerve
R. superficialis nervi radialis
Ulnar nerve
N. ulnaris
Superficial branch of ulnar nerve
R. superficialis nervi ulnaris
Common palmar digital branches of median nerve
Nn.digitales palmares communes
MAJOR NERVES OF THE WRIST

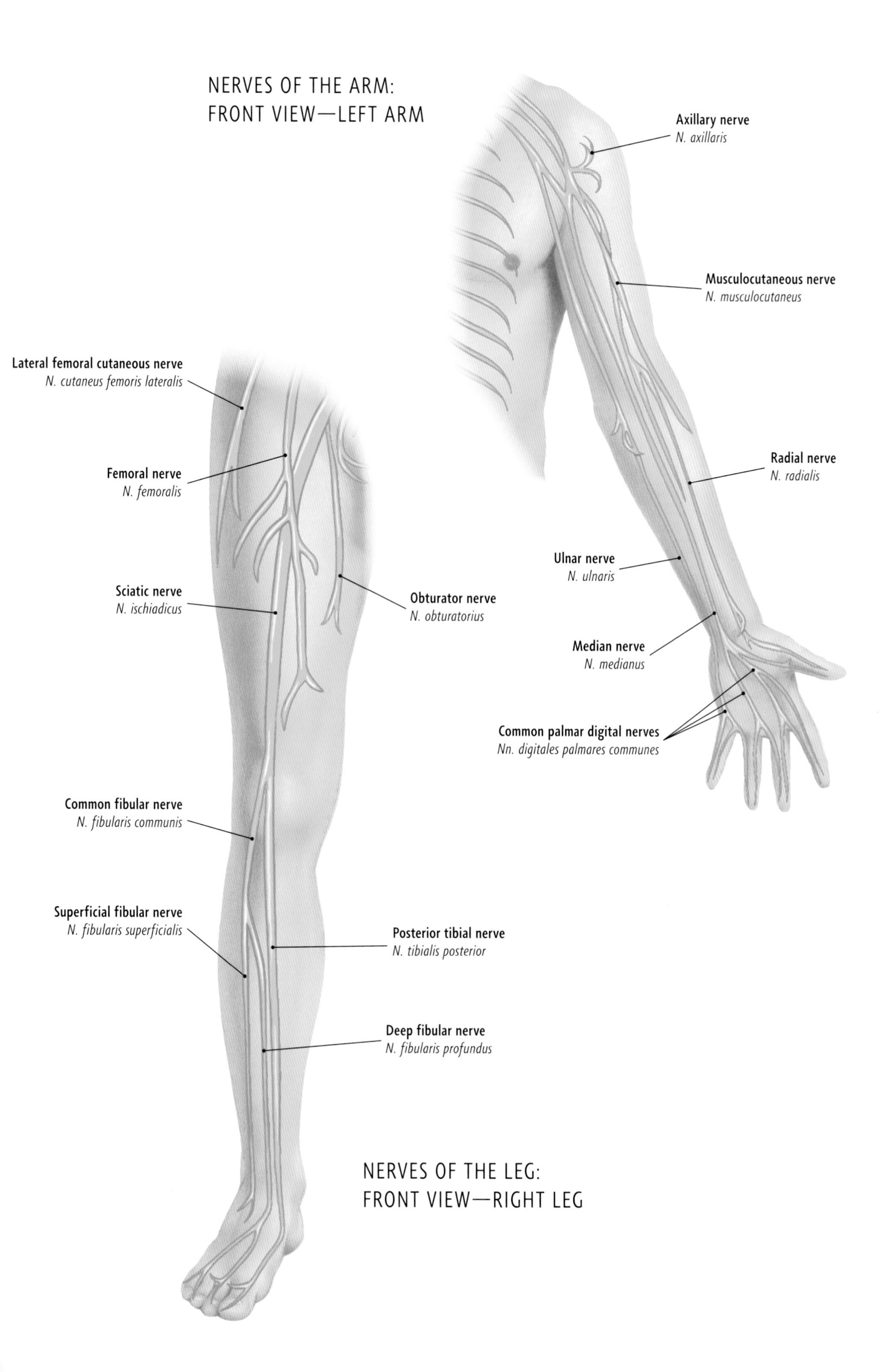
NERVES OF THE ARM:
FRONT VIEW—LEFT ARM
Axillary nerve
N. axillaris
Musculocutaneous nerve
N. musculocutaneus
Radial nerve
N. radialis
Ulnar nerve
N. ulnaris
Median nerve
N. medianus
Common palmar digital nerves
Nn. digitales palmares communes
Lateral femoral cutaneous nerve
N. cutaneus femoris lateralis
Femoral nerve
N. femoralis
Sciatic nerve
N. ischiadicus
Obturator nerve
N. obturatorius
Common fibular nerve
N. fibularis communis
Superficial fibular nerve
N. fibularis superficialis
Posterior tibial nerve
N. tibialis posterior
Deep fibular nerve
N. fibularis profundus
NERVES OF THE LEG:
FRONT VIEW—RIGHT LEG

Autonomic Nervous System

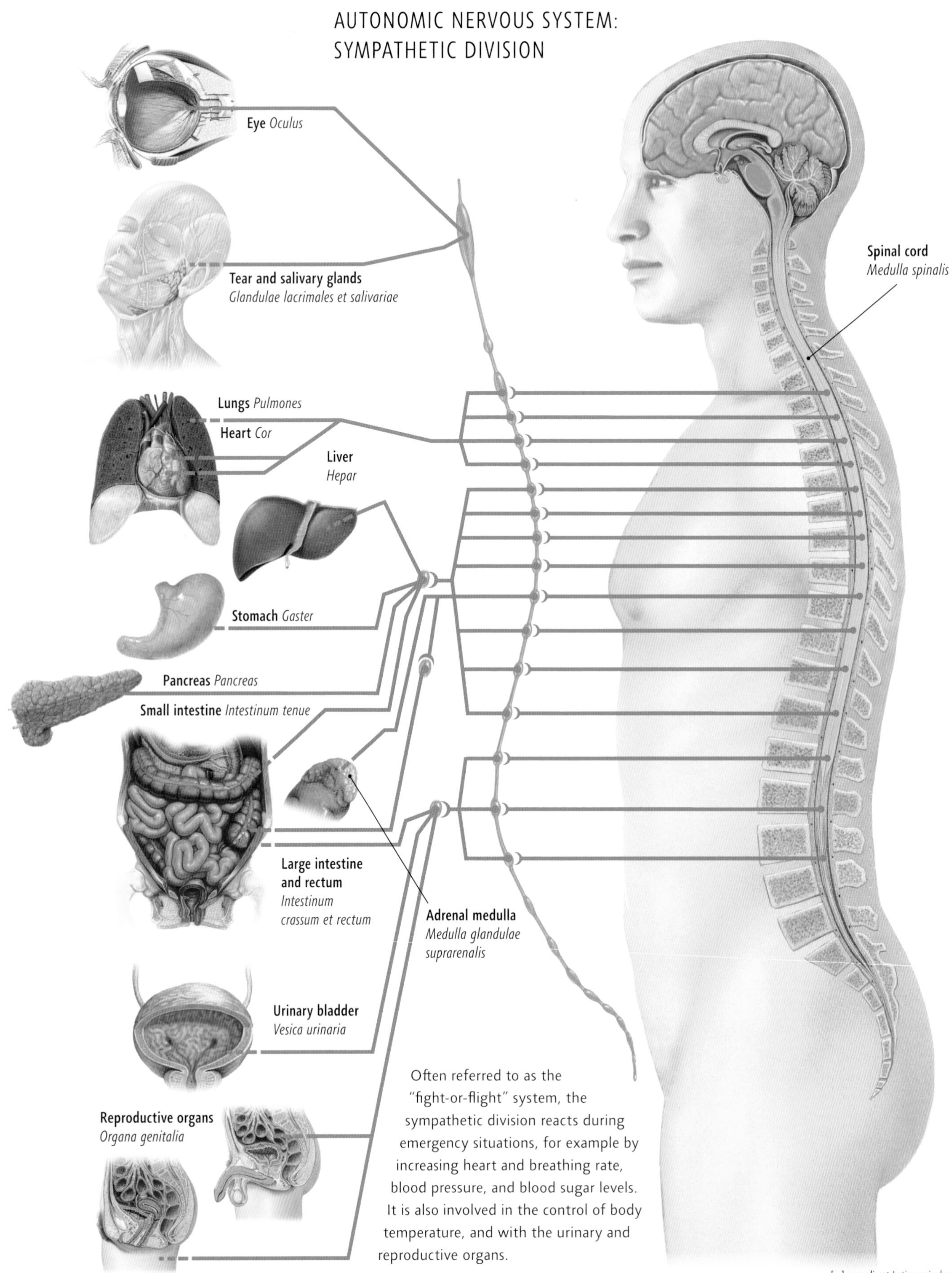
AUTONOMIC NERVOUS SYSTEM:
SYMPATHETIC DIVISION
Eye *Oculus*
Tear and salivary glands
Glandulae lacrimales et salivariae
Lungs *Pulmones*
Heart *Cor*
Liver
Hepar
Stomach *Gaster*
Pancreas *Pancreas*
Small intestine *Intestinum tenue*
Large intestine
and rectum
Intestinum
crassum et rectum
Adrenal medulla
Medulla glandulae
suprarenalis
Urinary bladder
Vesica urinaria
Reproductive organs
Organa genitalia
Spinal cord
Medulla spinalis

Often referred to as the "fight-or-flight" system, the sympathetic division reacts during emergency situations, for example by increasing heart and breathing rate, blood pressure, and blood sugar levels. It is also involved in the control of body temperature, and with the urinary and reproductive organs.

[~] = no direct Latin equivalent

AUTONOMIC NERVOUS SYSTEM: PARASYMPATHETIC DIVISION

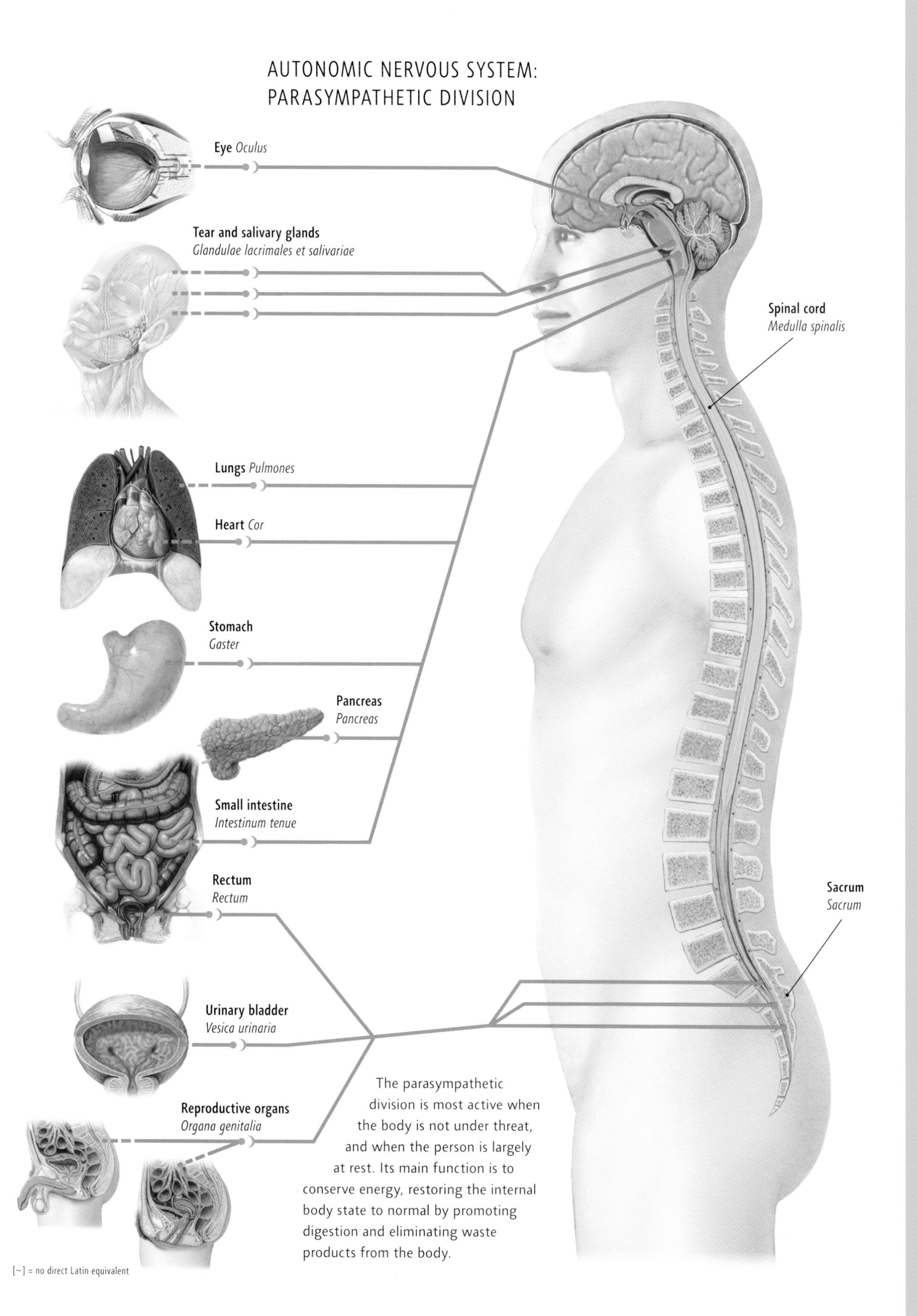
Eye *Oculus*
Tear and salivary glands *Glandulae lacrimales et salivariae*
Lungs *Pulmones*
Heart *Cor*
Stomach *Gaster*
Pancreas *Pancreas*
Small intestine *Intestinum tenue*
Rectum *Rectum*
Urinary bladder *Vesica urinaria*
Reproductive organs *Organa genitalia*
Spinal cord *Medulla spinalis*
Sacrum *Sacrum*

The parasympathetic division is most active when the body is not under threat, and when the person is largely at rest. Its main function is to conserve energy, restoring the internal body state to normal by promoting digestion and eliminating waste products from the body.

[~] = no direct Latin equivalent

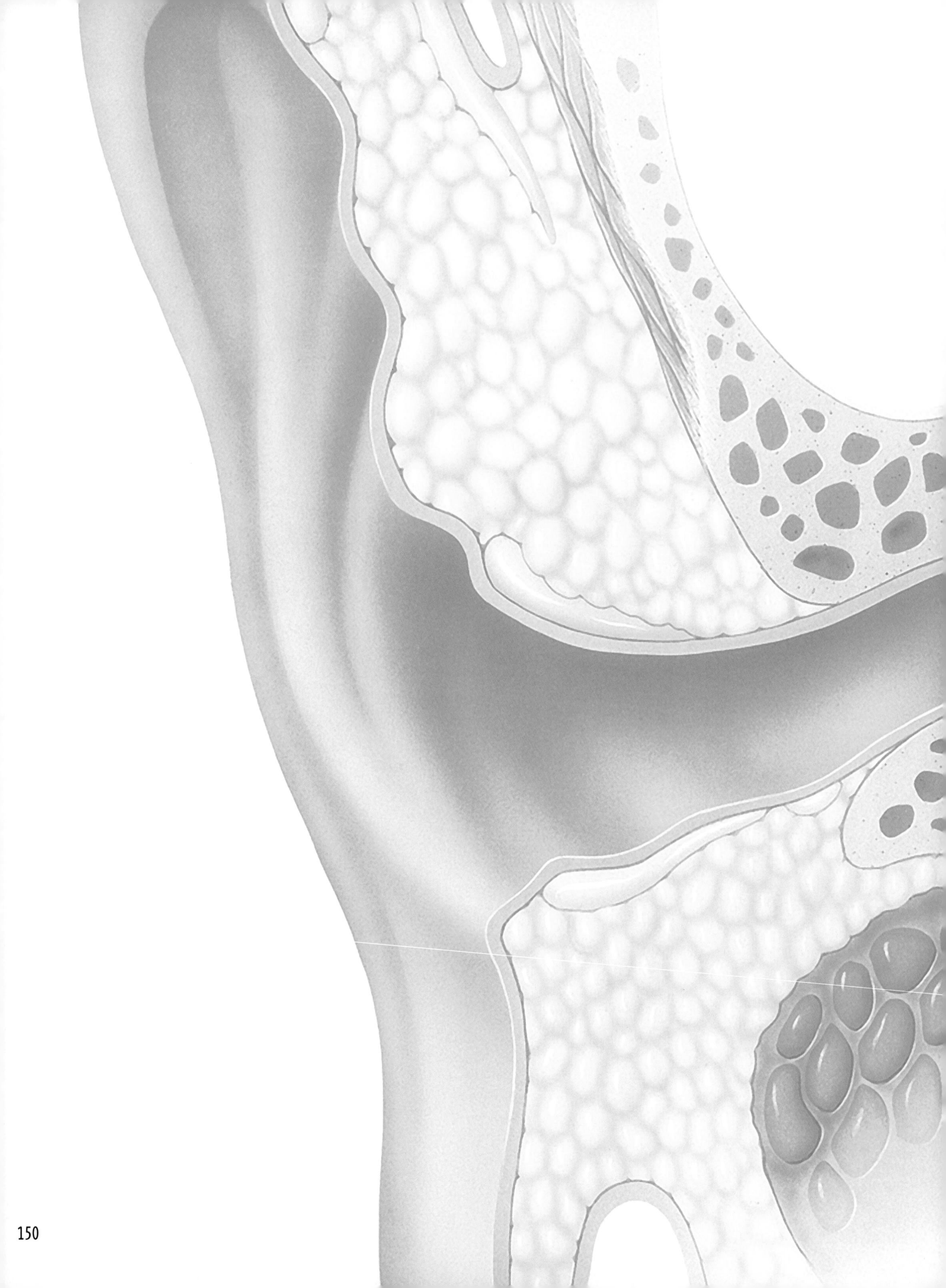

Special Sense Organs

The Eye

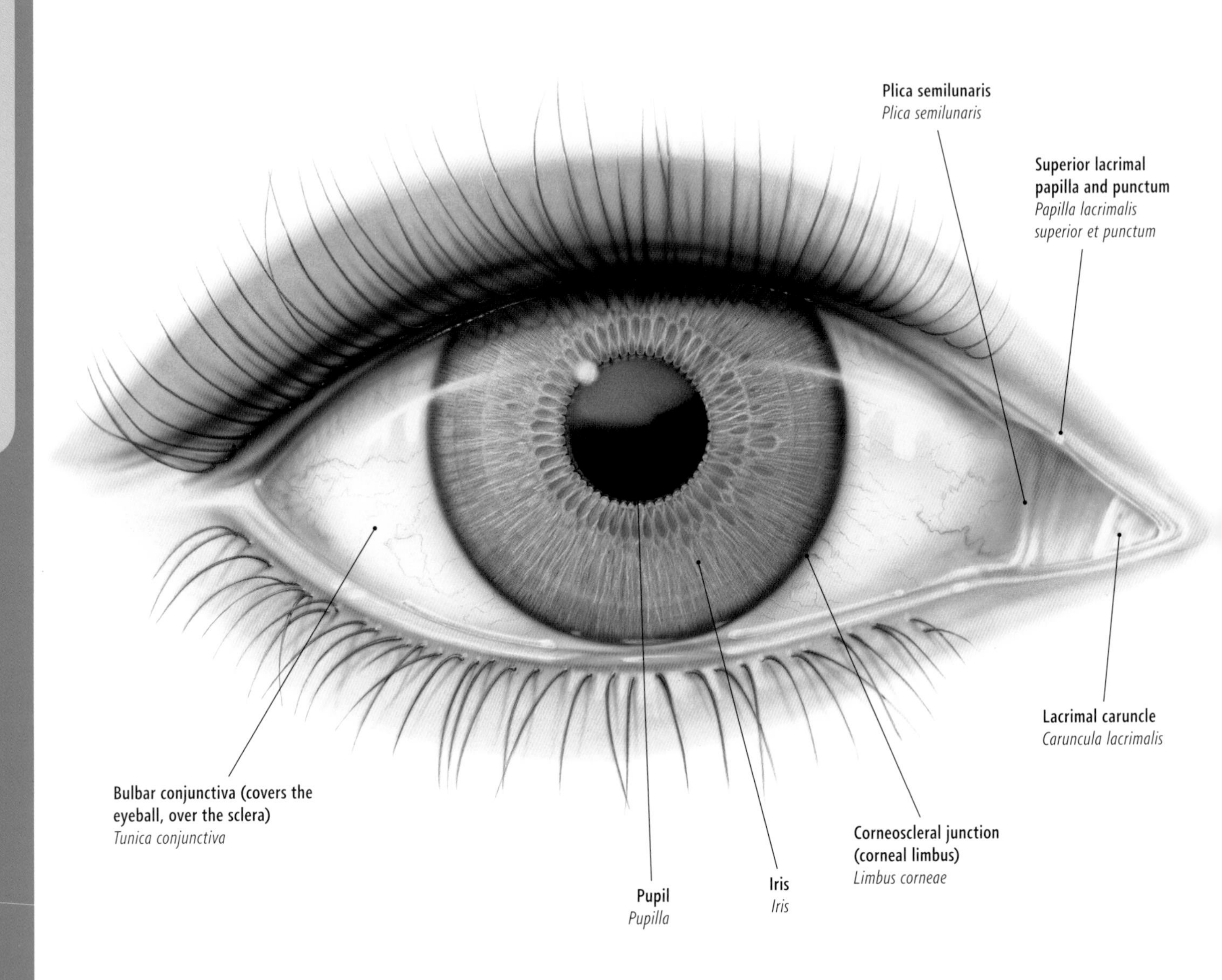
Plica semilunaris
Plica semilunaris
Superior lacrimal papilla and punctum
Papilla lacrimalis superior et punctum
Lacrimal caruncle
Caruncula lacrimalis
Bulbar conjunctiva (covers the eyeball, over the sclera)
Tunica conjunctiva
Corneoscleral junction (corneal limbus)
Limbus corneae
Pupil
Pupilla
Iris
Iris

FRONT VIEW

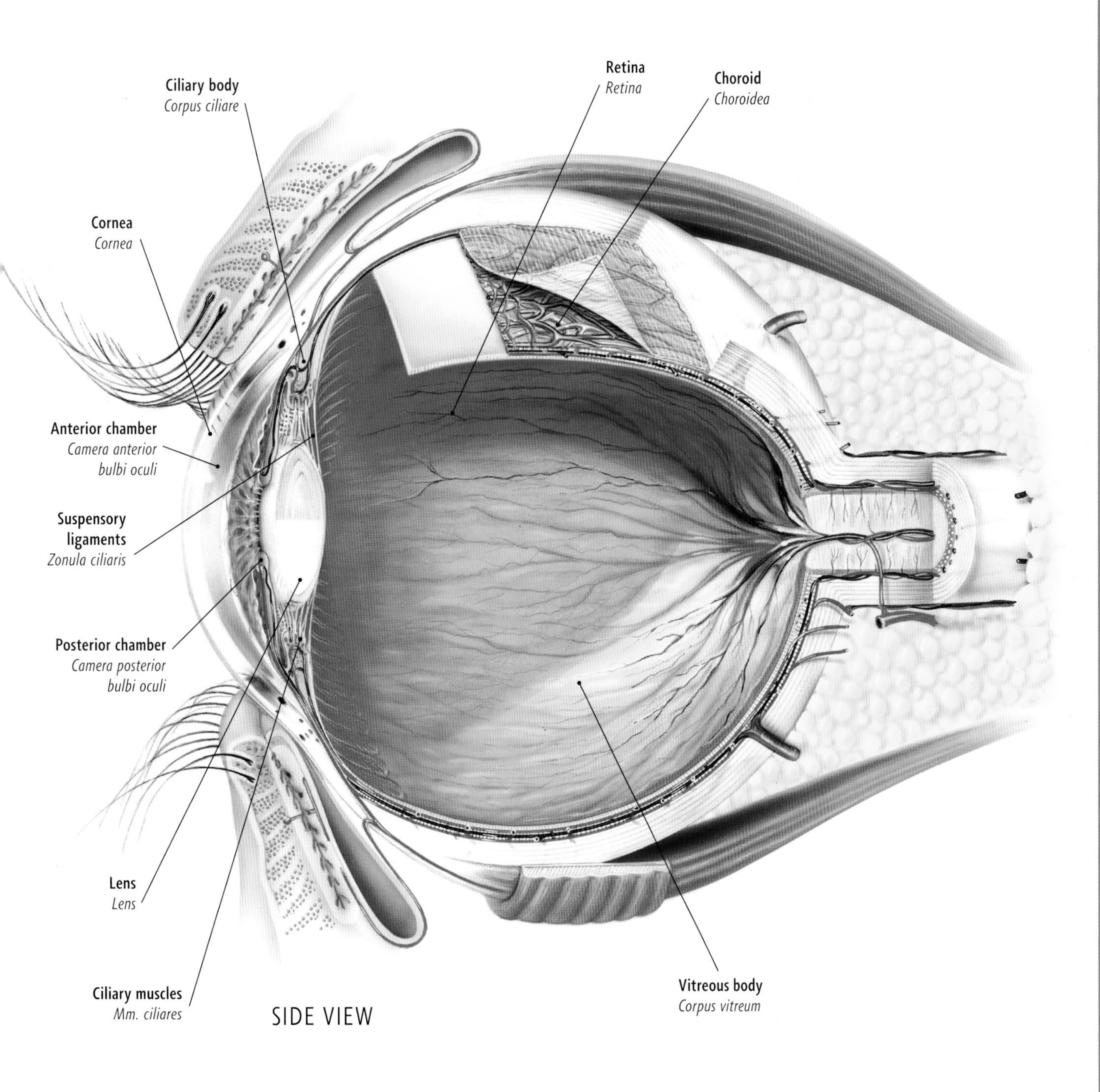
Ciliary body
Corpus ciliare
Retina
Retina
Choroid
Choroidea
Cornea
Cornea
Anterior chamber
Camera anterior bulbi oculi
Suspensory ligaments
Zonula ciliaris
Posterior chamber
Camera posterior bulbi oculi
Lens
Lens
Ciliary muscles
Mm. ciliares
Vitreous body
Corpus vitreum

SIDE VIEW

Sight

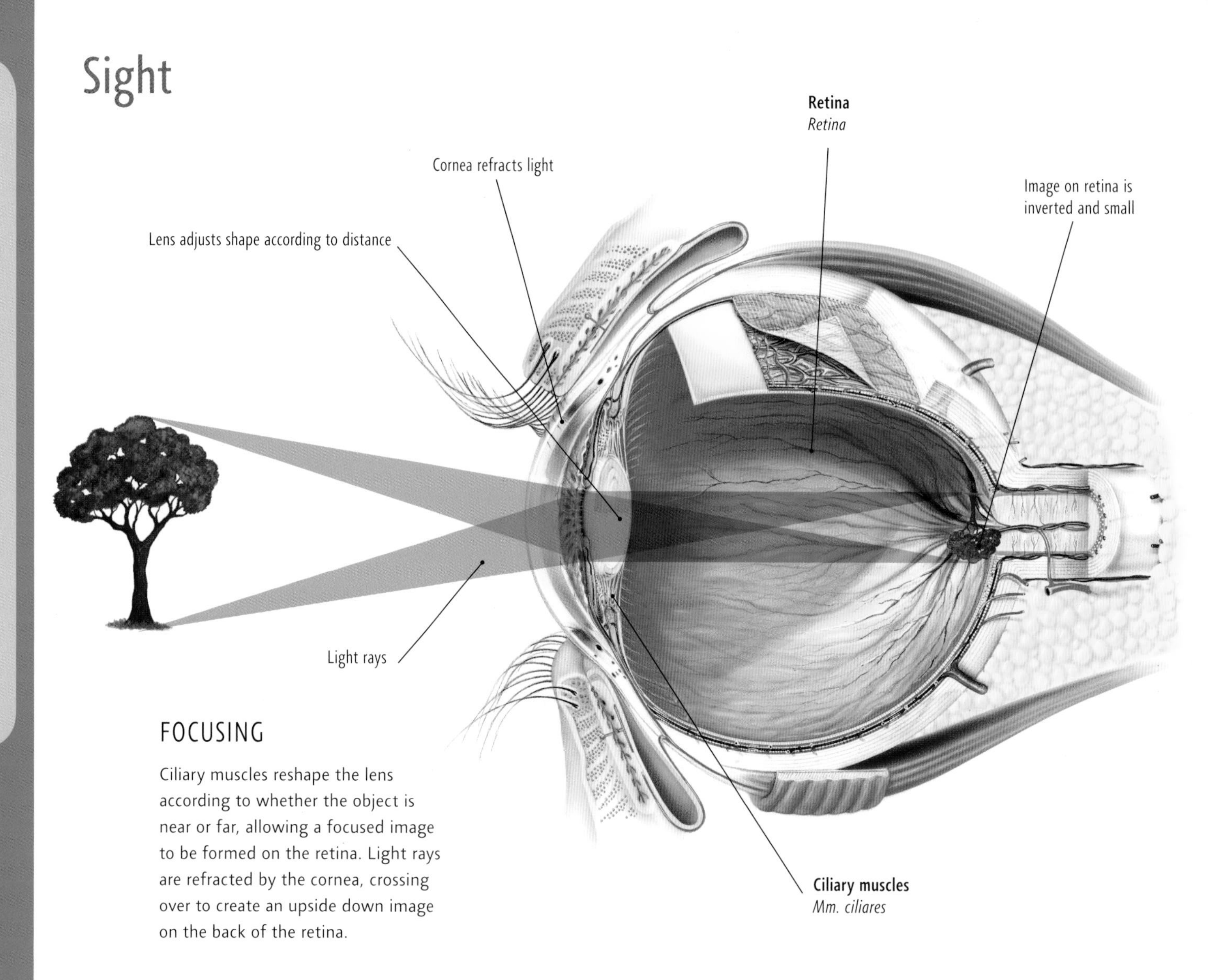
Retina
Retina
Cornea refracts light
Image on retina is inverted and small
Lens adjusts shape according to distance
Light rays
Ciliary muscles
Mm. ciliares

FOCUSING

Ciliary muscles reshape the lens according to whether the object is near or far, allowing a focused image to be formed on the retina. Light rays are refracted by the cornea, crossing over to create an upside down image on the back of the retina.

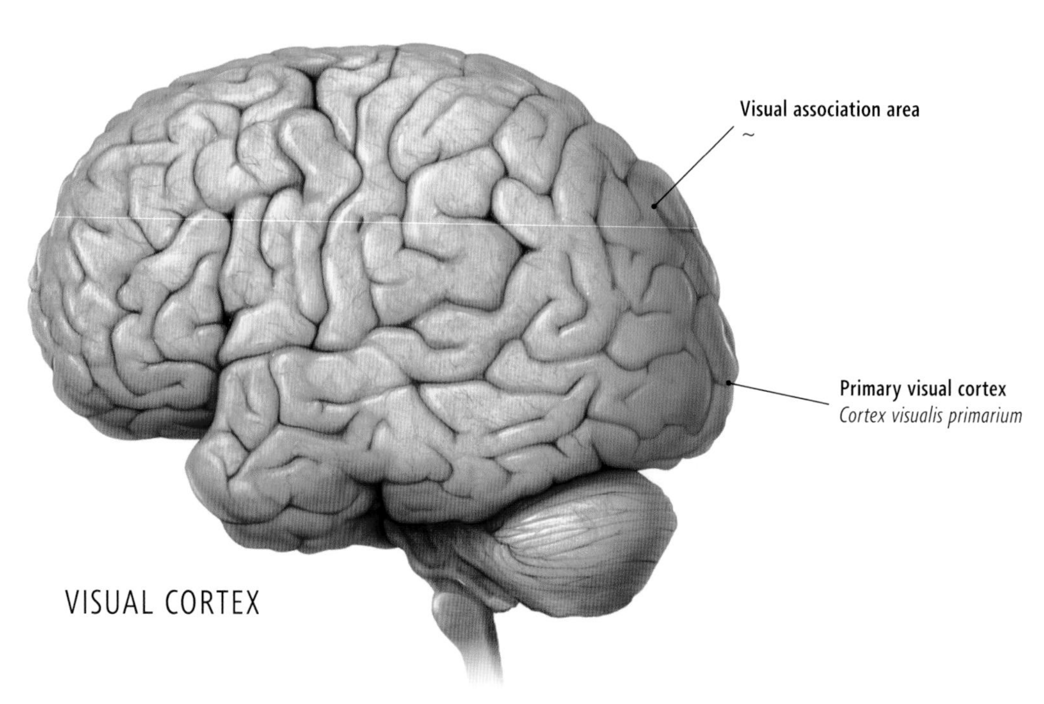
Visual association area
~
Primary visual cortex
Cortex visualis primarium

VISUAL CORTEX

[~] = no direct Latin equivalent

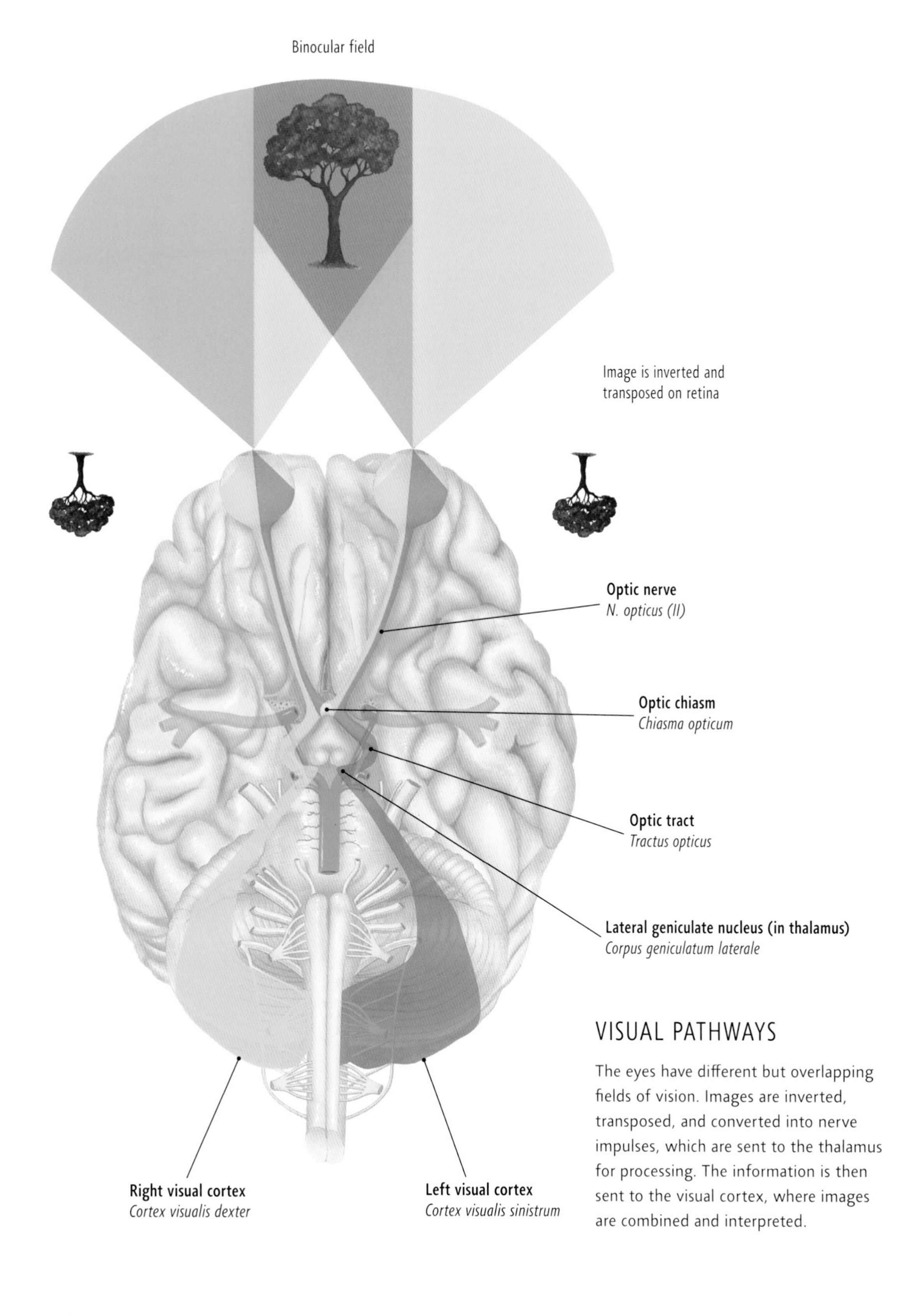
Binocular field
Image is inverted and transposed on retina
Optic nerve
N. opticus (II)
Optic chiasm
Chiasma opticum
Optic tract
Tractus opticus
Lateral geniculate nucleus (in thalamus)
Corpus geniculatum laterale
Right visual cortex
Cortex visualis dexter
Left visual cortex
Cortex visualis sinistrum

VISUAL PATHWAYS

The eyes have different but overlapping fields of vision. Images are inverted, transposed, and converted into nerve impulses, which are sent to the thalamus for processing. The information is then sent to the visual cortex, where images are combined and interpreted.

The Ear

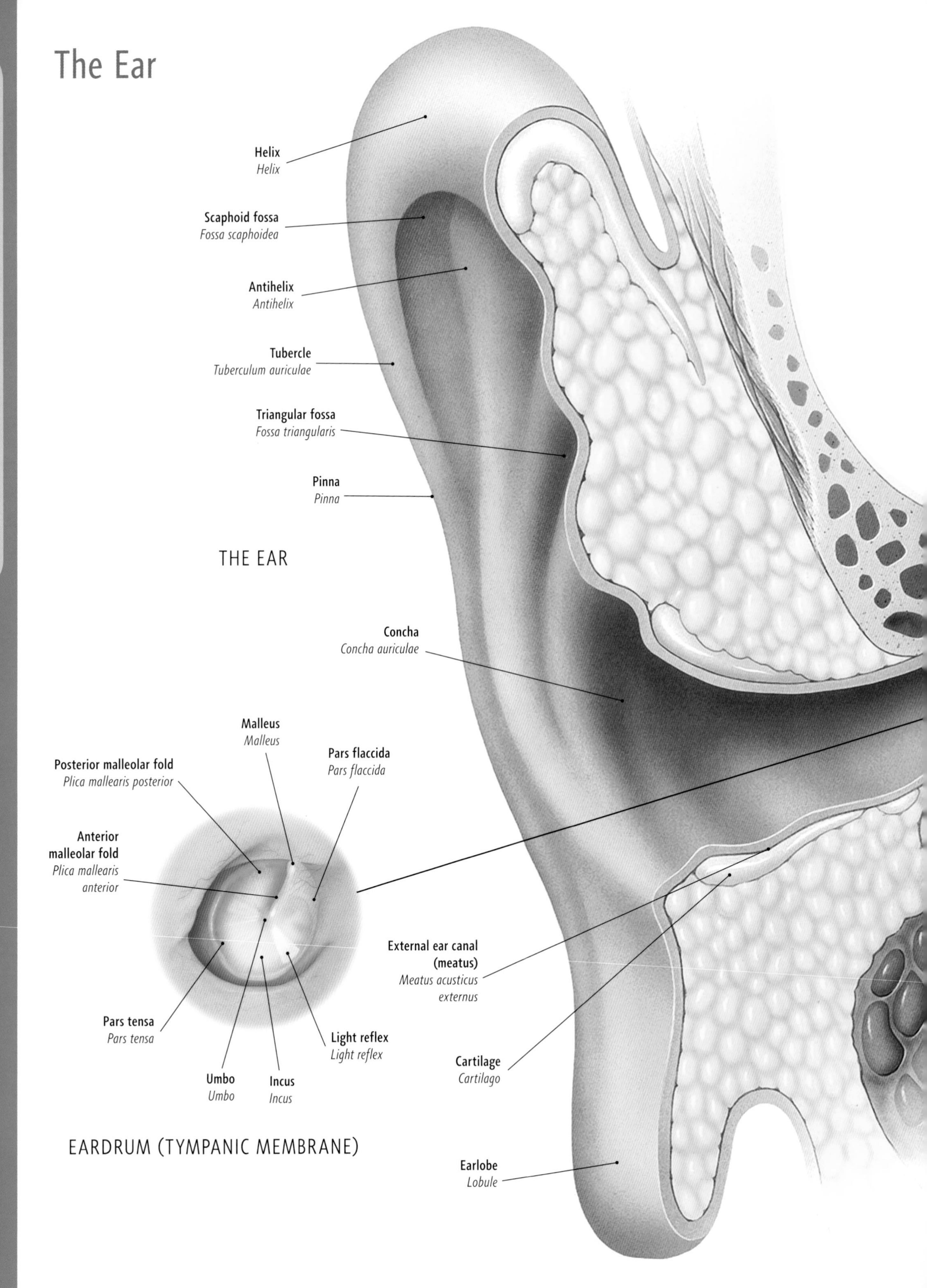

Helix
Helix
Scaphoid fossa
Fossa scaphoidea
Antihelix
Antihelix
Tubercle
Tuberculum auriculae
Triangular fossa
Fossa triangularis
Pinna
Pinna
THE EAR
Concha
Concha auriculae
Malleus
Malleus
Posterior malleolar fold
Plica mallearis posterior
Pars flaccida
Pars flaccida
Anterior malleolar fold
Plica mallearis anterior
External ear canal (meatus)
Meatus acusticus externus
Pars tensa
Pars tensa
Light reflex
Light reflex
Cartilage
Cartilago
Umbo
Umbo
Incus
Incus
EARDRUM (TYMPANIC MEMBRANE)
Earlobe
Lobule

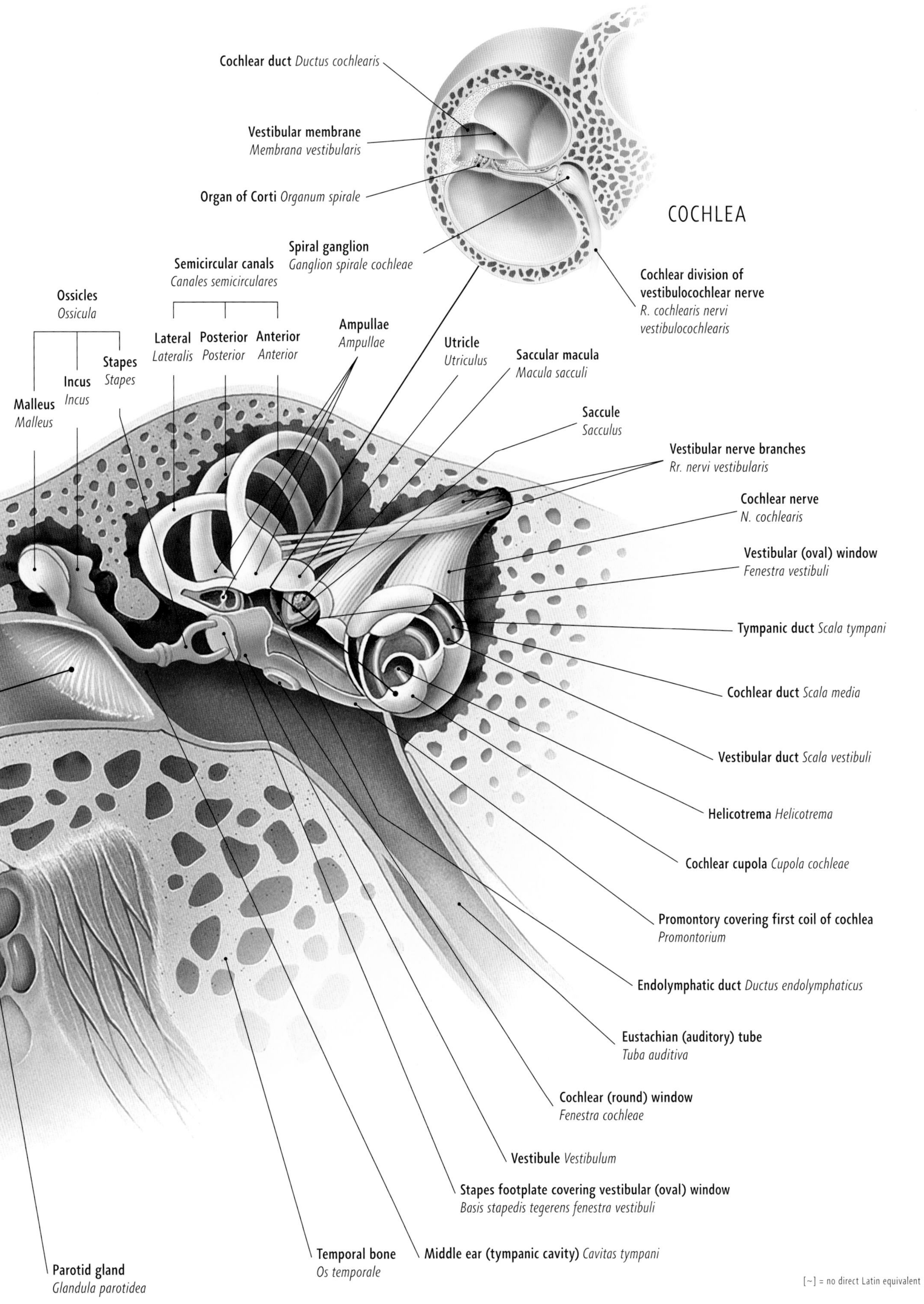
Cochlear duct Ductus cochlearis
Vestibular membrane Membrana vestibularis
Organ of Corti Organum spirale
Spiral ganglion Ganglion spirale cochleae
COCHLEA
Cochlear division of vestibulocochlear nerve R. cochlearis nervi vestibulocochlearis
Semicircular canals Canales semicirculares
Ossicles Ossicula
Lateral Lateralis
Posterior Posterior
Anterior Anterior
Ampullae Ampullae
Utricle Utriculus
Saccular macula Macula sacculi
Stapes Stapes
Incus Incus
Malleus Malleus
Saccule Sacculus
Vestibular nerve branches Rr. nervi vestibularis
Cochlear nerve N. cochlearis
Vestibular (oval) window Fenestra vestibuli
Tympanic duct Scala tympani
Cochlear duct Scala media
Vestibular duct Scala vestibuli
Helicotrema Helicotrema
Cochlear cupola Cupola cochleae
Promontory covering first coil of cochlea Promontorium
Endolymphatic duct Ductus endolymphaticus
Eustachian (auditory) tube Tuba auditiva
Cochlear (round) window Fenestra cochleae
Vestibule Vestibulum
Stapes footplate covering vestibular (oval) window Basis stapedis tegerens fenestra vestibuli
Temporal bone Os temporale
Middle ear (tympanic cavity) Cavitas tympani
Parotid gland Glandula parotidea

[~] = no direct Latin equivalent

Hearing and Balance

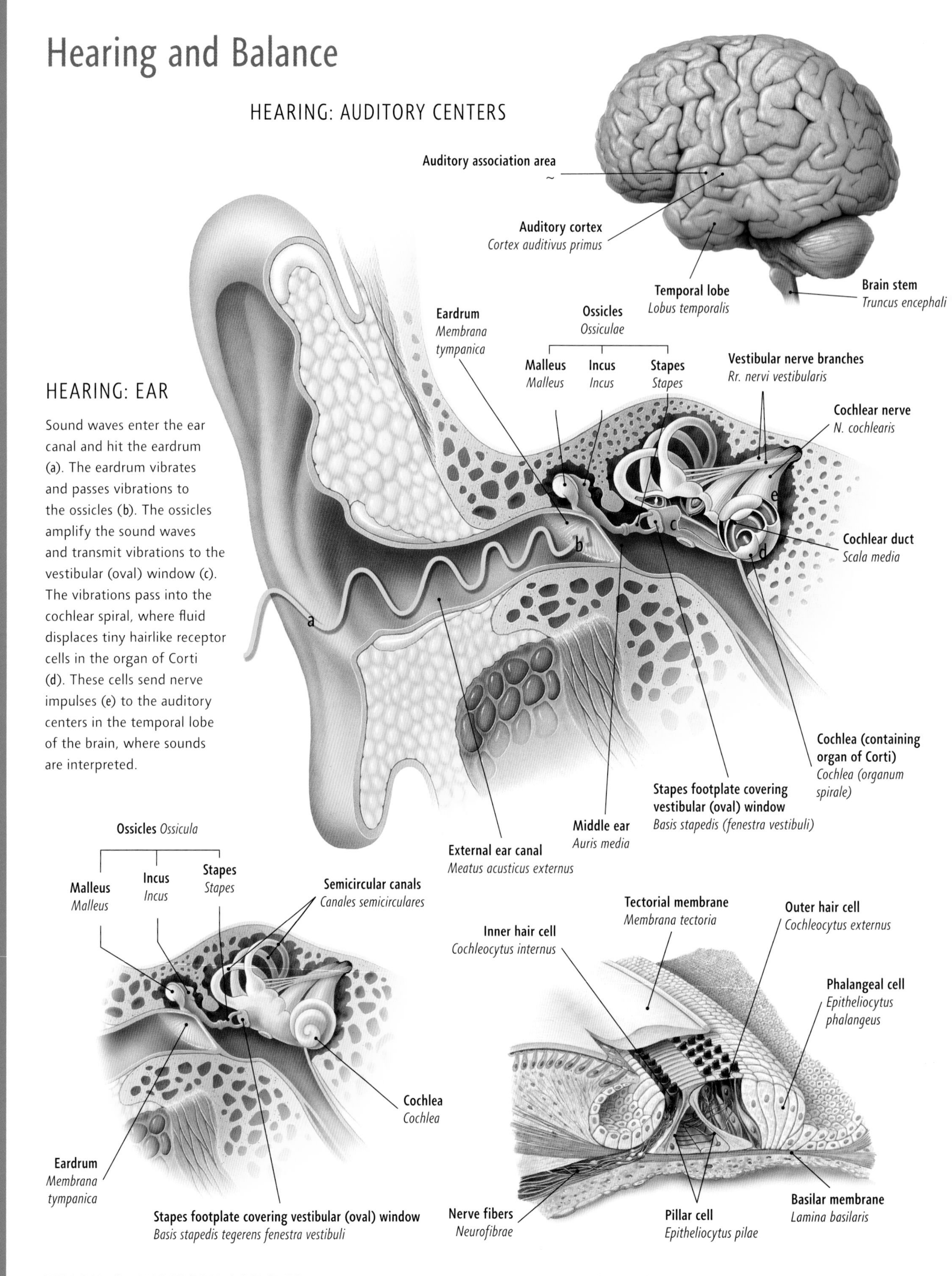
HEARING: AUDITORY CENTERS
Auditory association area
~
Auditory cortex
Cortex auditivus primus
Temporal lobe
Lobus temporalis
Brain stem
Truncus encephali
Eardrum
Membrana tympanica
Ossicles
Ossiculae
Malleus
Malleus
Incus
Incus
Stapes
Stapes
Vestibular nerve branches
Rr. nervi vestibularis
Cochlear nerve
N. cochlearis
Cochlear duct
Scala media
a
b
c
d
e
Cochlea (containing organ of Corti)
Cochlea (organum spirale)
Stapes footplate covering vestibular (oval) window
Basis stapedis (fenestra vestibuli)
Middle ear
Auris media
External ear canal
Meatus acusticus externus
Ossicles Ossicula
Malleus
Malleus
Incus
Incus
Stapes
Stapes
Semicircular canals
Canales semicirculares
Cochlea
Cochlea
Eardrum
Membrana tympanica
Stapes footplate covering vestibular (oval) window
Basis stapedis tegerens fenestra vestibuli
HEARING: AUDITORY OSSICLES
Tectorial membrane
Membrana tectoria
Outer hair cell
Cochleocytus externus
Inner hair cell
Cochleocytus internus
Phalangeal cell
Epitheliocytus phalangeus
Basilar membrane
Lamina basilaris
Nerve fibers
Neurofibrae
Pillar cell
Epitheliocytus pilae
HEARING: ORGAN OF CORTI
[~] = no direct Latin equivalent

HEARING: EAR

Sound waves enter the ear canal and hit the eardrum (a). The eardrum vibrates and passes vibrations to the ossicles (b). The ossicles amplify the sound waves and transmit vibrations to the vestibular (oval) window (c). The vibrations pass into the cochlear spiral, where fluid displaces tiny hairlike receptor cells in the organ of Corti (d). These cells send nerve impulses (e) to the auditory centers in the temporal lobe of the brain, where sounds are interpreted.

BALANCE: BALANCE MECHANISM IN THE EAR

Semicircular canals and otolith organs in the inner ear contain tiny hairs that are sensitive to the body's position. Changes in position cause the hairs to send nerve signals to the brain, which uses this information to help balance the body.

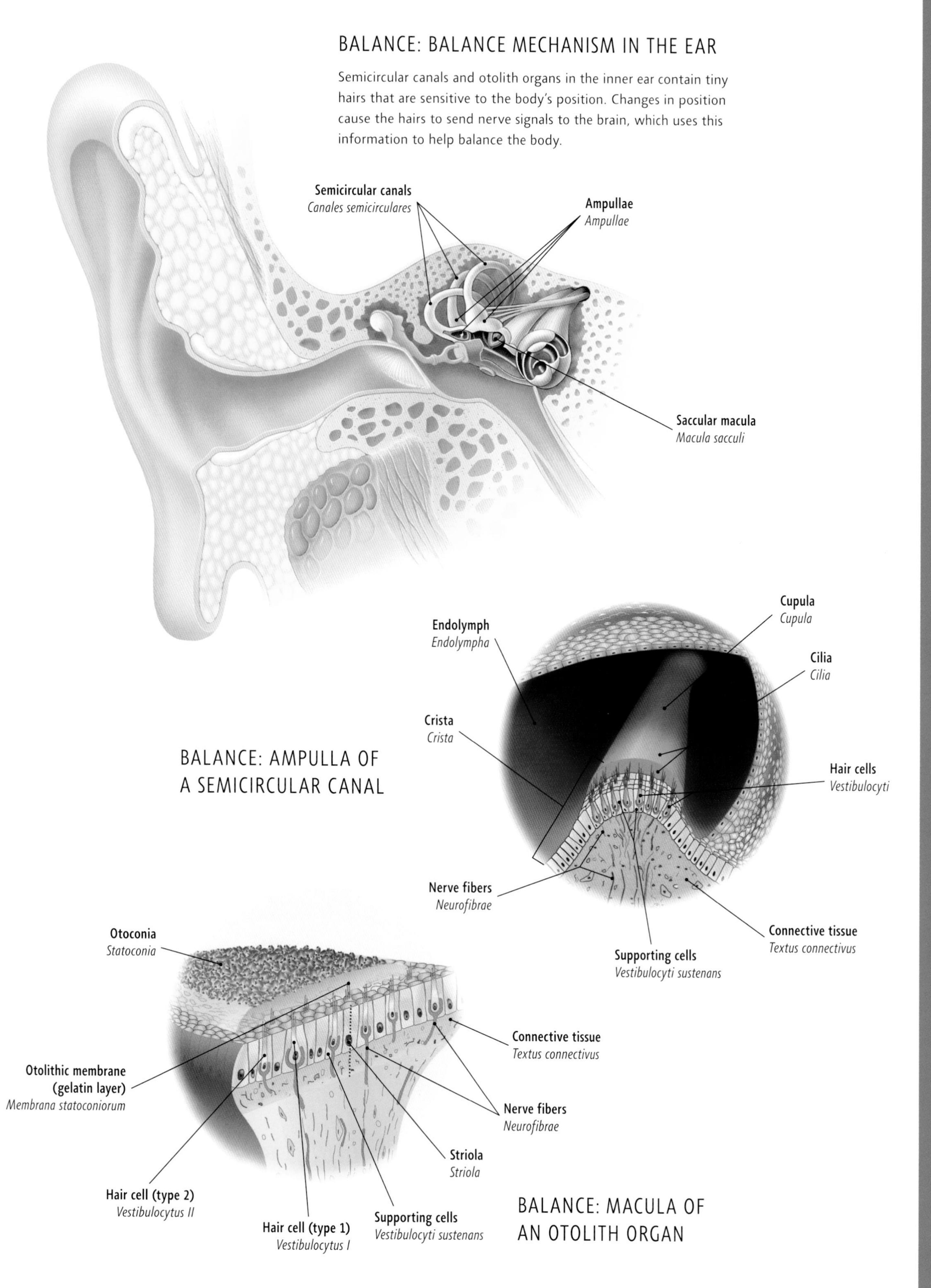

Semicircular canals
Canales semicirculares
Ampullae
Ampullae
Saccular macula
Macula sacculi
Endolymph
Endolympha
Cupula
Cupula
Cilia
Cilia
Crista
Crista
Hair cells
Vestibulocyti
Nerve fibers
Neurofibrae
Connective tissue
Textus connectivus
Supporting cells
Vestibulocyti sustenans
Otoconia
Statoconia
Otolithic membrane (gelatin layer)
Membrana statoconiorum
Connective tissue
Textus connectivus
Nerve fibers
Neurofibrae
Striola
Striola
Hair cell (type 2)
Vestibulocytus II
Hair cell (type 1)
Vestibulocytus I
Supporting cells
Vestibulocyti sustenans

BALANCE: AMPULLA OF A SEMICIRCULAR CANAL

BALANCE: MACULA OF AN OTOLITH ORGAN

Taste

TASTE PATHWAYS

Taste buds on the tongue and in the throat send nerve impulses via the medulla and thalamus to taste-receiving areas in the parietal lobe, where the taste is identified. The olfactory organs provide additional information vital for interpreting different tastes.

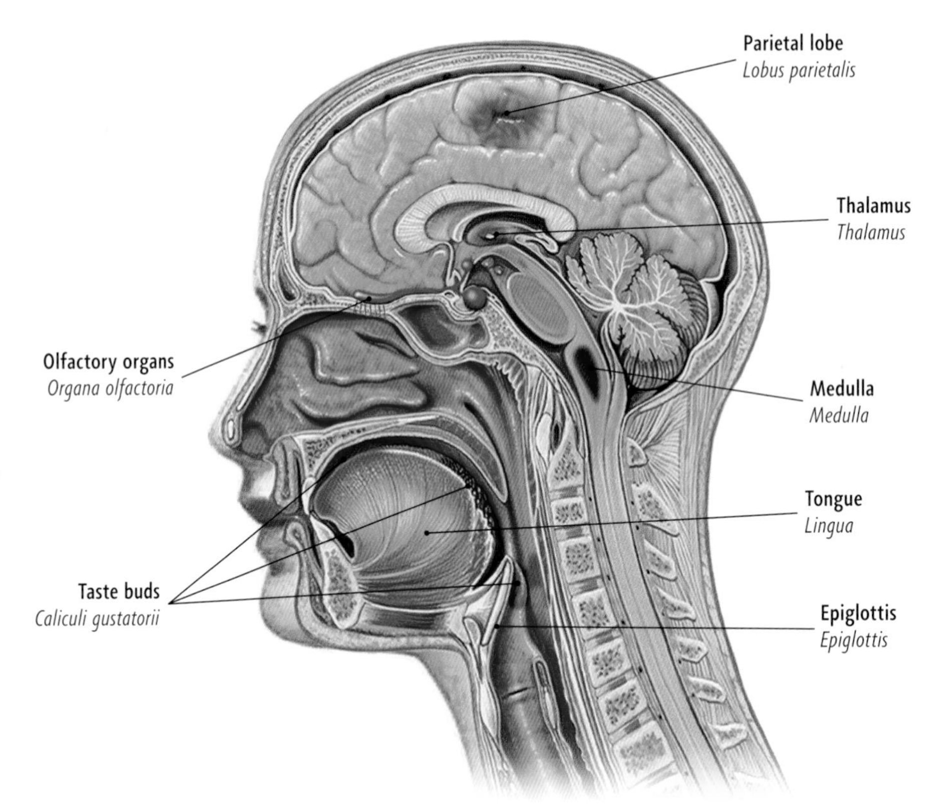
Parietal lobe *Lobus parietalis*
Thalamus *Thalamus*
Olfactory organs *Organa olfactoria*
Medulla *Medulla*
Tongue *Lingua*
Taste buds *Caliculi gustatorii*
Epiglottis *Epiglottis*

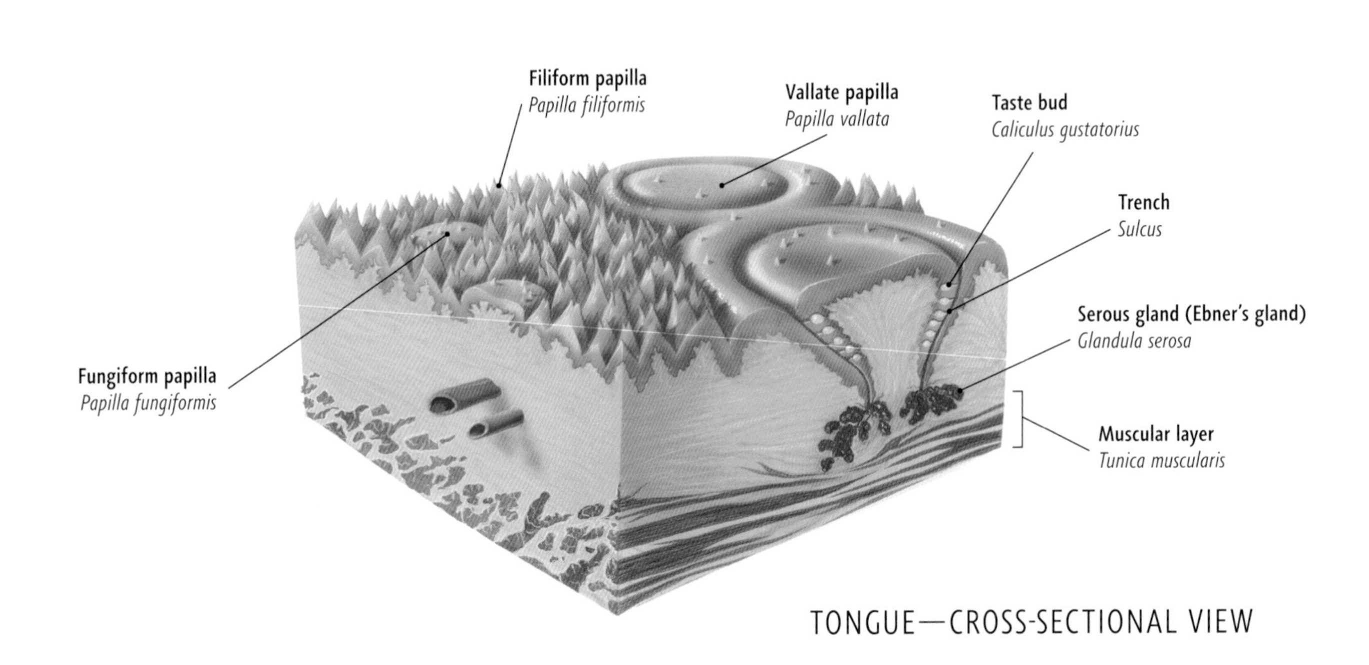
Filiform papilla *Papilla filiformis*
Vallate papilla *Papilla vallata*
Taste bud *Caliculus gustatorius*
Trench *Sulcus*
Serous gland (Ebner's gland) *Glandula serosa*
Fungiform papilla *Papilla fungiformis*
Muscular layer *Tunica muscularis*

TONGUE—CROSS-SECTIONAL VIEW

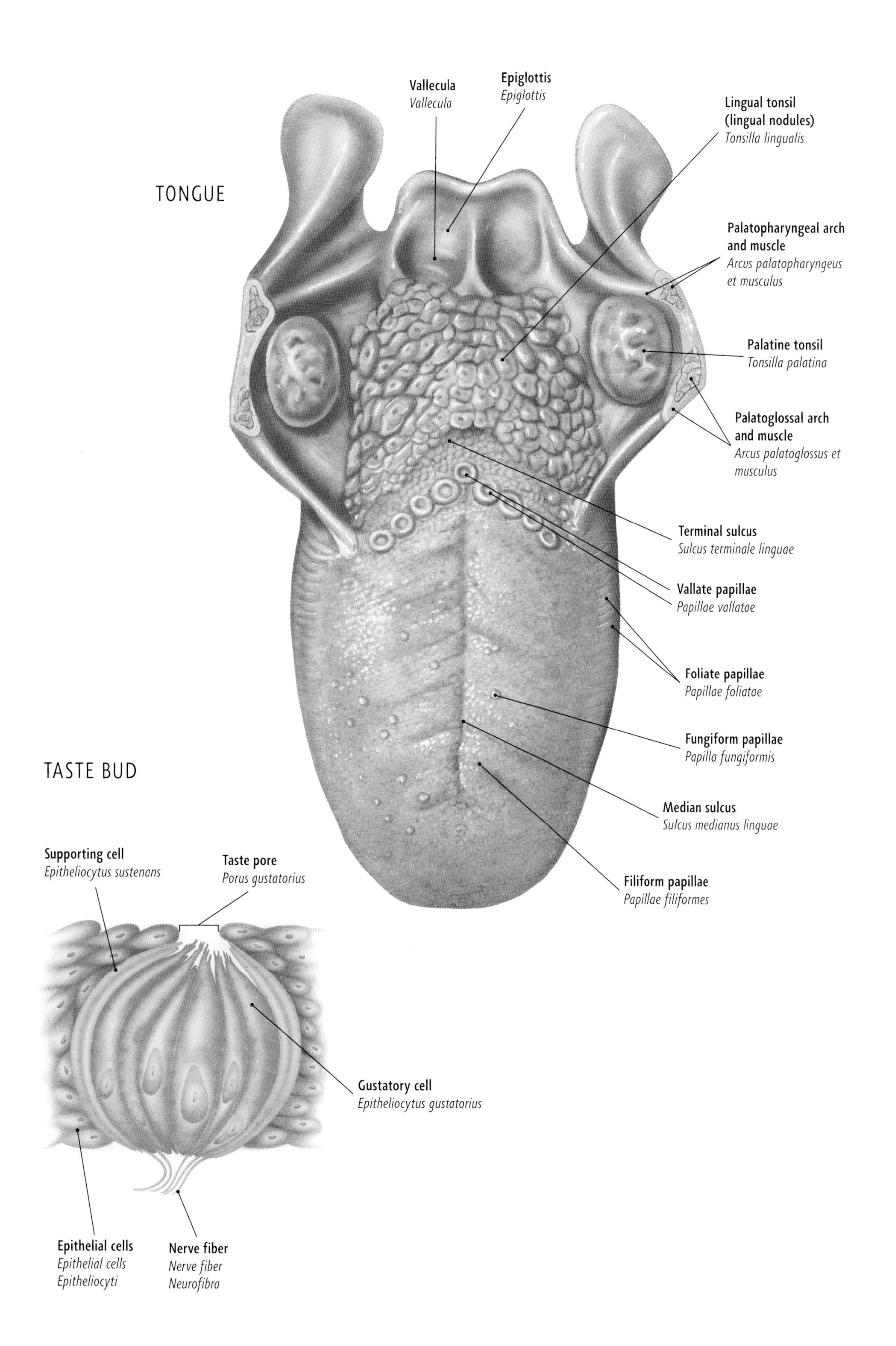
TONGUE
Vallecula
Vallecula
Epiglottis
Epiglottis
Lingual tonsil
(lingual nodules)
Tonsilla lingualis
Palatopharyngeal arch
and muscle
Arcus palatopharyngeus
et musculus
Palatine tonsil
Tonsilla palatina
Palatoglossal arch
and muscle
Arcus palatoglossus et
musculus
Terminal sulcus
Sulcus terminale linguae
Vallate papillae
Papillae vallatae
Foliate papillae
Papillae foliatae
Fungiform papillae
Papilla fungiformis
Median sulcus
Sulcus medianus linguae
Filiform papillae
Papillae filiformes
TASTE BUD
Supporting cell
Epitheliocytus sustenans
Taste pore
Porus gustatorius
Gustatory cell
Epitheliocytus gustatorius
Epithelial cells
Epithelial cells
Epitheliocyti
Nerve fiber
Nerve fiber
Neurofibra

Smell

OLFACTORY PATHWAY

Odor molecules enter the nostrils and pass into the nasal cavity, where they dissolve in olfactory mucosa in the nasal lining. Receptors send nerve signals
via the olfactory bulb to the olfactory center in the brain, where smells are identified.

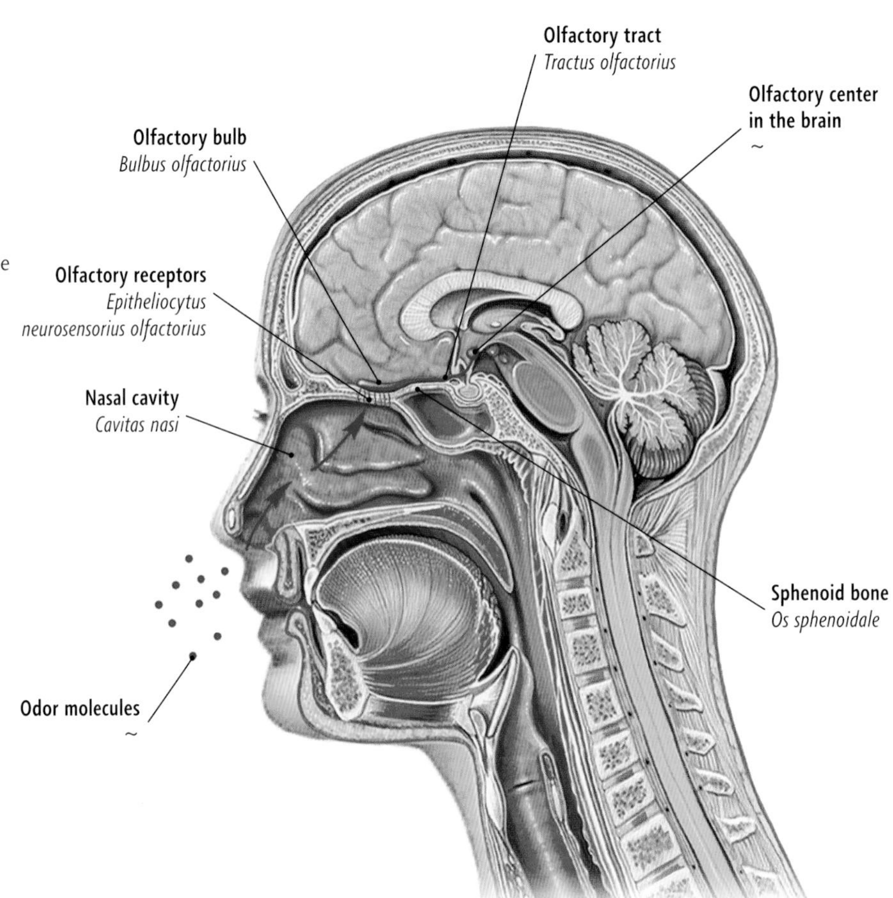
Olfactory tract
Tractus olfactorius
Olfactory center in the brain
~
Olfactory bulb
Bulbus olfactorius
Olfactory receptors
Epitheliocytus neurosensorius olfactorius
Nasal cavity
Cavitas nasi
Sphenoid bone
Os sphenoidale
Odor molecules
~

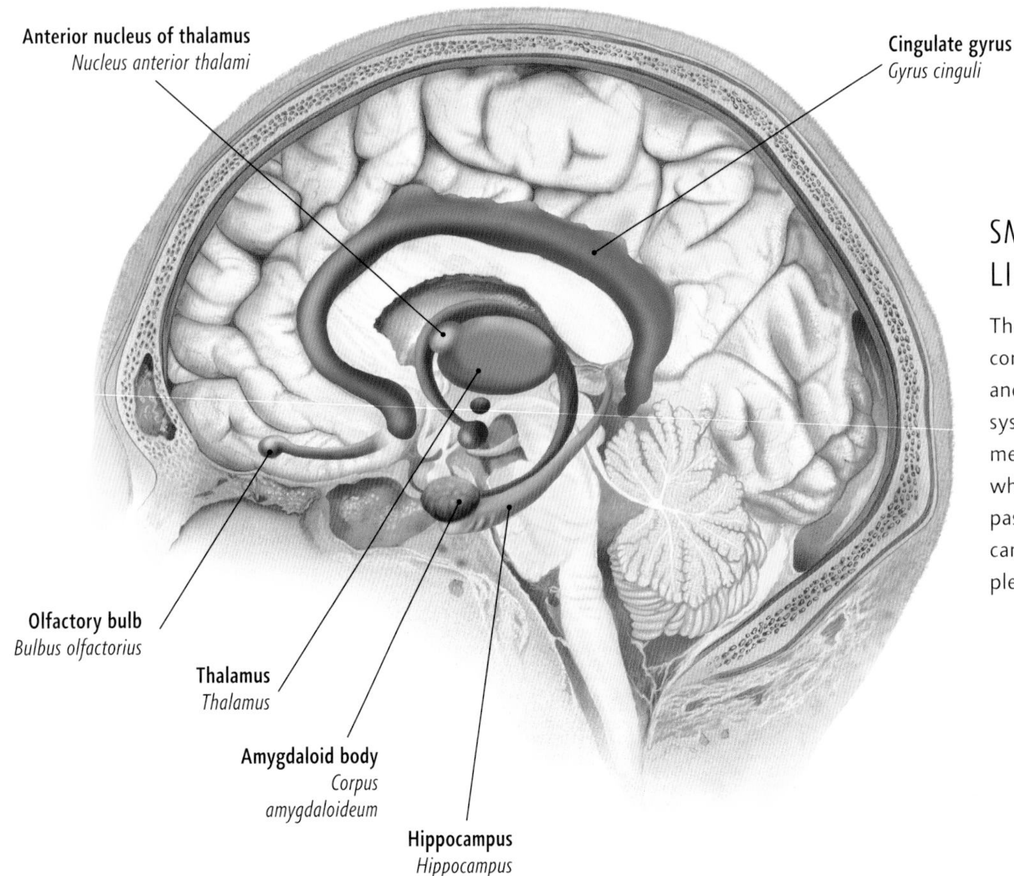
Anterior nucleus of thalamus
Nucleus anterior thalami
Cingulate gyrus
Gyrus cinguli
Olfactory bulb
Bulbus olfactorius
Thalamus
Thalamus
Amygdaloid body
Corpus amygdaloideum
Hippocampus
Hippocampus

SMELL AND THE LIMBIC SYSTEM

The olfactory bulb is directly connected to the hippocampus and amygdala in the limbic system, which is important for memory and emotion. This is why smells are evocative of past places and feelings, and can trigger responses like pleasure and fear.

[~] = no direct Latin equivalent

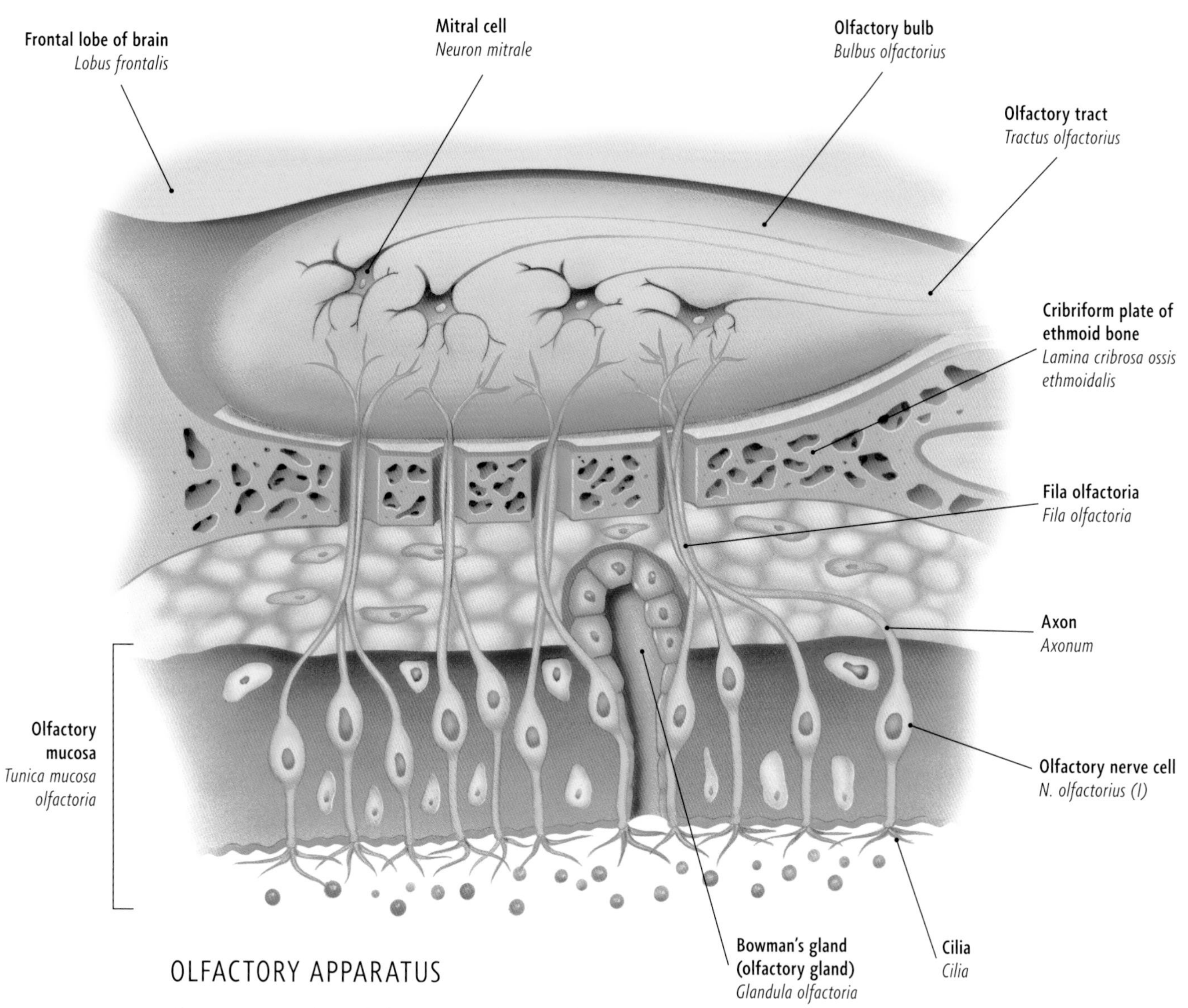

Frontal lobe of brain
Lobus frontalis
Mitral cell
Neuron mitrale
Olfactory bulb
Bulbus olfactorius
Olfactory tract
Tractus olfactorius
Cribriform plate of ethmoid bone
Lamina cribrosa ossis ethmoidalis
Fila olfactoria
Fila olfactoria
Axon
Axonum
Olfactory mucosa
Tunica mucosa olfactoria
Olfactory nerve cell
N. olfactorius (I)
Bowman's gland (olfactory gland)
Glandula olfactoria
Cilia
Cilia

OLFACTORY APPARATUS

The mucous membrane (olfactory epithelium) of the olfactory receptors contains millions of nerve cells bearing cilia. Odor molecules stimulate the nerve cells to send impulses along the nerve fibers through holes in the cribriform plate of the ethmoid bone to the olfactory bulb.

Pain

PAIN PATHWAYS

When sensory neurons (nociceptors) in skin, muscles, joints, or internal organs are activated by pain, they send signals along nerve fibers to the spinal cord and on to the brain's sensory cortex, where the sensation of pain is registered.

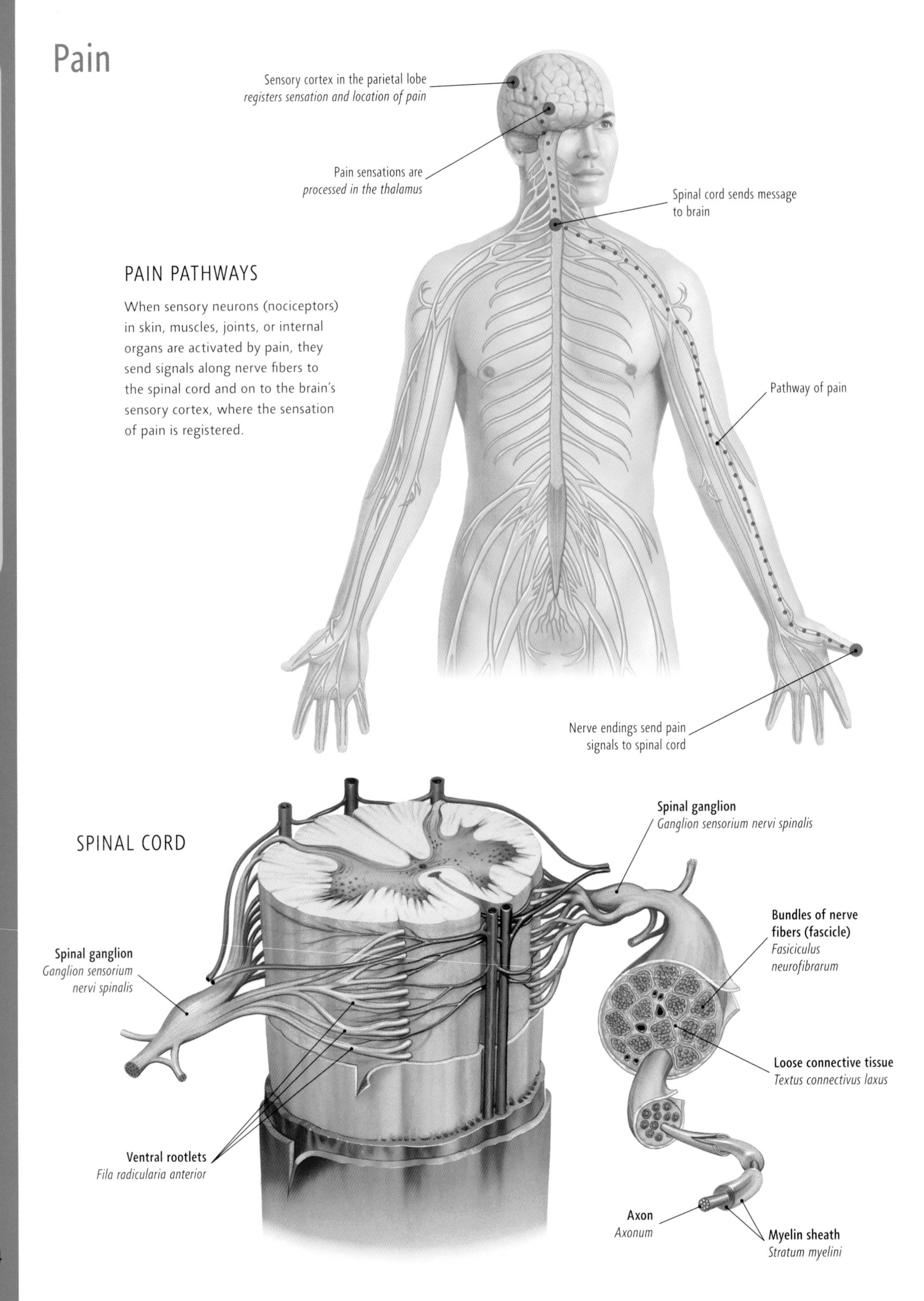
Sensory cortex in the parietal lobe
registers sensation and location of pain
Pain sensations are
processed in the thalamus
Spinal cord sends message to brain
Pathway of pain
Nerve endings send pain signals to spinal cord
SPINAL CORD
Spinal ganglion
Ganglion sensorium nervi spinalis
Bundles of nerve fibers (fascicle)
Fasiciculus neurofibrarum
Spinal ganglion
Ganglion sensorium nervi spinalis
Loose connective tissue
Textus connectivus laxus
Ventral rootlets
Fila radicularia anterior
Axon
Axonum
Myelin sheath
Stratum myelini

REFERRED PAIN

The pain from internal organs may be felt on the surface of the skin as well as internally. This is because the skin and internal organs may share the same pain pathways.

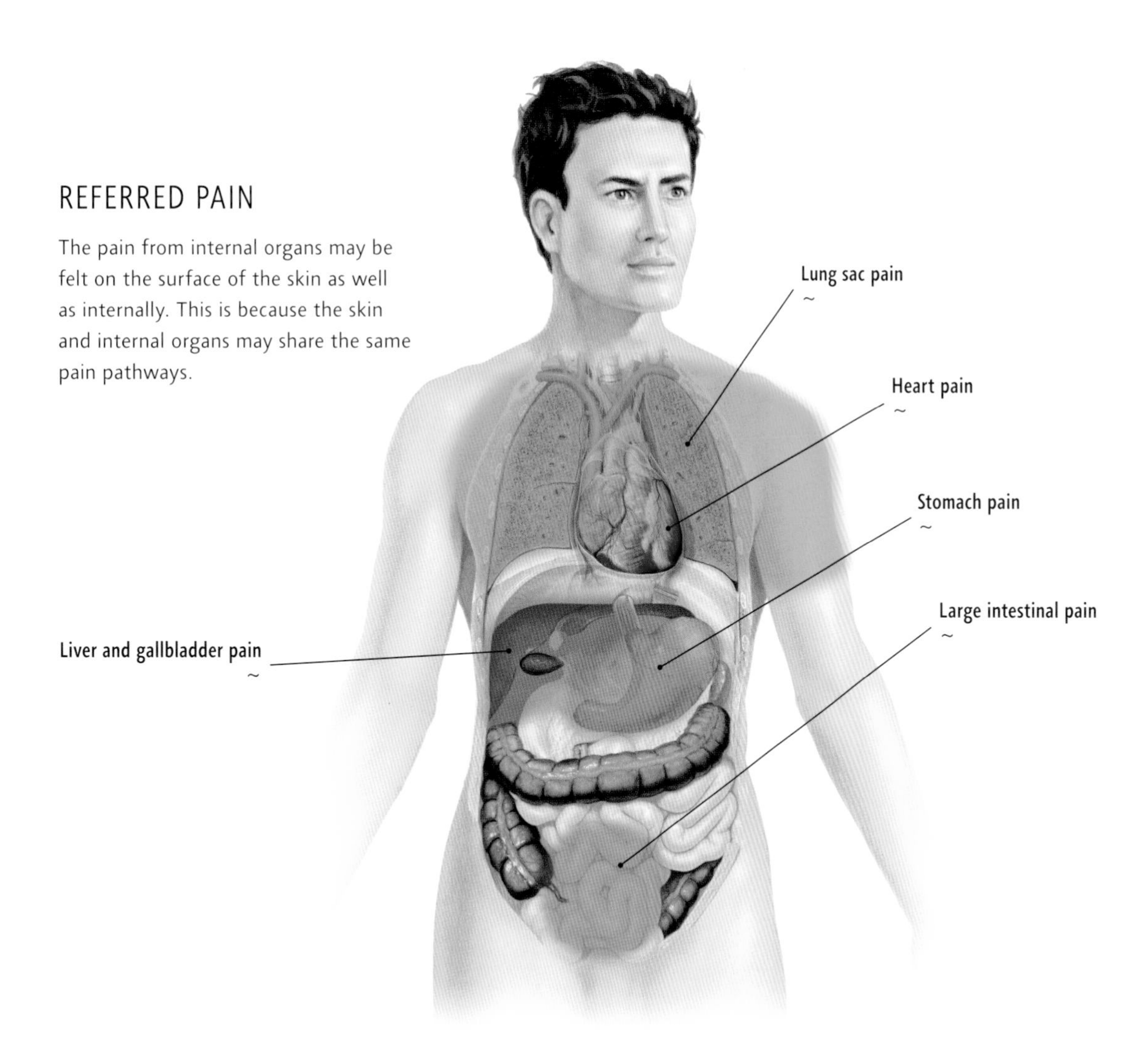
Lung sac pain
~
Heart pain
~
Stomach pain
~
Large intestinal pain
~
Liver and gallbladder pain
~

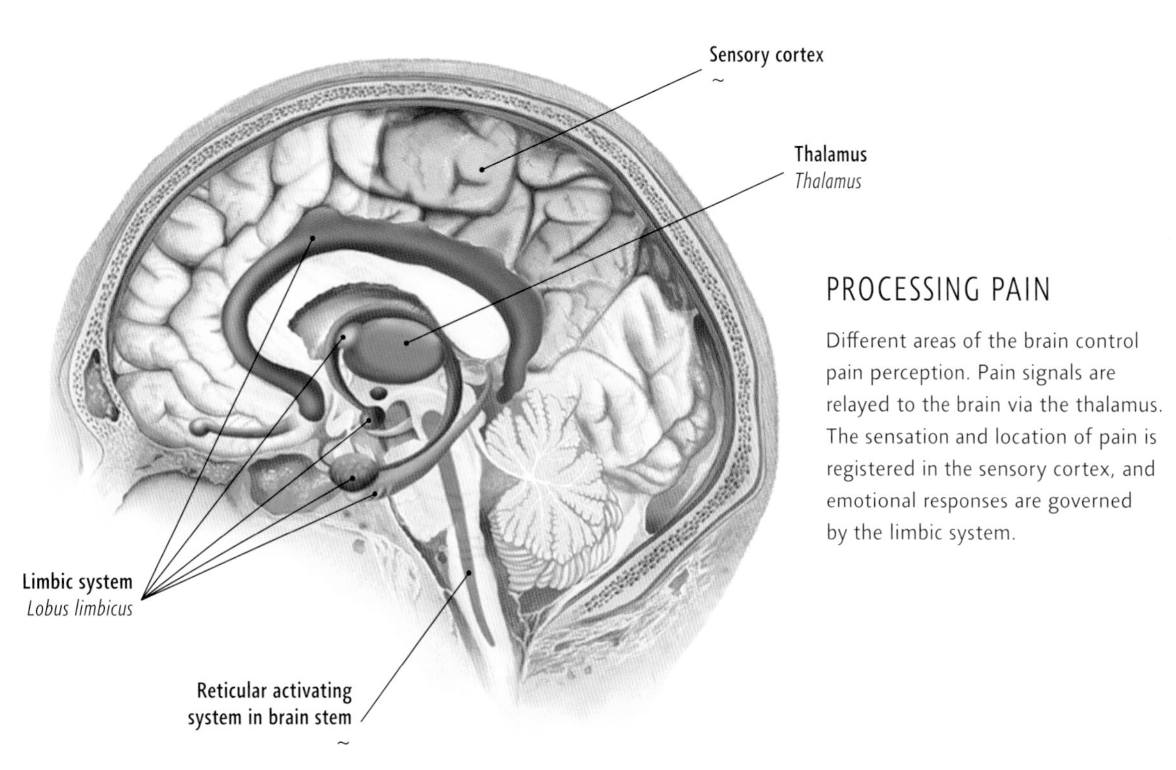
Sensory cortex
~
Thalamus
Thalamus
Limbic system
Lobus limbicus
Reticular activating system in brain stem
~

PROCESSING PAIN

Different areas of the brain control pain perception. Pain signals are relayed to the brain via the thalamus. The sensation and location of pain is registered in the sensory cortex, and emotional responses are governed by the limbic system.

[~] = no direct Latin equivalent

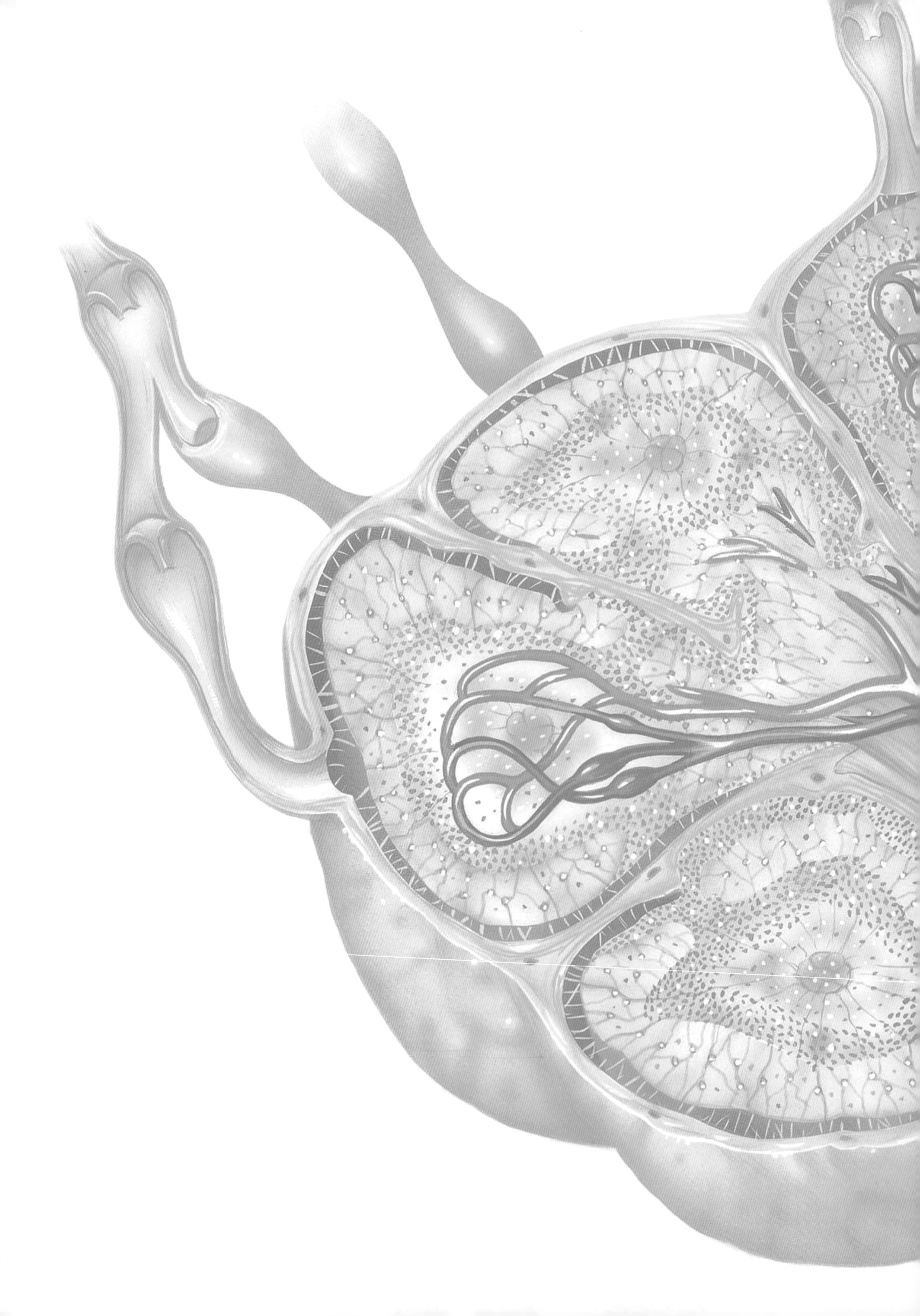

The Lymphatic System

The Lymphatic System

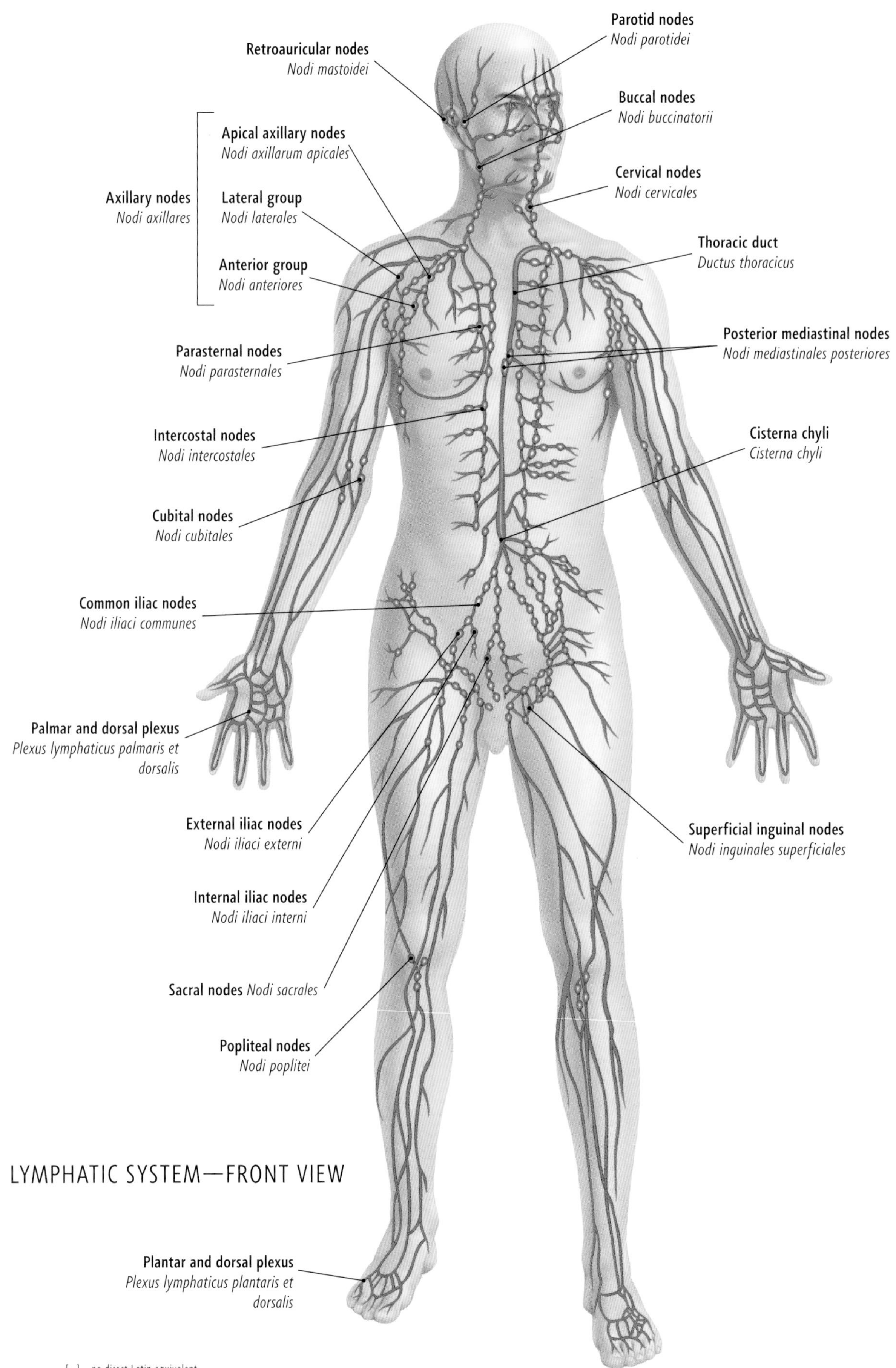
Parotid nodes
Nodi parotidei
Retroauricular nodes
Nodi mastoidei
Buccal nodes
Nodi buccinatorii
Apical axillary nodes
Nodi axillarum apicales
Cervical nodes
Nodi cervicales
Axillary nodes
Nodi axillares
Lateral group
Nodi laterales
Thoracic duct
Ductus thoracicus
Anterior group
Nodi anteriores
Posterior mediastinal nodes
Nodi mediastinales posteriores
Parasternal nodes
Nodi parasternales
Intercostal nodes
Nodi intercostales
Cisterna chyli
Cisterna chyli
Cubital nodes
Nodi cubitales
Common iliac nodes
Nodi iliaci communes
Palmar and dorsal plexus
Plexus lymphaticus palmaris et dorsalis
External iliac nodes
Nodi iliaci externi
Superficial inguinal nodes
Nodi inguinales superficiales
Internal iliac nodes
Nodi iliaci interni
Sacral nodes Nodi sacrales
Popliteal nodes
Nodi poplitei
Plantar and dorsal plexus
Plexus lymphaticus plantaris et dorsalis

LYMPHATIC SYSTEM—FRONT VIEW

[~] = no direct Latin equivalent

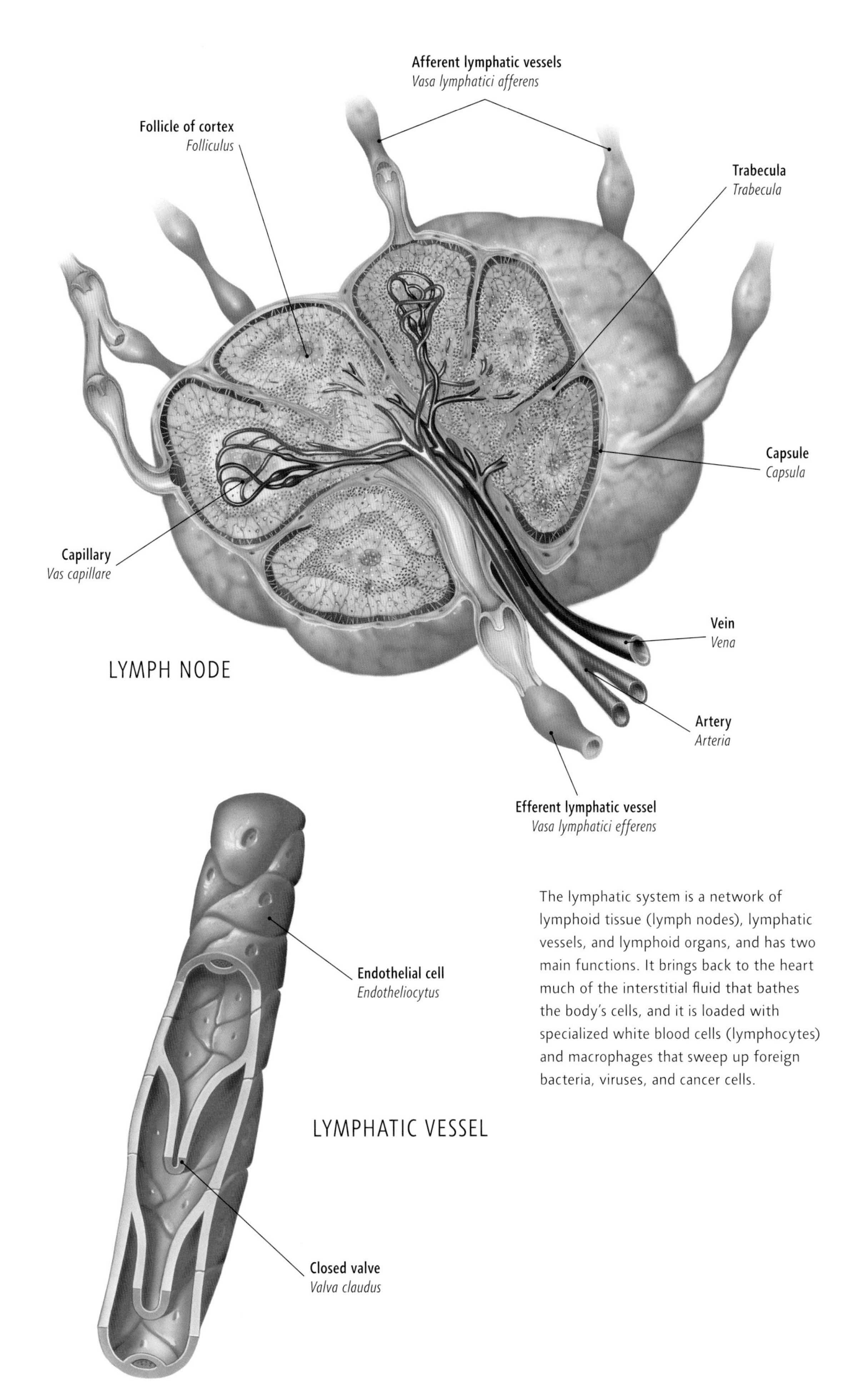
Afferent lymphatic vessels
Vasa lymphatici afferens
Follicle of cortex
Folliculus
Trabecula
Trabecula
Capsule
Capsula
Capillary
Vas capillare
Vein
Vena
LYMPH NODE
Artery
Arteria
Efferent lymphatic vessel
Vasa lymphatici efferens
Endothelial cell
Endotheliocytus
LYMPHATIC VESSEL
Closed valve
Valva claudus

The lymphatic system is a network of lymphoid tissue (lymph nodes), lymphatic vessels, and lymphoid organs, and has two main functions. It brings back to the heart much of the interstitial fluid that bathes the body's cells, and it is loaded with specialized white blood cells (lymphocytes) and macrophages that sweep up foreign bacteria, viruses, and cancer cells.

[~] = no direct Latin equivalent

Lymphoid Organs

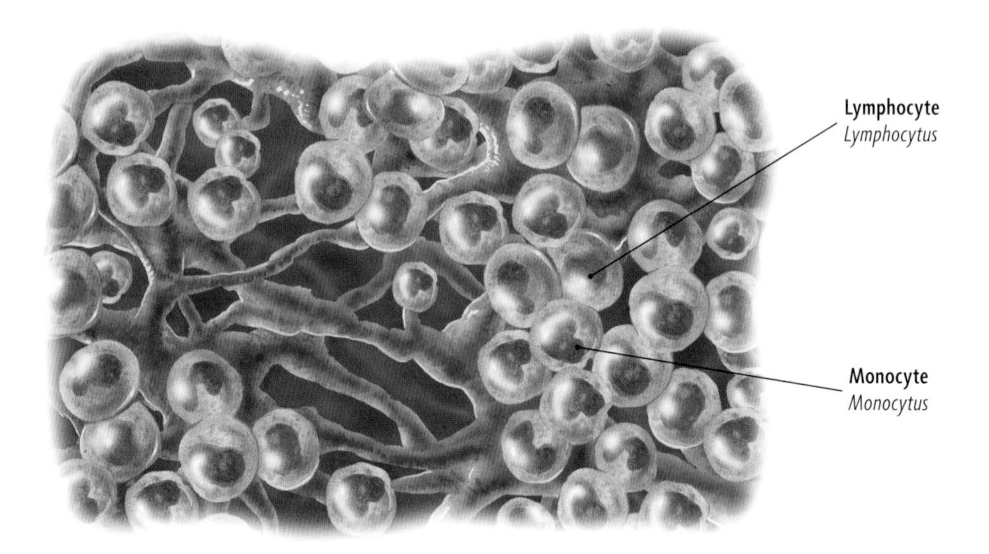
Lymphocyte
Lymphocytus
Monocyte
Monocytus

LYMPHATIC TISSUE

The spleen is the largest concentration of lymphatic tissue in the body; lymphatic tissue is also found in the lymphatic nodules of the gut and tonsils, and in the linings (mucosa) of the respiratory system, urogenital tract, and digestive tract.

LYMPHOCYTE

A lymphocyte is a type of white blood cell that plays an important role in the immune response. There are several types, including B cells, T cells, and natural killer cells.

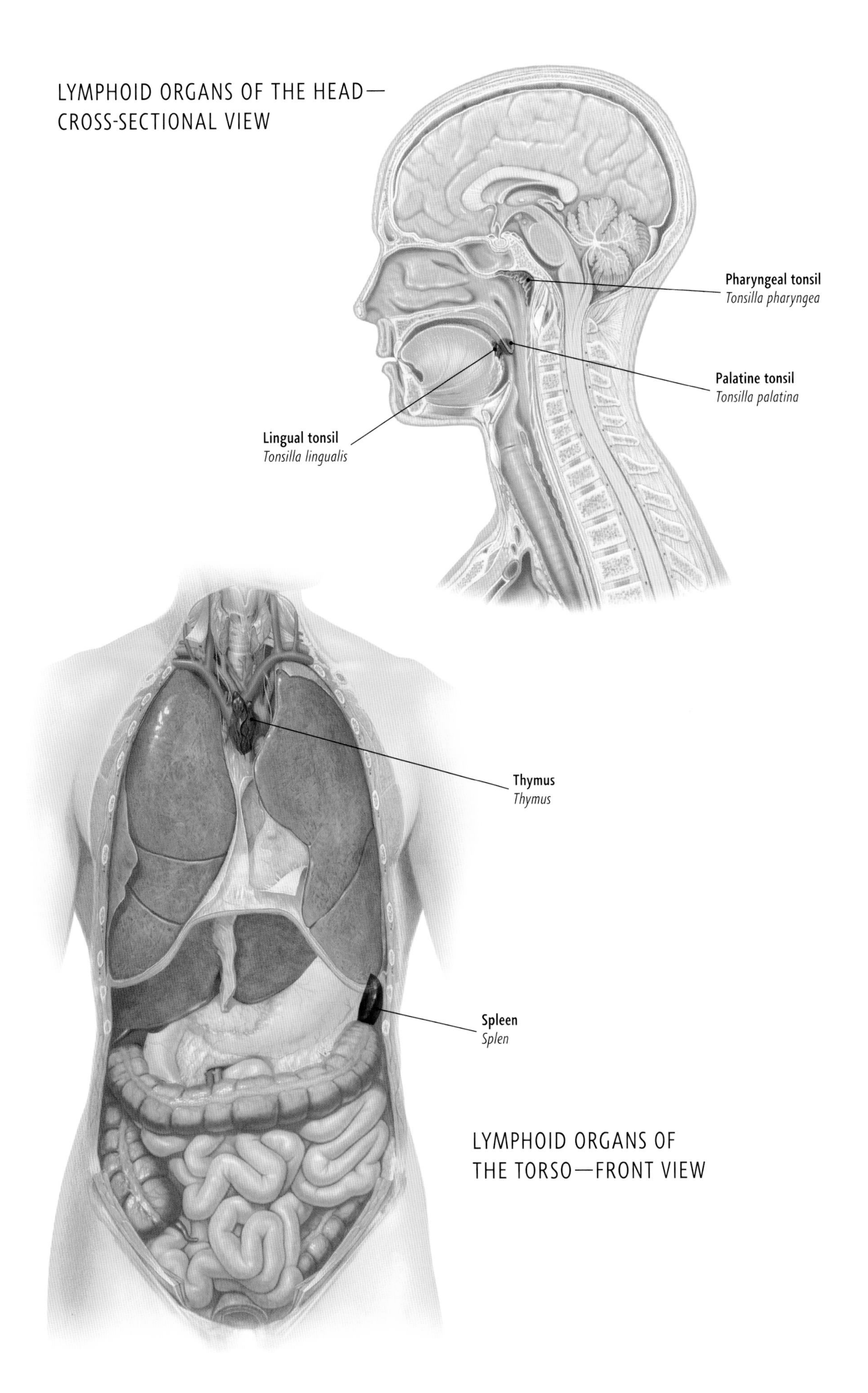
LYMPHOID ORGANS OF THE HEAD—
CROSS-SECTIONAL VIEW
Pharyngeal tonsil
Tonsilla pharyngea
Palatine tonsil
Tonsilla palatina
Lingual tonsil
Tonsilla lingualis
Thymus
Thymus
Spleen
Splen
LYMPHOID ORGANS OF
THE TORSO—FRONT VIEW

Lymphoid Organs

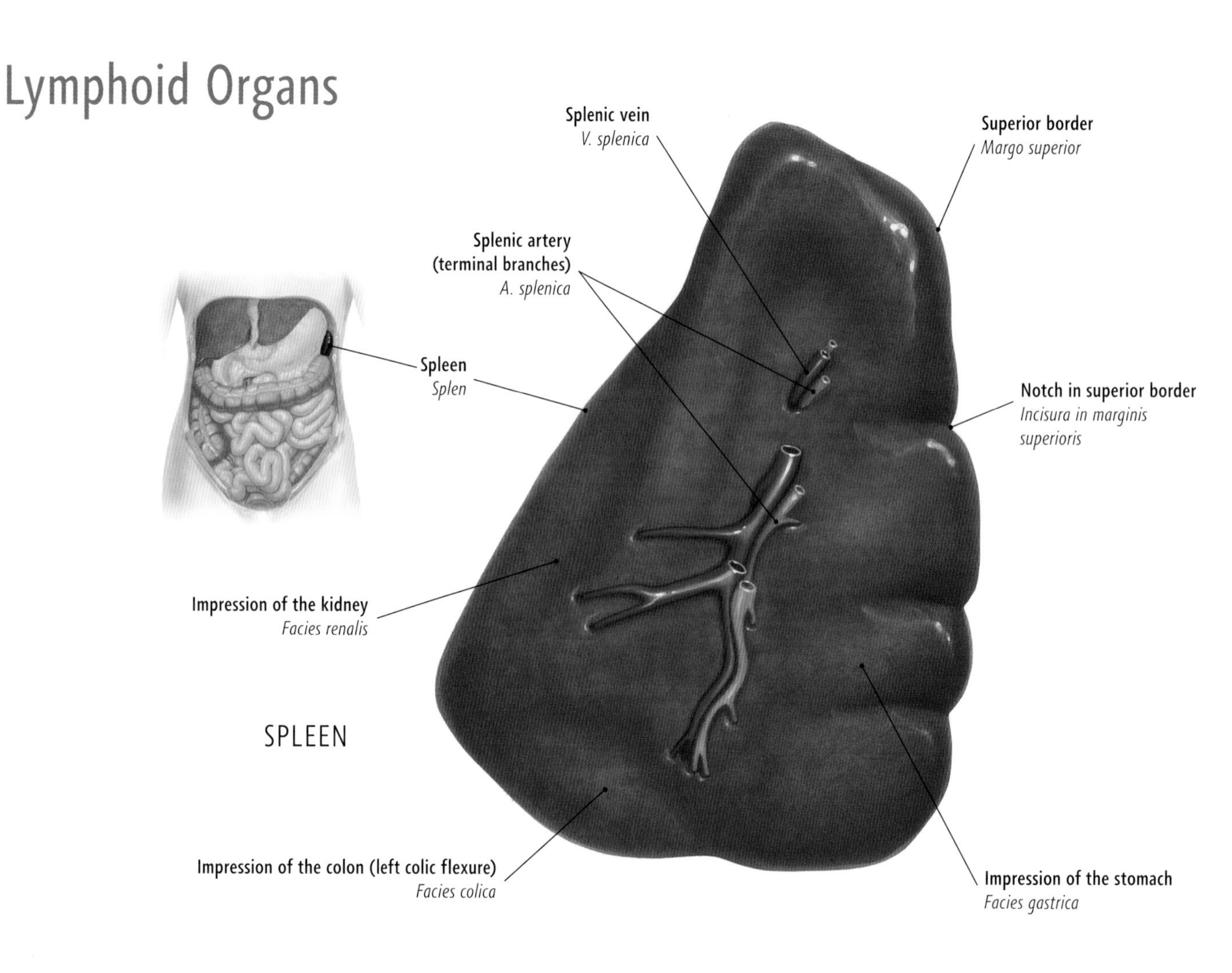
Splenic vein
V. splenica
Superior border
Margo superior
Splenic artery
(terminal branches)
A. splenica
Spleen
Splen
Notch in superior border
Incisura in marginis
superioris
Impression of the kidney
Facies renalis
Impression of the colon (left colic flexure)
Facies colica
Impression of the stomach
Facies gastrica

SPLEEN

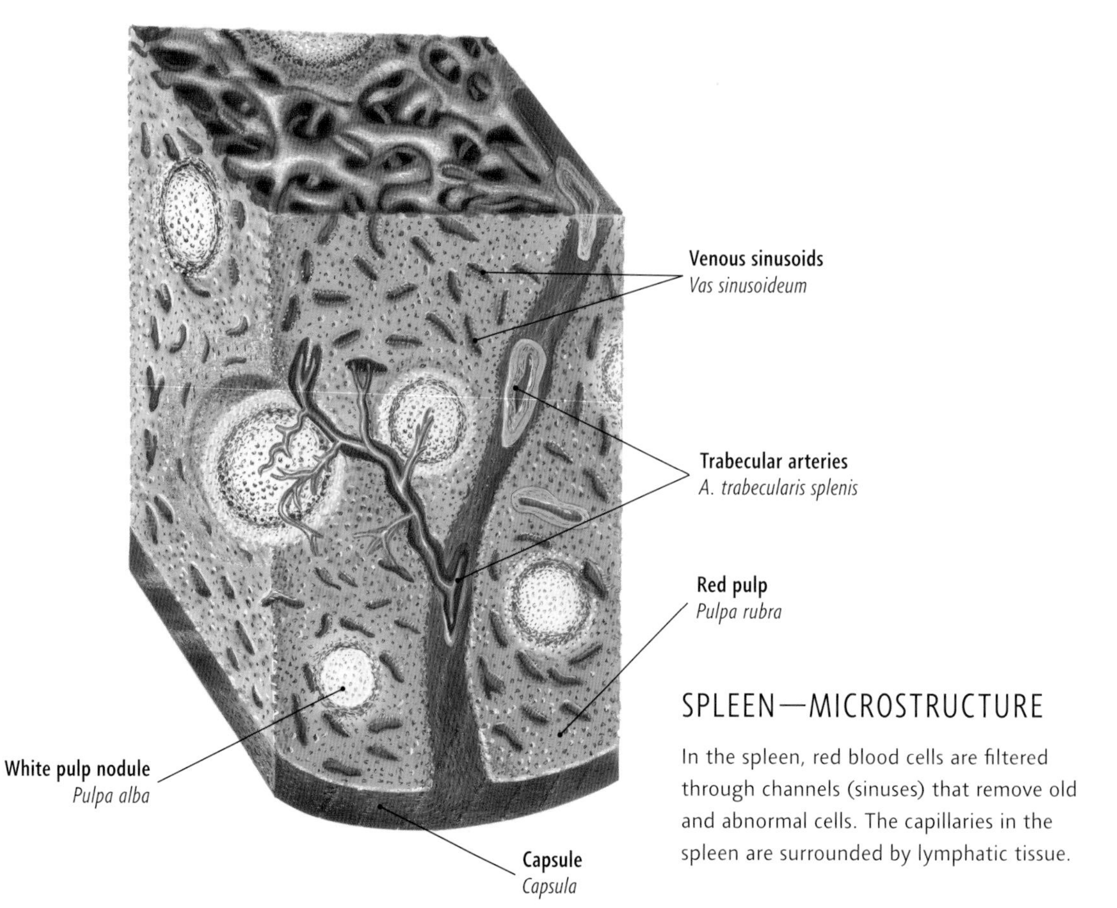
Venous sinusoids
Vas sinusoideum
Trabecular arteries
A. trabecularis splenis
Red pulp
Pulpa rubra
White pulp nodule
Pulpa alba
Capsule
Capsula

SPLEEN—MICROSTRUCTURE

In the spleen, red blood cells are filtered through channels (sinuses) that remove old and abnormal cells. The capillaries in the spleen are surrounded by lymphatic tissue.

THYMUS

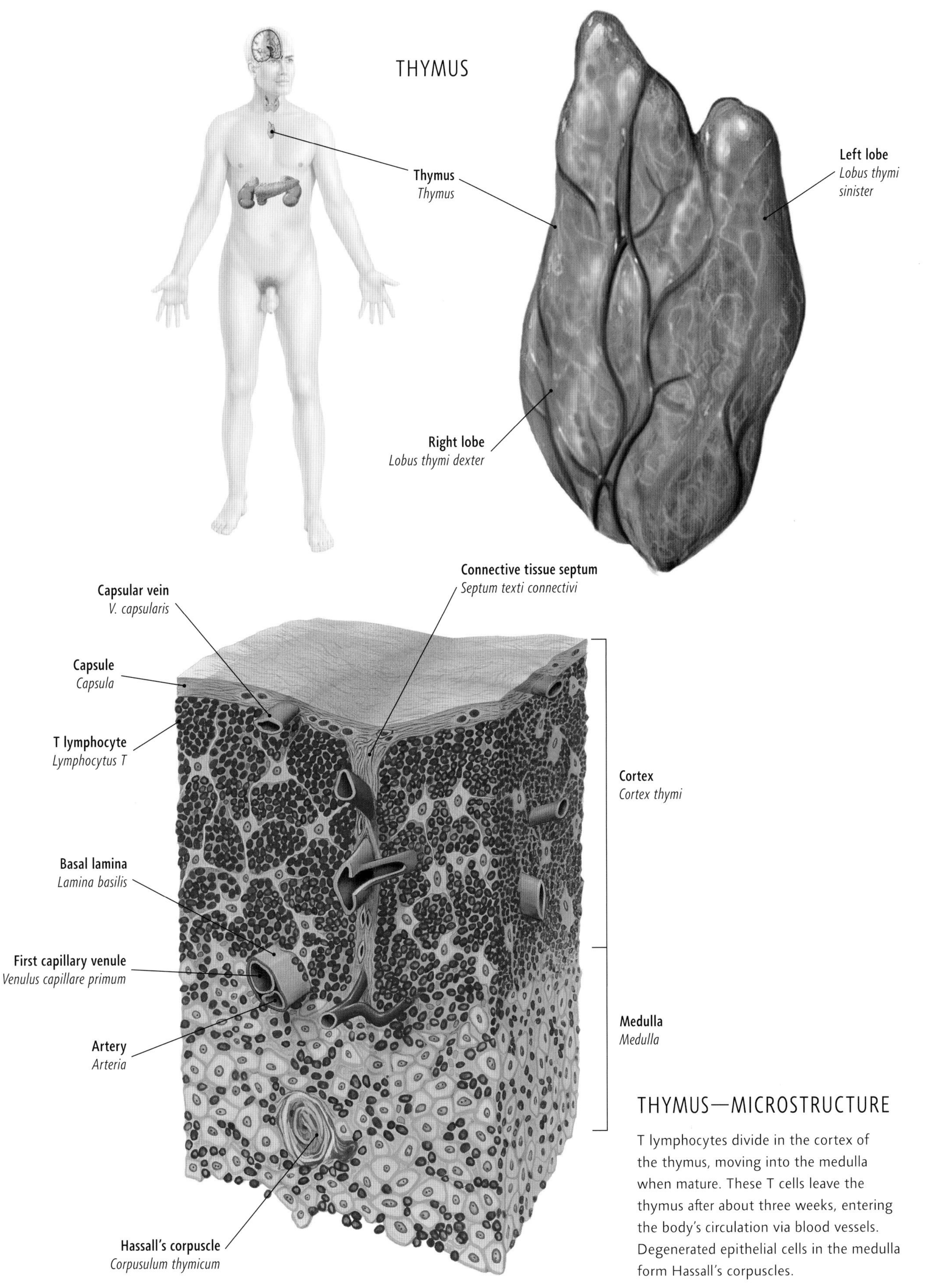
Thymus
Thymus
Left lobe
Lobus thymi sinister
Right lobe
Lobus thymi dexter
Connective tissue septum
Septum texti connectivi
Capsular vein
V. capsularis
Capsule
Capsula
T lymphocyte
Lymphocytus T
Cortex
Cortex thymi
Basal lamina
Lamina basilis
First capillary venule
Venulus capillare primum
Medulla
Medulla
Artery
Arteria
Hassall's corpuscle
Corpusulum thymicum

THYMUS—MICROSTRUCTURE

T lymphocytes divide in the cortex of the thymus, moving into the medulla when mature. These T cells leave the thymus after about three weeks, entering the body's circulation via blood vessels. Degenerated epithelial cells in the medulla form Hassall's corpuscles.

[~] = no direct Latin equivalent

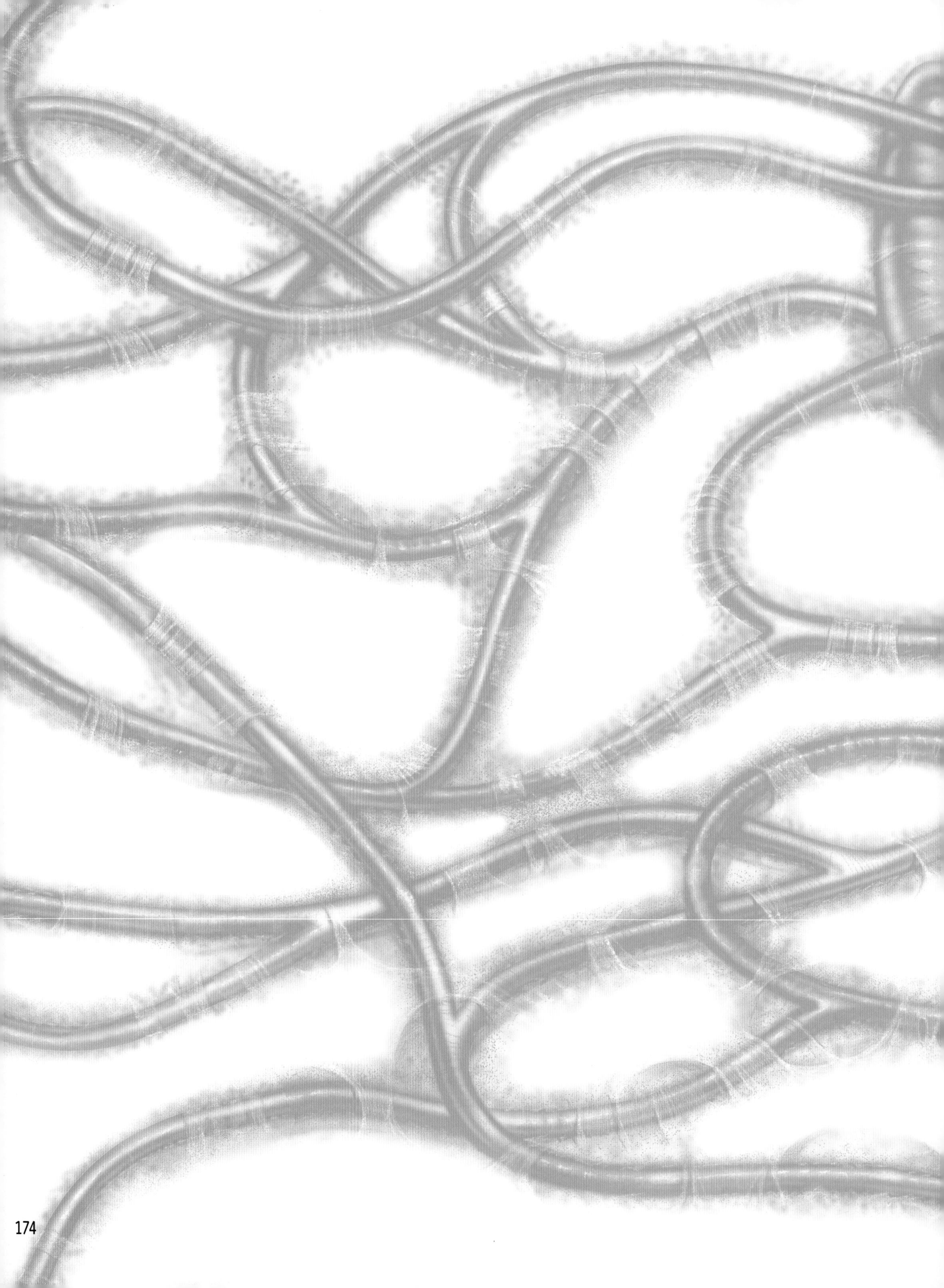

The Circulatory System

The Circulatory System

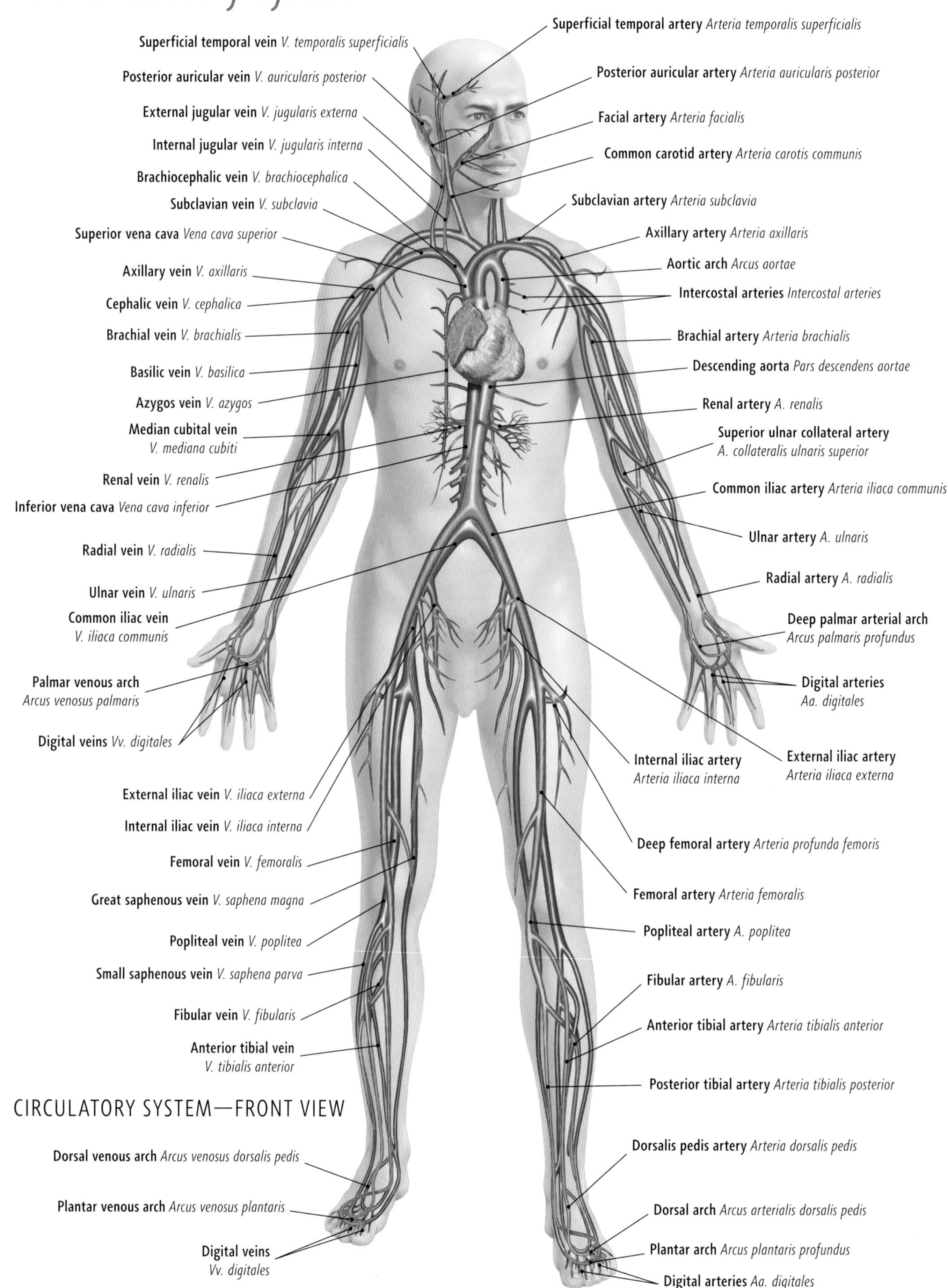
Superficial temporal vein V. temporalis superficialis
Posterior auricular vein V. auricularis posterior
External jugular vein V. jugularis externa
Internal jugular vein V. jugularis interna
Brachiocephalic vein V. brachiocephalica
Subclavian vein V. subclavia
Superior vena cava Vena cava superior
Axillary vein V. axillaris
Cephalic vein V. cephalica
Brachial vein V. brachialis
Basilic vein V. basilica
Azygos vein V. azygos
Median cubital vein V. mediana cubiti
Renal vein V. renalis
Inferior vena cava Vena cava inferior
Radial vein V. radialis
Ulnar vein V. ulnaris
Common iliac vein V. iliaca communis
Palmar venous arch Arcus venosus palmaris
Digital veins Vv. digitales
External iliac vein V. iliaca externa
Internal iliac vein V. iliaca interna
Femoral vein V. femoralis
Great saphenous vein V. saphena magna
Popliteal vein V. poplitea
Small saphenous vein V. saphena parva
Fibular vein V. fibularis
Anterior tibial vein V. tibialis anterior
Dorsal venous arch Arcus venosus dorsalis pedis
Plantar venous arch Arcus venosus plantaris
Digital veins Vv. digitales
Superficial temporal artery Arteria temporalis superficialis
Posterior auricular artery Arteria auricularis posterior
Facial artery Arteria facialis
Common carotid artery Arteria carotis communis
Subclavian artery Arteria subclavia
Axillary artery Arteria axillaris
Aortic arch Arcus aortae
Intercostal arteries Intercostal arteries
Brachial artery Arteria brachialis
Descending aorta Pars descendens aortae
Renal artery A. renalis
Superior ulnar collateral artery A. collateralis ulnaris superior
Common iliac artery Arteria iliaca communis
Ulnar artery A. ulnaris
Radial artery A. radialis
Deep palmar arterial arch Arcus palmaris profundus
Digital arteries Aa. digitales
Internal iliac artery Arteria iliaca interna
External iliac artery Arteria iliaca externa
Deep femoral artery Arteria profunda femoris
Femoral artery Arteria femoralis
Popliteal artery A. poplitea
Fibular artery A. fibularis
Anterior tibial artery Arteria tibialis anterior
Posterior tibial artery Arteria tibialis posterior
Dorsalis pedis artery Arteria dorsalis pedis
Dorsal arch Arcus arterialis dorsalis pedis
Plantar arch Arcus plantaris profundus
Digital arteries Aa. digitales

CIRCULATORY SYSTEM—FRONT VIEW

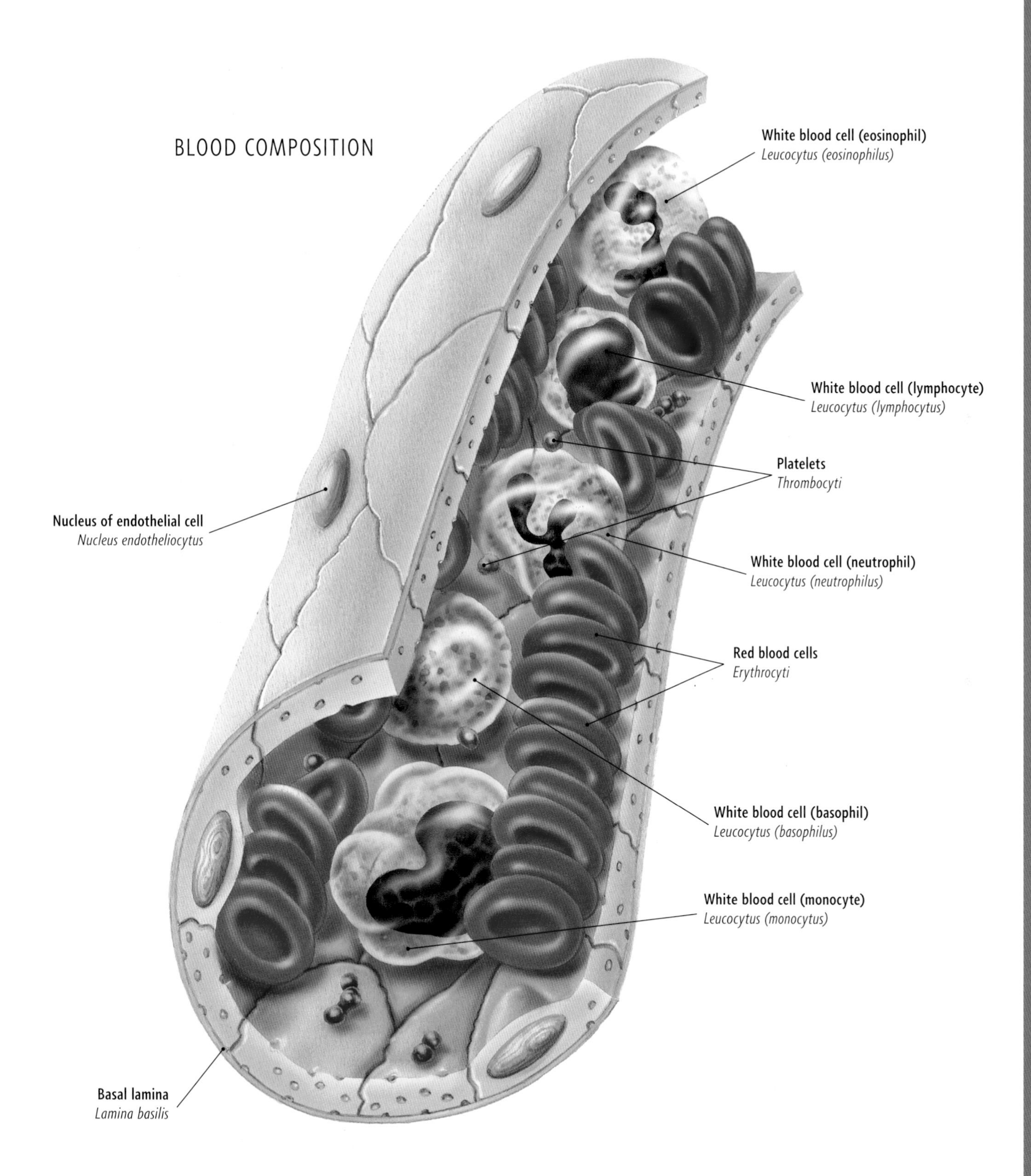
BLOOD COMPOSITION
White blood cell (eosinophil)
Leucocytus (eosinophilus)
White blood cell (lymphocyte)
Leucocytus (lymphocytus)
Platelets
Thrombocyti
Nucleus of endothelial cell
Nucleus endotheliocytus
White blood cell (neutrophil)
Leucocytus (neutrophilus)
Red blood cells
Erythrocyti
White blood cell (basophil)
Leucocytus (basophilus)
White blood cell (monocyte)
Leucocytus (monocytus)
Basal lamina
Lamina basilis

Blood Vessels

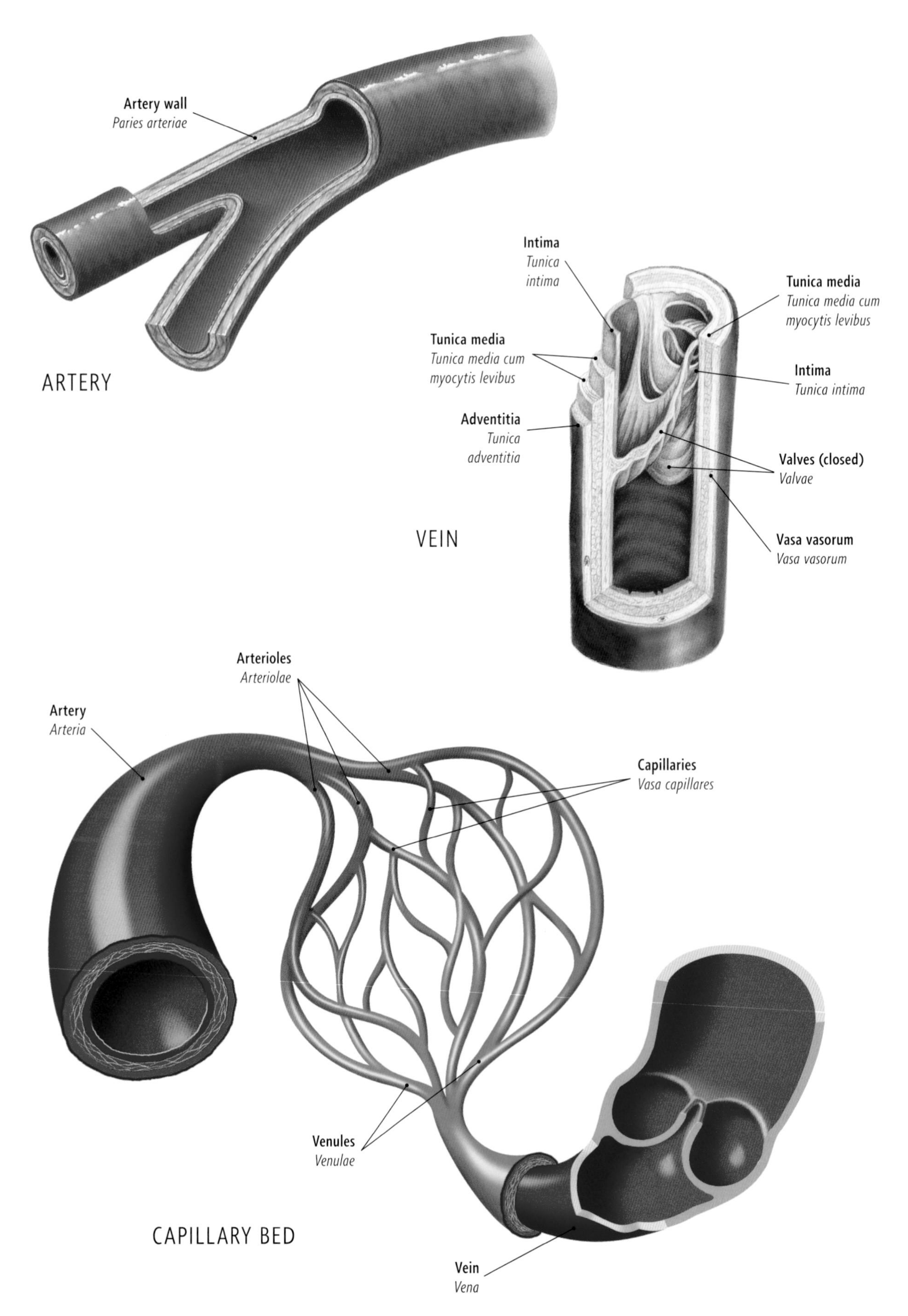
Artery wall
Paries arteriae
ARTERY
Intima
Tunica intima
Tunica media
Tunica media cum myocytis levibus
Tunica media
Tunica media cum myocytis levibus
Intima
Tunica intima
Adventitia
Tunica adventitia
Valves (closed)
Valvae
VEIN
Vasa vasorum
Vasa vasorum
Arterioles
Arteriolae
Artery
Arteria
Capillaries
Vasa capillares
Venules
Venulae
CAPILLARY BED
Vein
Vena

[~] = no direct Latin equivalent

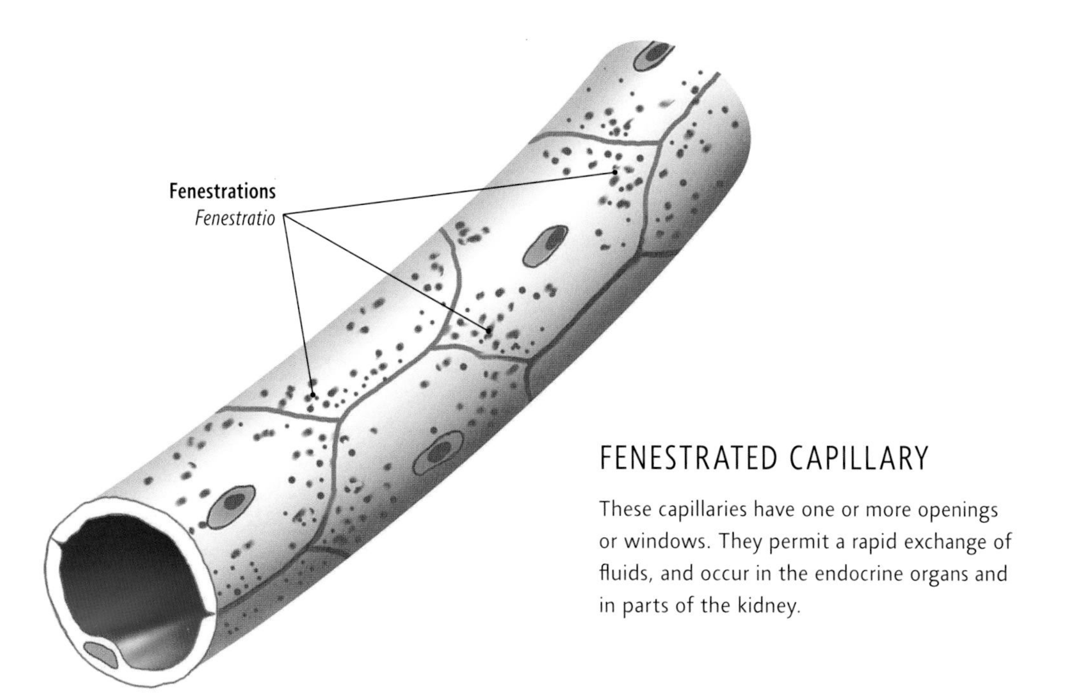
Fenestrations
Fenestratio

FENESTRATED CAPILLARY

These capillaries have one or more openings or windows. They permit a rapid exchange of fluids, and occur in the endocrine organs and in parts of the kidney.

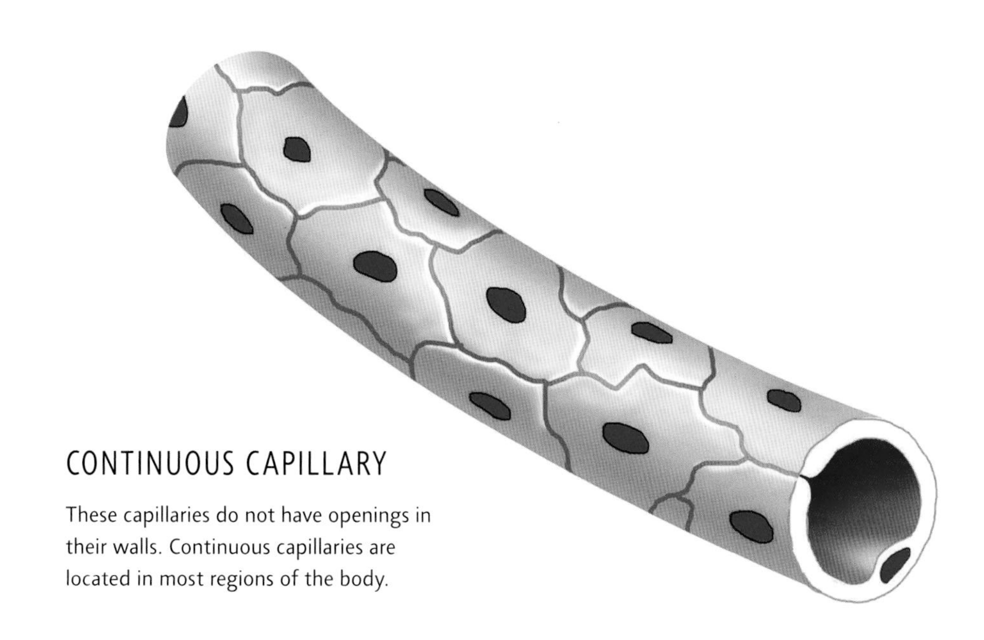

CONTINUOUS CAPILLARY

These capillaries do not have openings in their walls. Continuous capillaries are located in most regions of the body.

Major Arteries and Veins of the Body

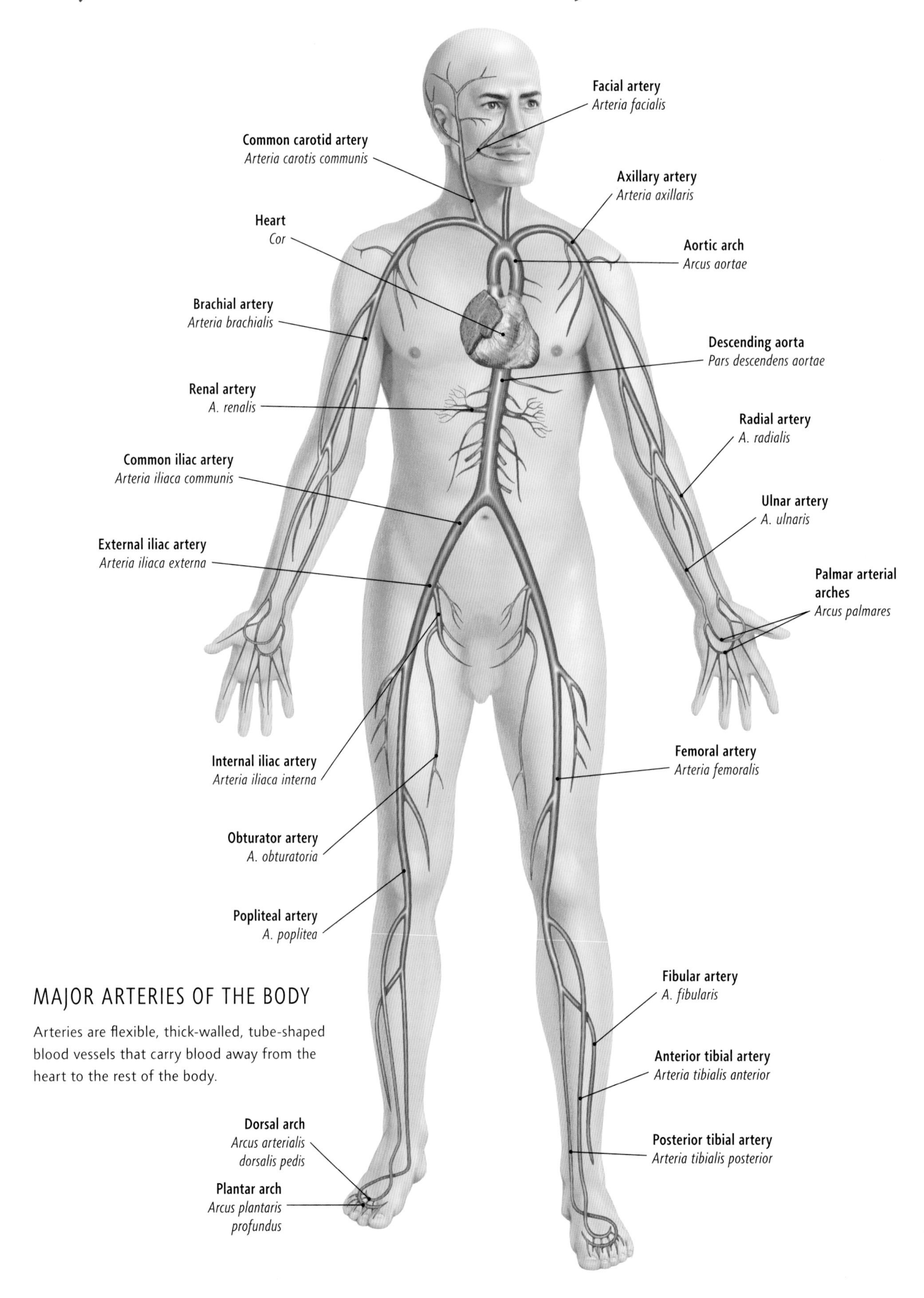
Facial artery
Arteria facialis
Common carotid artery
Arteria carotis communis
Axillary artery
Arteria axillaris
Heart
Cor
Aortic arch
Arcus aortae
Brachial artery
Arteria brachialis
Descending aorta
Pars descendens aortae
Renal artery
A. renalis
Radial artery
A. radialis
Common iliac artery
Arteria iliaca communis
Ulnar artery
A. ulnaris
External iliac artery
Arteria iliaca externa
Palmar arterial arches
Arcus palmares
Internal iliac artery
Arteria iliaca interna
Femoral artery
Arteria femoralis
Obturator artery
A. obturatoria
Popliteal artery
A. poplitea
Fibular artery
A. fibularis
Anterior tibial artery
Arteria tibialis anterior
Dorsal arch
Arcus arterialis dorsalis pedis
Posterior tibial artery
Arteria tibialis posterior
Plantar arch
Arcus plantaris profundus

MAJOR ARTERIES OF THE BODY

Arteries are flexible, thick-walled, tube-shaped blood vessels that carry blood away from the heart to the rest of the body.

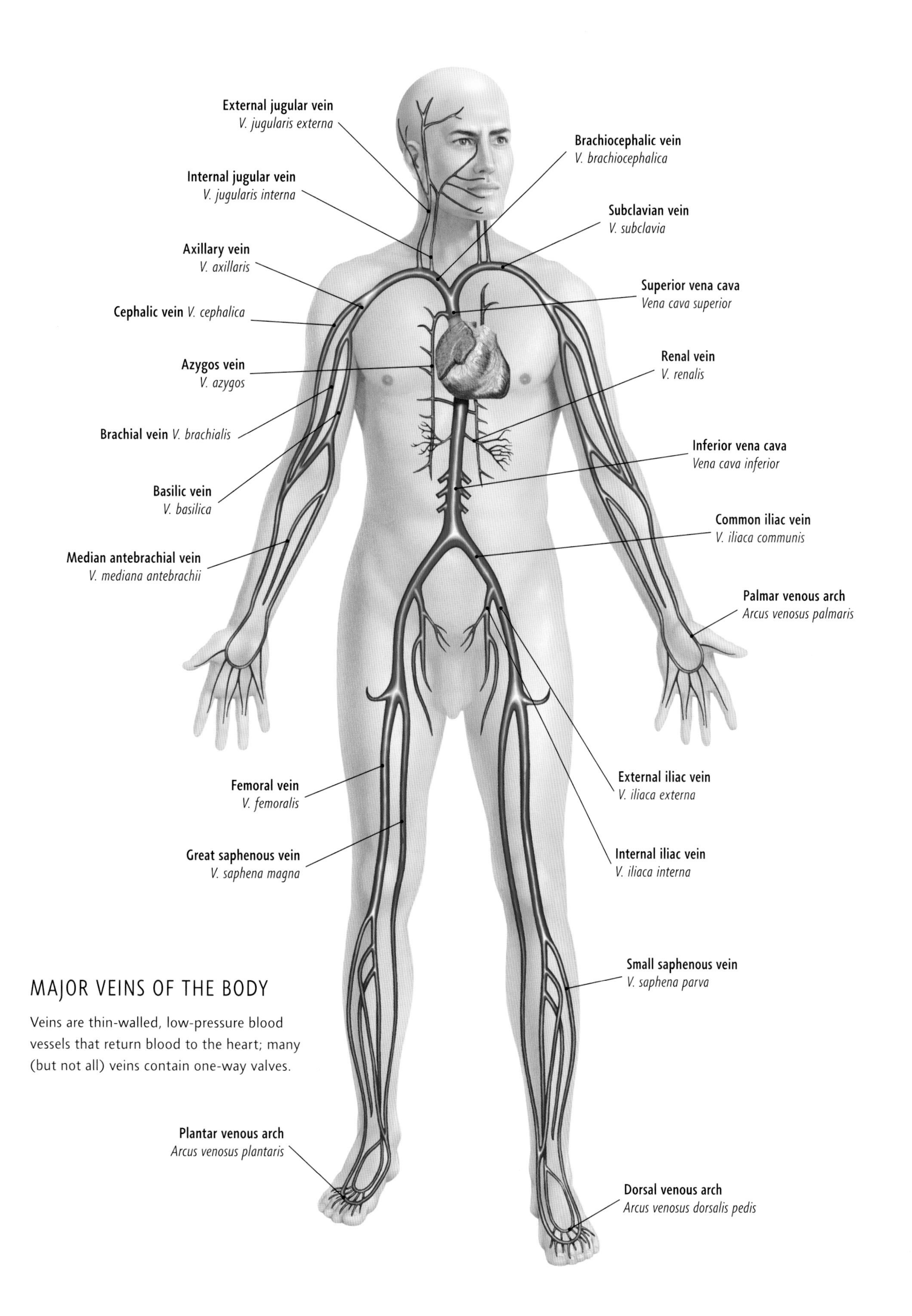
External jugular vein
V. jugularis externa
Internal jugular vein
V. jugularis interna
Axillary vein
V. axillaris
Cephalic vein V. cephalica
Azygos vein
V. azygos
Brachial vein V. brachialis
Basilic vein
V. basilica
Median antebrachial vein
V. mediana antebrachii
Femoral vein
V. femoralis
Great saphenous vein
V. saphena magna
Plantar venous arch
Arcus venosus plantaris
Brachiocephalic vein
V. brachiocephalica
Subclavian vein
V. subclavia
Superior vena cava
Vena cava superior
Renal vein
V. renalis
Inferior vena cava
Vena cava inferior
Common iliac vein
V. iliaca communis
Palmar venous arch
Arcus venosus palmaris
External iliac vein
V. iliaca externa
Internal iliac vein
V. iliaca interna
Small saphenous vein
V. saphena parva
Dorsal venous arch
Arcus venosus dorsalis pedis

MAJOR VEINS OF THE BODY

Veins are thin-walled, low-pressure blood vessels that return blood to the heart; many (but not all) veins contain one-way valves.

Arteries of the Brain

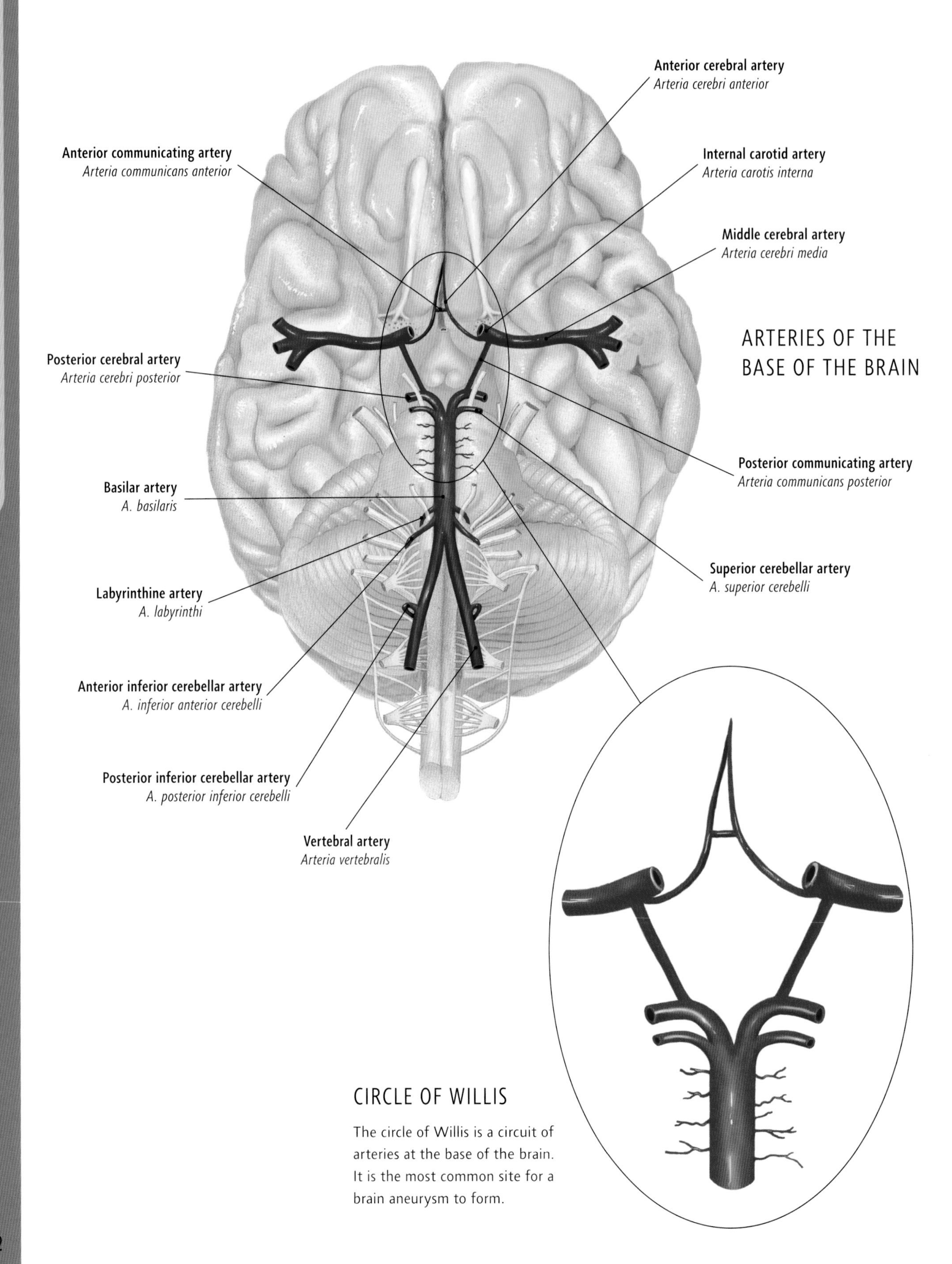
Anterior cerebral artery
Arteria cerebri anterior
Anterior communicating artery
Arteria communicans anterior
Internal carotid artery
Arteria carotis interna
Middle cerebral artery
Arteria cerebri media
ARTERIES OF THE BASE OF THE BRAIN
Posterior cerebral artery
Arteria cerebri posterior
Posterior communicating artery
Arteria communicans posterior
Basilar artery
A. basilaris
Superior cerebellar artery
A. superior cerebelli
Labyrinthine artery
A. labyrinthi
Anterior inferior cerebellar artery
A. inferior anterior cerebelli
Posterior inferior cerebellar artery
A. posterior inferior cerebelli
Vertebral artery
Arteria vertebralis

CIRCLE OF WILLIS

The circle of Willis is a circuit of arteries at the base of the brain. It is the most common site for a brain aneurysm to form.

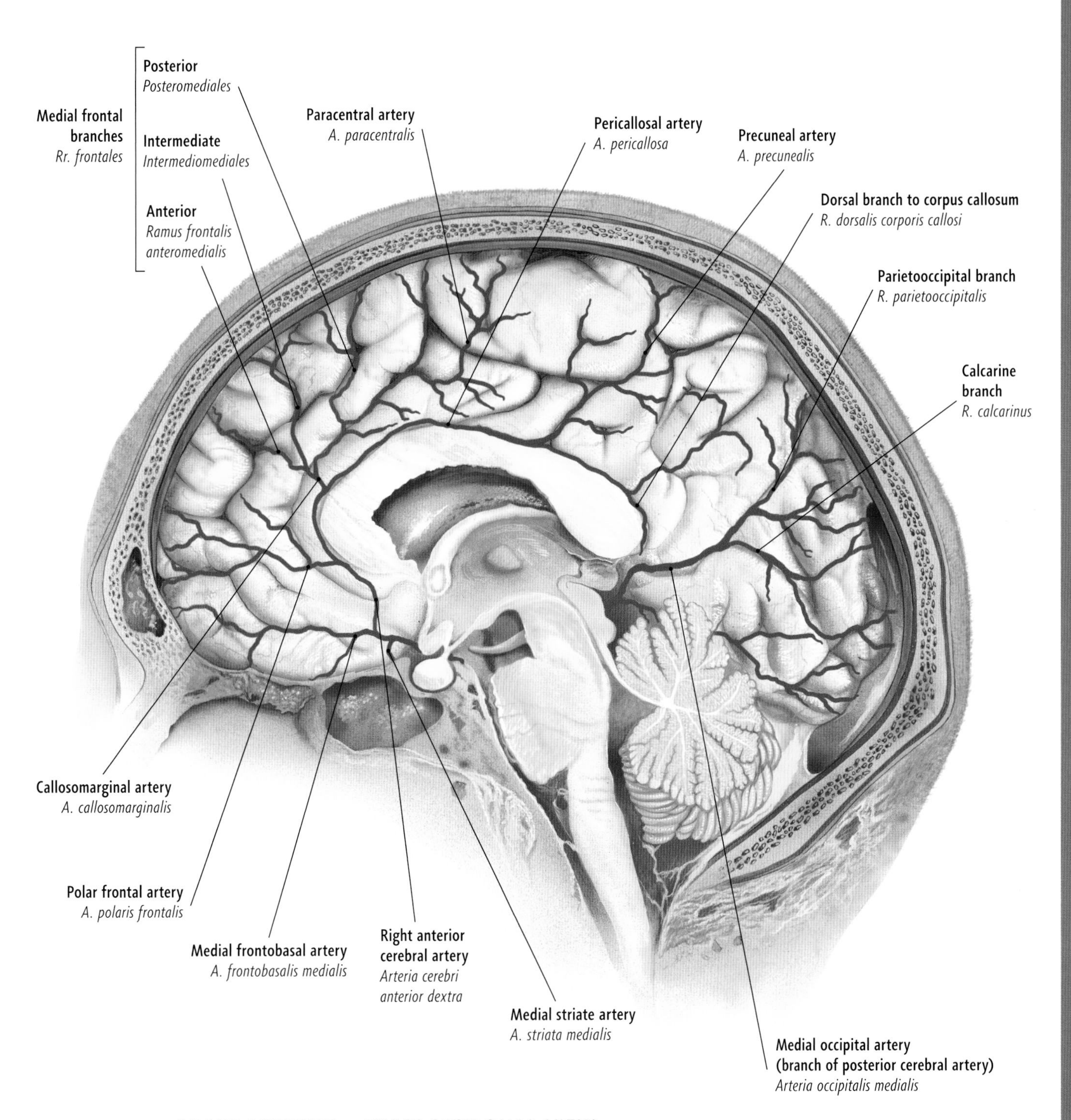
Medial frontal branches
Rr. frontales
Posterior
Posteromediales
Intermediate
Intermediomediales
Anterior
Ramus frontalis anteromedialis
Paracentral artery
A. paracentralis
Pericallosal artery
A. pericallosa
Precuneal artery
A. precunealis
Dorsal branch to corpus callosum
R. dorsalis corporis callosi
Parietooccipital branch
R. parietooccipitalis
Calcarine branch
R. calcarinus
Callosomarginal artery
A. callosomarginalis
Polar frontal artery
A. polaris frontalis
Medial frontobasal artery
A. frontobasalis medialis
Right anterior cerebral artery
Arteria cerebri anterior dextra
Medial striate artery
A. striata medialis
Medial occipital artery
(branch of posterior cerebral artery)
Arteria occipitalis medialis

BRAIN ARTERIES—CROSS-SECTIONAL VIEW

Blood Vessels of the Head and Neck

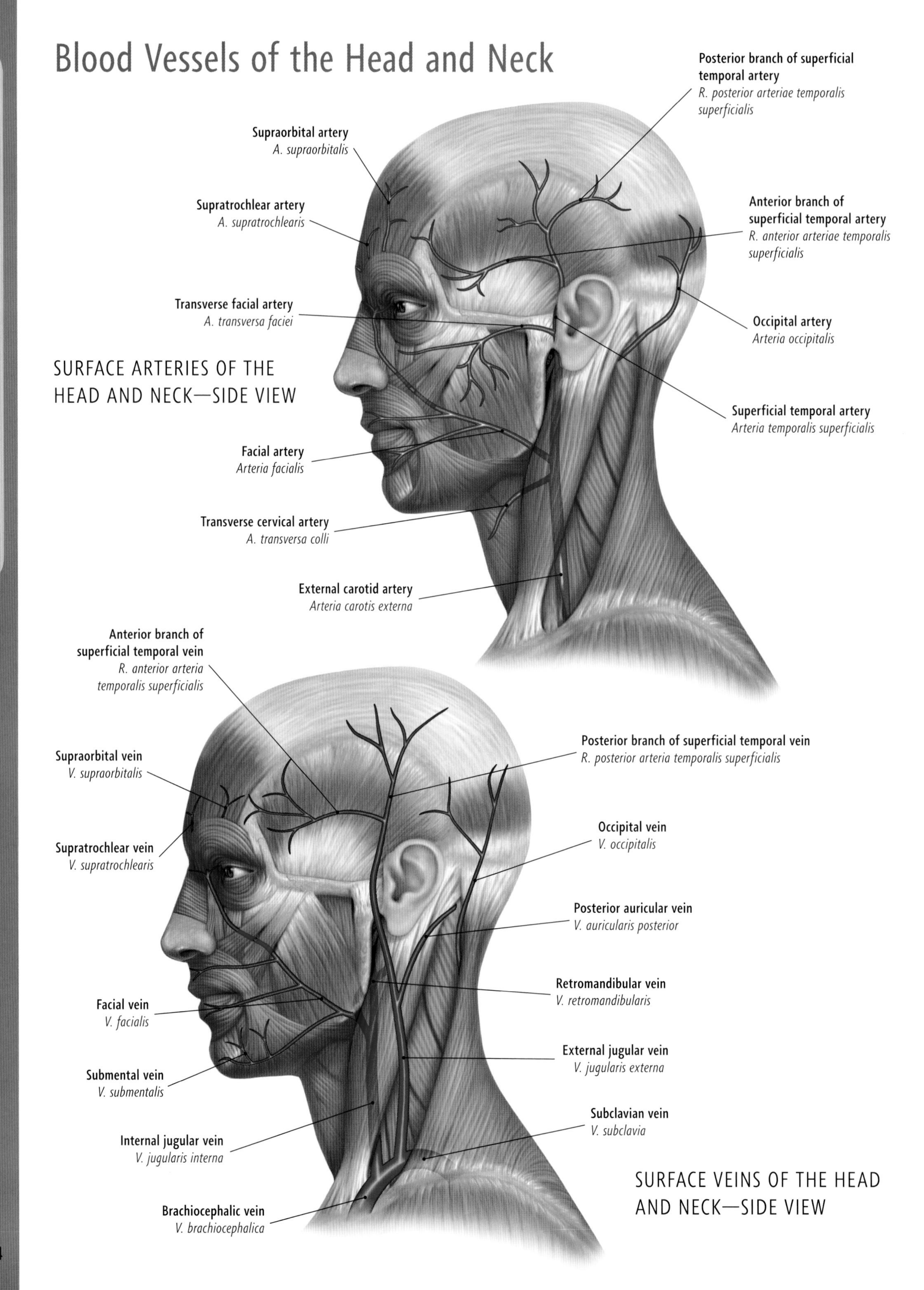
Posterior branch of superficial temporal artery
R. posterior arteriae temporalis superficialis
Supraorbital artery
A. supraorbitalis
Supratrochlear artery
A. supratrochlearis
Anterior branch of superficial temporal artery
R. anterior arteriae temporalis superficialis
Transverse facial artery
A. transversa faciei
Occipital artery
Arteria occipitalis
SURFACE ARTERIES OF THE HEAD AND NECK—SIDE VIEW
Superficial temporal artery
Arteria temporalis superficialis
Facial artery
Arteria facialis
Transverse cervical artery
A. transversa colli
External carotid artery
Arteria carotis externa
Anterior branch of superficial temporal vein
R. anterior arteria temporalis superficialis
Posterior branch of superficial temporal vein
R. posterior arteria temporalis superficialis
Supraorbital vein
V. supraorbitalis
Occipital vein
V. occipitalis
Supratrochlear vein
V. supratrochlearis
Posterior auricular vein
V. auricularis posterior
Retromandibular vein
V. retromandibularis
Facial vein
V. facialis
External jugular vein
V. jugularis externa
Submental vein
V. submentalis
Subclavian vein
V. subclavia
Internal jugular vein
V. jugularis interna
SURFACE VEINS OF THE HEAD AND NECK—SIDE VIEW
Brachiocephalic vein
V. brachiocephalica

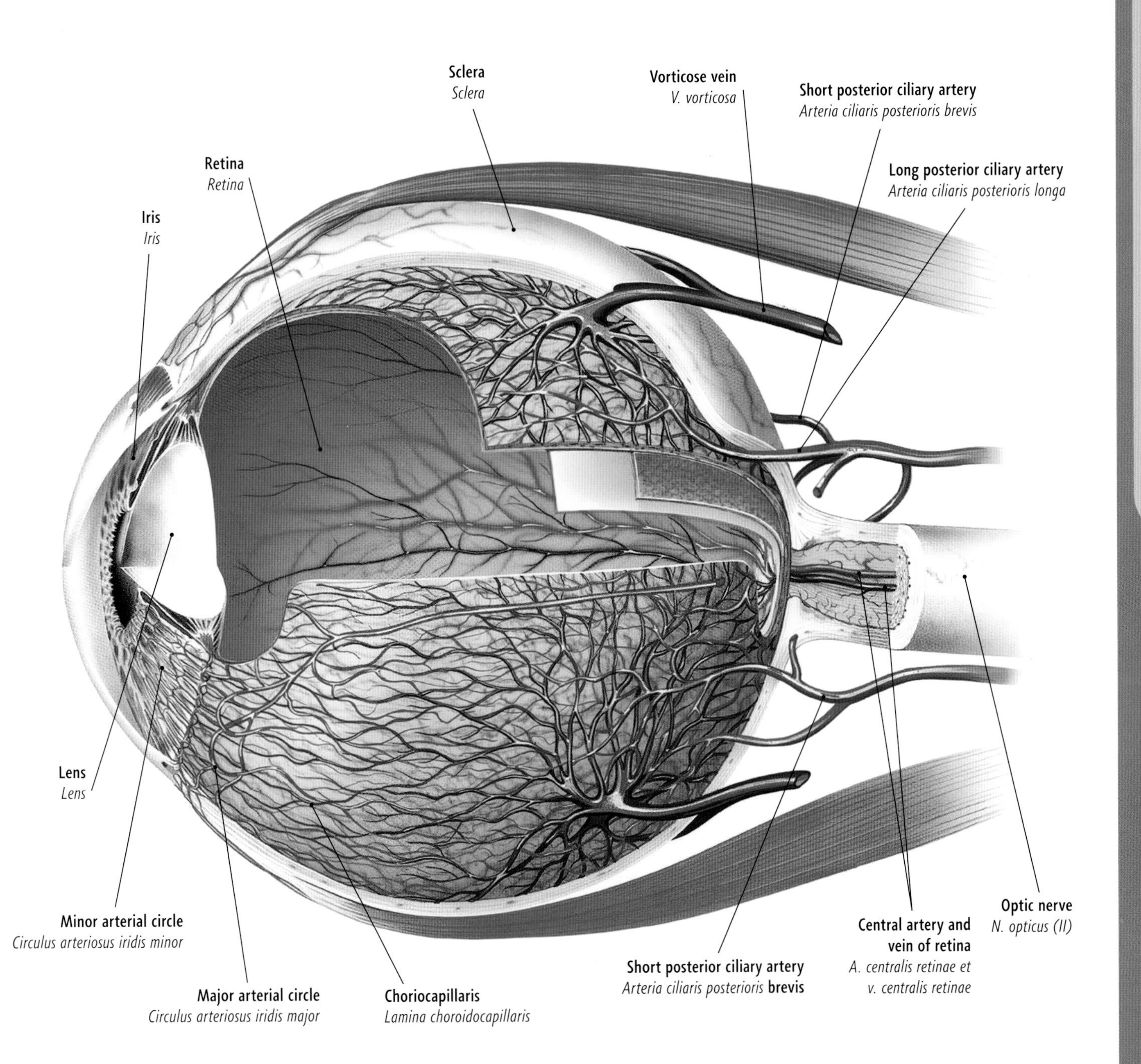
Sclera
Sclera
Vorticose vein
V. vorticosa
Short posterior ciliary artery
Arteria ciliaris posterioris brevis
Retina
Retina
Long posterior ciliary artery
Arteria ciliaris posterioris longa
Iris
Iris
Lens
Lens
Minor arterial circle
Circulus arteriosus iridis minor
Major arterial circle
Circulus arteriosus iridis major
Choriocapillaris
Lamina choroidocapillaris
Short posterior ciliary artery
Arteria ciliaris posterioris brevis
Central artery and
vein of retina
A. centralis retinae et
v. centralis retinae
Optic nerve
N. opticus (II)

SURFACE ARTERIES OF THE EYE—SIDE VIEW

The Heart

FRONT VIEW

REAR VIEW

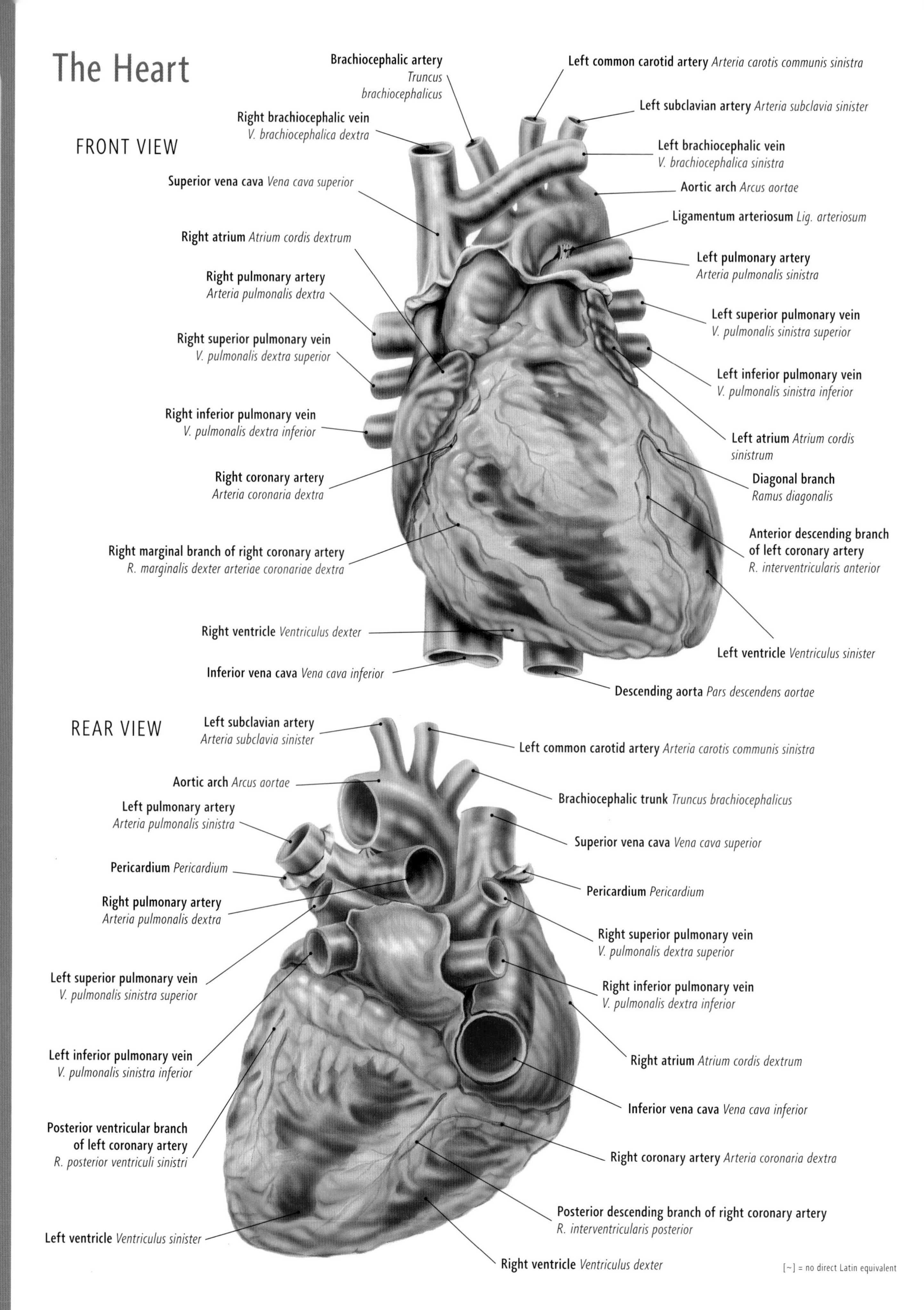

Brachiocephalic artery
Truncus brachiocephalicus
Left common carotid artery Arteria carotis communis sinistra
Left subclavian artery Arteria subclavia sinister
Right brachiocephalic vein
V. brachiocephalica dextra
Left brachiocephalic vein
V. brachiocephalica sinistra
Superior vena cava Vena cava superior
Aortic arch Arcus aortae
Ligamentum arteriosum Lig. arteriosum
Right atrium Atrium cordis dextrum
Left pulmonary artery
Arteria pulmonalis sinistra
Right pulmonary artery
Arteria pulmonalis dextra
Left superior pulmonary vein
V. pulmonalis sinistra superior
Right superior pulmonary vein
V. pulmonalis dextra superior
Left inferior pulmonary vein
V. pulmonalis sinistra inferior
Right inferior pulmonary vein
V. pulmonalis dextra inferior
Left atrium Atrium cordis sinistrum
Right coronary artery
Arteria coronaria dextra
Diagonal branch
Ramus diagonalis
Anterior descending branch of left coronary artery
R. interventricularis anterior
Right marginal branch of right coronary artery
R. marginalis dexter arteriae coronariae dextra
Right ventricle Ventriculus dexter
Left ventricle Ventriculus sinister
Inferior vena cava Vena cava inferior
Descending aorta Pars descendens aortae
Left subclavian artery
Arteria subclavia sinister
Left common carotid artery Arteria carotis communis sinistra
Aortic arch Arcus aortae
Brachiocephalic trunk Truncus brachiocephalicus
Left pulmonary artery
Arteria pulmonalis sinistra
Superior vena cava Vena cava superior
Pericardium Pericardium
Pericardium Pericardium
Right pulmonary artery
Arteria pulmonalis dextra
Right superior pulmonary vein
V. pulmonalis dextra superior
Left superior pulmonary vein
V. pulmonalis sinistra superior
Right inferior pulmonary vein
V. pulmonalis dextra inferior
Left inferior pulmonary vein
V. pulmonalis sinistra inferior
Right atrium Atrium cordis dextrum
Inferior vena cava Vena cava inferior
Posterior ventricular branch of left coronary artery
R. posterior ventriculi sinistri
Right coronary artery Arteria coronaria dextra
Posterior descending branch of right coronary artery
R. interventricularis posterior
Left ventricle Ventriculus sinister
Right ventricle Ventriculus dexter

[~] = no direct Latin equivalent

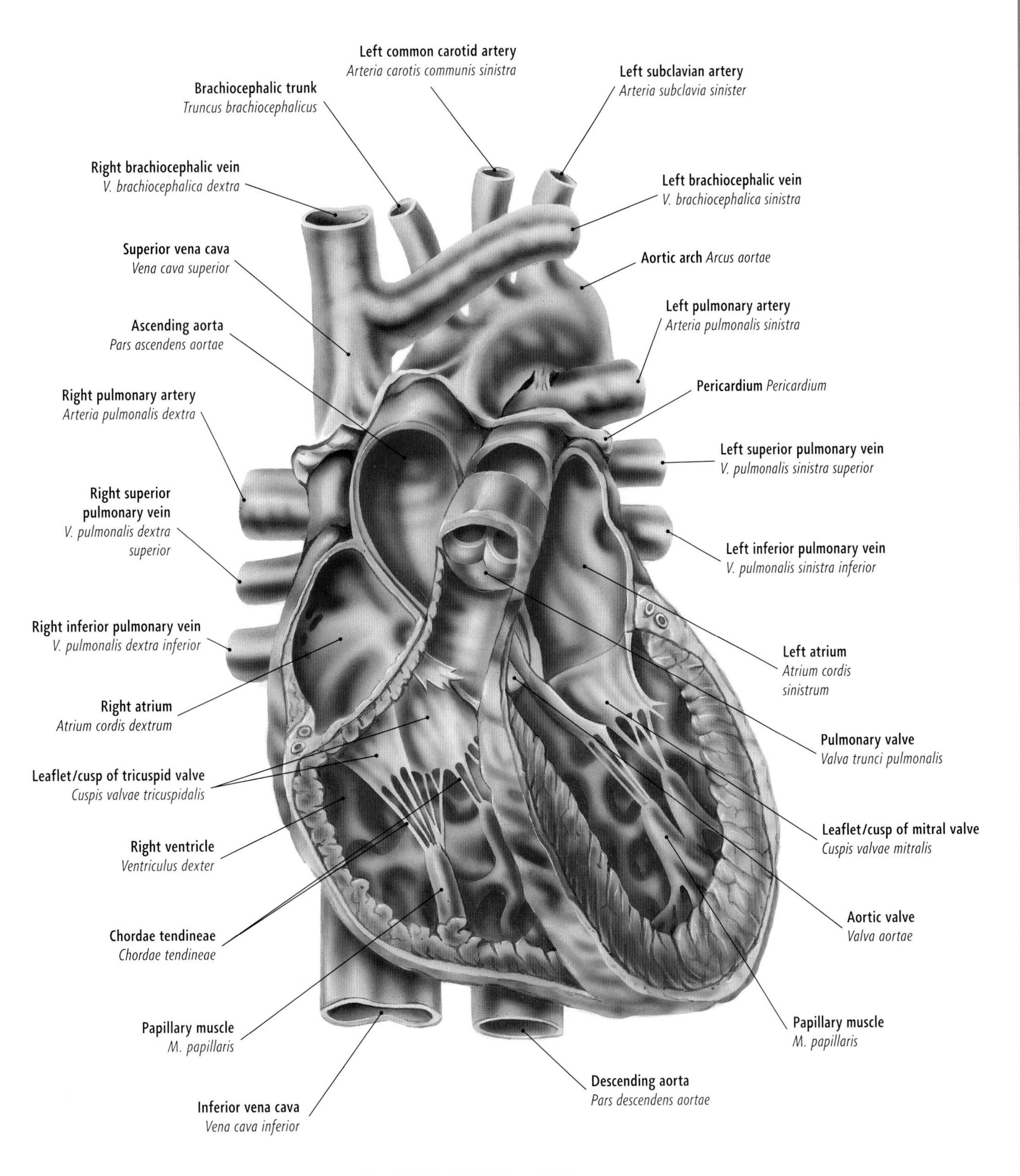

Left common carotid artery
Arteria carotis communis sinistra
Left subclavian artery
Arteria subclavia sinister
Brachiocephalic trunk
Truncus brachiocephalicus
Right brachiocephalic vein
V. brachiocephalica dextra
Left brachiocephalic vein
V. brachiocephalica sinistra
Superior vena cava
Vena cava superior
Aortic arch Arcus aortae
Ascending aorta
Pars ascendens aortae
Left pulmonary artery
Arteria pulmonalis sinistra
Pericardium Pericardium
Right pulmonary artery
Arteria pulmonalis dextra
Left superior pulmonary vein
V. pulmonalis sinistra superior
Right superior pulmonary vein
V. pulmonalis dextra superior
Left inferior pulmonary vein
V. pulmonalis sinistra inferior
Right inferior pulmonary vein
V. pulmonalis dextra inferior
Left atrium
Atrium cordis sinistrum
Right atrium
Atrium cordis dextrum
Pulmonary valve
Valva trunci pulmonalis
Leaflet/cusp of tricuspid valve
Cuspis valvae tricuspidalis
Leaflet/cusp of mitral valve
Cuspis valvae mitralis
Right ventricle
Ventriculus dexter
Aortic valve
Valva aortae
Chordae tendineae
Chordae tendineae
Papillary muscle
M. papillaris
Papillary muscle
M. papillaris
Descending aorta
Pars descendens aortae
Inferior vena cava
Vena cava inferior

CROSS-SECTIONAL VIEW

Heart Valves

HEART VALVES: VENTRICULAR SYSTOLE

The ventricles of the heart contract, pushing oxygen-rich blood into the aorta (for circulation around the body) and oxygen-poor blood into the pulmonary artery (to be sent to the lungs).

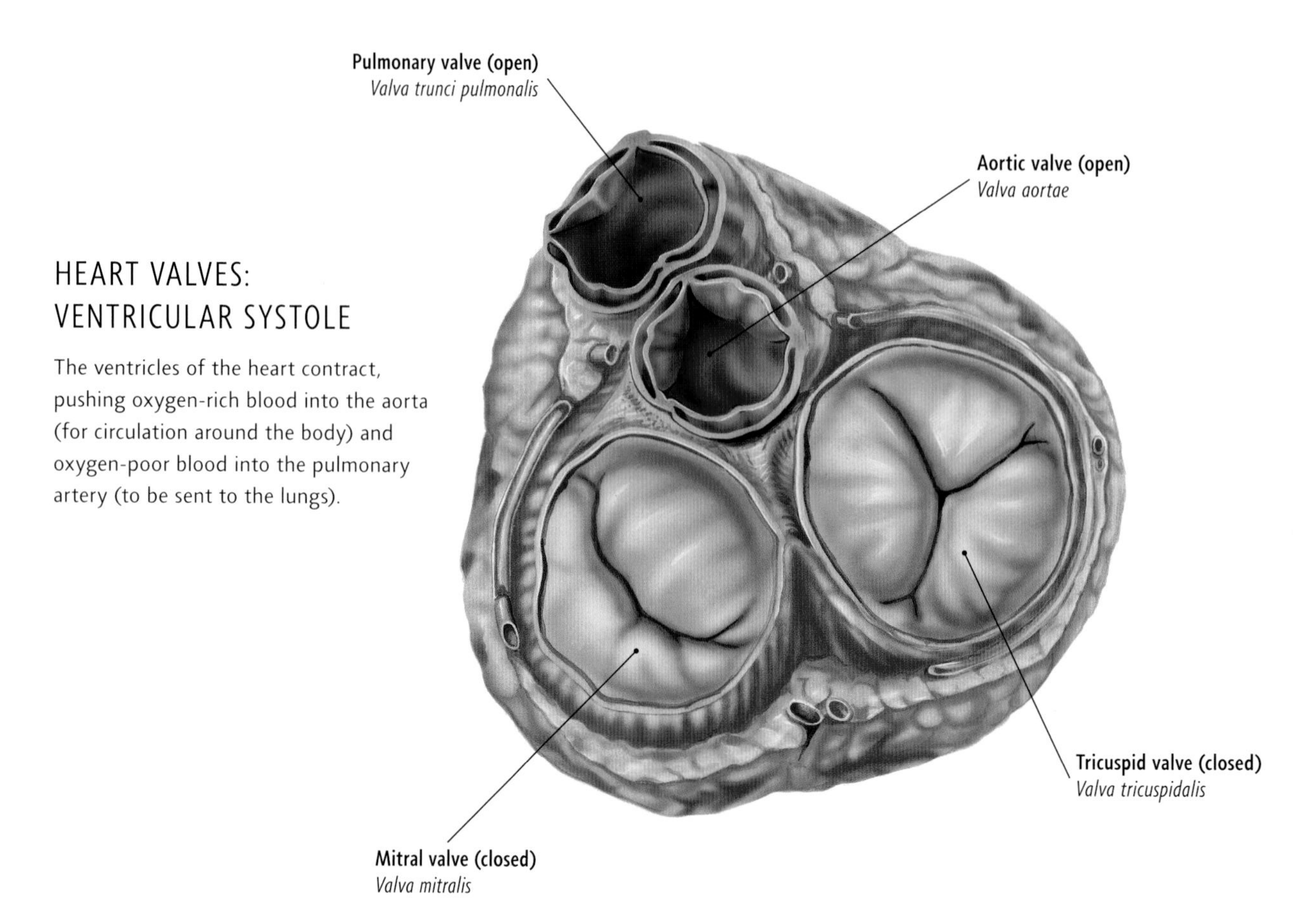
Pulmonary valve (open)
Valva trunci pulmonalis
Aortic valve (open)
Valva aortae
Tricuspid valve (closed)
Valva tricuspidalis
Mitral valve (closed)
Valva mitralis

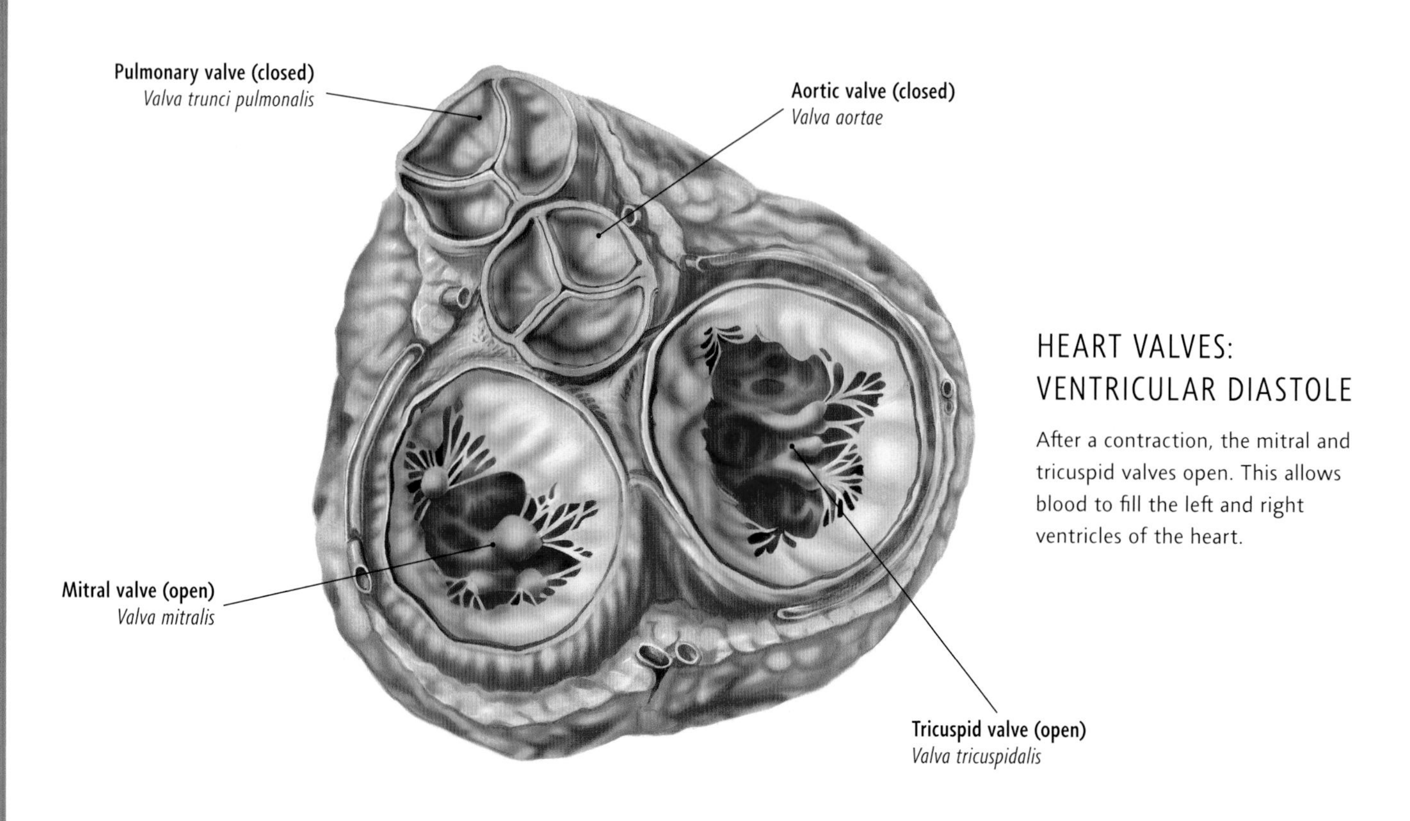
Pulmonary valve (closed)
Valva trunci pulmonalis
Aortic valve (closed)
Valva aortae
Mitral valve (open)
Valva mitralis
Tricuspid valve (open)
Valva tricuspidalis

HEART VALVES: VENTRICULAR DIASTOLE

After a contraction, the mitral and tricuspid valves open. This allows blood to fill the left and right ventricles of the heart.

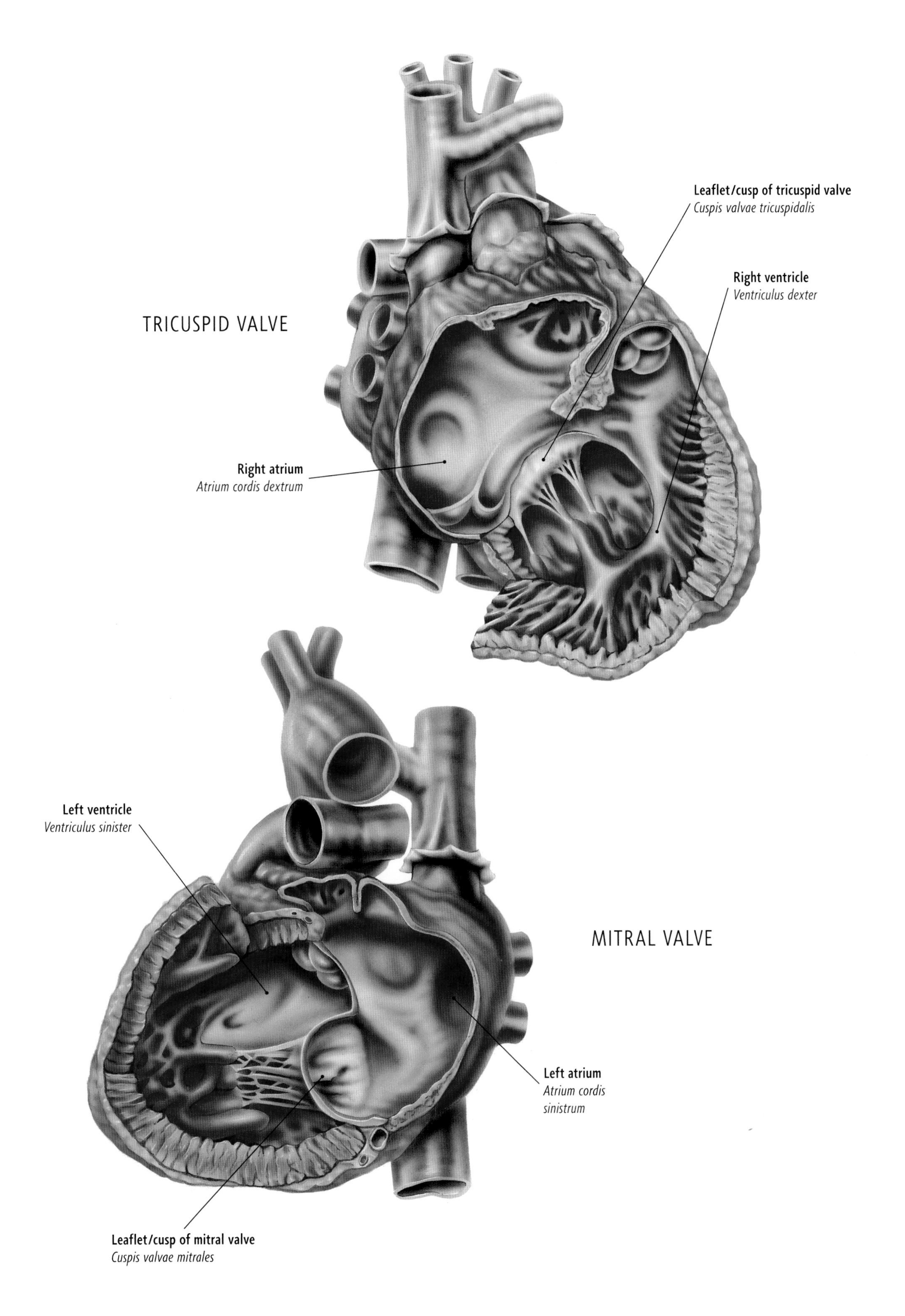
TRICUSPID VALVE
Leaflet/cusp of tricuspid valve
Cuspis valvae tricuspidalis
Right ventricle
Ventriculus dexter
Right atrium
Atrium cordis dextrum
MITRAL VALVE
Left ventricle
Ventriculus sinister
Left atrium
Atrium cordis
sinistrum
Leaflet/cusp of mitral valve
Cuspis valvae mitrales

Heart Cycle

HEART CYCLE 1

Oxygen-poor blood (shown in blue) flows from the body into the right atrium. Oxygen-rich blood (shown in red) enters the left atrium from the lungs.

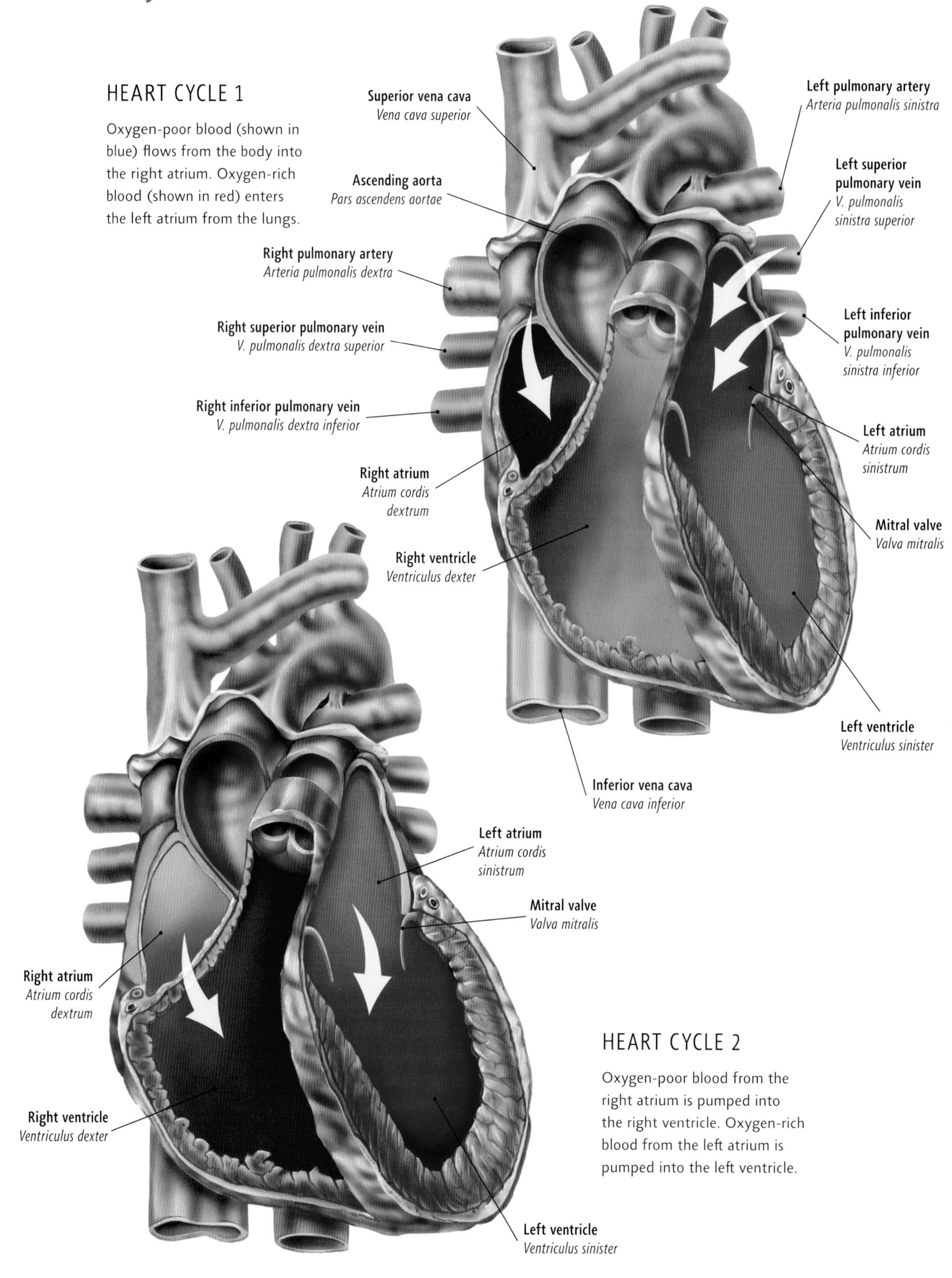
Superior vena cava
Vena cava superior
Left pulmonary artery
Arteria pulmonalis sinistra
Ascending aorta
Pars ascendens aortae
Left superior pulmonary vein
V. pulmonalis sinistra superior
Right pulmonary artery
Arteria pulmonalis dextra
Right superior pulmonary vein
V. pulmonalis dextra superior
Left inferior pulmonary vein
V. pulmonalis sinistra inferior
Right inferior pulmonary vein
V. pulmonalis dextra inferior
Left atrium
Atrium cordis sinistrum
Right atrium
Atrium cordis dextrum
Mitral valve
Valva mitralis
Right ventricle
Ventriculus dexter
Left ventricle
Ventriculus sinister
Inferior vena cava
Vena cava inferior
Left atrium
Atrium cordis sinistrum
Mitral valve
Valva mitralis
Right atrium
Atrium cordis dextrum
Right ventricle
Ventriculus dexter
Left ventricle
Ventriculus sinister

HEART CYCLE 2

Oxygen-poor blood from the right atrium is pumped into the right ventricle. Oxygen-rich blood from the left atrium is pumped into the left ventricle.

HEART CYCLE 3

The left ventricle pumps oxygen-rich blood through the ascending aorta to all parts of the body. The right ventricle pumps oxygen-poor blood to the lungs through the pulmonary arteries.

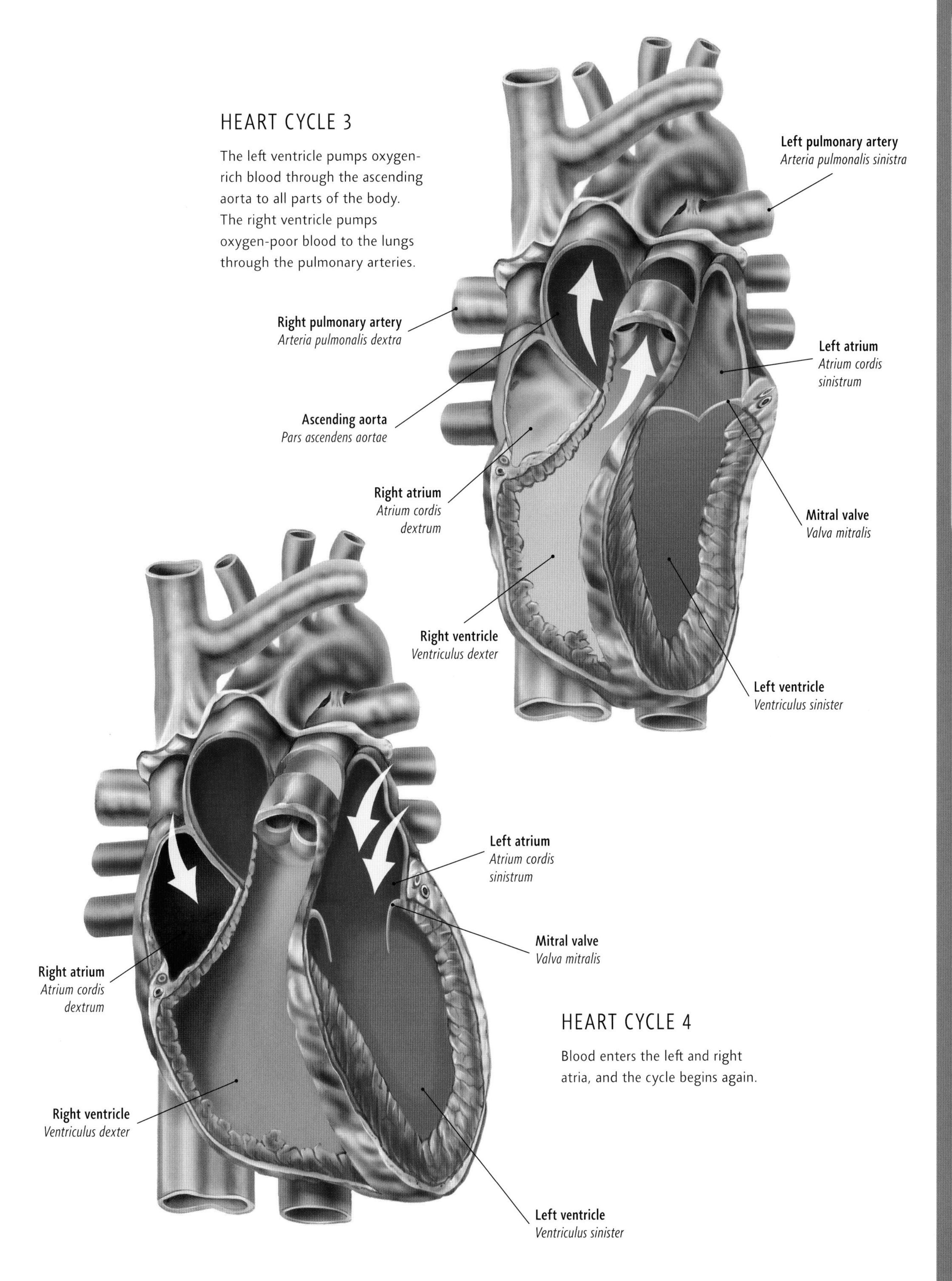

Left pulmonary artery
Arteria pulmonalis sinistra
Right pulmonary artery
Arteria pulmonalis dextra
Left atrium
Atrium cordis sinistrum
Ascending aorta
Pars ascendens aortae
Right atrium
Atrium cordis dextrum
Mitral valve
Valva mitralis
Right ventricle
Ventriculus dexter
Left ventricle
Ventriculus sinister
Left atrium
Atrium cordis sinistrum
Mitral valve
Valva mitralis
Right atrium
Atrium cordis dextrum
Right ventricle
Ventriculus dexter
Left ventricle
Ventriculus sinister

HEART CYCLE 4

Blood enters the left and right atria, and the cycle begins again.

Blood Vessels of the Abdomen

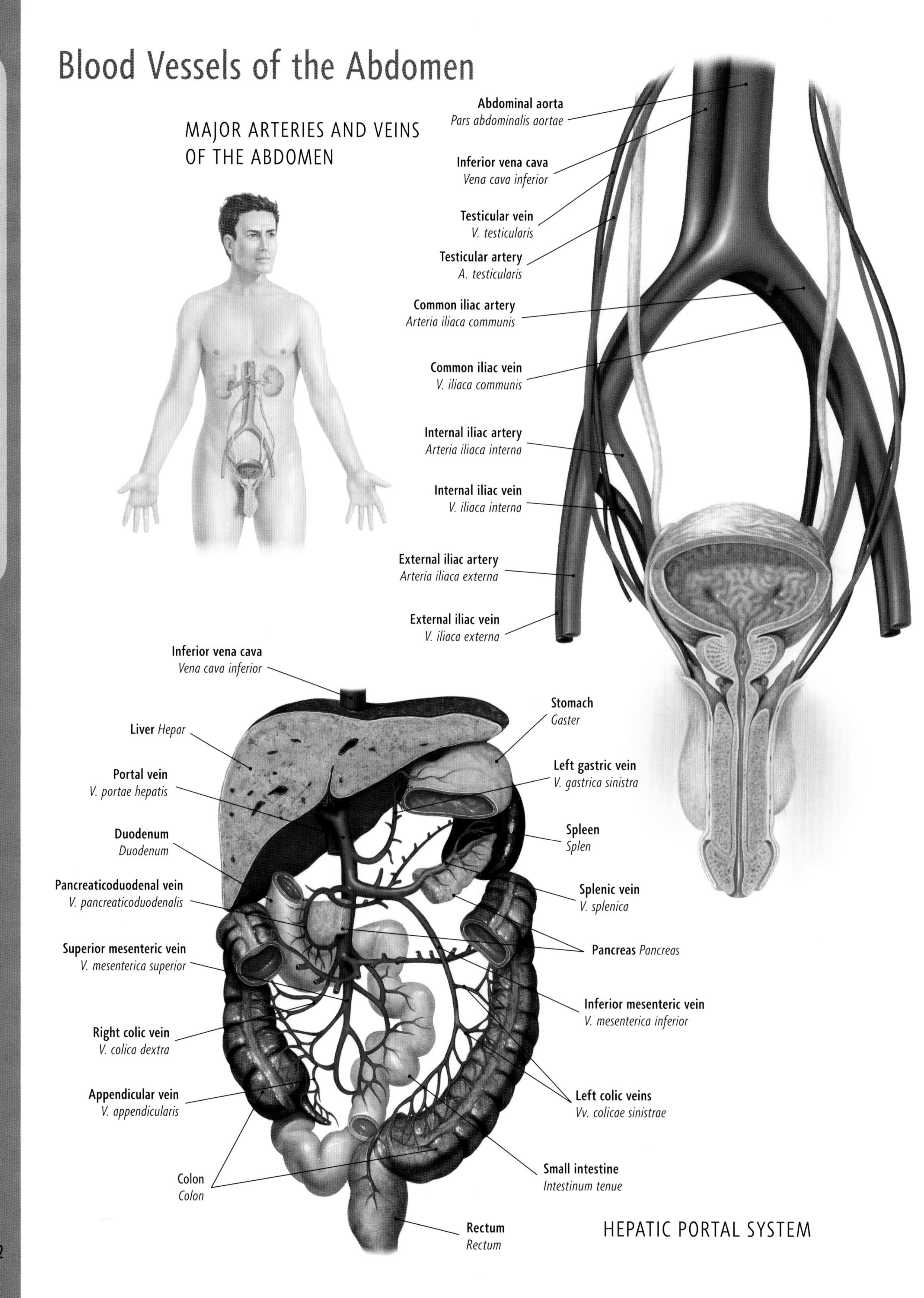

MAJOR ARTERIES AND VEINS OF THE ABDOMEN
Abdominal aorta
Pars abdominalis aortae
Inferior vena cava
Vena cava inferior
Testicular vein
V. testicularis
Testicular artery
A. testicularis
Common iliac artery
Arteria iliaca communis
Common iliac vein
V. iliaca communis
Internal iliac artery
Arteria iliaca interna
Internal iliac vein
V. iliaca interna
External iliac artery
Arteria iliaca externa
External iliac vein
V. iliaca externa
Inferior vena cava
Vena cava inferior
Liver Hepar
Portal vein
V. portae hepatis
Duodenum
Duodenum
Pancreaticoduodenal vein
V. pancreaticoduodenalis
Superior mesenteric vein
V. mesenterica superior
Right colic vein
V. colica dextra
Appendicular vein
V. appendicularis
Colon
Colon
Stomach
Gaster
Left gastric vein
V. gastrica sinistra
Spleen
Splen
Splenic vein
V. splenica
Pancreas Pancreas
Inferior mesenteric vein
V. mesenterica inferior
Left colic veins
Vv. colicae sinistrae
Small intestine
Intestinum tenue
Rectum
Rectum
HEPATIC PORTAL SYSTEM

HEPATIC ARTERY

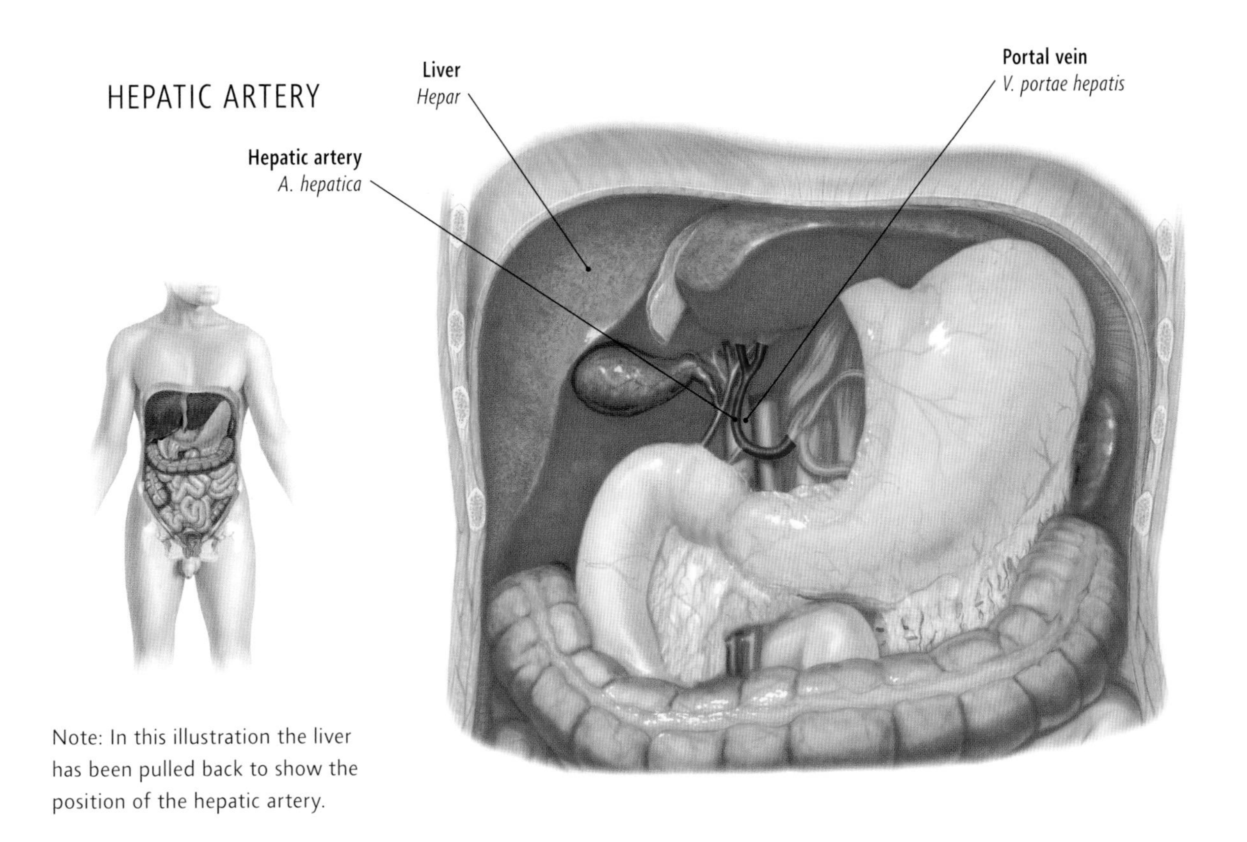
Liver
Hepar
Portal vein
V. portae hepatis
Hepatic artery
A. hepatica

Note: In this illustration the liver has been pulled back to show the position of the hepatic artery.

RENAL ARTERY

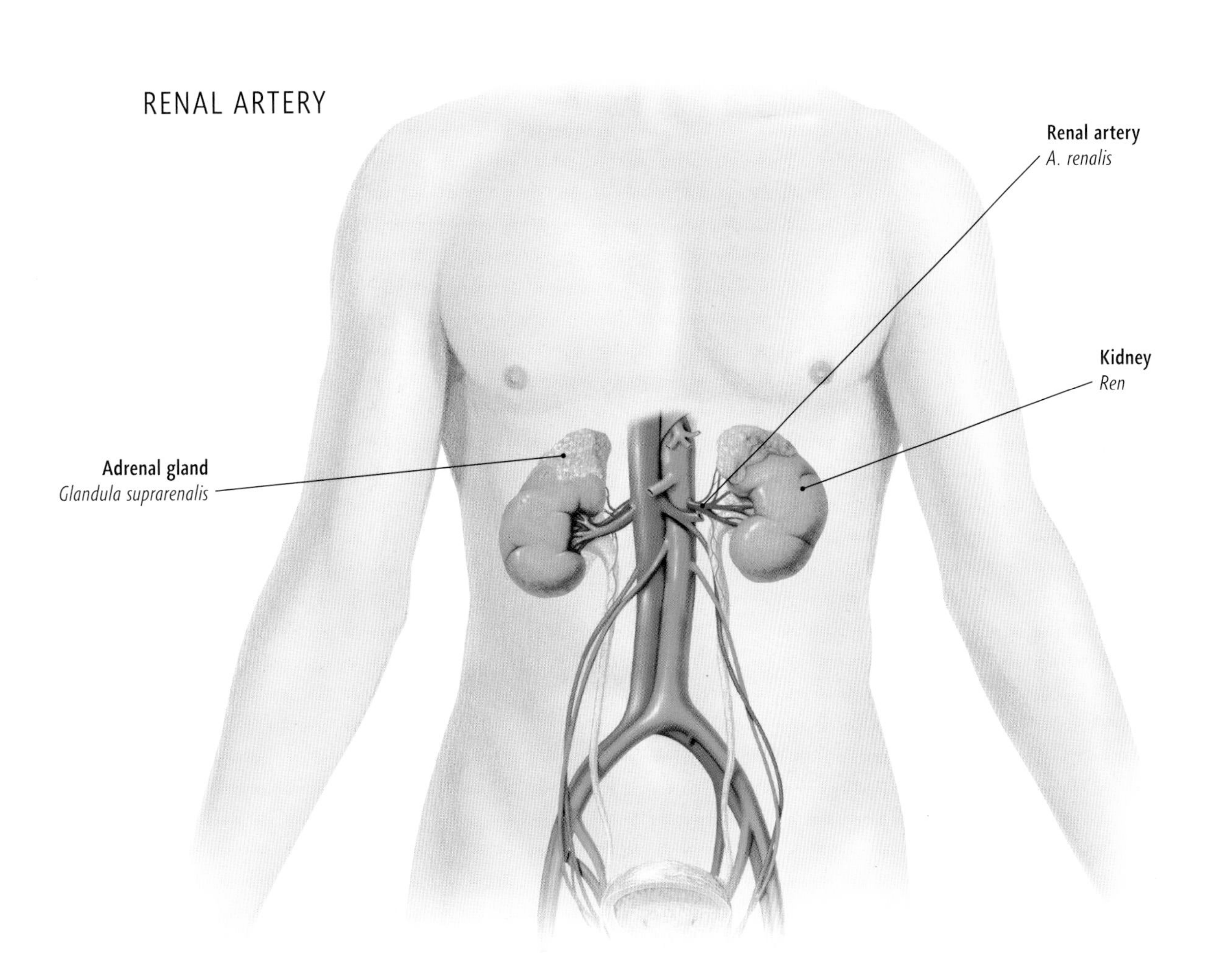
Renal artery
A. renalis
Kidney
Ren
Adrenal gland
Glandula suprarenalis

Blood Vessels of the Arms and Legs

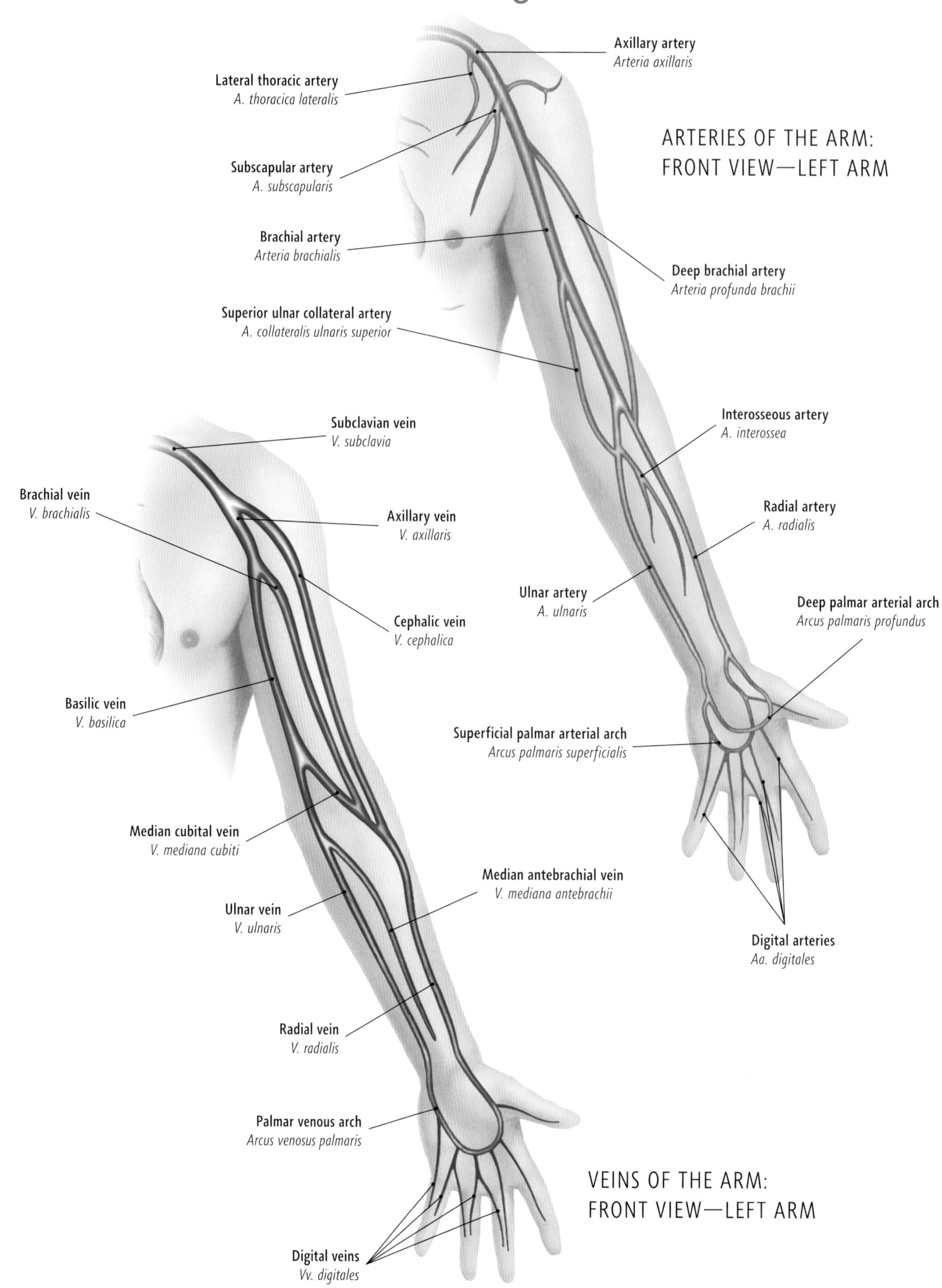

Axillary artery
Arteria axillaris
Lateral thoracic artery
A. thoracica lateralis
ARTERIES OF THE ARM:
FRONT VIEW—LEFT ARM
Subscapular artery
A. subscapularis
Brachial artery
Arteria brachialis
Deep brachial artery
Arteria profunda brachii
Superior ulnar collateral artery
A. collateralis ulnaris superior
Interosseous artery
A. interossea
Radial artery
A. radialis
Ulnar artery
A. ulnaris
Deep palmar arterial arch
Arcus palmaris profundus
Superficial palmar arterial arch
Arcus palmaris superficialis
Digital arteries
Aa. digitales
Subclavian vein
V. subclavia
Brachial vein
V. brachialis
Axillary vein
V. axillaris
Cephalic vein
V. cephalica
Basilic vein
V. basilica
Median cubital vein
V. mediana cubiti
Median antebrachial vein
V. mediana antebrachii
Ulnar vein
V. ulnaris
Radial vein
V. radialis
Palmar venous arch
Arcus venosus palmaris
VEINS OF THE ARM:
FRONT VIEW—LEFT ARM
Digital veins
Vv. digitales

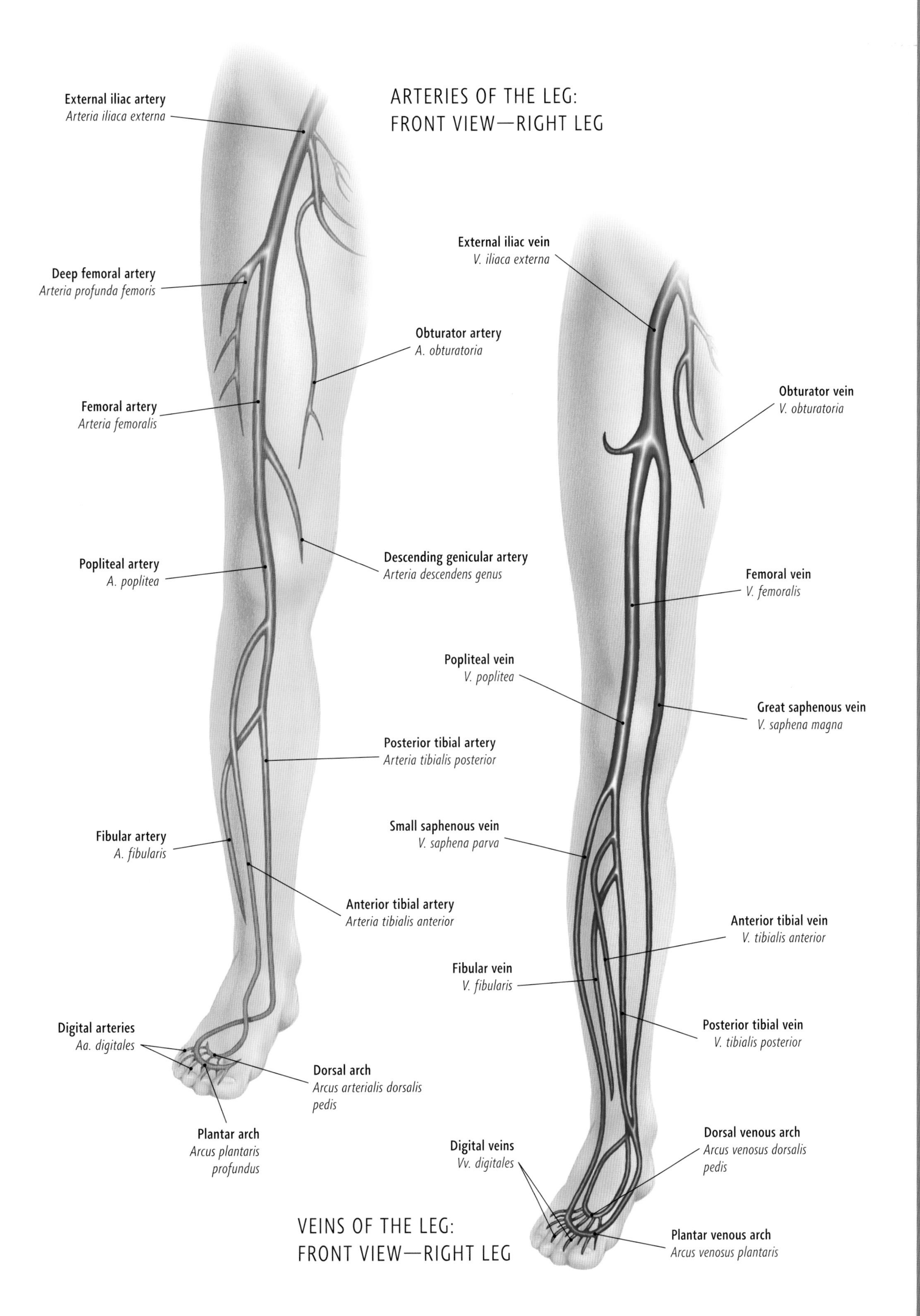
ARTERIES OF THE LEG:
FRONT VIEW—RIGHT LEG
External iliac artery
Arteria iliaca externa
Deep femoral artery
Arteria profunda femoris
Obturator artery
A. obturatoria
Femoral artery
Arteria femoralis
Popliteal artery
A. poplitea
Descending genicular artery
Arteria descendens genus
Posterior tibial artery
Arteria tibialis posterior
Fibular artery
A. fibularis
Anterior tibial artery
Arteria tibialis anterior
Digital arteries
Aa. digitales
Dorsal arch
Arcus arterialis dorsalis pedis
Plantar arch
Arcus plantaris profundus
External iliac vein
V. iliaca externa
Obturator vein
V. obturatoria
Femoral vein
V. femoralis
Popliteal vein
V. poplitea
Great saphenous vein
V. saphena magna
Small saphenous vein
V. saphena parva
Anterior tibial vein
V. tibialis anterior
Fibular vein
V. fibularis
Posterior tibial vein
V. tibialis posterior
Digital veins
Vv. digitales
Dorsal venous arch
Arcus venosus dorsalis pedis
Plantar venous arch
Arcus venosus plantaris
VEINS OF THE LEG:
FRONT VIEW—RIGHT LEG

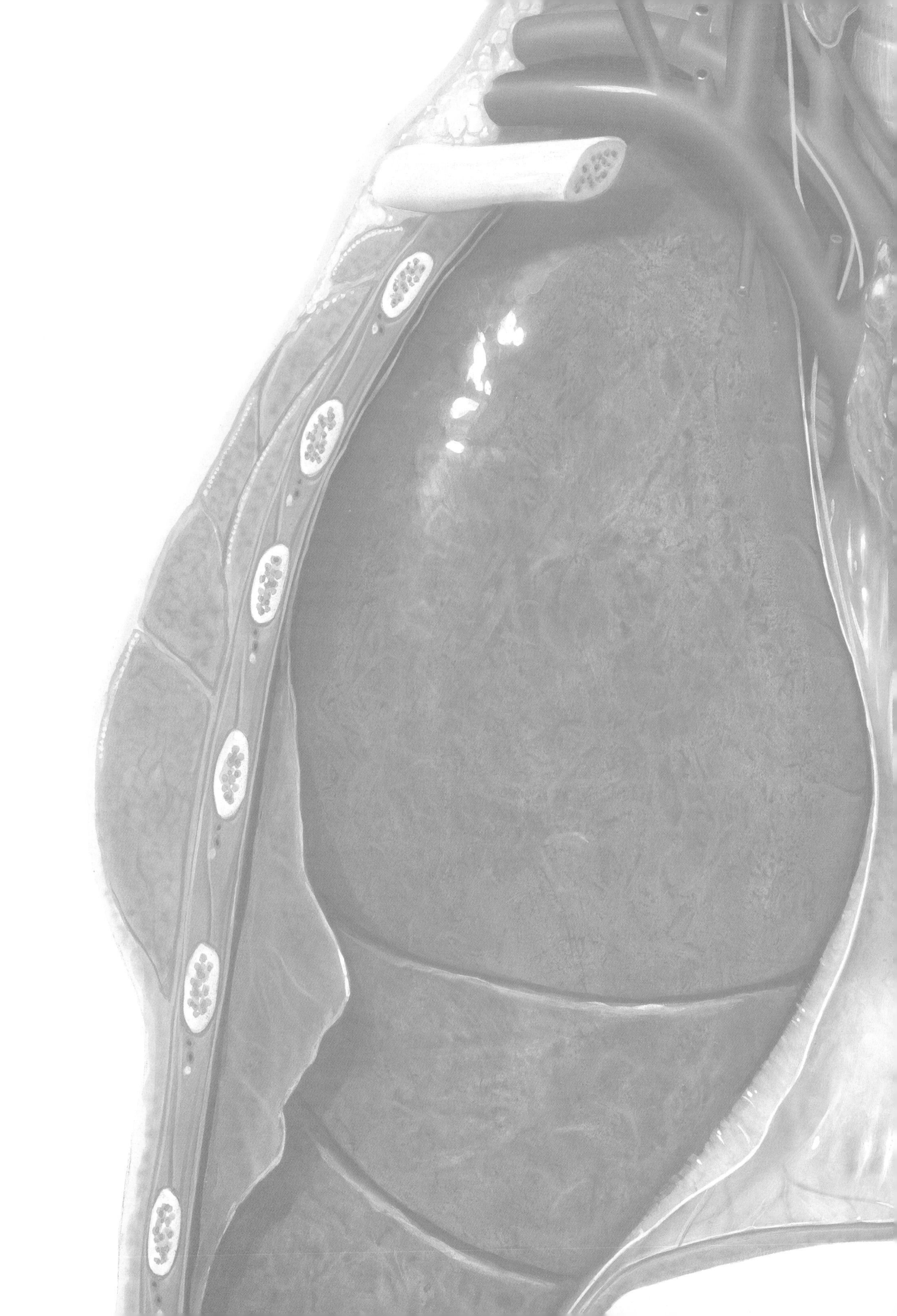

The Respiratory System

The Respiratory System

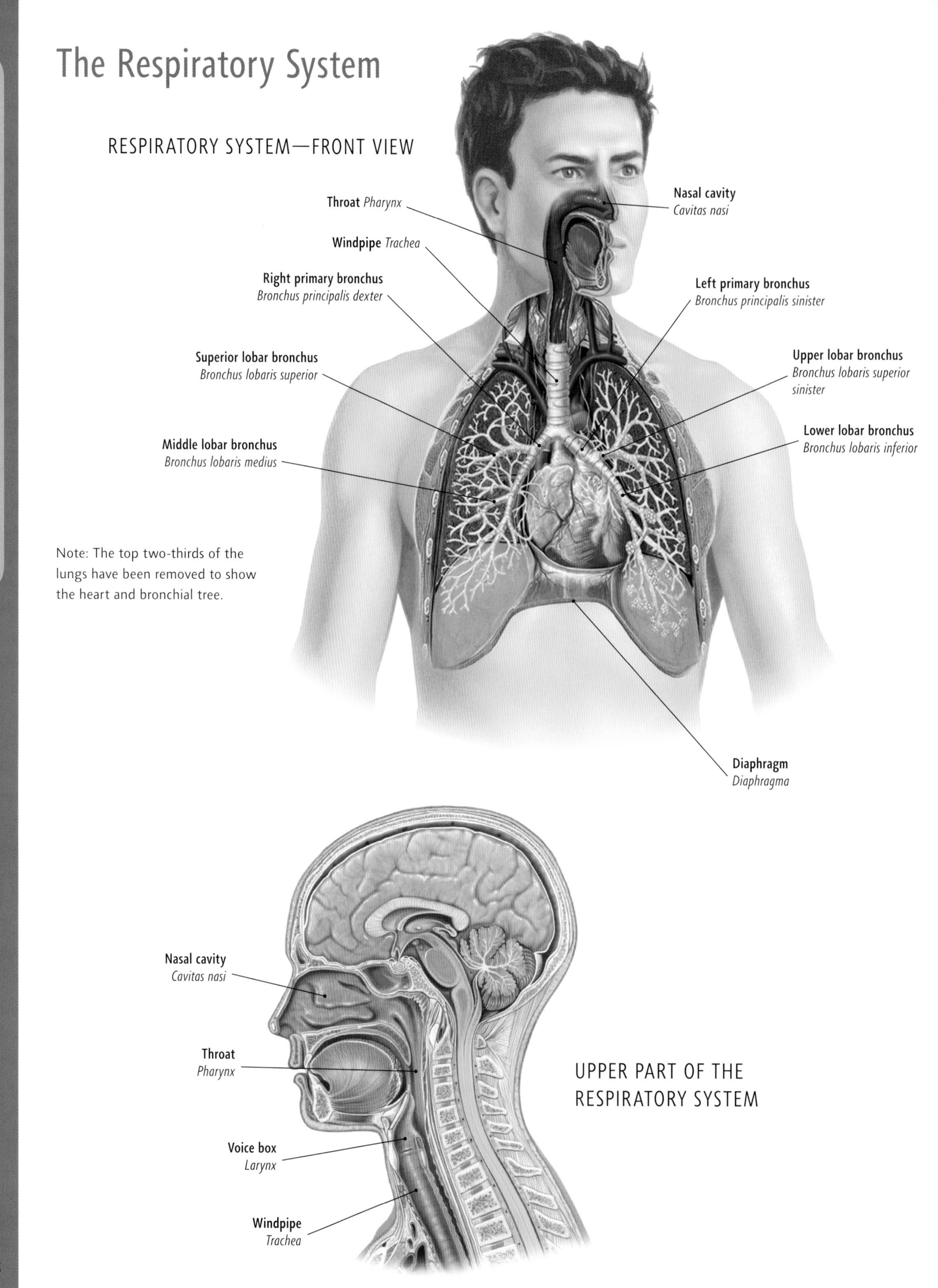
RESPIRATORY SYSTEM—FRONT VIEW
Throat Pharynx
Windpipe Trachea
Right primary bronchus
Bronchus principalis dexter
Superior lobar bronchus
Bronchus lobaris superior
Middle lobar bronchus
Bronchus lobaris medius
Nasal cavity
Cavitas nasi
Left primary bronchus
Bronchus principalis sinister
Upper lobar bronchus
Bronchus lobaris superior sinister
Lower lobar bronchus
Bronchus lobaris inferior
Diaphragm
Diaphragma
Nasal cavity
Cavitas nasi
Throat
Pharynx
Voice box
Larynx
Windpipe
Trachea
UPPER PART OF THE RESPIRATORY SYSTEM

Note: The top two-thirds of the lungs have been removed to show the heart and bronchial tree.

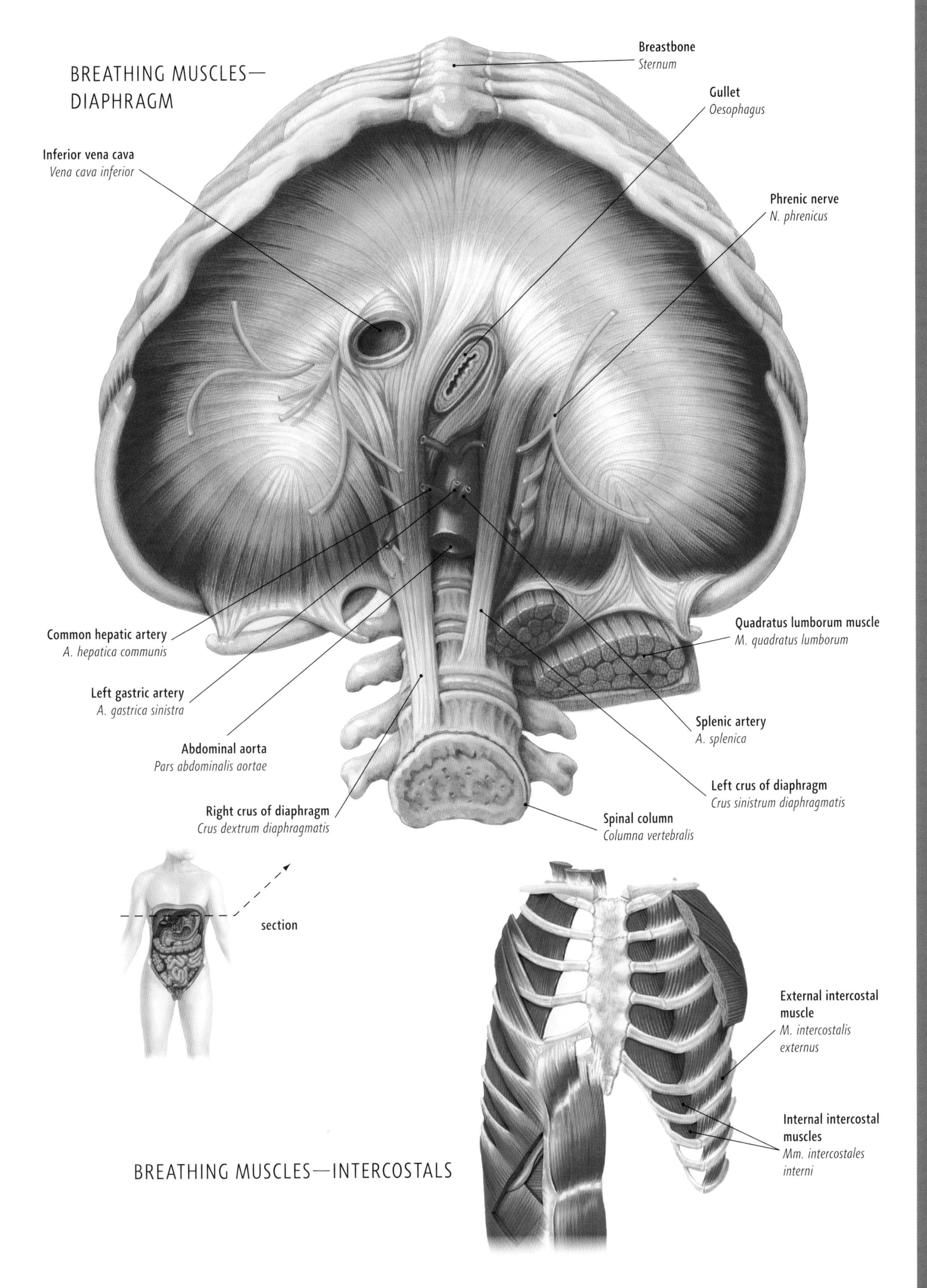
BREATHING MUSCLES—
DIAPHRAGM
Breastbone
Sternum
Gullet
Oesophagus
Inferior vena cava
Vena cava inferior
Phrenic nerve
N. phrenicus
Common hepatic artery
A. hepatica communis
Quadratus lumborum muscle
M. quadratus lumborum
Left gastric artery
A. gastrica sinistra
Splenic artery
A. splenica
Abdominal aorta
Pars abdominalis aortae
Left crus of diaphragm
Crus sinistrum diaphragmatis
Right crus of diaphragm
Crus dextrum diaphragmatis
Spinal column
Columna vertebralis
section
External intercostal
muscle
M. intercostalis
externus
Internal intercostal
muscles
Mm. intercostales
interni
BREATHING MUSCLES—INTERCOSTALS

Nose and Throat

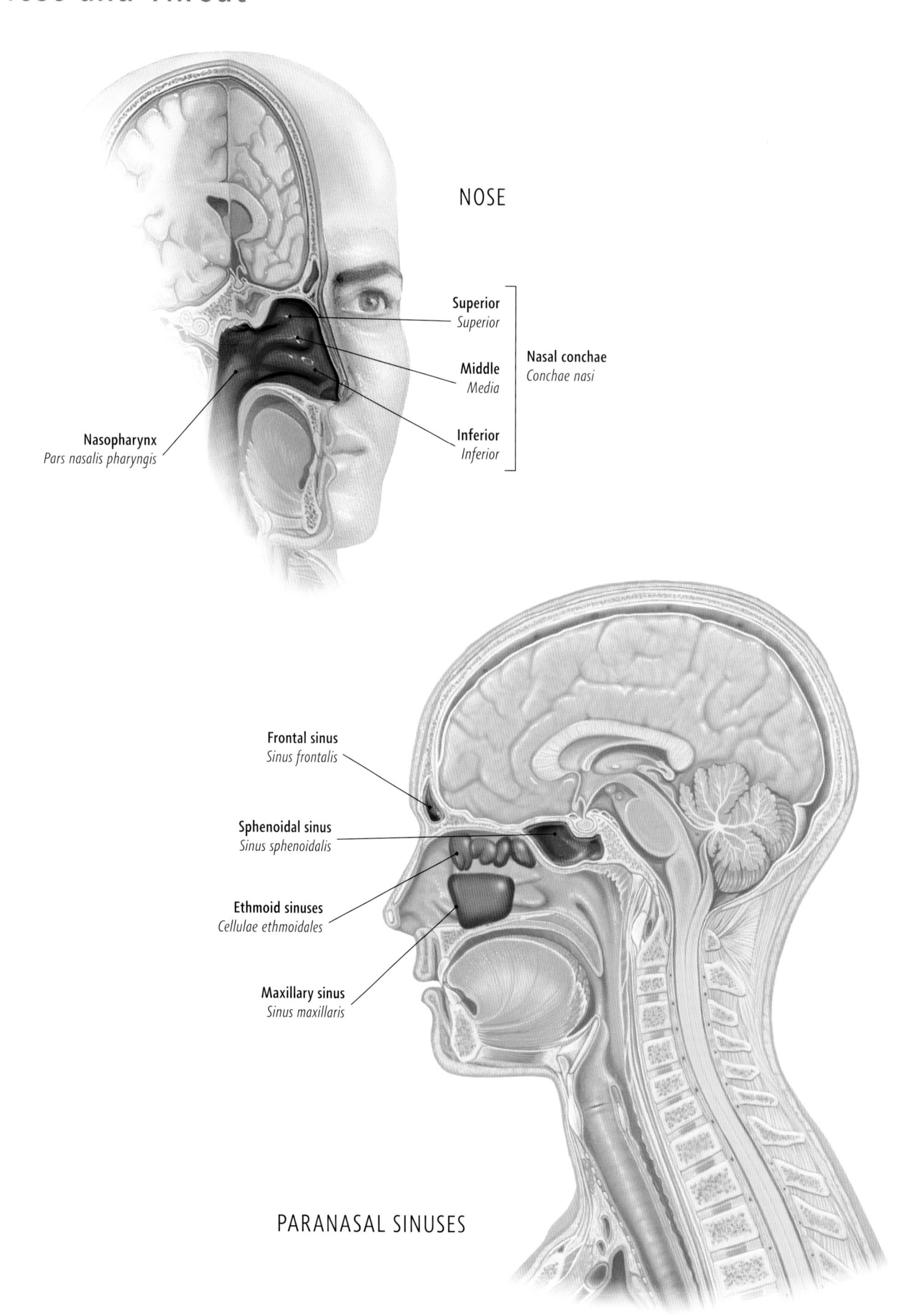
NOSE
Superior
Superior
Middle
Media
Inferior
Inferior
Nasal conchae
Conchae nasi
Nasopharynx
Pars nasalis pharyngis
Frontal sinus
Sinus frontalis
Sphenoidal sinus
Sinus sphenoidalis
Ethmoid sinuses
Cellulae ethmoidales
Maxillary sinus
Sinus maxillaris
PARANASAL SINUSES

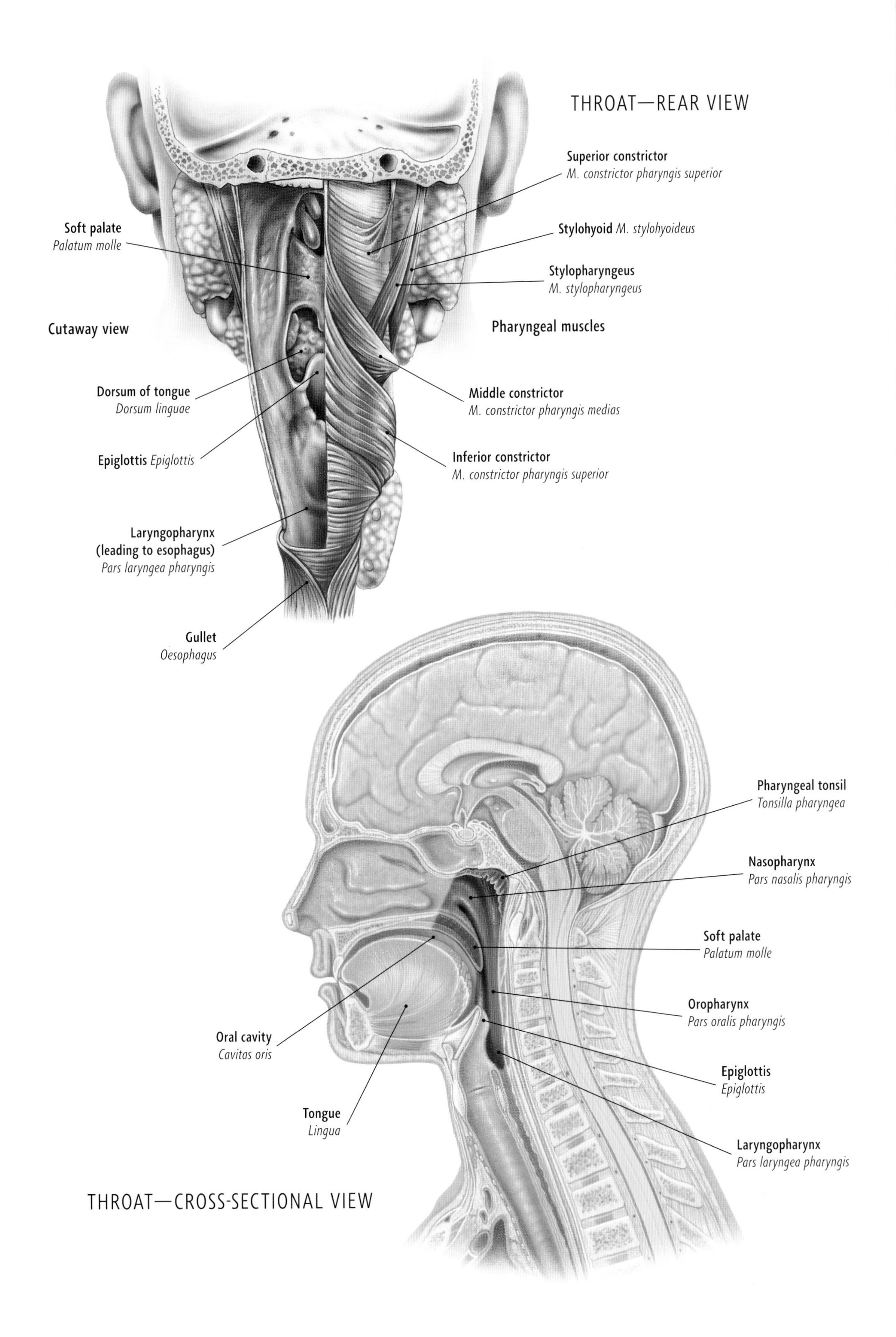
THROAT—REAR VIEW
Superior constrictor
M. constrictor pharyngis superior
Soft palate
Palatum molle
Stylohyoid M. stylohyoideus
Stylopharyngeus
M. stylopharyngeus
Cutaway view
Pharyngeal muscles
Dorsum of tongue
Dorsum linguae
Middle constrictor
M. constrictor pharyngis medias
Epiglottis Epiglottis
Inferior constrictor
M. constrictor pharyngis superior
Laryngopharynx
(leading to esophagus)
Pars laryngea pharyngis
Gullet
Oesophagus
Pharyngeal tonsil
Tonsilla pharyngea
Nasopharynx
Pars nasalis pharyngis
Soft palate
Palatum molle
Oropharynx
Pars oralis pharyngis
Oral cavity
Cavitas oris
Epiglottis
Epiglottis
Tongue
Lingua
Laryngopharynx
Pars laryngea pharyngis
THROAT—CROSS-SECTIONAL VIEW

Larynx

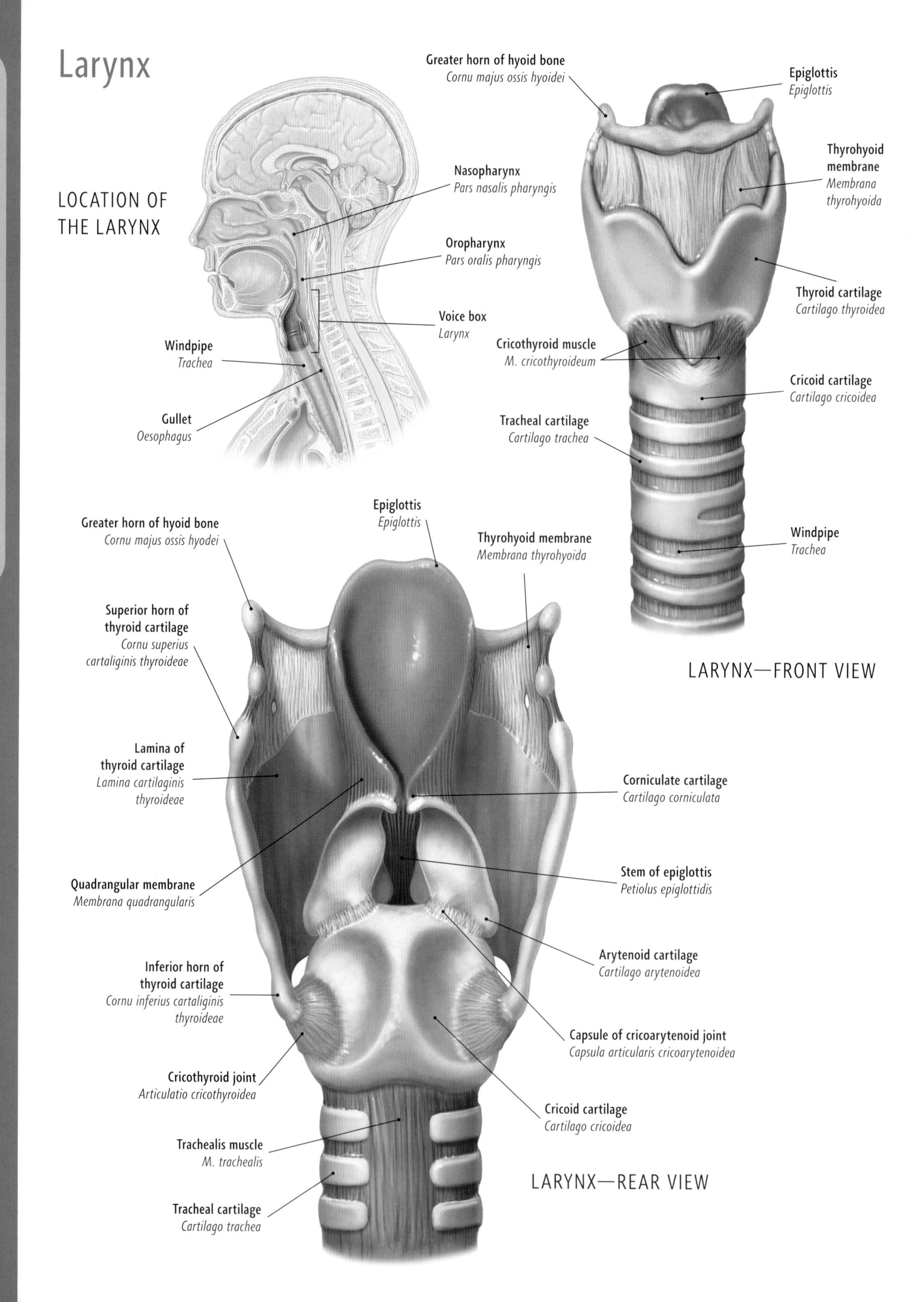
LOCATION OF THE LARYNX
Nasopharynx
Pars nasalis pharyngis
Oropharynx
Pars oralis pharyngis
Voice box
Larynx
Windpipe
Trachea
Gullet
Oesophagus
Greater horn of hyoid bone
Cornu majus ossis hyoidei
Epiglottis
Epiglottis
Thyrohyoid membrane
Membrana thyrohyoida
Thyroid cartilage
Cartilago thyroidea
Cricothyroid muscle
M. cricothyroideum
Cricoid cartilage
Cartilago cricoidea
Tracheal cartilage
Cartilago trachea
Windpipe
Trachea
LARYNX—FRONT VIEW
Greater horn of hyoid bone
Cornu majus ossis hyodei
Epiglottis
Epiglottis
Thyrohyoid membrane
Membrana thyrohyoida
Superior horn of thyroid cartilage
Cornu superius cartaliginis thyroideae
Lamina of thyroid cartilage
Lamina cartilaginis thyroideae
Corniculate cartilage
Cartilago corniculata
Stem of epiglottis
Petiolus epiglottidis
Quadrangular membrane
Membrana quadrangularis
Arytenoid cartilage
Cartilago arytenoidea
Inferior horn of thyroid cartilage
Cornu inferius cartaliginis thyroideae
Capsule of cricoarytenoid joint
Capsula articularis cricoarytenoidea
Cricothyroid joint
Articulatio cricothyroidea
Cricoid cartilage
Cartilago cricoidea
Trachealis muscle
M. trachealis
Tracheal cartilage
Cartilago trachea
LARYNX—REAR VIEW

EPIGLOTTIS: SWALLOWING

The epiglottis is a flap of elastic cartilage in the larynx. During swallowing it folds down over the glottis to prevent food and drink passing into the airway.

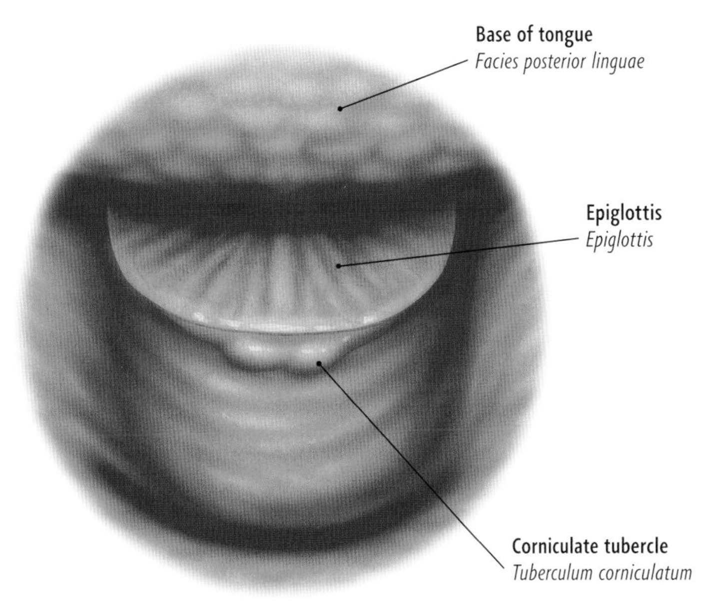
Base of tongue
Facies posterior linguae
Epiglottis
Epiglottis
Corniculate tubercle
Tuberculum corniculatum

Epiglottis
Epiglottis

Root of tongue
Radix linguae

facies inferior linguae

Vocal cords
Plicae vocales

Vocal process of arytenoid cartilage
Processus vocalis cartilaginis arytenoideae

EPIGLOTTIS: SPEAKING

During speaking, the epiglottis remains upright so that exhaled air can flow through the larynx and vibrate the vocal folds (vocal cords), producing sound. The tension and length of the cords determines the pitch of the sound.

EPIGLOTTIS: BREATHING

During breathing, the epiglottis remains upright, and the vocal folds (vocal cords) are moved apart by muscles in the larynx, allowing air to flow downward into the trachea.

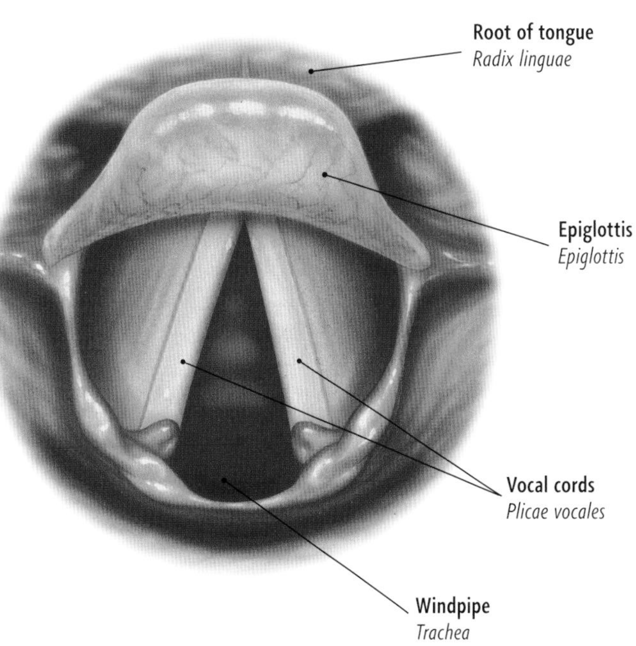
Root of tongue
Radix linguae
Epiglottis
Epiglottis
Vocal cords
Plicae vocales
Windpipe
Trachea

Trachea and Bronchi

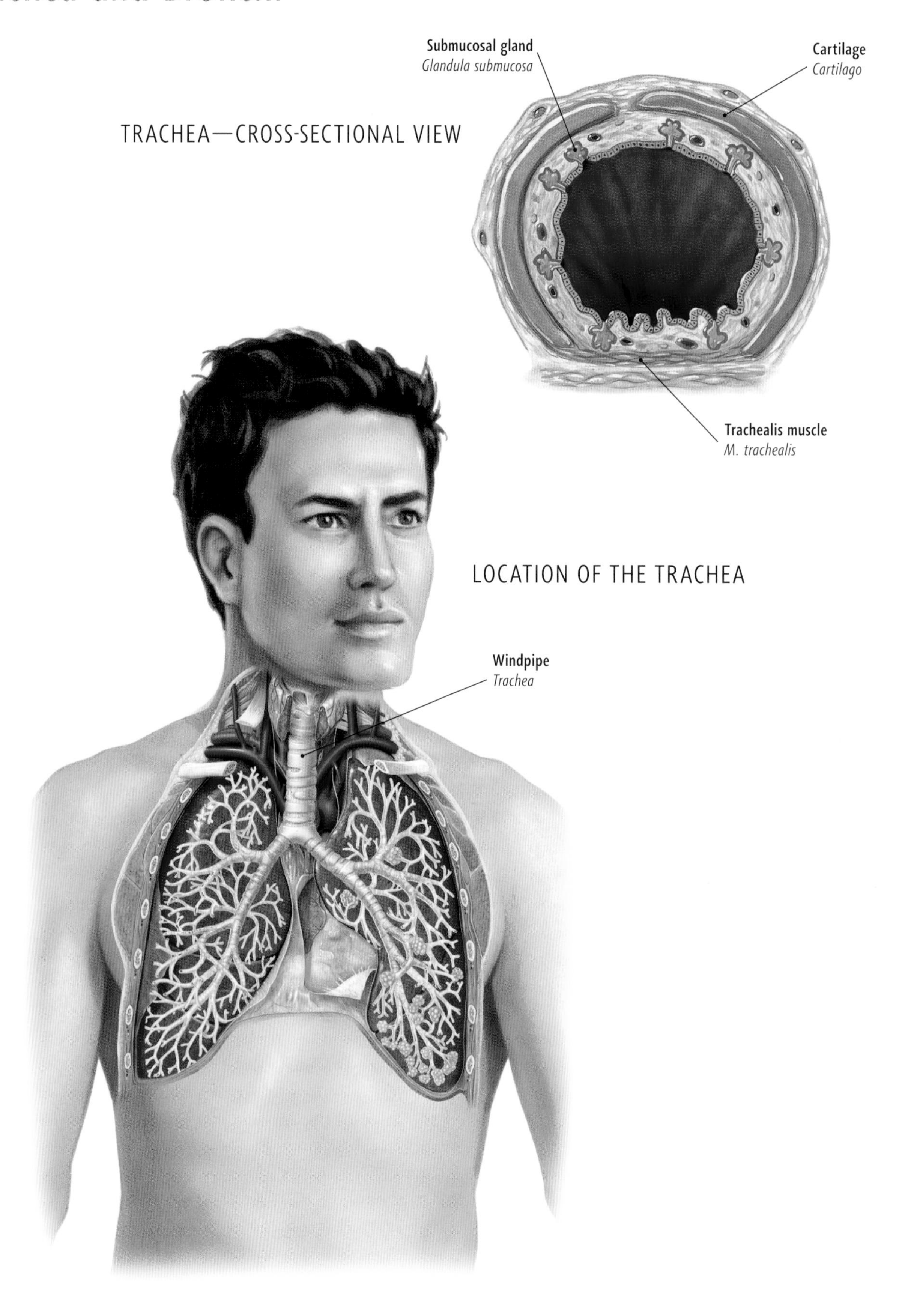
TRACHEA—CROSS-SECTIONAL VIEW
Submucosal gland
Glandula submucosa
Cartilage
Cartilago
Trachealis muscle
M. trachealis
LOCATION OF THE TRACHEA
Windpipe
Trachea

BRONCHUS

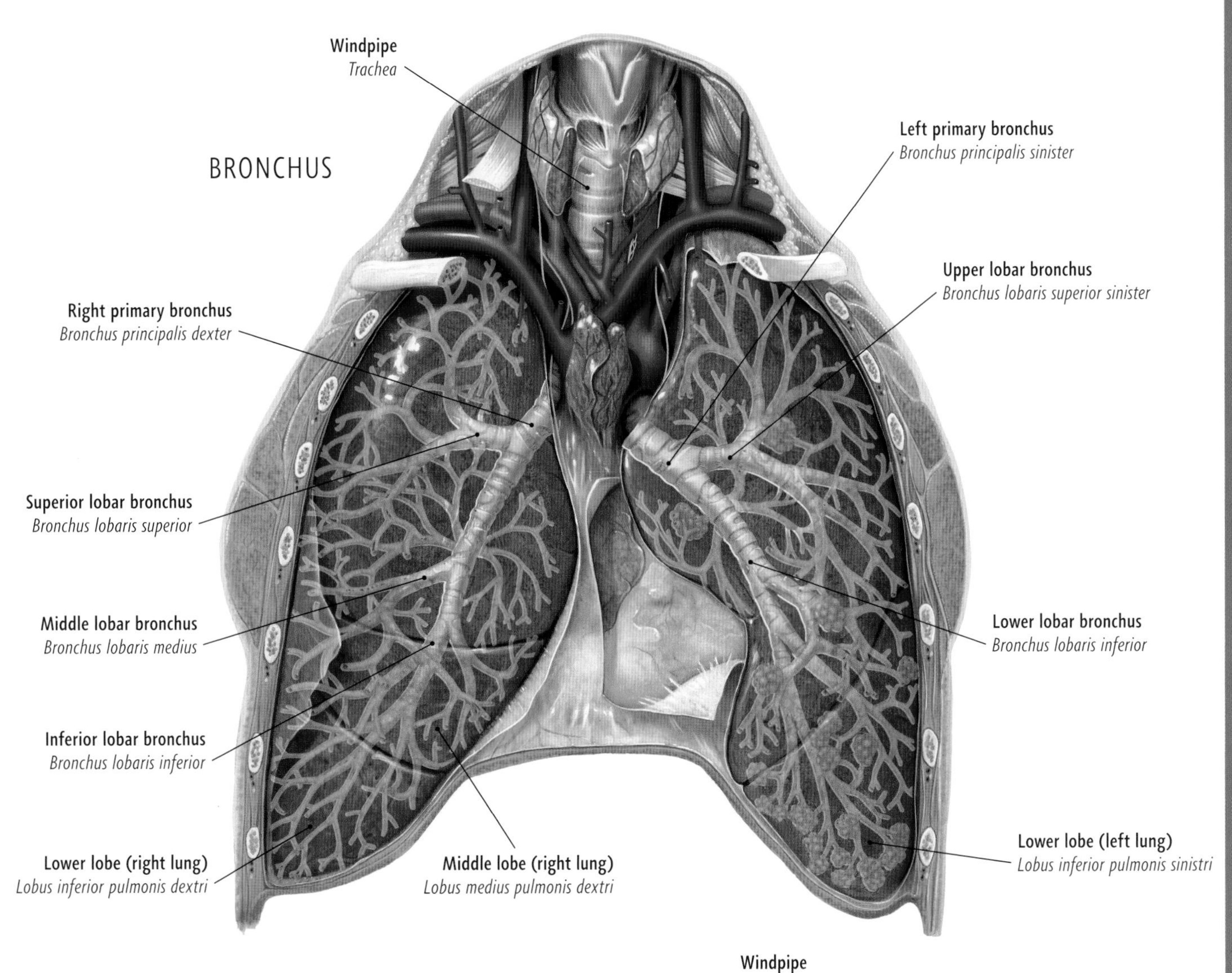
Windpipe
Trachea
Left primary bronchus
Bronchus principalis sinister
Upper lobar bronchus
Bronchus lobaris superior sinister
Right primary bronchus
Bronchus principalis dexter
Superior lobar bronchus
Bronchus lobaris superior
Middle lobar bronchus
Bronchus lobaris medius
Lower lobar bronchus
Bronchus lobaris inferior
Inferior lobar bronchus
Bronchus lobaris inferior
Lower lobe (left lung)
Lobus inferior pulmonis sinistri
Lower lobe (right lung)
Lobus inferior pulmonis dextri
Middle lobe (right lung)
Lobus medius pulmonis dextri

BRONCHUS—CROSS-SECTIONAL VIEW

BRONCHIAL TREE

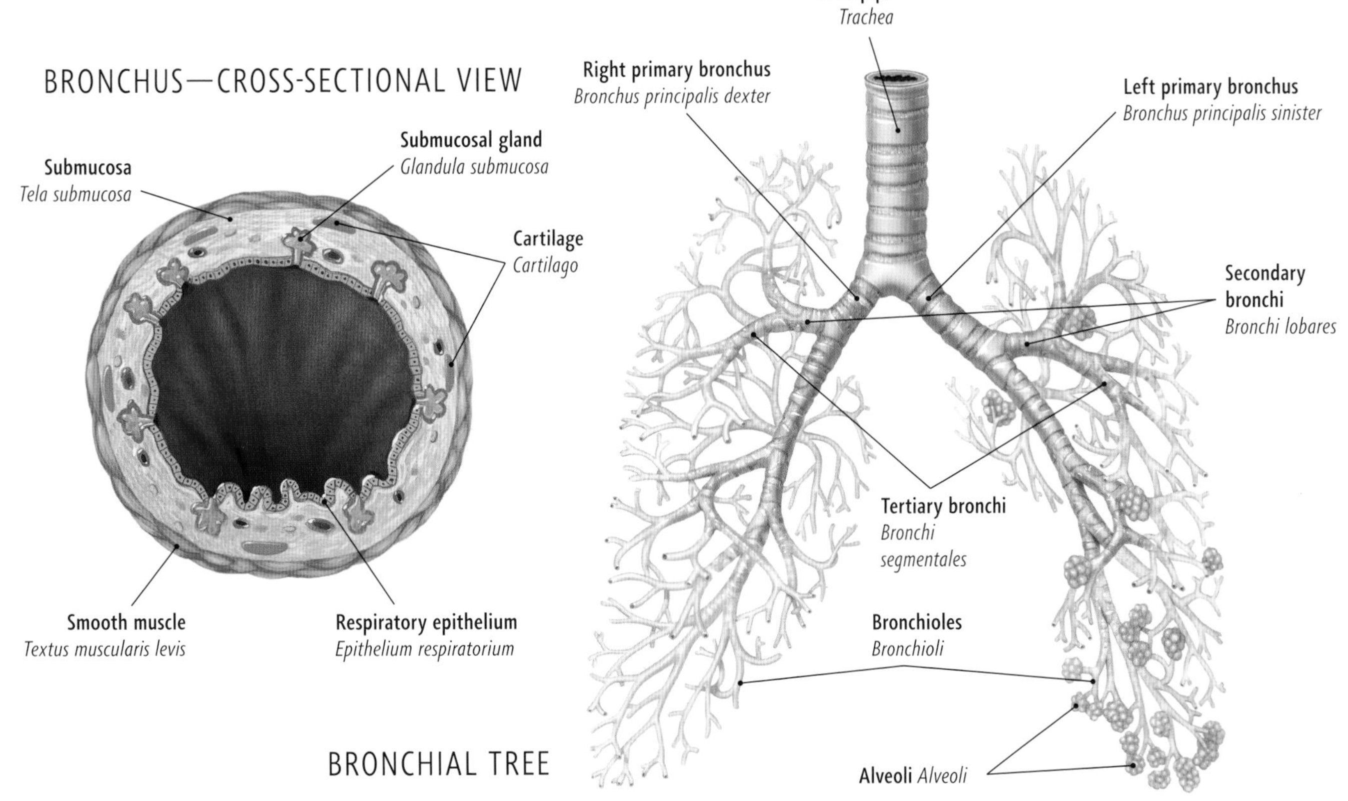
Submucosal gland
Glandula submucosa
Submucosa
Tela submucosa
Cartilage
Cartilago
Smooth muscle
Textus muscularis levis
Respiratory epithelium
Epithelium respiratorium
Windpipe
Trachea
Right primary bronchus
Bronchus principalis dexter
Left primary bronchus
Bronchus principalis sinister
Secondary bronchi
Bronchi lobares
Tertiary bronchi
Bronchi segmentales
Bronchioles
Bronchioli
Alveoli Alveoli

Lungs

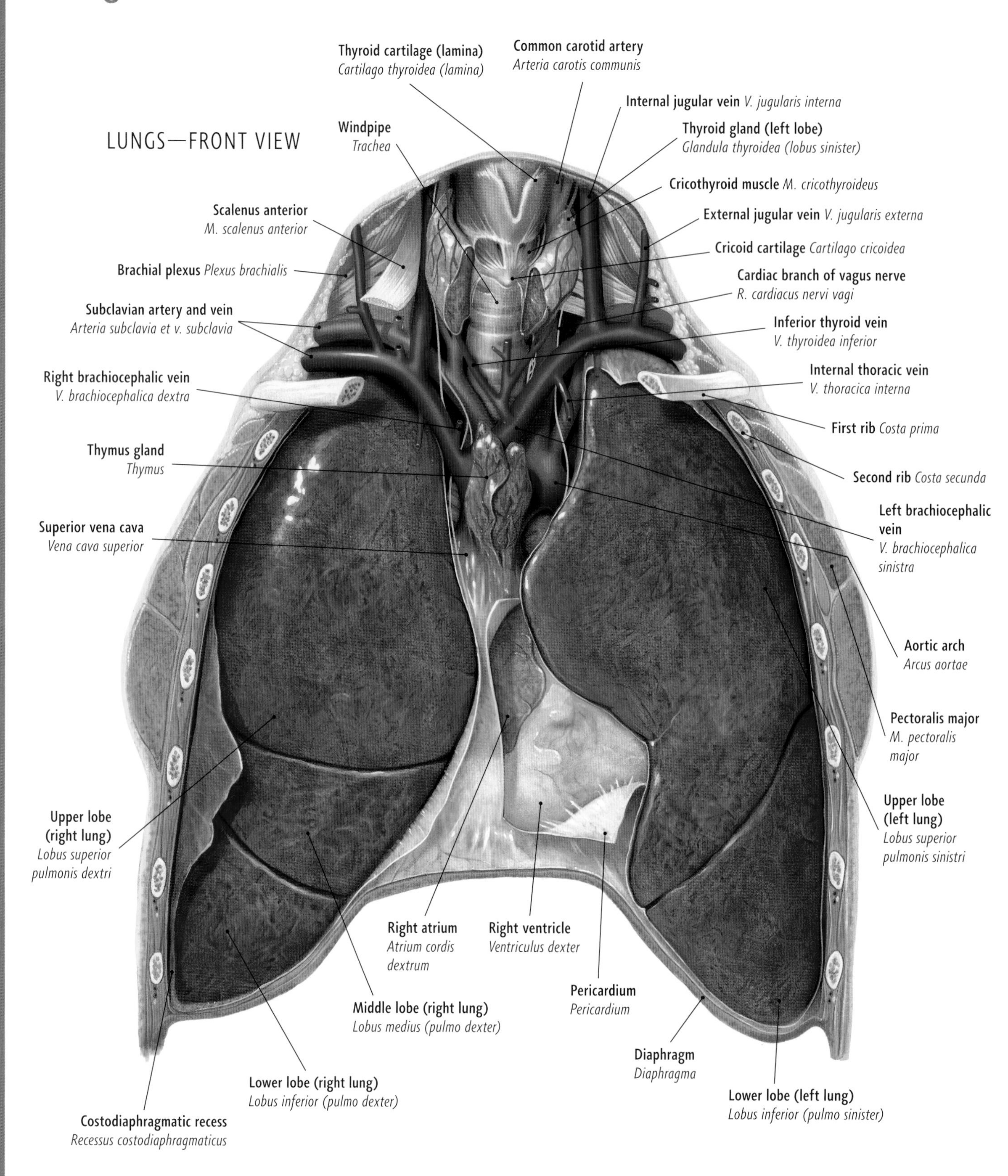
LUNGS—FRONT VIEW
Thyroid cartilage (lamina) Cartilago thyroidea (lamina)
Common carotid artery Arteria carotis communis
Internal jugular vein V. jugularis interna
Windpipe Trachea
Thyroid gland (left lobe) Glandula thyroidea (lobus sinister)
Cricothyroid muscle M. cricothyroideus
Scalenus anterior M. scalenus anterior
External jugular vein V. jugularis externa
Cricoid cartilage Cartilago cricoidea
Brachial plexus Plexus brachialis
Cardiac branch of vagus nerve R. cardiacus nervi vagi
Subclavian artery and vein Arteria subclavia et v. subclavia
Inferior thyroid vein V. thyroidea inferior
Right brachiocephalic vein V. brachiocephalica dextra
Internal thoracic vein V. thoracica interna
First rib Costa prima
Thymus gland Thymus
Second rib Costa secunda
Left brachiocephalic vein V. brachiocephalica sinistra
Superior vena cava Vena cava superior
Aortic arch Arcus aortae
Pectoralis major M. pectoralis major
Upper lobe (right lung) Lobus superior pulmonis dextri
Upper lobe (left lung) Lobus superior pulmonis sinistri
Right atrium Atrium cordis dextrum
Right ventricle Ventriculus dexter
Pericardium Pericardium
Middle lobe (right lung) Lobus medius (pulmo dexter)
Diaphragm Diaphragma
Lower lobe (right lung) Lobus inferior (pulmo dexter)
Lower lobe (left lung) Lobus inferior (pulmo sinister)
Costodiaphragmatic recess Recessus costodiaphragmaticus

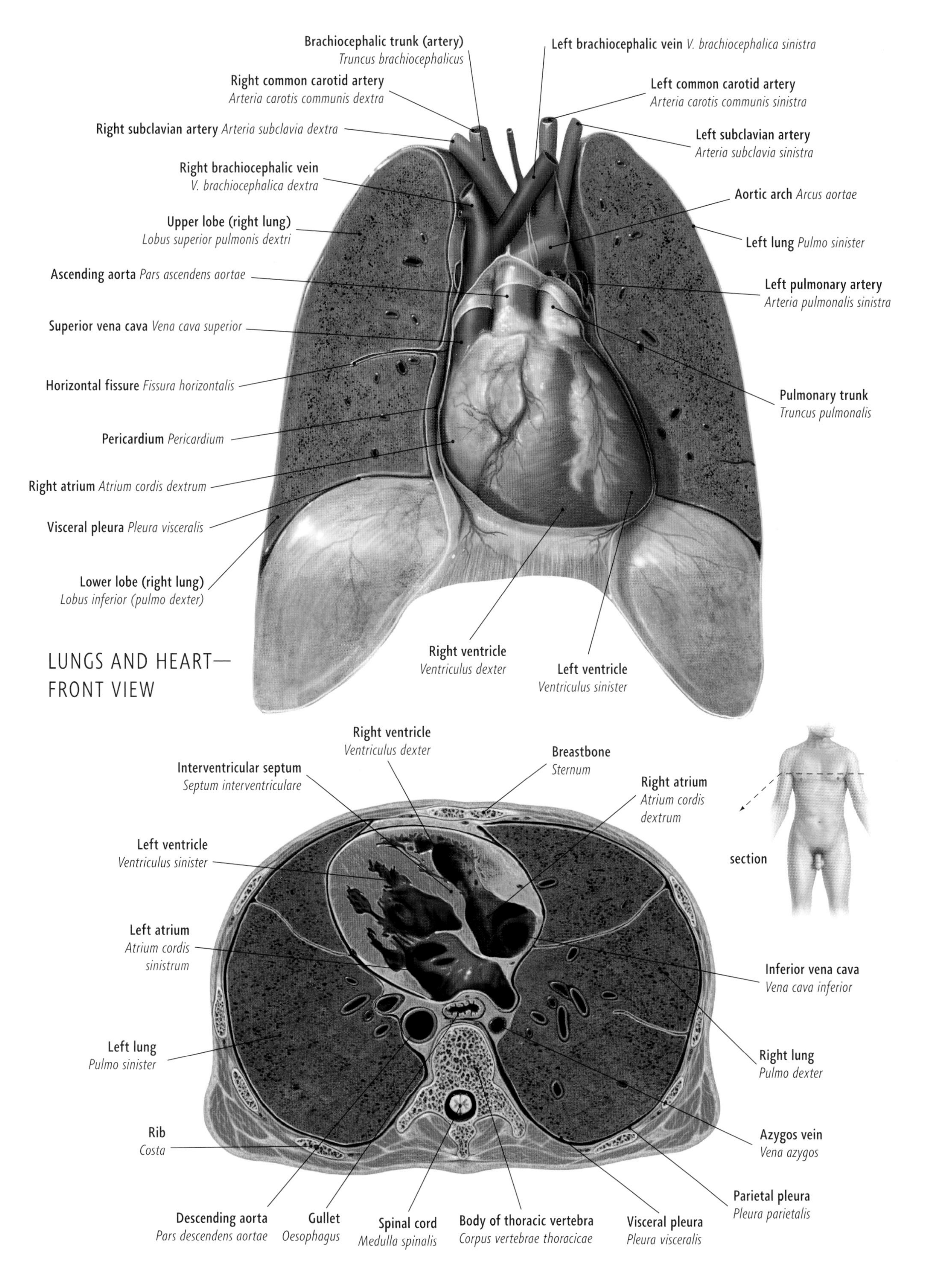
Brachiocephalic trunk (artery)
Truncus brachiocephalicus
Left brachiocephalic vein V. brachiocephalica sinistra
Right common carotid artery
Arteria carotis communis dextra
Left common carotid artery
Arteria carotis communis sinistra
Right subclavian artery Arteria subclavia dextra
Left subclavian artery
Arteria subclavia sinistra
Right brachiocephalic vein
V. brachiocephalica dextra
Aortic arch Arcus aortae
Upper lobe (right lung)
Lobus superior pulmonis dextri
Left lung Pulmo sinister
Ascending aorta Pars ascendens aortae
Left pulmonary artery
Arteria pulmonalis sinistra
Superior vena cava Vena cava superior
Horizontal fissure Fissura horizontalis
Pulmonary trunk
Truncus pulmonalis
Pericardium Pericardium
Right atrium Atrium cordis dextrum
Visceral pleura Pleura visceralis
Lower lobe (right lung)
Lobus inferior (pulmo dexter)
Right ventricle
Ventriculus dexter
Left ventricle
Ventriculus sinister
Right ventricle
Ventriculus dexter
Breastbone
Sternum
Interventricular septum
Septum interventriculare
Right atrium
Atrium cordis
dextrum
section
Left ventricle
Ventriculus sinister
Left atrium
Atrium cordis
sinistrum
Inferior vena cava
Vena cava inferior
Left lung
Pulmo sinister
Right lung
Pulmo dexter
Azygos vein
Vena azygos
Rib
Costa
Parietal pleura
Pleura parietalis
Descending aorta
Pars descendens aortae
Gullet
Oesophagus
Spinal cord
Medulla spinalis
Body of thoracic vertebra
Corpus vertebrae thoracicae
Visceral pleura
Pleura visceralis

LUNGS AND HEART—FRONT VIEW

LUNGS AND HEART—CROSS-SECTIONAL VIEW FROM ABOVE

Breathing

HOW WE BREATHE

Respiratory centers in the brain control the rate and depth of breathing, depending on oxygen and carbon dioxide levels in the blood. Contraction of the diaphragm and chest muscles draws air into the lungs via the nasal passage and trachea; the lungs transfer oxygen into the bloodstream.

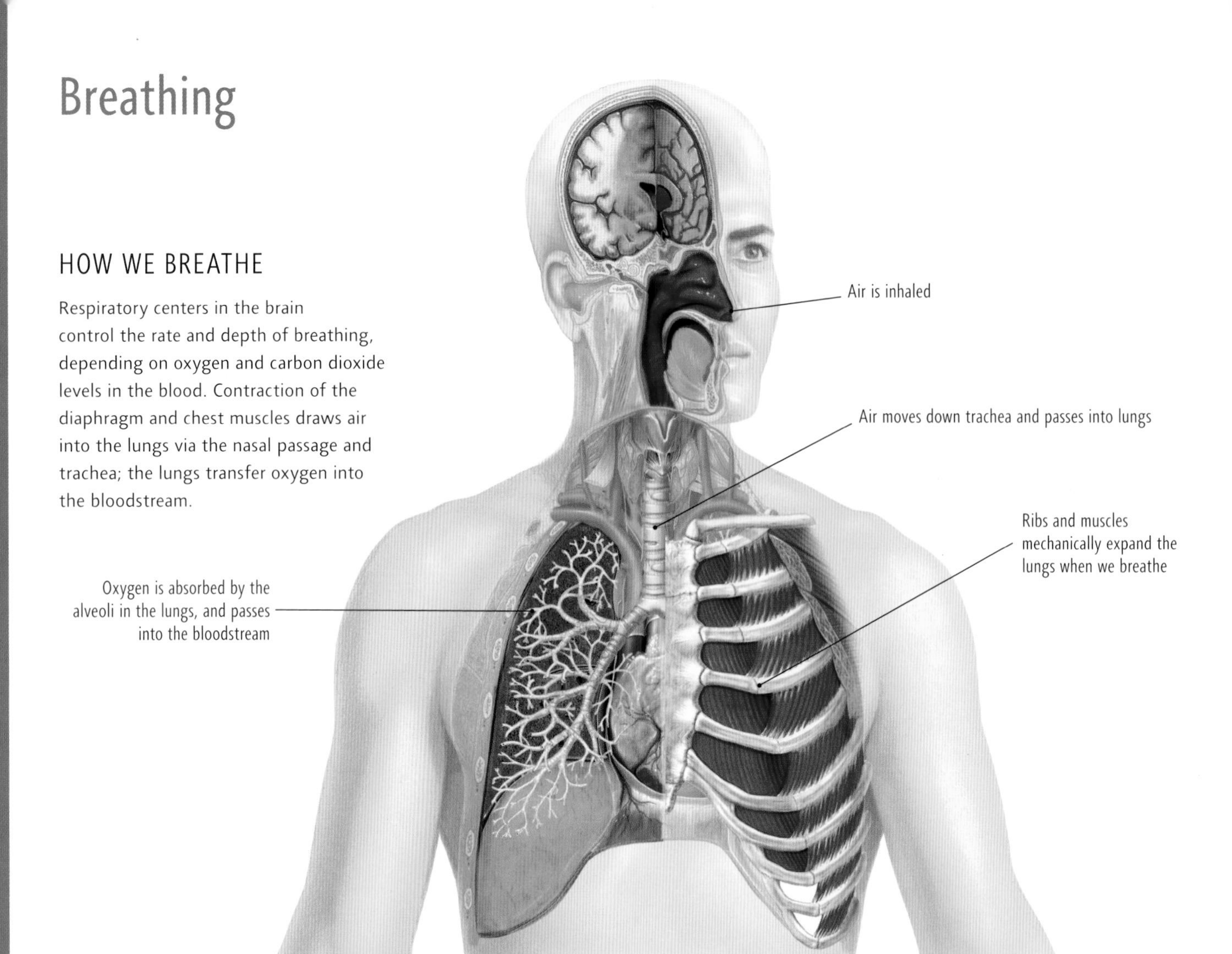
Air is inhaled
Air moves down trachea and passes into lungs
Ribs and muscles mechanically expand the lungs when we breathe
Oxygen is absorbed by the alveoli in the lungs, and passes into the bloodstream

RESPIRATORY CENTERS IN THE BRAIN

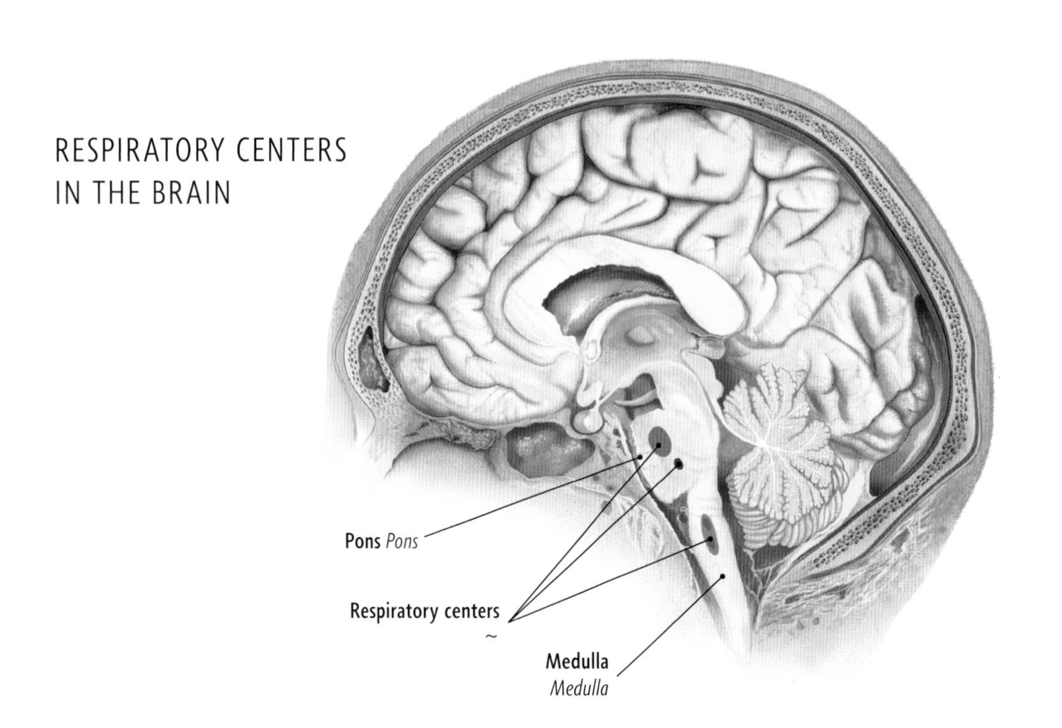
Pons *Pons*
Respiratory centers
~
Medulla
Medulla

[~] = no direct Latin equivalent

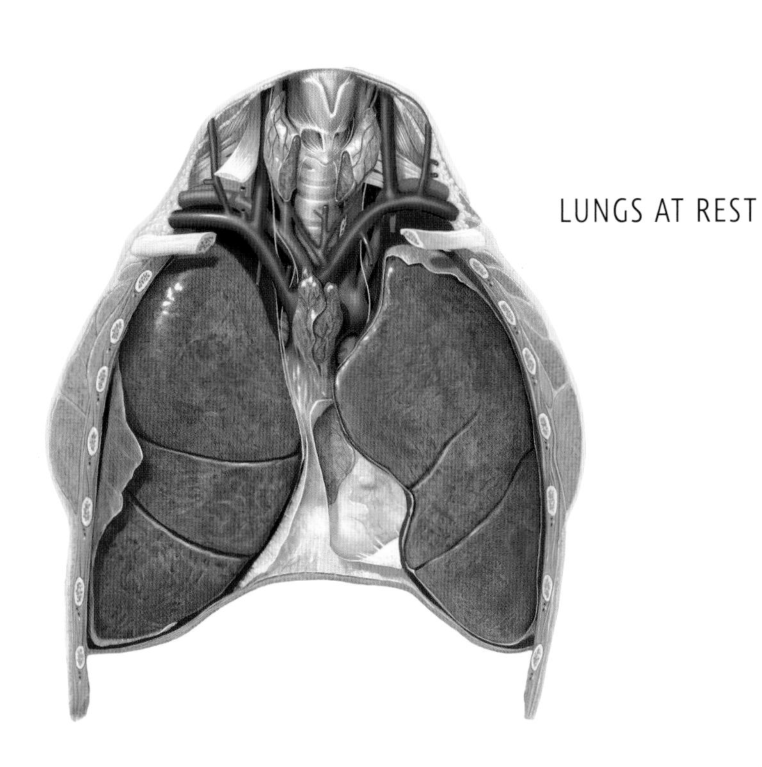

LUNGS AT REST

When we breathe in, the intercostal muscles move the ribs upward and outward, and the diaphragm pushes downward, drawing air into the expanded lungs. Breathing out depends mainly on the passive recoil of elastic tension built up in the lungs and chest during inspiration.

INSPIRATION
(BREATHING IN)

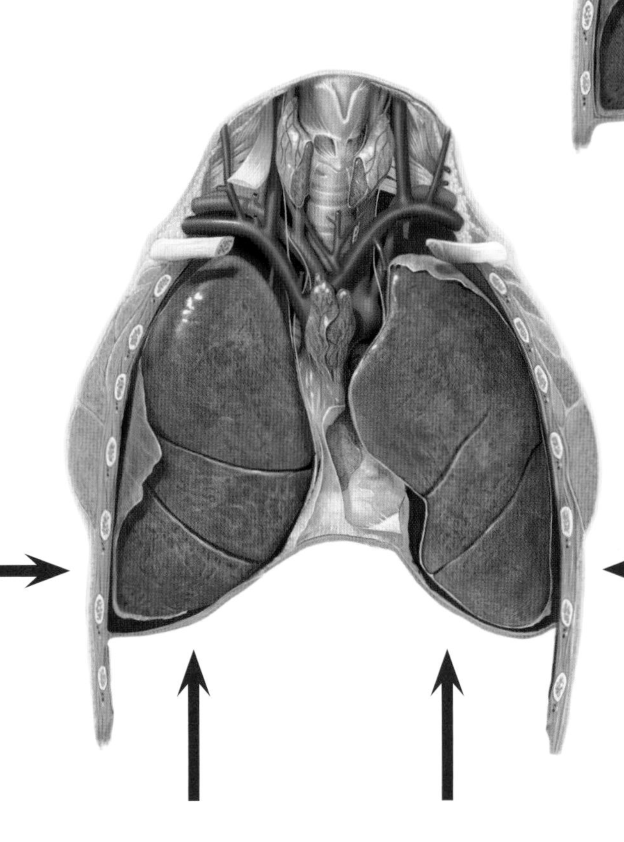

EXPIRATION
(BREATHING OUT)

Breathing

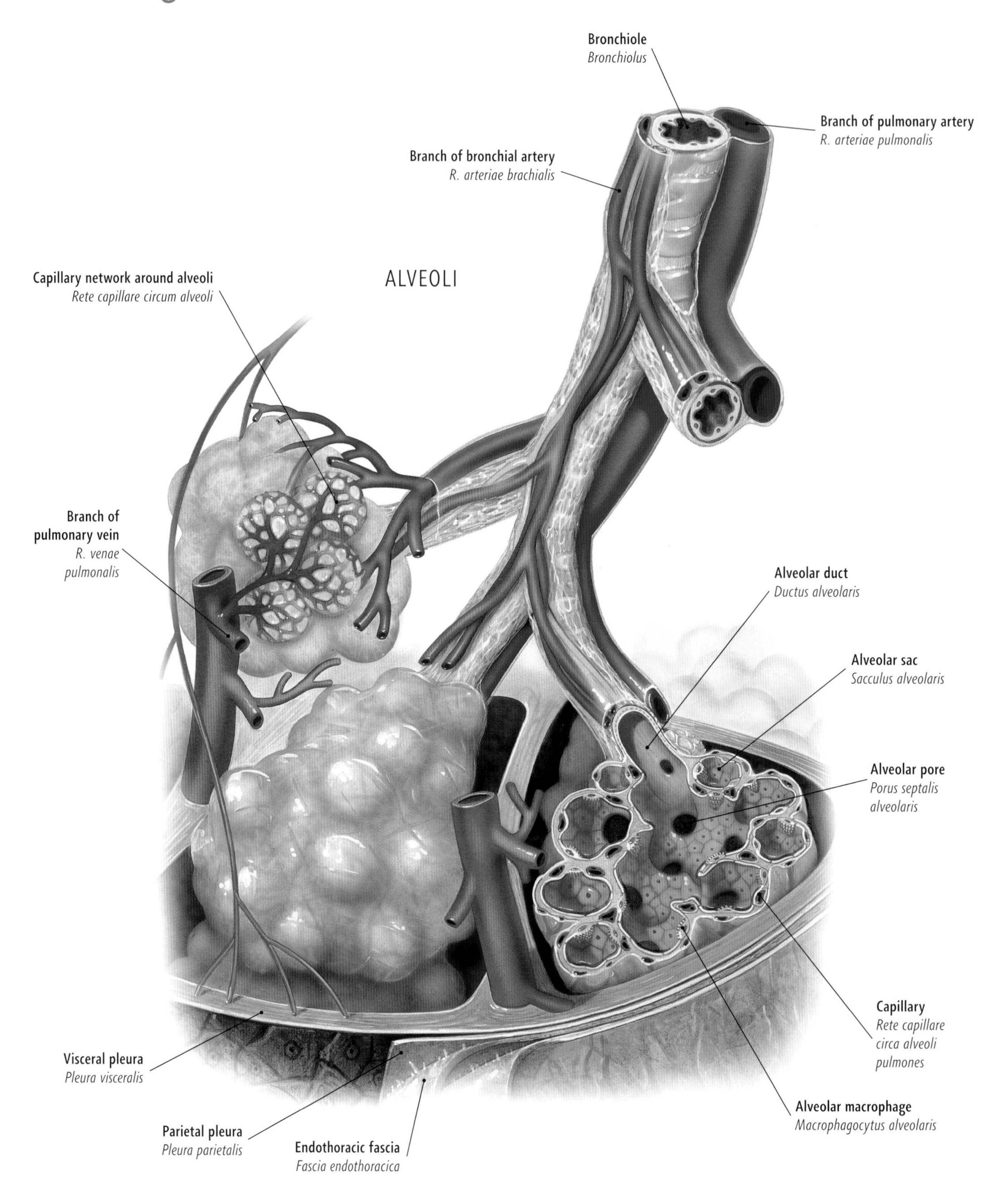
ALVEOLI
Bronchiole
Bronchiolus
Branch of pulmonary artery
R. arteriae pulmonalis
Branch of bronchial artery
R. arteriae brachialis
Capillary network around alveoli
Rete capillare circum alveoli
Branch of
pulmonary vein
R. venae
pulmonalis
Alveolar duct
Ductus alveolaris
Alveolar sac
Sacculus alveolaris
Alveolar pore
Porus septalis
alveolaris
Capillary
Rete capillare
circa alveoli
pulmones
Alveolar macrophage
Macrophagocytus alveolaris
Visceral pleura
Pleura visceralis
Parietal pleura
Pleura parietalis
Endothoracic fascia
Fascia endothoracica

[~] = no direct Latin equivalent

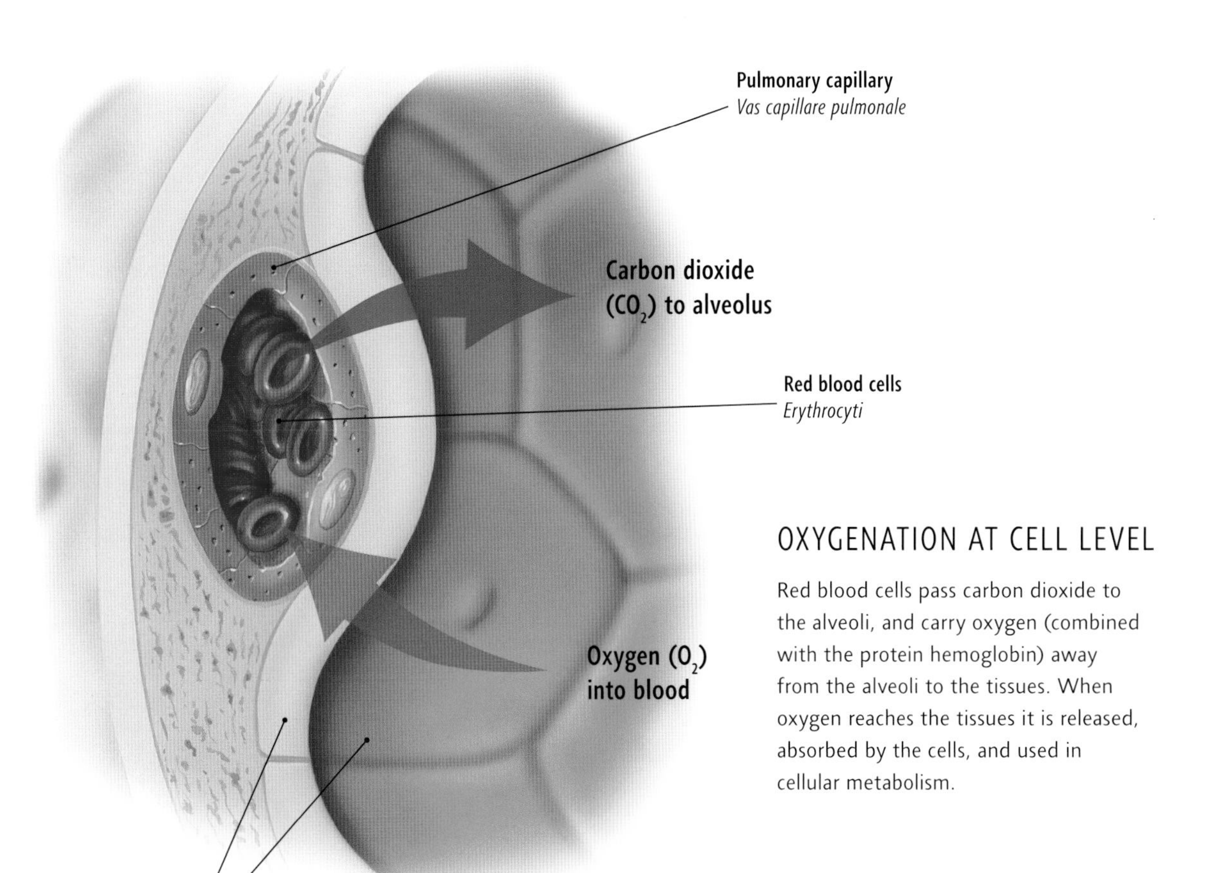
Pulmonary capillary
Vas capillare pulmonale
Carbon dioxide (CO_2) to alveolus
Red blood cells
Erythrocyti
Oxygen (O_2) into blood
Alveolar epithelium
Epithelium alveolare

OXYGENATION AT CELL LEVEL

Red blood cells pass carbon dioxide to the alveoli, and carry oxygen (combined with the protein hemoglobin) away from the alveoli to the tissues. When oxygen reaches the tissues it is released, absorbed by the cells, and used in cellular metabolism.

MOVING OXYGEN AROUND THE BODY

Blood travels continuously through two circulatory systems: the pulmonary (lung) and the systemic (body) circulations. The heart pumps oxygen-poor blood from the systemic circulation into the pulmonary circulation. This blood is oxygenated by the lungs and then pumped by the heart into the systemic circulation.

Note: Part of the lungs has been removed to reveal the pleura covering the diaphragm.

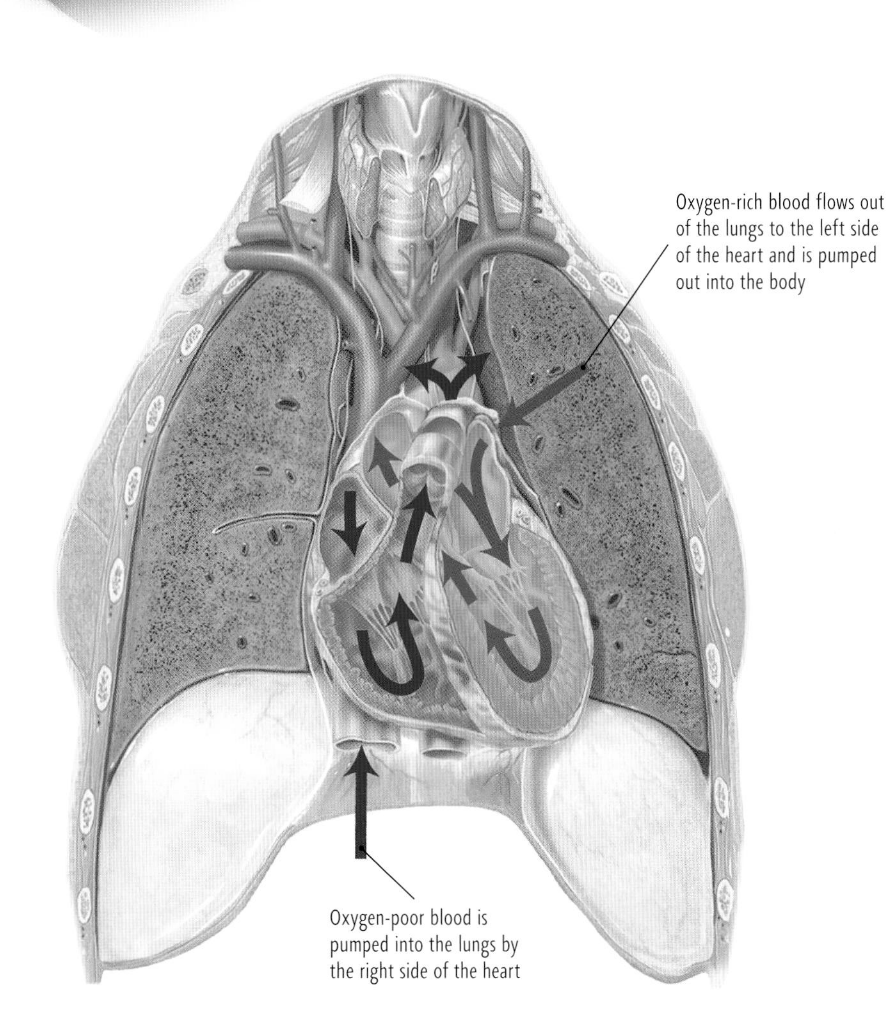
Oxygen-rich blood flows out of the lungs to the left side of the heart and is pumped out into the body
Oxygen-poor blood is pumped into the lungs by the right side of the heart

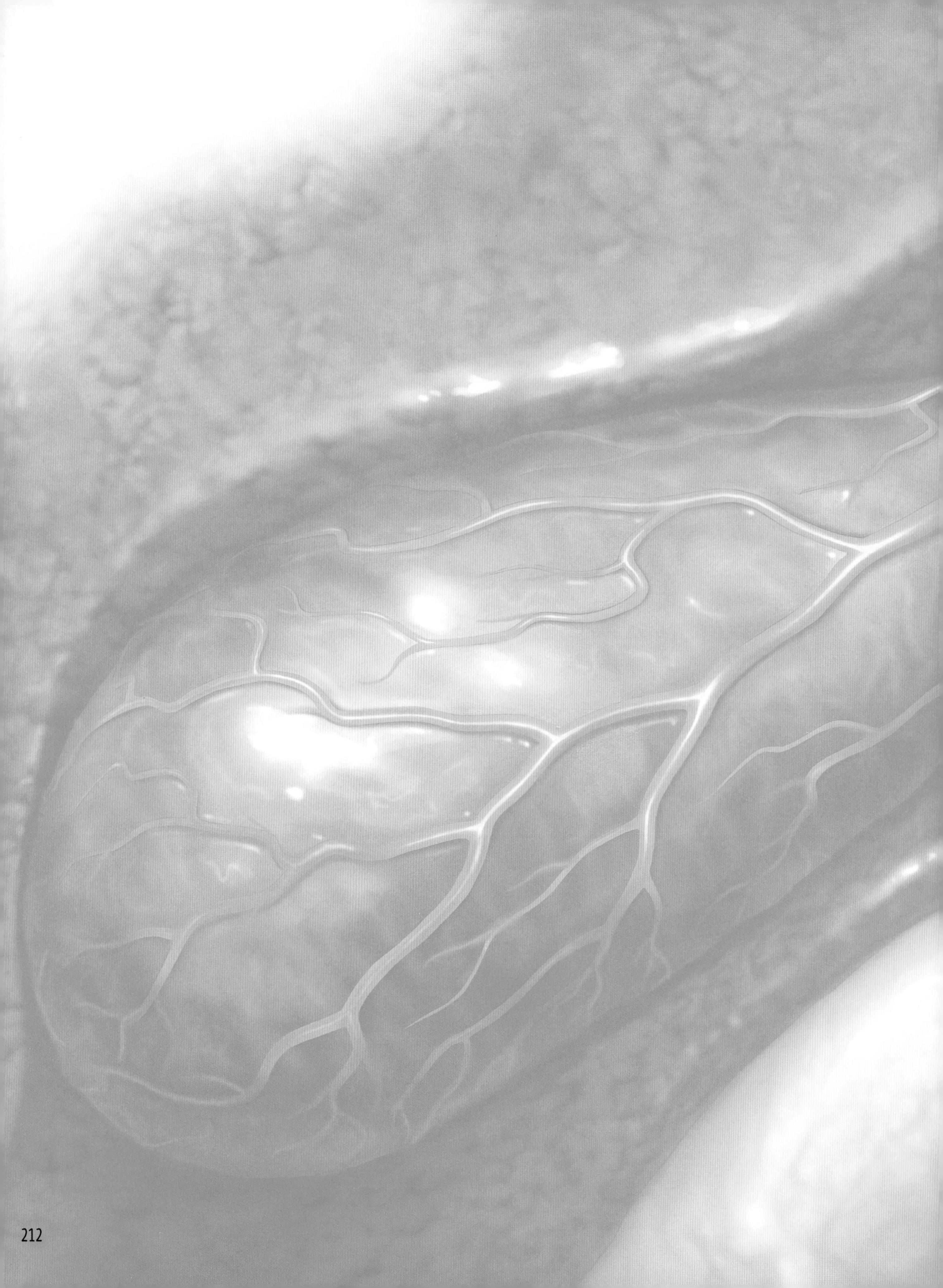

The Digestive System

The Digestive System

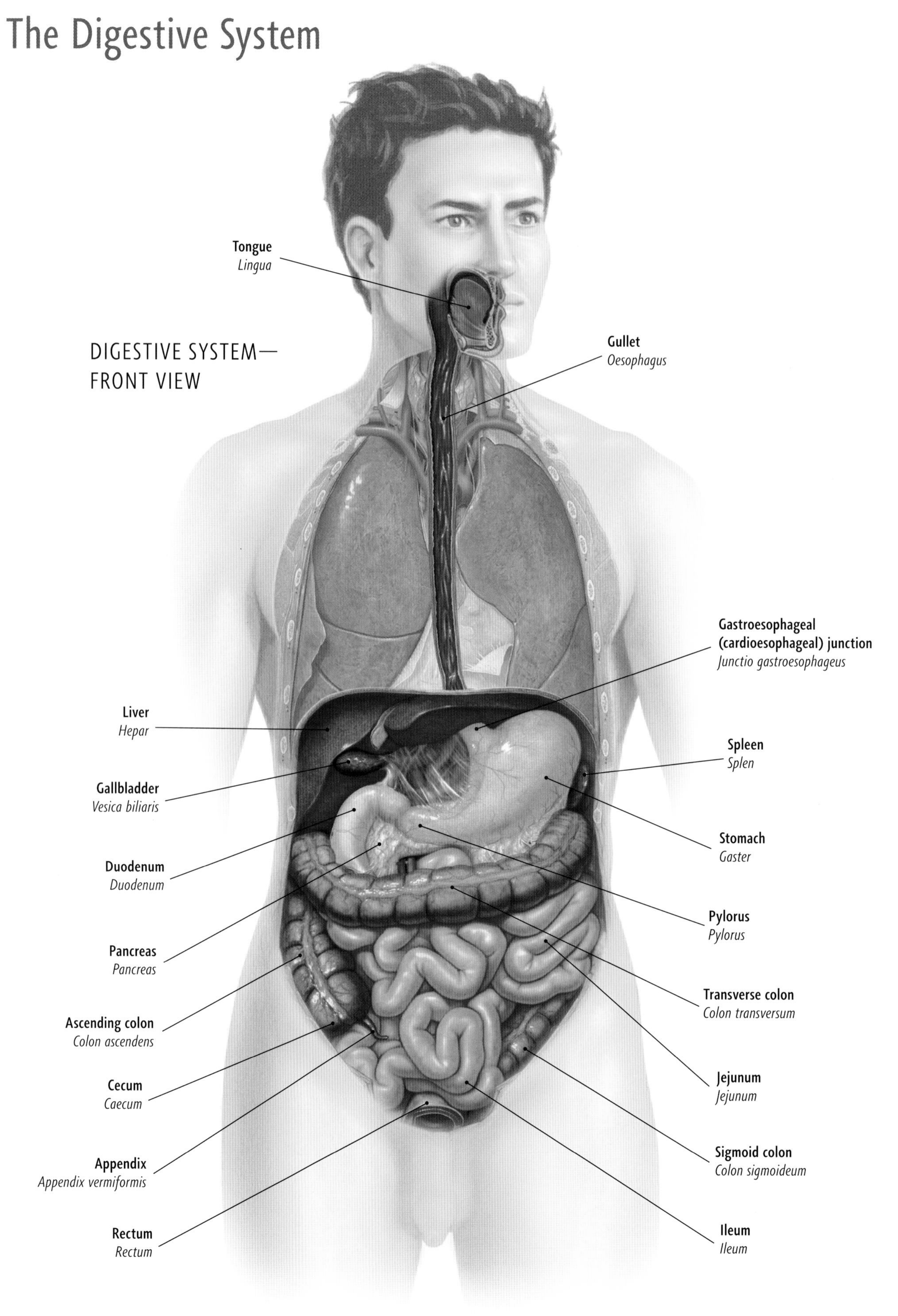
DIGESTIVE SYSTEM—
FRONT VIEW
Tongue
Lingua
Gullet
Oesophagus
Gastroesophageal
(cardioesophageal) junction
Junctio gastroesophageus
Liver
Hepar
Spleen
Splen
Gallbladder
Vesica biliaris
Stomach
Gaster
Duodenum
Duodenum
Pylorus
Pylorus
Pancreas
Pancreas
Transverse colon
Colon transversum
Ascending colon
Colon ascendens
Jejunum
Jejunum
Cecum
Caecum
Sigmoid colon
Colon sigmoideum
Appendix
Appendix vermiformis
Rectum
Rectum
Ileum
Ileum

[~] = no direct Latin equivalent

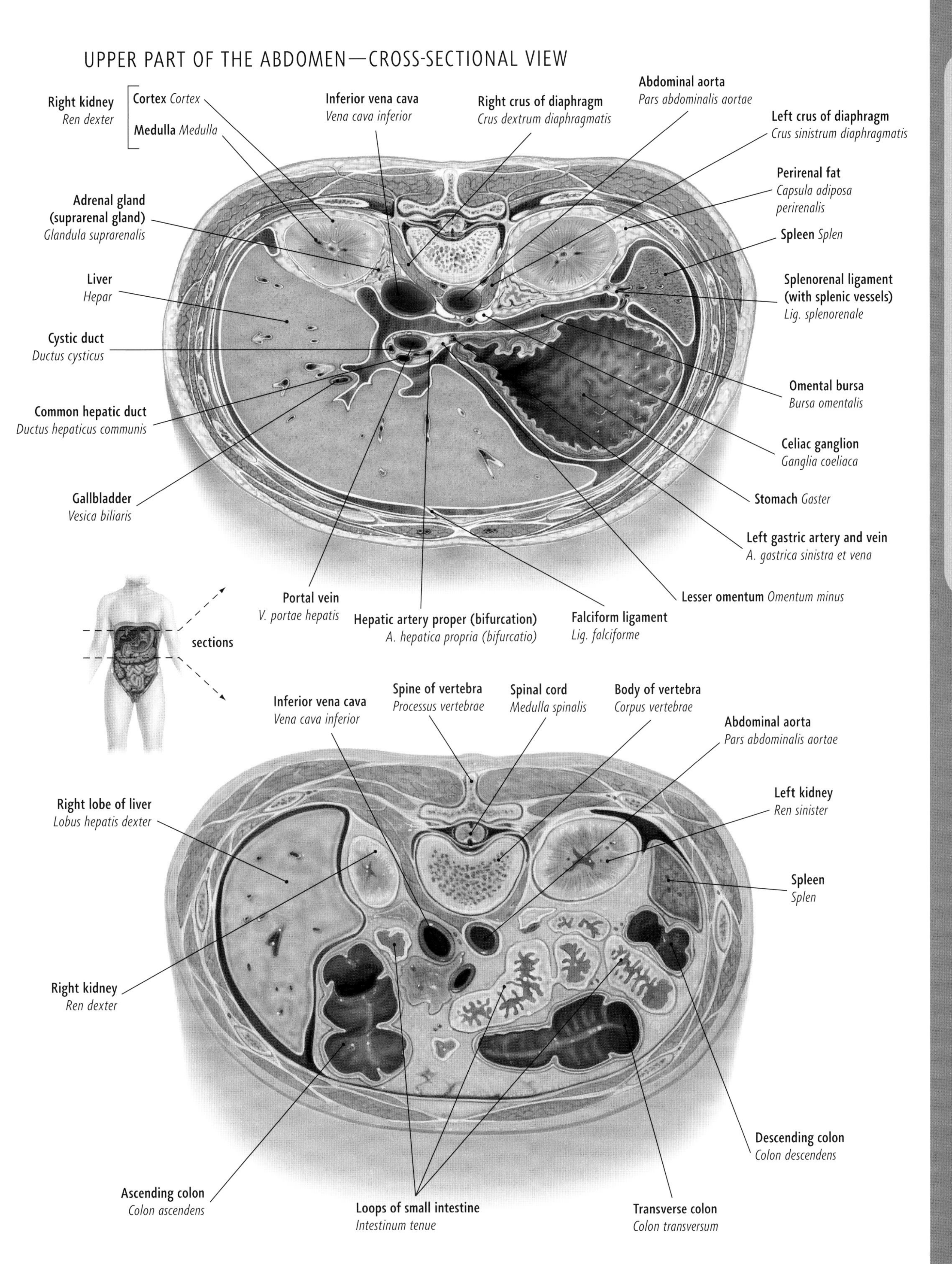

UPPER PART OF THE ABDOMEN—CROSS-SECTIONAL VIEW
Right kidney
Ren dexter
Cortex Cortex
Medulla Medulla
Inferior vena cava
Vena cava inferior
Right crus of diaphragm
Crus dextrum diaphragmatis
Abdominal aorta
Pars abdominalis aortae
Left crus of diaphragm
Crus sinistrum diaphragmatis
Perirenal fat
Capsula adiposa perirenalis
Adrenal gland (suprarenal gland)
Glandula suprarenalis
Spleen Splen
Liver
Hepar
Splenorenal ligament (with splenic vessels)
Lig. splenorenale
Cystic duct
Ductus cysticus
Omental bursa
Bursa omentalis
Common hepatic duct
Ductus hepaticus communis
Celiac ganglion
Ganglia coeliaca
Gallbladder
Vesica biliaris
Stomach Gaster
Left gastric artery and vein
A. gastrica sinistra et vena
Lesser omentum Omentum minus
Portal vein
V. portae hepatis
Hepatic artery proper (bifurcation)
A. hepatica propria (bifurcatio)
Falciform ligament
Lig. falciforme
sections
Inferior vena cava
Vena cava inferior
Spine of vertebra
Processus vertebrae
Spinal cord
Medulla spinalis
Body of vertebra
Corpus vertebrae
Abdominal aorta
Pars abdominalis aortae
Left kidney
Ren sinister
Right lobe of liver
Lobus hepatis dexter
Spleen
Splen
Right kidney
Ren dexter
Descending colon
Colon descendens
Ascending colon
Colon ascendens
Loops of small intestine
Intestinum tenue
Transverse colon
Colon transversum
MIDDLE PART OF THE ABDOMEN—CROSS-SECTIONAL VIEW

The Digestive System

DIGESTIVE ORGANS AND GREATER OMENTUM

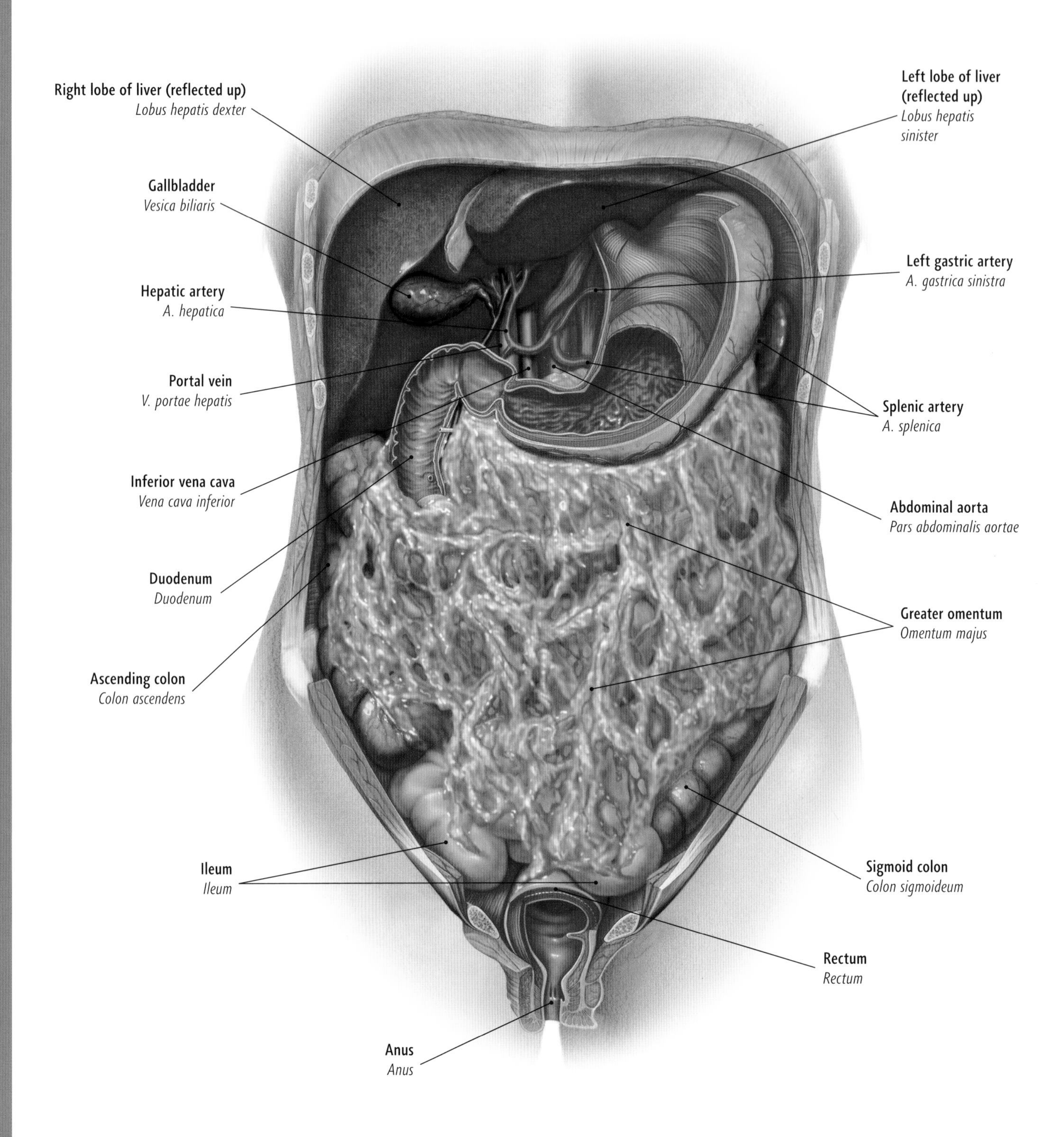

Right lobe of liver (reflected up)
Lobus hepatis dexter
Gallbladder
Vesica biliaris
Hepatic artery
A. hepatica
Portal vein
V. portae hepatis
Inferior vena cava
Vena cava inferior
Duodenum
Duodenum
Ascending colon
Colon ascendens
Ileum
Ileum
Anus
Anus
Left lobe of liver (reflected up)
Lobus hepatis sinister
Left gastric artery
A. gastrica sinistra
Splenic artery
A. splenica
Abdominal aorta
Pars abdominalis aortae
Greater omentum
Omentum majus
Sigmoid colon
Colon sigmoideum
Rectum
Rectum

ABDOMINAL ORGANS

Falciform ligament *Lig. falciforme*

Common hepatic duct *Ductus hepaticus communis*

Right lobe of liver *Lobus hepatis dexter*

Left hepatic artery *R. arteriae hepatica sinister*

Inferior vena cava *Vena cava inferior*

Right crus of diaphragm *Crus dextrum diaphragmae*

Common hepatic artery *A. hepatica communis*

Gallbladder *Vesica biliaris*

Rib *Costa*

Intercostal muscles *Mm. intercostales*

Cystic duct *Ductus cysticus*

Bile duct *Ductus biliaris*

Portal vein *V. portae hepatis*

Pyloric sphincter *M. sphincter pyloricus*

Pylorus *Pylorus*

Plica circularis of duodenum *Plica circularis duodeni*

Accessory pancreatic duct *Ductus pancreaticus accessorius*

Hepatic (right colic) flexure *Flexura coli dextra*

Main pancreatic duct *Ductus pancreaticus*

Greater duodenal papilla *Papilla duodeni major*

Head of pancreas *Caput pancreatis*

Uncinate process of pancreas *Processus uncinatus pancreatis*

Ascending colon *Colon ascendens*

Iliac bone *Os ilium*

External abdominal oblique *M. obliquus externus abdominis*

Cecum *Caecum*

Appendix *Appendix vermiformis*

Ileum *Ileum*

Transversus abdominis *M. transversus abdominis*

Internal oblique *M. obliquus internus abdominis*

Pubic bone *Os pubis*

Anal column *Columna analis*

Anus *Anus*

External anal sphincter *M. sphincter ani externus*

Left lobe of liver *Lobus hepatis sinister*

Diaphragm *Diaphragma*

Left gastric artery *A. gastrica sinistra*

Stomach *Gaster*

Left crus of diaphragm *Crus sinistrum diaphragmae*

Splenic artery *A. splenica*

Spleen *Splen*

Abdominal aorta *Pars abdominalis aortae*

Body of pancreas *Corpus pancreatis*

Left gastroepiploic artery *A. gastroomentalis sinistra*

Greater omentum *Omentum majus*

Transverse colon *Colon transversum*

Jejunum *Jejunum*

Duodenojejunal junction *Flexura duodenojejunalis*

Superior mesenteric artery *Arteria mesenterica superior*

Superior mesenteric vein *V. mesenterica superior*

Sigmoid colon *Colon sigmoideum*

Ileum *Ileum*

Rectum *Rectum*

Levator ani *M. levator ani*

Internal anal sphincter *M. sphincter ani internus*

Note: In this illustration the liver has been pulled back to reveal the other abdominal organs.

Digestion and Absorption of Nutrients

STOMACH

Food is swallowed via the mouth, travels down through the esophagus, and enters the stomach, where it is mixed with acid and pepsin (an enzyme that breaks down proteins into amino acids).

SMALL INTESTINE

From the stomach, semiliquid food matter (known as chyme) passes into the small intestine, where it is broken down into simple sugars, amino acids, and fatty acids that can be absorbed through the lining of the small intestine.

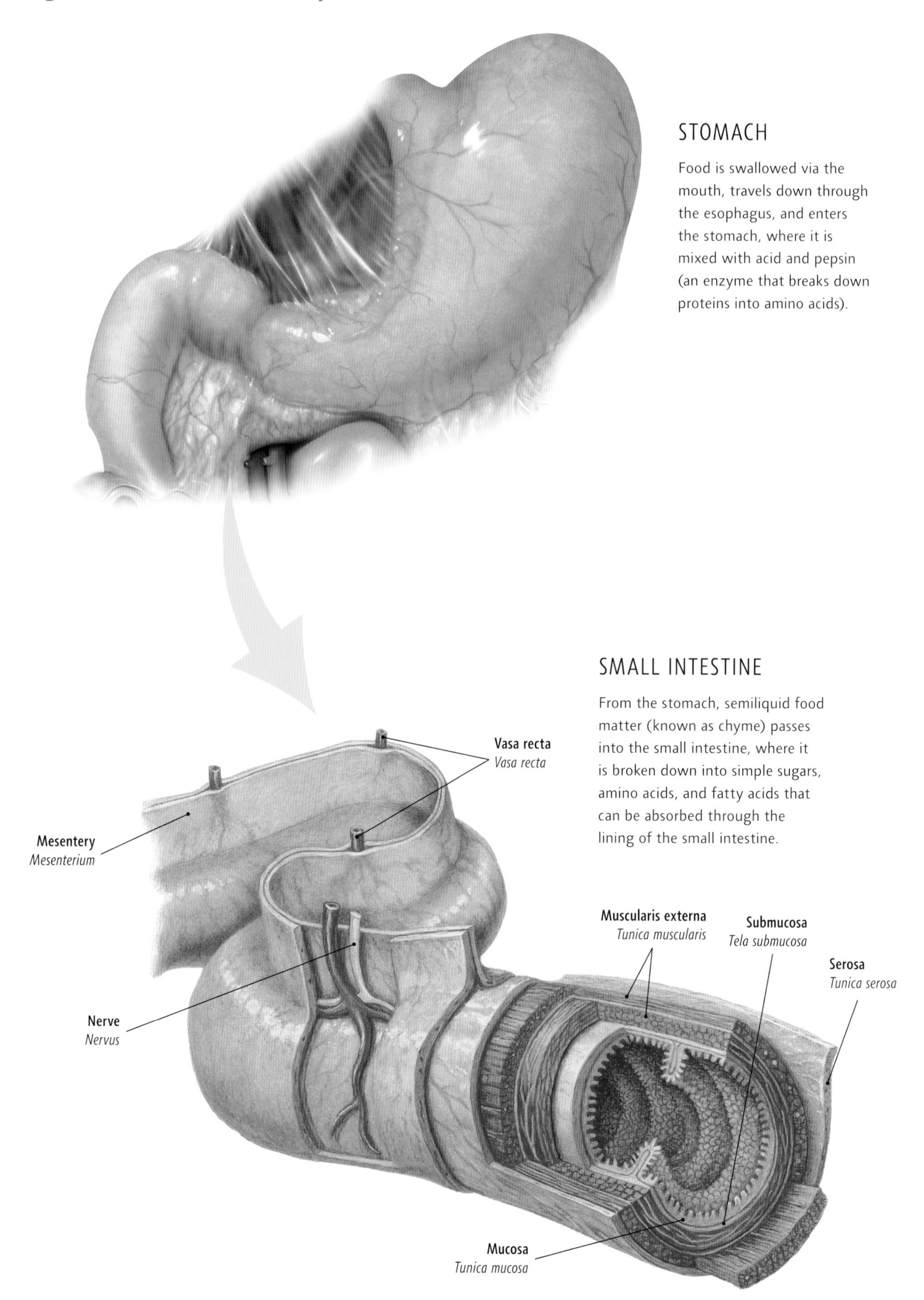

Vasa recta
Vasa recta
Mesentery
Mesenterium
Muscularis externa
Tunica muscularis
Submucosa
Tela submucosa
Serosa
Tunica serosa
Nerve
Nervus
Mucosa
Tunica mucosa

LIVER

A liver lobule is full of hepatic cells arranged around a branch of the portal vein. The hepatic cells extract nutrients from the blood that flows along the branches of this vein and extract oxygen and fat from the branches of the hepatic arteries.

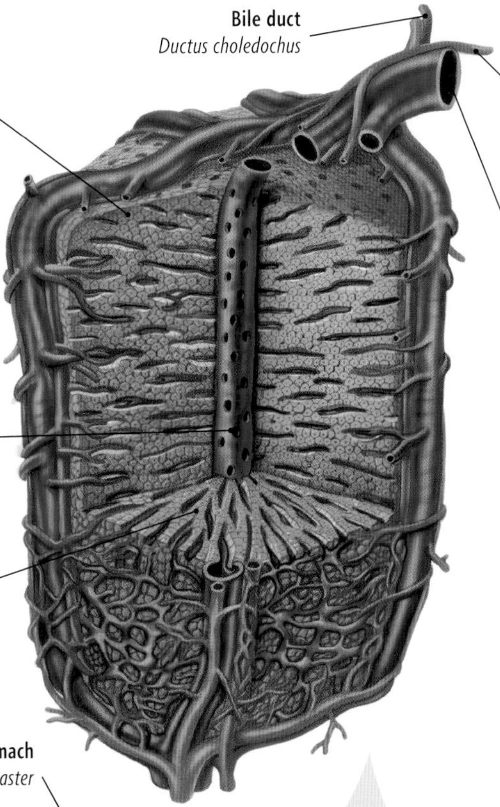

Bile duct *Ductus choledochus*

Liver cells (hepatocytes) *Hepatocyti*

Artery *Arteria*

Branch of portal vein *R. venae portae hepatis*

Central vein *V. centralis*

Liver plate *Lamina hepatica*

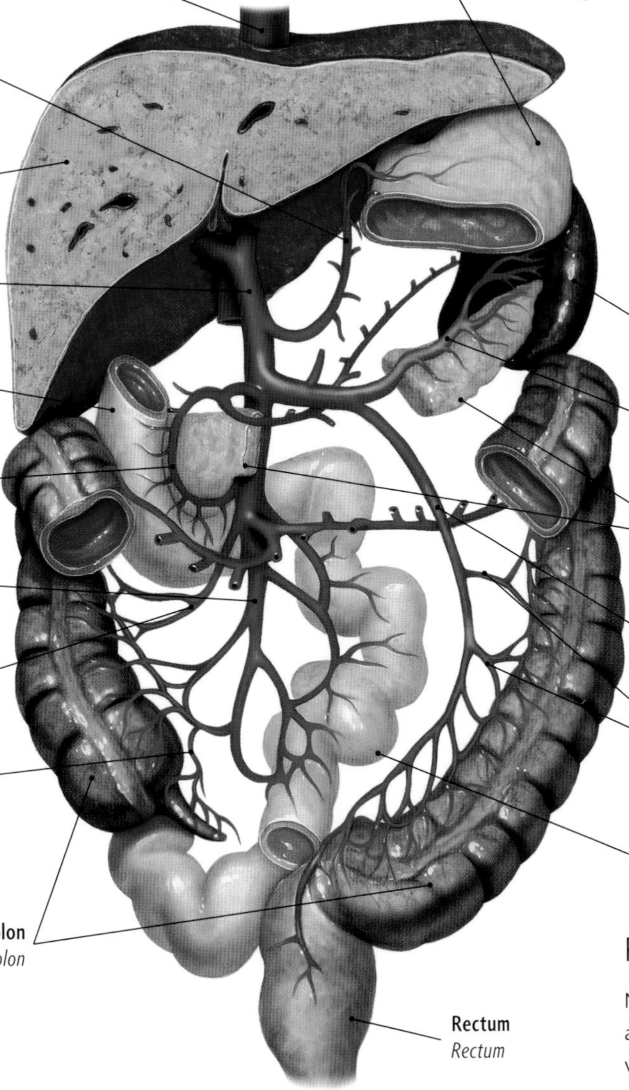

Inferior vena cava *Vena cava inferior*

Stomach *Gaster*

Left gastric vein *V. gastrica sinistra*

Liver *Hepar*

Portal vein *V. portae hepatis*

Duodenum *Duodenum*

Spleen *Splen*

Splenic vein *V. splenica*

Pancreaticoduodenal vein *V. pancreaticoduodenalis*

Pancreas *Pancreas*

Superior mesenteric vein *V. mesenterica superior*

Inferior mesenteric vein *V. mesenterica inferior*

Right colic vein *V. colica dextra*

Left colic veins *Vv. colicae sinistrae*

Appendicular vein *V. appendicularis*

Small intestine *Intestinum tenue*

Colon *Colon*

Rectum *Rectum*

PORTAL SYSTEM

Nutrients extracted by the digestive system are transported to the liver by a network of veins, known as the hepatic portal system, entering the liver via the portal vein.

Mouth and Tongue

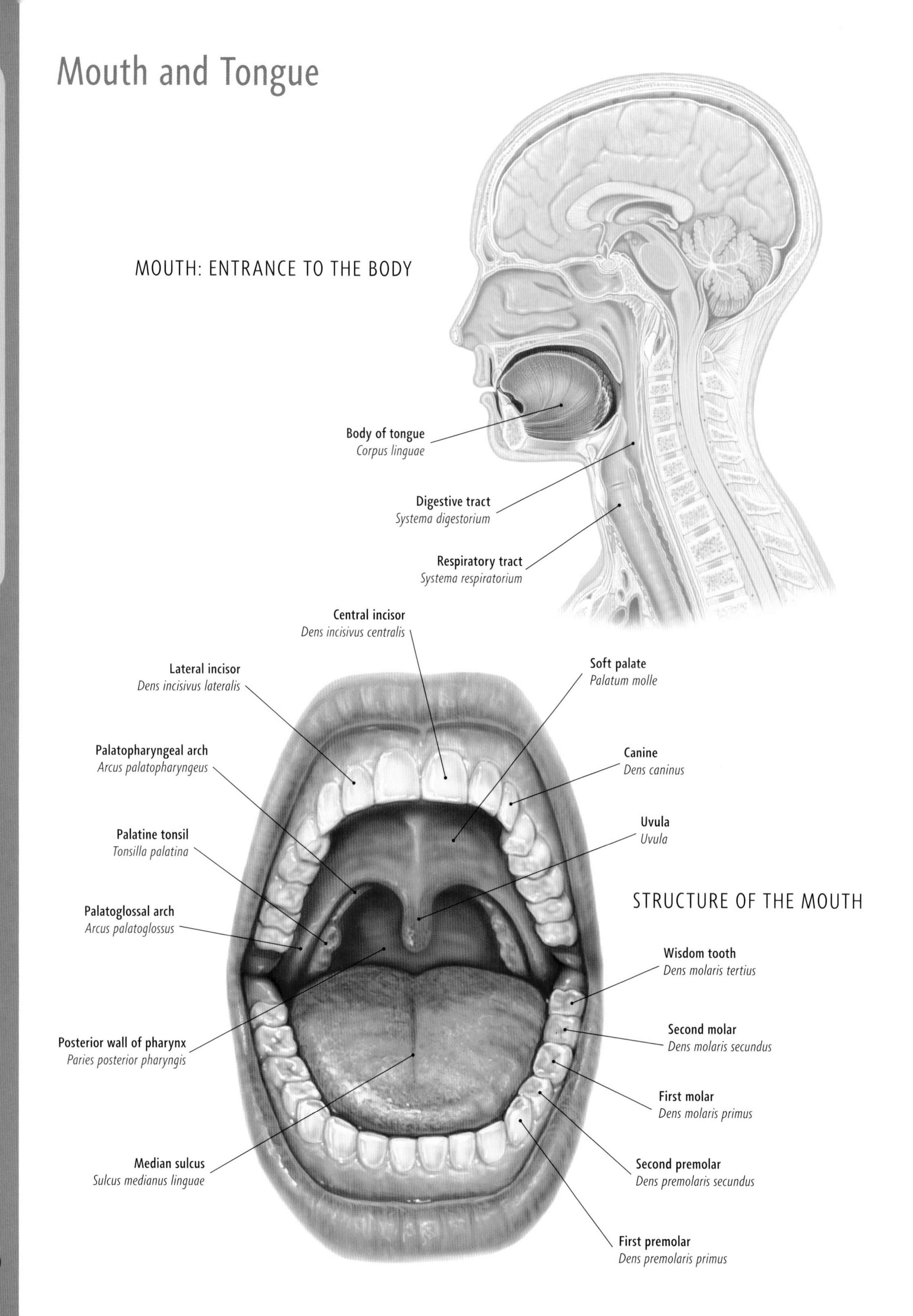
THE DIGESTIVE SYSTEM
MOUTH: ENTRANCE TO THE BODY
Body of tongue
Corpus linguae
Digestive tract
Systema digestorium
Respiratory tract
Systema respiratorium
Central incisor
Dens incisivus centralis
Lateral incisor
Dens incisivus lateralis
Soft palate
Palatum molle
Palatopharyngeal arch
Arcus palatopharyngeus
Canine
Dens caninus
Uvula
Uvula
Palatine tonsil
Tonsilla palatina
STRUCTURE OF THE MOUTH
Palatoglossal arch
Arcus palatoglossus
Wisdom tooth
Dens molaris tertius
Second molar
Dens molaris secundus
Posterior wall of pharynx
Paries posterior pharyngis
First molar
Dens molaris primus
Second premolar
Dens premolaris secundus
Median sulcus
Sulcus medianus linguae
First premolar
Dens premolaris primus

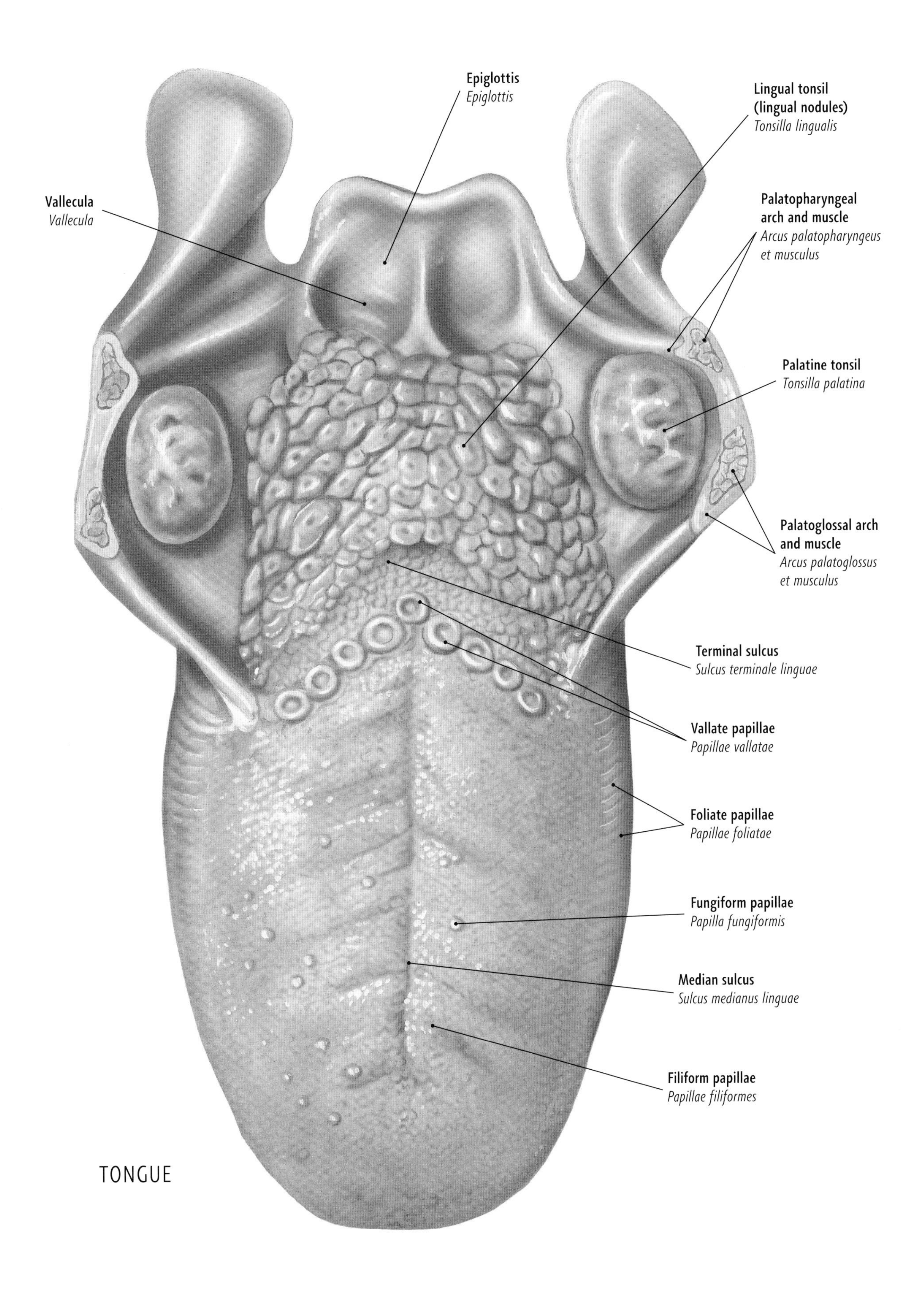
Epiglottis
Epiglottis
Lingual tonsil (lingual nodules)
Tonsilla lingualis
Vallecula
Vallecula
Palatopharyngeal arch and muscle
Arcus palatopharyngeus et musculus
Palatine tonsil
Tonsilla palatina
Palatoglossal arch and muscle
Arcus palatoglossus et musculus
Terminal sulcus
Sulcus terminale linguae
Vallate papillae
Papillae vallatae
Foliate papillae
Papillae foliatae
Fungiform papillae
Papilla fungiformis
Median sulcus
Sulcus medianus linguae
Filiform papillae
Papillae filiformes

TONGUE

Teeth

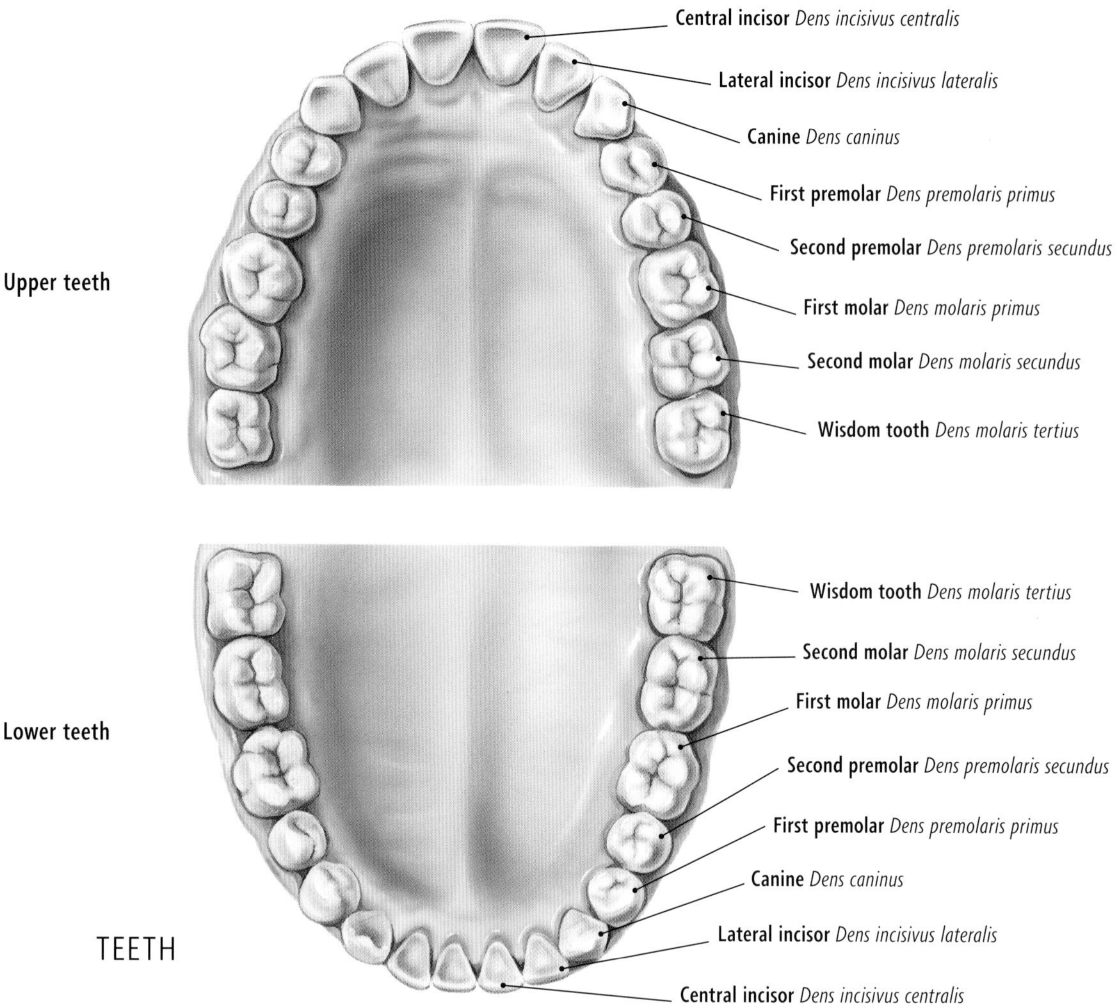

Upper teeth

Central incisor *Dens incisivus centralis*

Lateral incisor *Dens incisivus lateralis*

Canine *Dens caninus*

First premolar *Dens premolaris primus*

Second premolar *Dens premolaris secundus*

First molar *Dens molaris primus*

Second molar *Dens molaris secundus*

Wisdom tooth *Dens molaris tertius*

Lower teeth

Wisdom tooth *Dens molaris tertius*

Second molar *Dens molaris secundus*

First molar *Dens molaris primus*

Second premolar *Dens premolaris secundus*

First premolar *Dens premolaris primus*

Canine *Dens caninus*

Lateral incisor *Dens incisivus lateralis*

Central incisor *Dens incisivus centralis*

TEETH

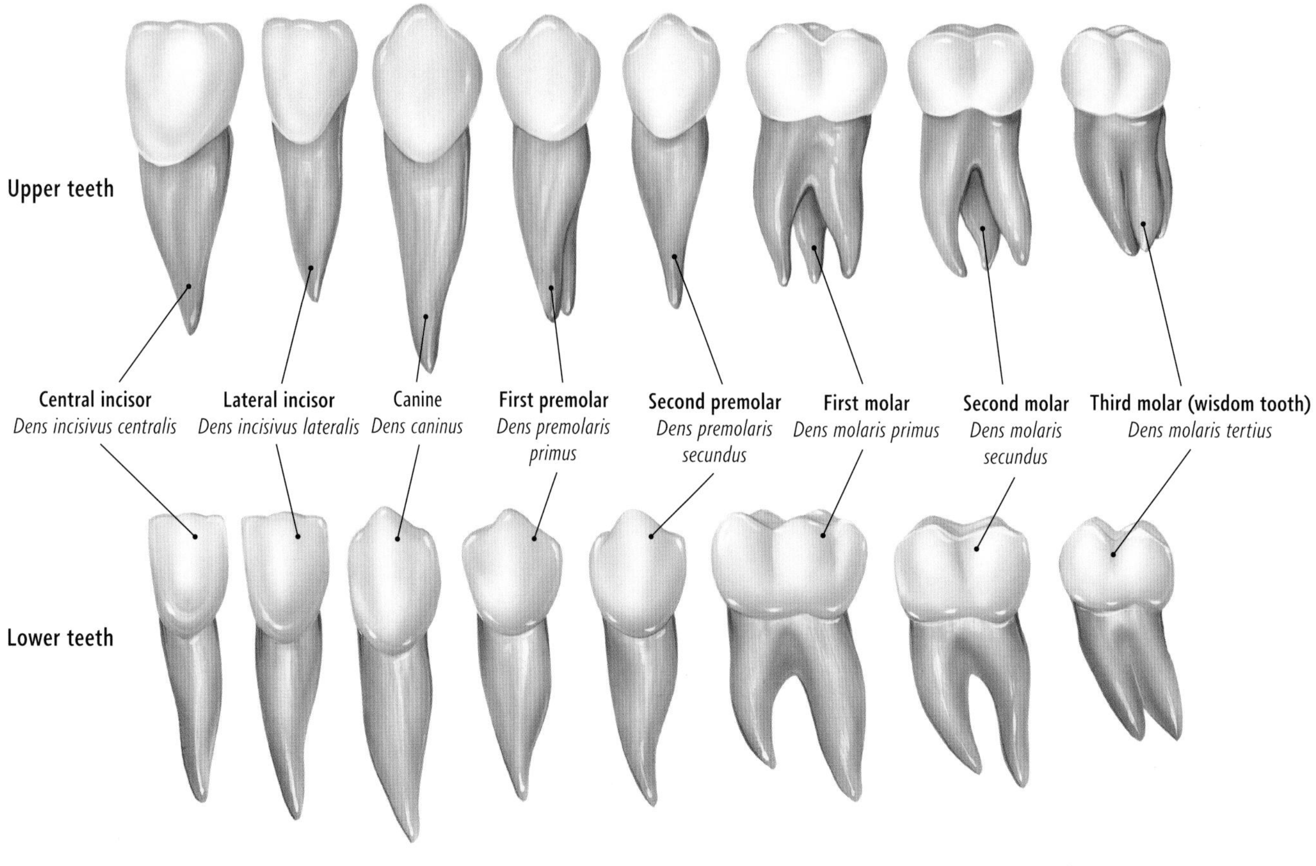

Upper teeth

Central incisor *Dens incisivus centralis*

Lateral incisor *Dens incisivus lateralis*

Canine *Dens caninus*

First premolar *Dens premolaris primus*

Second premolar *Dens premolaris secundus*

First molar *Dens molaris primus*

Second molar *Dens molaris secundus*

Third molar (wisdom tooth) *Dens molaris tertius*

Lower teeth

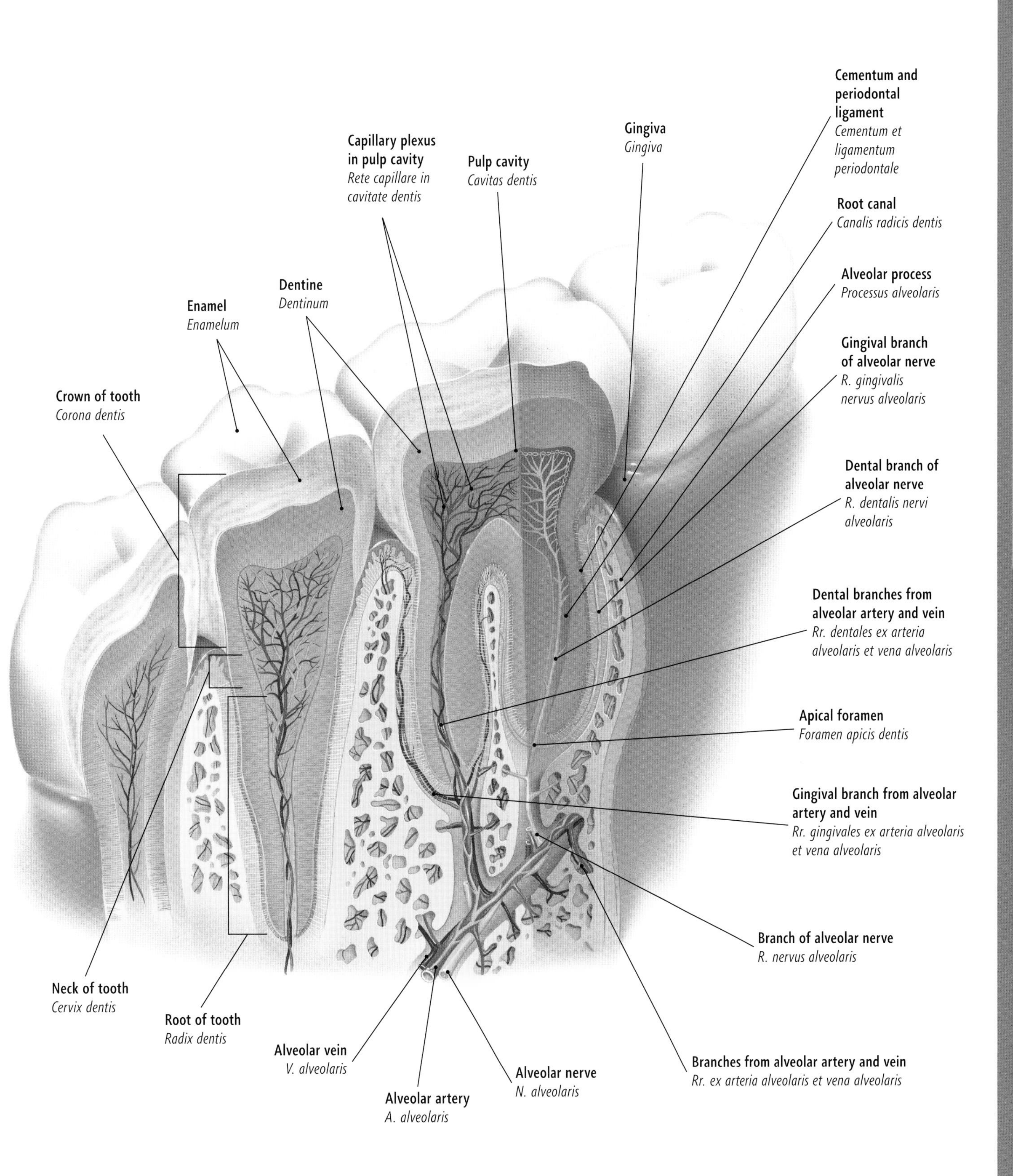

Cementum and periodontal ligament
Cementum et ligamentum periodontale
Gingiva
Gingiva
Capillary plexus in pulp cavity
Rete capillare in cavitate dentis
Pulp cavity
Cavitas dentis
Root canal
Canalis radicis dentis
Dentine
Dentinum
Enamel
Enamelum
Alveolar process
Processus alveolaris
Gingival branch of alveolar nerve
R. gingivalis nervus alveolaris
Crown of tooth
Corona dentis
Dental branch of alveolar nerve
R. dentalis nervi alveolaris
Dental branches from alveolar artery and vein
Rr. dentales ex arteria alveolaris et vena alveolaris
Apical foramen
Foramen apicis dentis
Gingival branch from alveolar artery and vein
Rr. gingivales ex arteria alveolaris et vena alveolaris
Branch of alveolar nerve
R. nervus alveolaris
Neck of tooth
Cervix dentis
Root of tooth
Radix dentis
Alveolar vein
V. alveolaris
Alveolar nerve
N. alveolaris
Branches from alveolar artery and vein
Rr. ex arteria alveolaris et vena alveolaris
Alveolar artery
A. alveolaris

STRUCTURE OF TEETH—CROSS-SECTIONAL VIEW

[~] = no direct Latin equivalent

Salivary Glands

ORAL CAVITY AND SALIVARY GLANDS

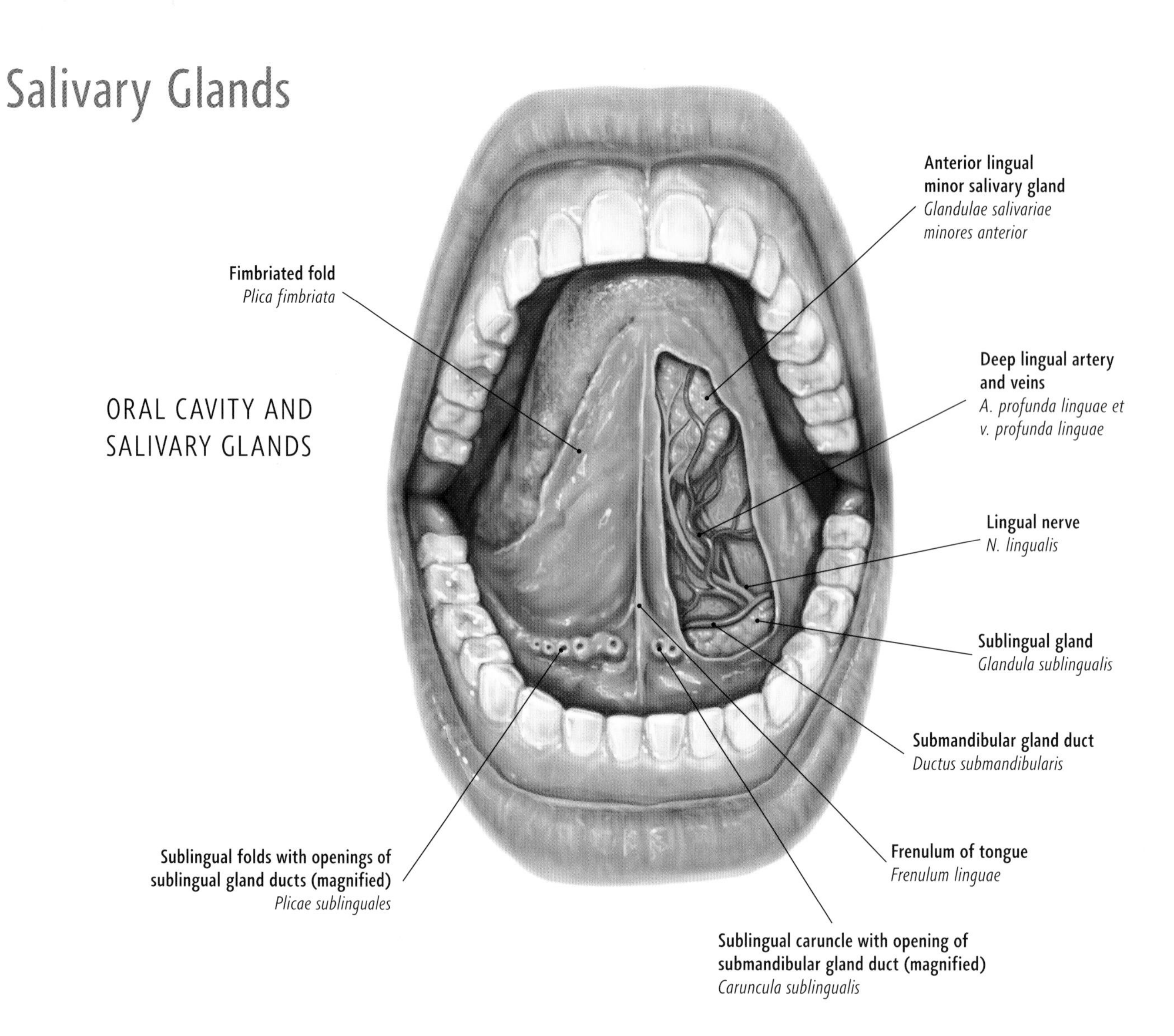

Anterior lingual minor salivary gland
Glandulae salivariae minores anterior
Fimbriated fold
Plica fimbriata
Deep lingual artery and veins
A. profunda linguae et v. profunda linguae
Lingual nerve
N. lingualis
Sublingual gland
Glandula sublingualis
Submandibular gland duct
Ductus submandibularis
Sublingual folds with openings of sublingual gland ducts (magnified)
Plicae sublinguales
Frenulum of tongue
Frenulum linguae
Sublingual caruncle with opening of submandibular gland duct (magnified)
Caruncula sublingualis

SALIVARY GLANDS

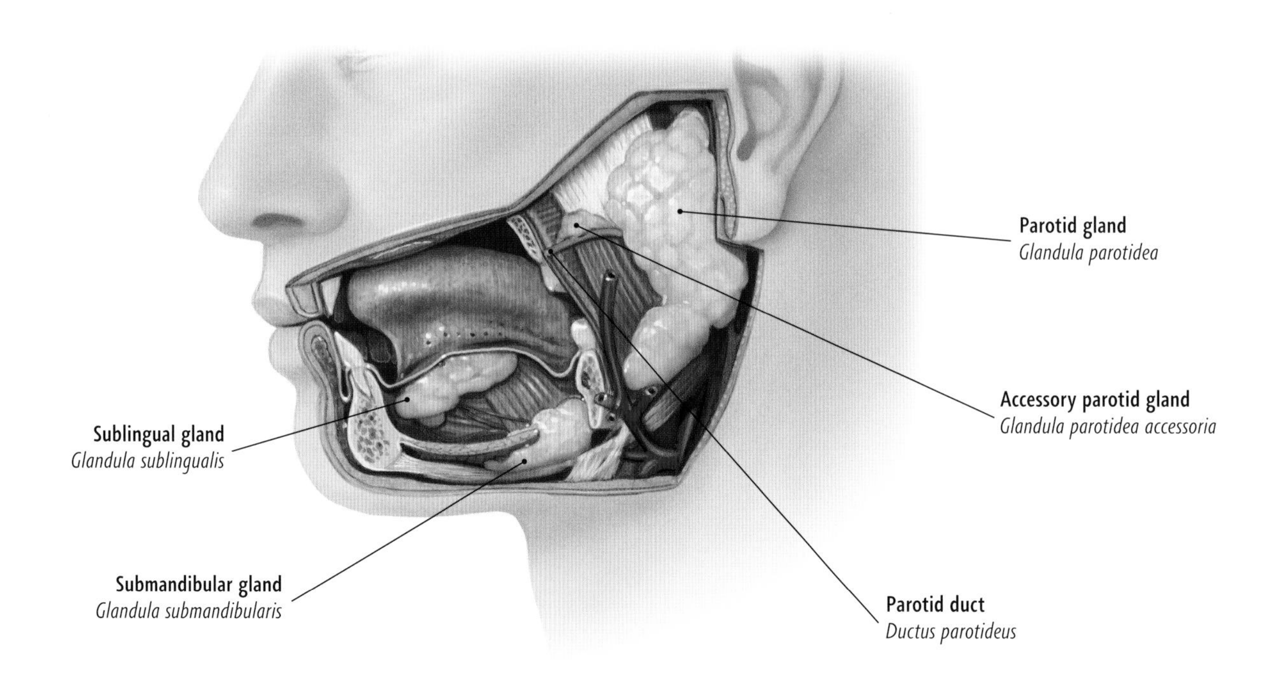

Parotid gland
Glandula parotidea
Accessory parotid gland
Glandula parotidea accessoria
Sublingual gland
Glandula sublingualis
Submandibular gland
Glandula submandibularis
Parotid duct
Ductus parotideus

PAROTID GLAND—MICROSTRUCTURE

The thin saliva produced by the parotid glands contains enzymes specially designed to break down starch.

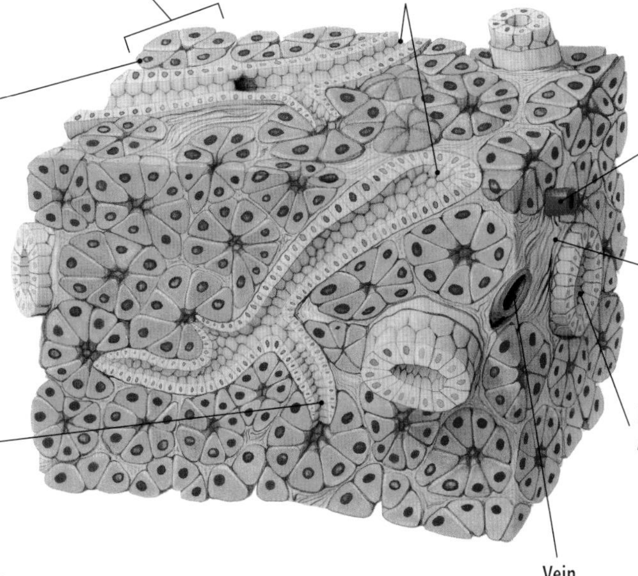

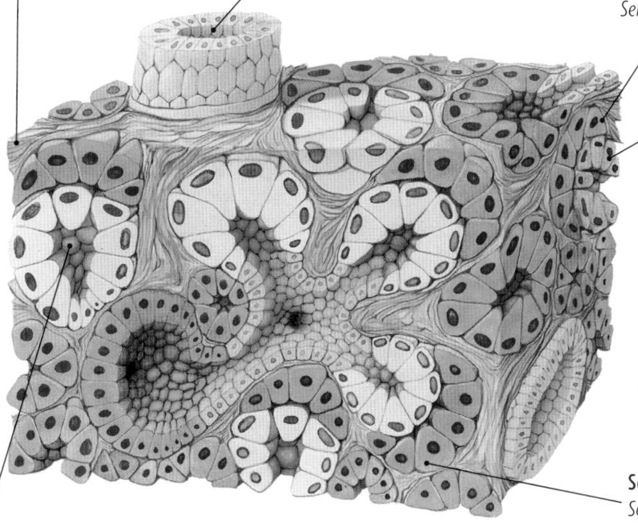

Acinar cell
Acinocytus

Intercalated ducts
Ducti intercalati

Serous cell
Serocytus

Artery
Arteria

Septum
Septum

Striated duct
Ductus striatus

Interlobular duct
Ductus interlobularis

Vein
Vena

SUBMANDIBULAR GLAND—MICROSTRUCTURE

The submandibular glands comprise a mixture of enzyme-producing serous cells and mucus-producing cells. Their saliva is predominantly water.

Interlobular duct
Ductus interlobularis

Septum of connective tissue
Septum texti connectivi

Serous cell (forming a serous crescent)
Serocytus

Mucous cell (forming a mucous acinus)
Mucocytus

Serous crescent (serous demilune)
Semiluna serosa

Mucous tubule
Tubulus mucosae

SUBLINGUAL GLAND—MICROSTRUCTURE

The sublingual glands produce thicker, watery mucus, particularly in response to the presence of milk or cream, which helps to lubricate the mouth.

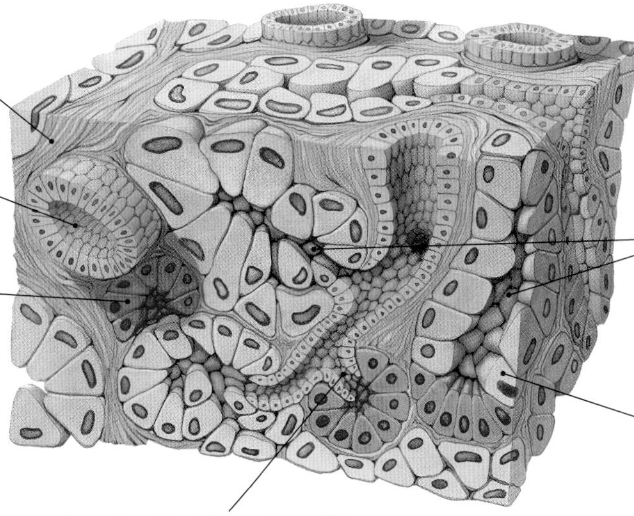

Septum
Septum

Interlobular duct
Ductus interlobularis

Acinar cell
Acinocytus

Mucous tubules
Tubuli mucosae

Mucous cell
Mucocytus

Intercalated duct (from acinar cells)
Ductus intercalatus

Liver and Gallbladder

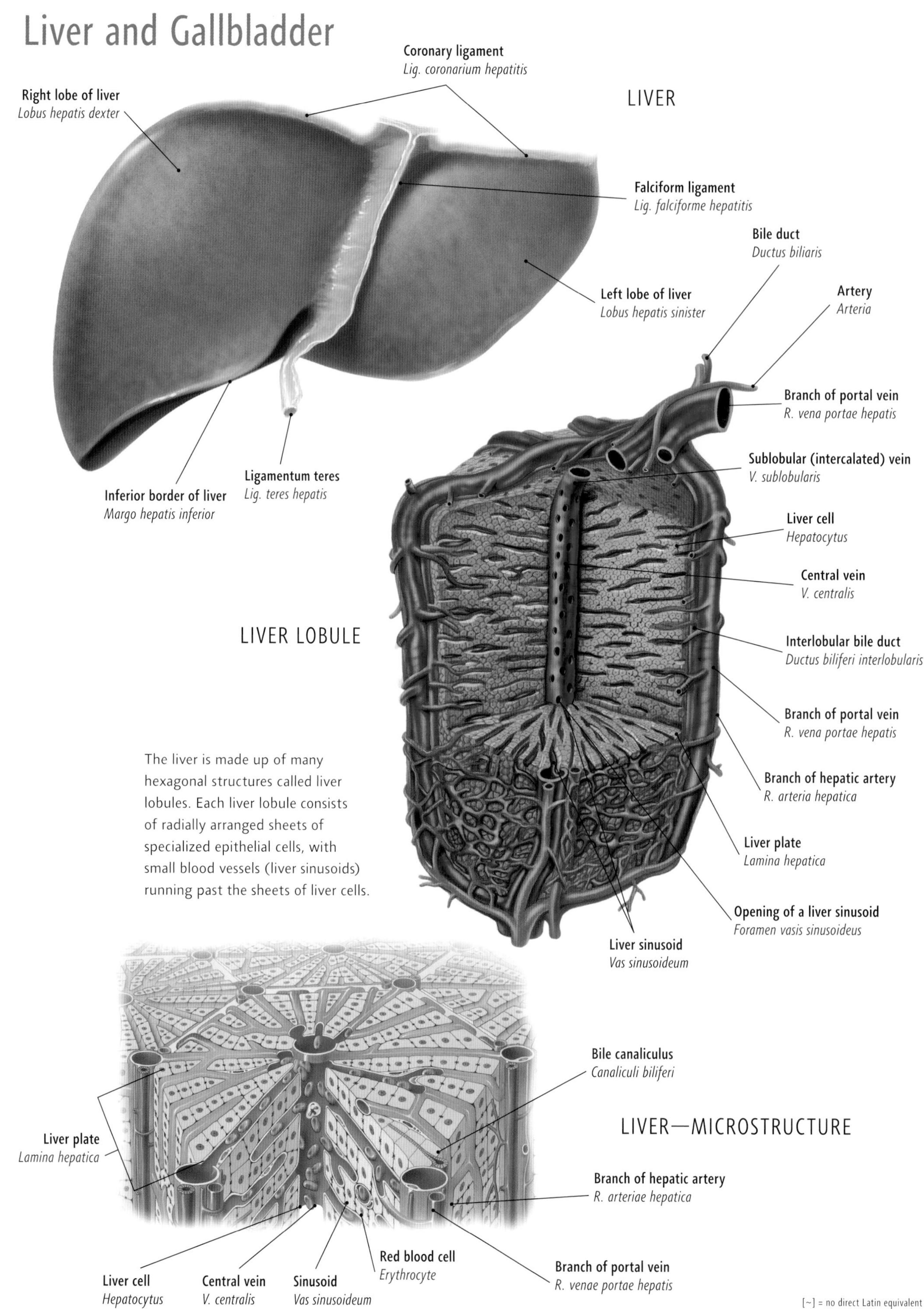
LIVER
Coronary ligament
Lig. coronarium hepatitis
Right lobe of liver
Lobus hepatis dexter
Falciform ligament
Lig. falciforme hepatitis
Left lobe of liver
Lobus hepatis sinister
Ligamentum teres
Lig. teres hepatis
Inferior border of liver
Margo hepatis inferior
LIVER LOBULE
Bile duct
Ductus biliaris
Artery
Arteria
Branch of portal vein
R. vena portae hepatis
Sublobular (intercalated) vein
V. sublobularis
Liver cell
Hepatocytus
Central vein
V. centralis
Interlobular bile duct
Ductus biliferi interlobularis
Branch of portal vein
R. vena portae hepatis
Branch of hepatic artery
R. arteria hepatica
Liver plate
Lamina hepatica
Opening of a liver sinusoid
Foramen vasis sinusoideus
Liver sinusoid
Vas sinusoideum
LIVER—MICROSTRUCTURE
Bile canaliculus
Canaliculi biliferi
Liver plate
Lamina hepatica
Branch of hepatic artery
R. arteriae hepatica
Red blood cell
Erythrocyte
Branch of portal vein
R. venae portae hepatis
Liver cell
Hepatocytus
Central vein
V. centralis
Sinusoid
Vas sinusoideum

The liver is made up of many hexagonal structures called liver lobules. Each liver lobule consists of radially arranged sheets of specialized epithelial cells, with small blood vessels (liver sinusoids) running past the sheets of liver cells.

[~] = no direct Latin equivalent

GALLBLADDER

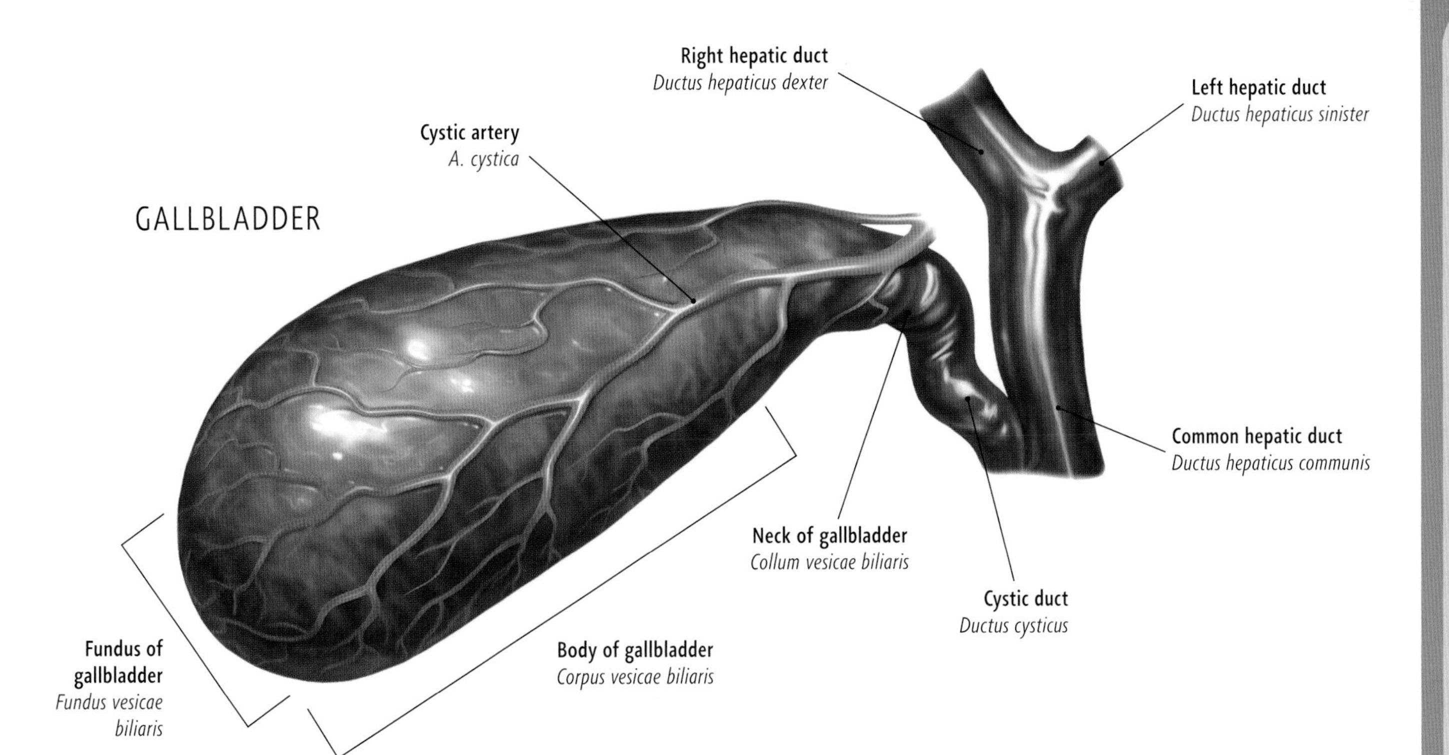
Right hepatic duct
Ductus hepaticus dexter
Left hepatic duct
Ductus hepaticus sinister
Cystic artery
A. cystica
Common hepatic duct
Ductus hepaticus communis
Neck of gallbladder
Collum vesicae biliaris
Cystic duct
Ductus cysticus
Fundus of gallbladder
Fundus vesicae biliaris
Body of gallbladder
Corpus vesicae biliaris

BILE DUCTS

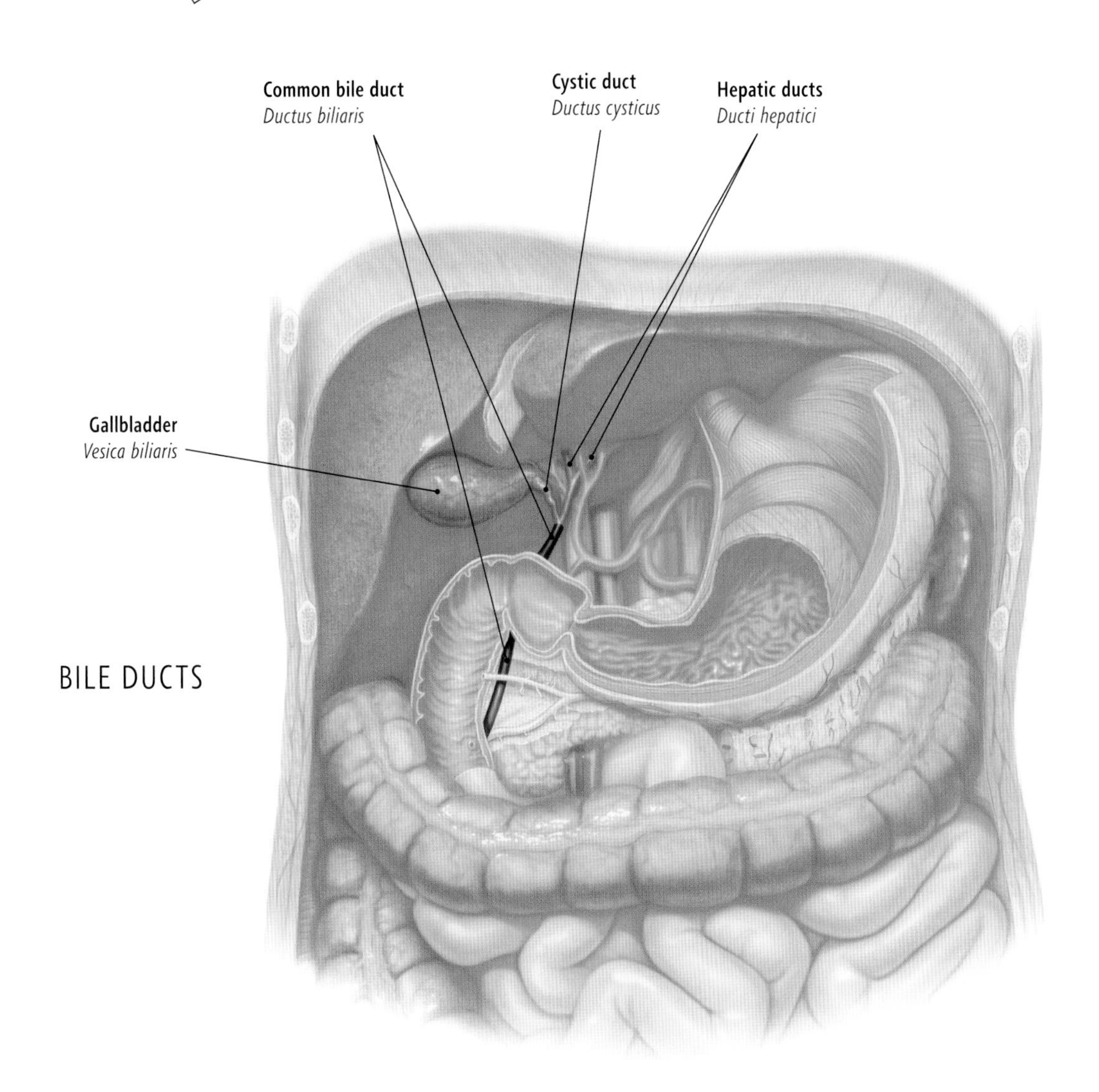
Common bile duct
Ductus biliaris
Cystic duct
Ductus cysticus
Hepatic ducts
Ducti hepatici
Gallbladder
Vesica biliaris

Stomach and Pancreas

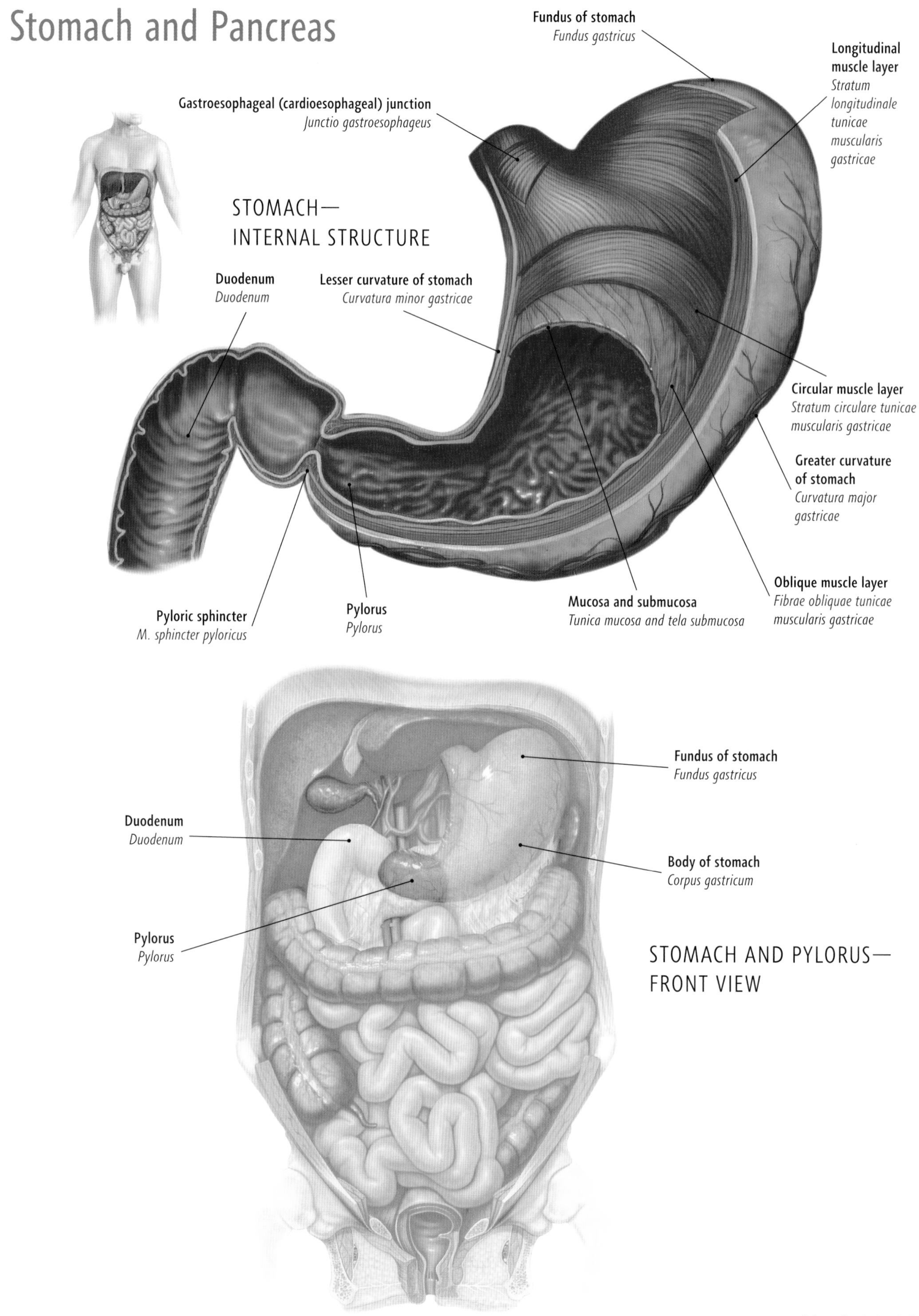
STOMACH—
INTERNAL STRUCTURE
Fundus of stomach
Fundus gastricus
Longitudinal muscle layer
Stratum longitudinale tunicae muscularis gastricae
Gastroesophageal (cardioesophageal) junction
Junctio gastroesophageus
Duodenum
Duodenum
Lesser curvature of stomach
Curvatura minor gastricae
Circular muscle layer
Stratum circulare tunicae muscularis gastricae
Greater curvature of stomach
Curvatura major gastricae
Oblique muscle layer
Fibrae obliquae tunicae muscularis gastricae
Mucosa and submucosa
Tunica mucosa and tela submucosa
Pylorus
Pylorus
Pyloric sphincter
M. sphincter pyloricus
STOMACH AND PYLORUS—
FRONT VIEW
Fundus of stomach
Fundus gastricus
Duodenum
Duodenum
Body of stomach
Corpus gastricum
Pylorus
Pylorus

[~] = no direct Latin equivalent

PANCREAS

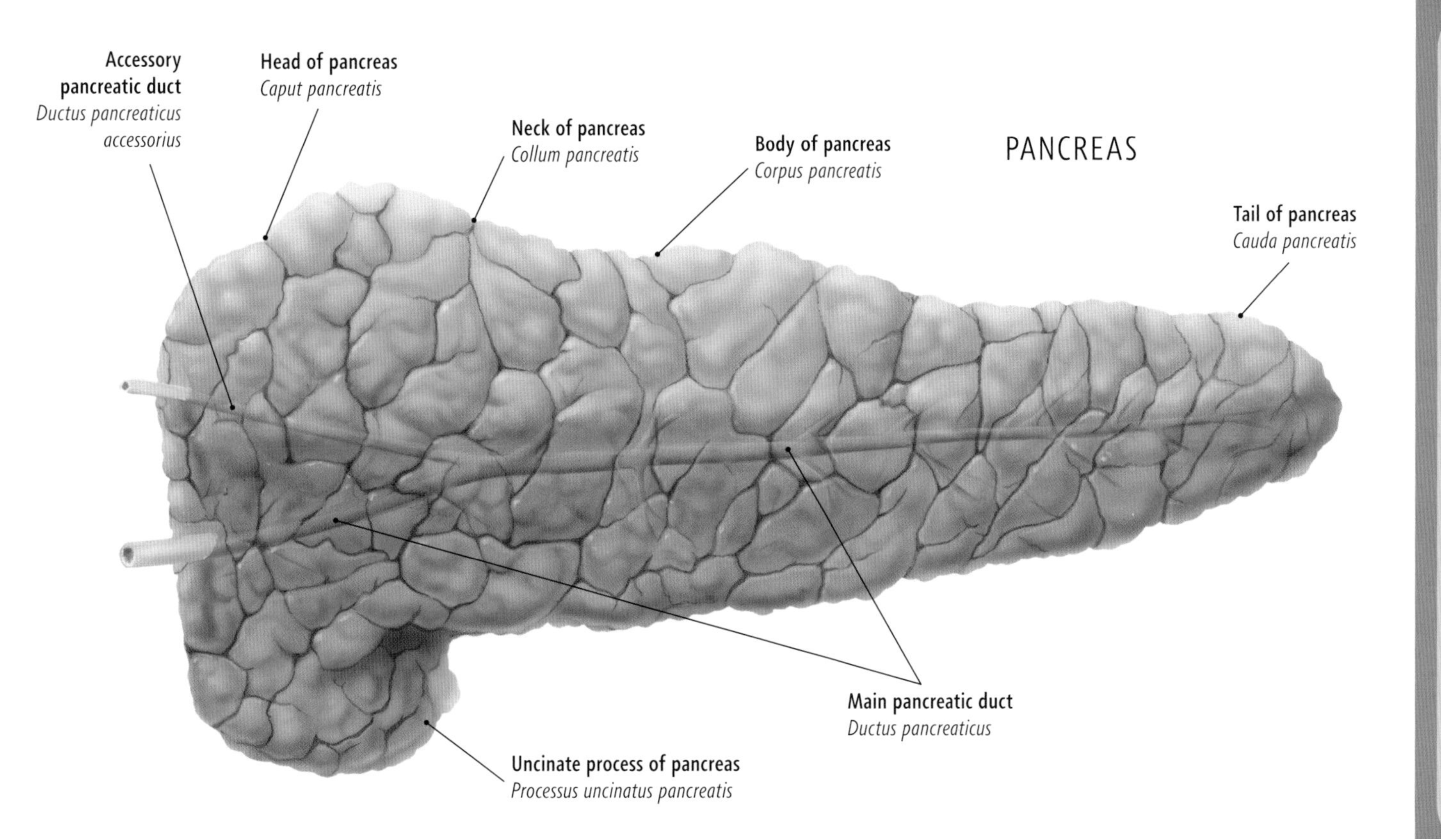
Accessory pancreatic duct
Ductus pancreaticus accessorius
Head of pancreas
Caput pancreatis
Neck of pancreas
Collum pancreatis
Body of pancreas
Corpus pancreatis
Tail of pancreas
Cauda pancreatis
Main pancreatic duct
Ductus pancreaticus
Uncinate process of pancreas
Processus uncinatus pancreatis

PANCREAS: EXOCRINE CELLS

Most of the pancreas consists of acinar cells, which secrete digestive enzymes that aid in food processing. The enzymes flow from the cells into the small intestine along a network of attached ducts.

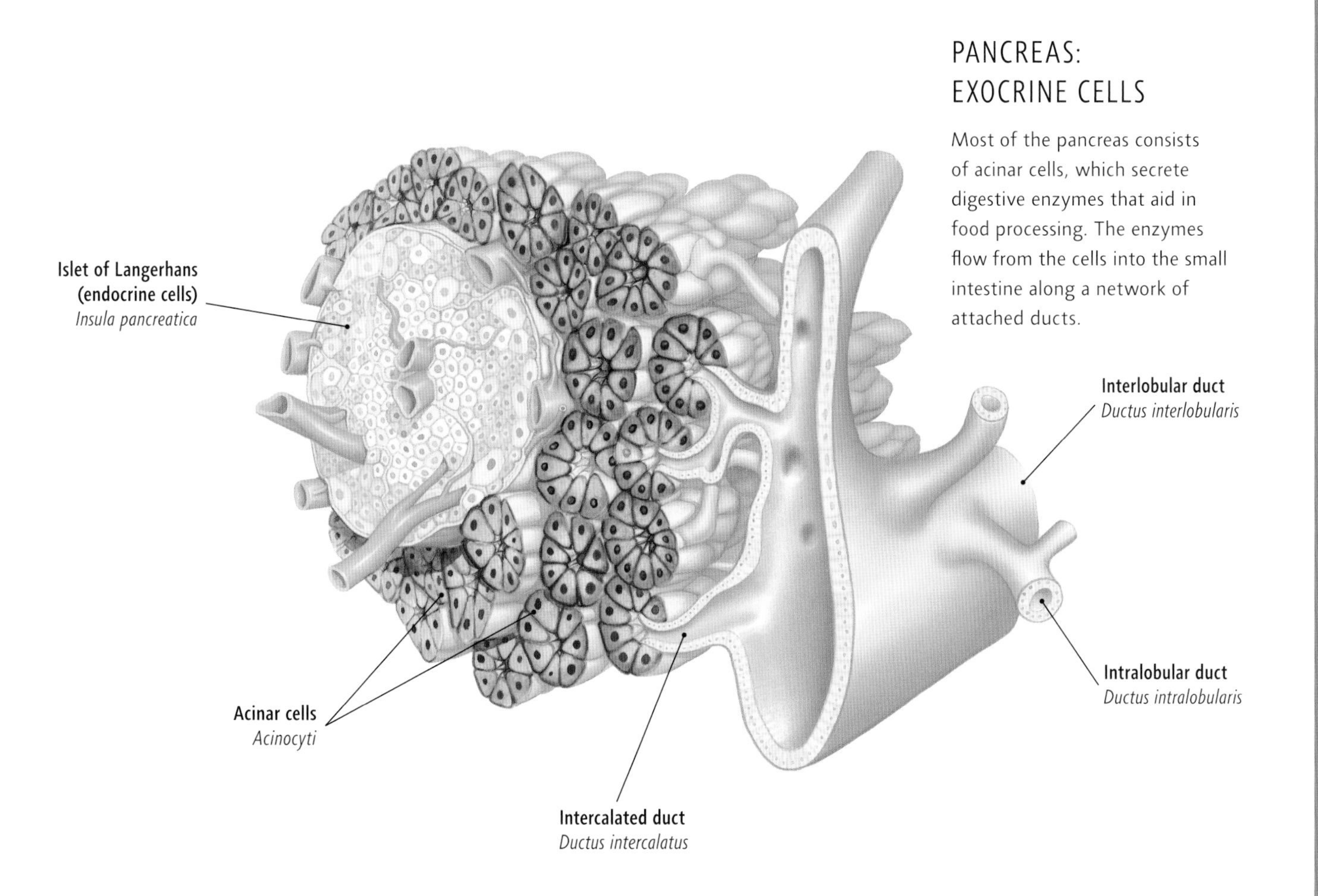
Islet of Langerhans (endocrine cells)
Insula pancreatica
Interlobular duct
Ductus interlobularis
Intralobular duct
Ductus intralobularis
Acinar cells
Acinocyti
Intercalated duct
Ductus intercalatus

Alimentary Canal and Peristalsis

ALIMENTARY CANAL

The alimentary canal is the body's "food processor." The first part breaks food down into fats, carbohydrates, and proteins, which are absorbed into the blood further along the canal. Water is absorbed along the final part, and anything left over is excreted.

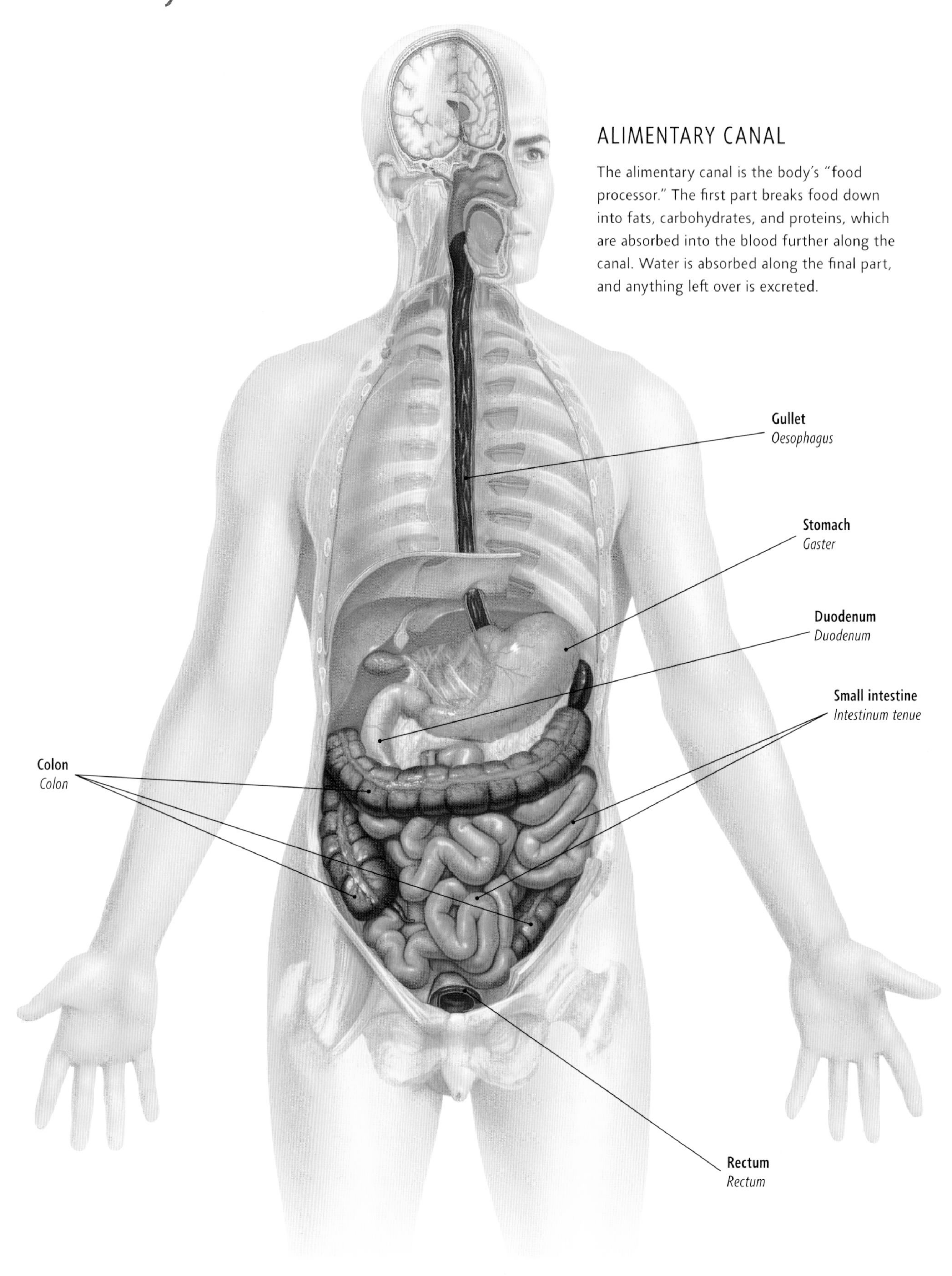
Gullet
Oesophagus
Stomach
Gaster
Duodenum
Duodenum
Small intestine
Intestinum tenue
Colon
Colon
Rectum
Rectum

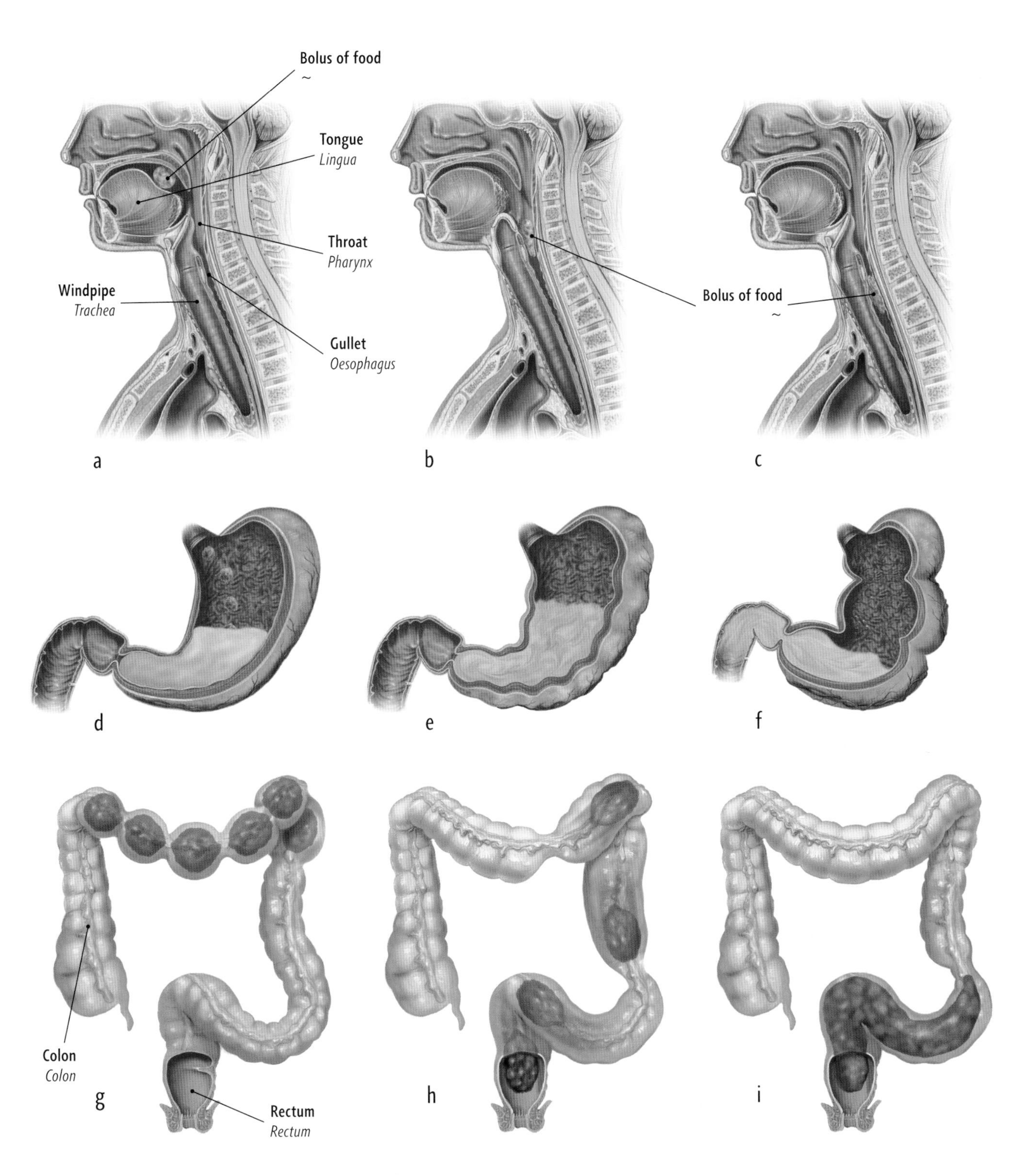
Bolus of food
~
Tongue
Lingua
Throat
Pharynx
Windpipe
Trachea
Gullet
Oesophagus
Bolus of food
~
a
b
c
d
e
f
Colon
Colon
g
Rectum
Rectum
h
i

PERISTALSIS

Peristalsis consists of wavelike contractions in the muscular walls of the esophagus, stomach, and intestines that propel the contents along. After food is swallowed, it enters the esophagus (a, b, c) and travels toward the stomach (d, e, f) before moving to the small intestine and then the colon. Water and bile salts are absorbed from the colon before peristaltic contractions push the waste matter along to the rectum (g, h, i).

[~] = no direct Latin equivalent

Intestines

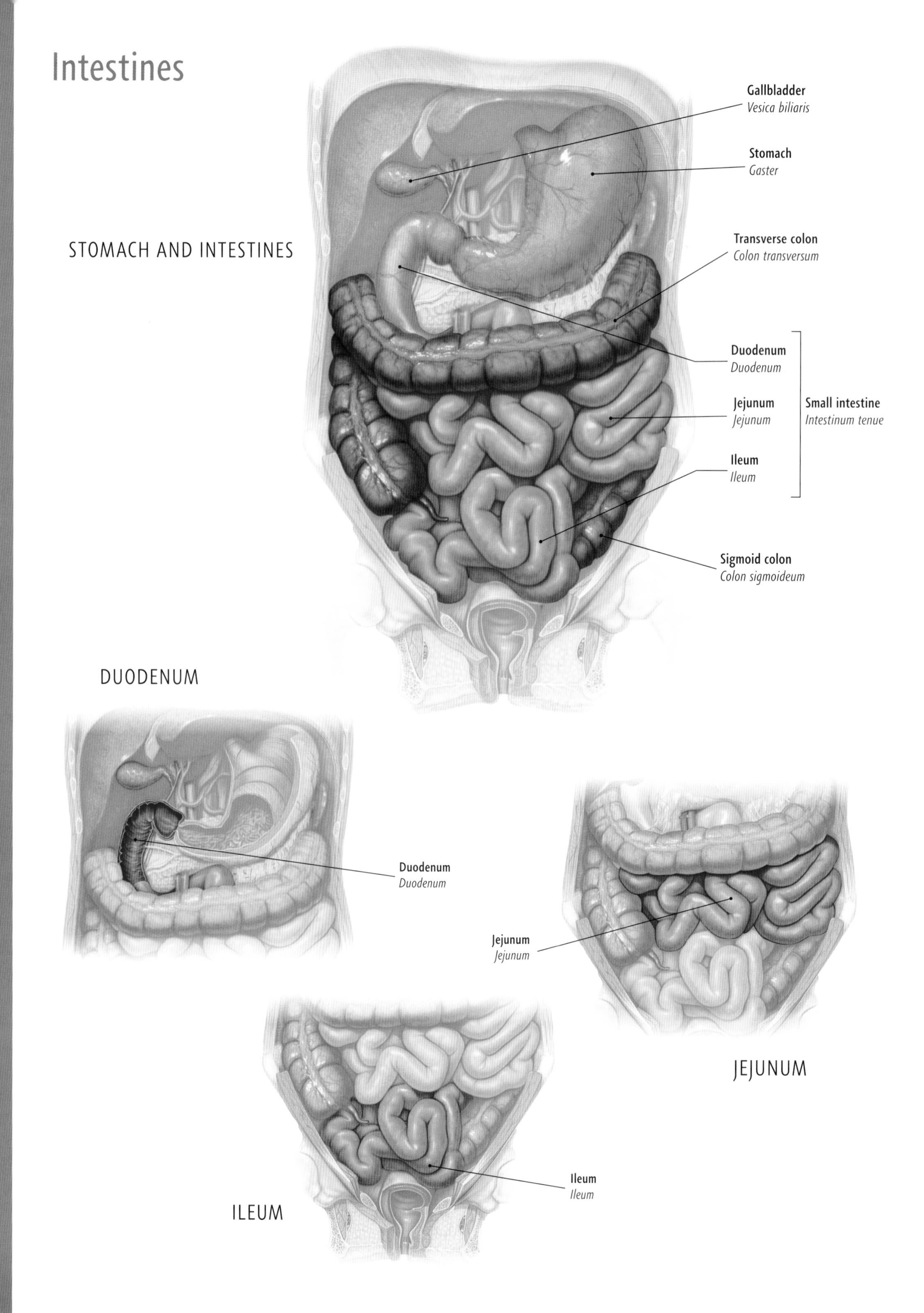
STOMACH AND INTESTINES
Gallbladder
Vesica biliaris
Stomach
Gaster
Transverse colon
Colon transversum
Duodenum
Duodenum
Jejunum
Jejunum
Ileum
Ileum
Small intestine
Intestinum tenue
Sigmoid colon
Colon sigmoideum
DUODENUM
Duodenum
Duodenum
Jejunum
Jejunum
JEJUNUM
Ileum
Ileum
ILEUM

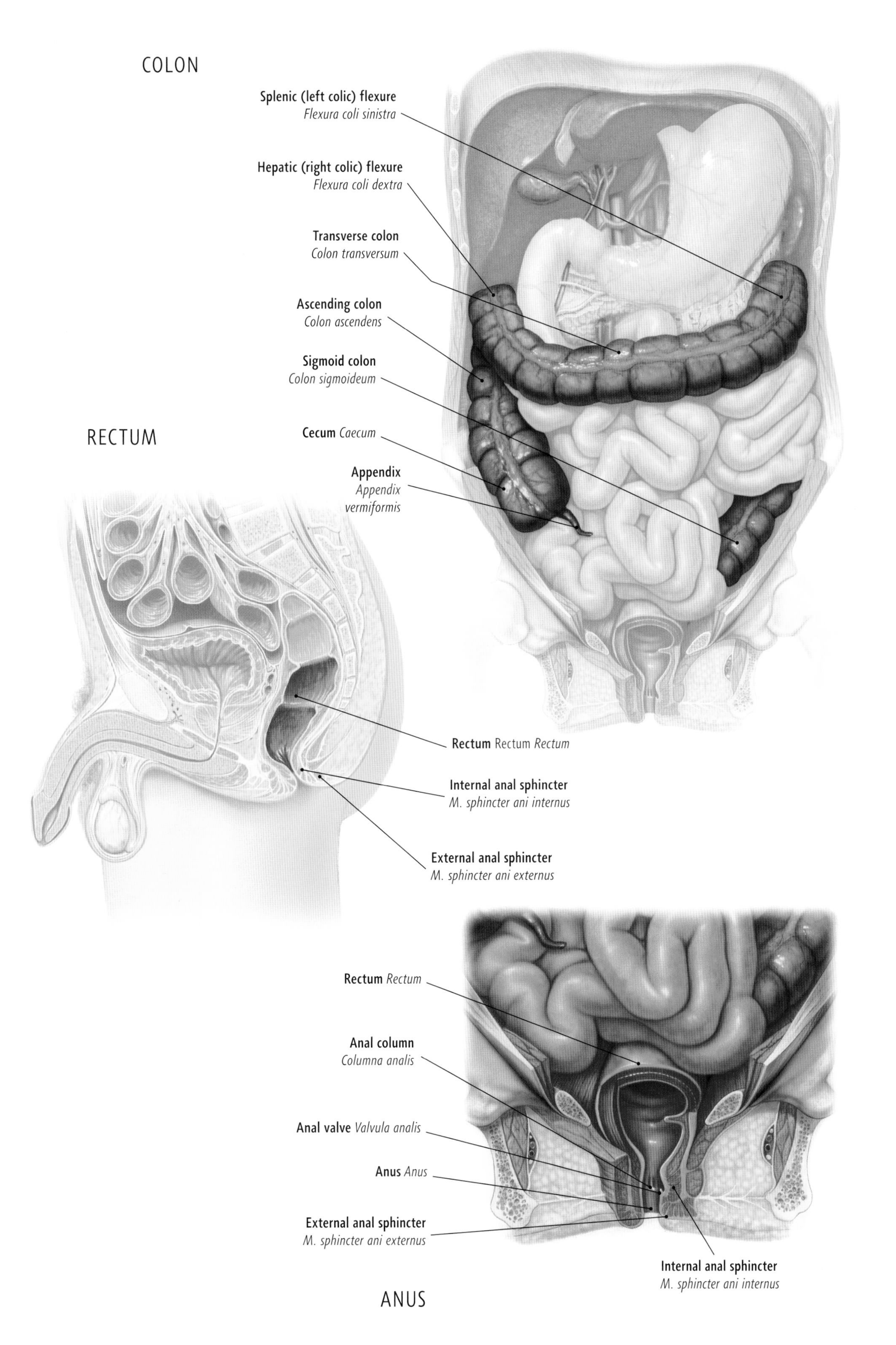
COLON
Splenic (left colic) flexure
Flexura coli sinistra
Hepatic (right colic) flexure
Flexura coli dextra
Transverse colon
Colon transversum
Ascending colon
Colon ascendens
Sigmoid colon
Colon sigmoideum
Cecum Caecum
Appendix
Appendix vermiformis
RECTUM
Rectum Rectum Rectum
Internal anal sphincter
M. sphincter ani internus
External anal sphincter
M. sphincter ani externus
Rectum Rectum
Anal column
Columna analis
Anal valve Valvula analis
Anus Anus
External anal sphincter
M. sphincter ani externus
Internal anal sphincter
M. sphincter ani internus
ANUS

Intestines

INTESTINAL JEJUNUM—CUT-AWAY VIEW

The jejunum is the part of the small intestine between the duodenum and the ileum. It digests and absorbs food, sending nutrients to the lymphatic vessels and liver.

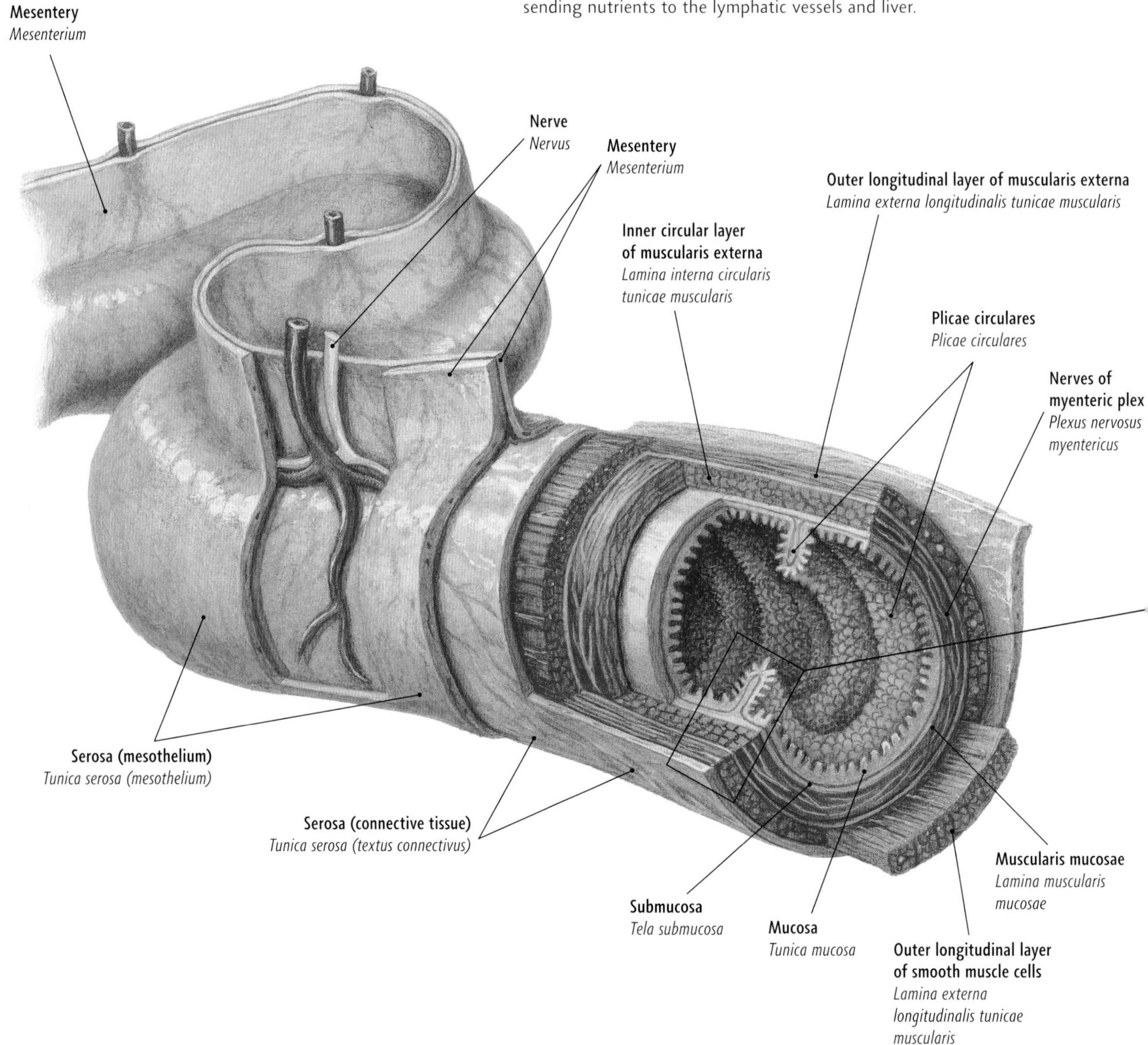
Mesentery
Mesenterium
Nerve
Nervus
Mesentery
Mesenterium
Outer longitudinal layer of muscularis externa
Lamina externa longitudinalis tunicae muscularis
Inner circular layer of muscularis externa
Lamina interna circularis tunicae muscularis
Plicae circulares
Plicae circulares
Nerves of myenteric plex
Plexus nervosus myentericus
Serosa (mesothelium)
Tunica serosa (mesothelium)
Serosa (connective tissue)
Tunica serosa (textus connectivus)
Muscularis mucosae
Lamina muscularis mucosae
Submucosa
Tela submucosa
Mucosa
Tunica mucosa
Outer longitudinal layer of smooth muscle cells
Lamina externa longitudinalis tunicae muscularis

[~] = no direct Latin equivalent

INTESTINAL JEJUNUM PLICA—CROSS-SECTIONAL VIEW

The lining of the jejunum has many small folds (known as plicae circulares) that feature tiny fingerlike projections called villi. The plicae circulares and villi greatly increase the surface area available for the absorption of nutrients.

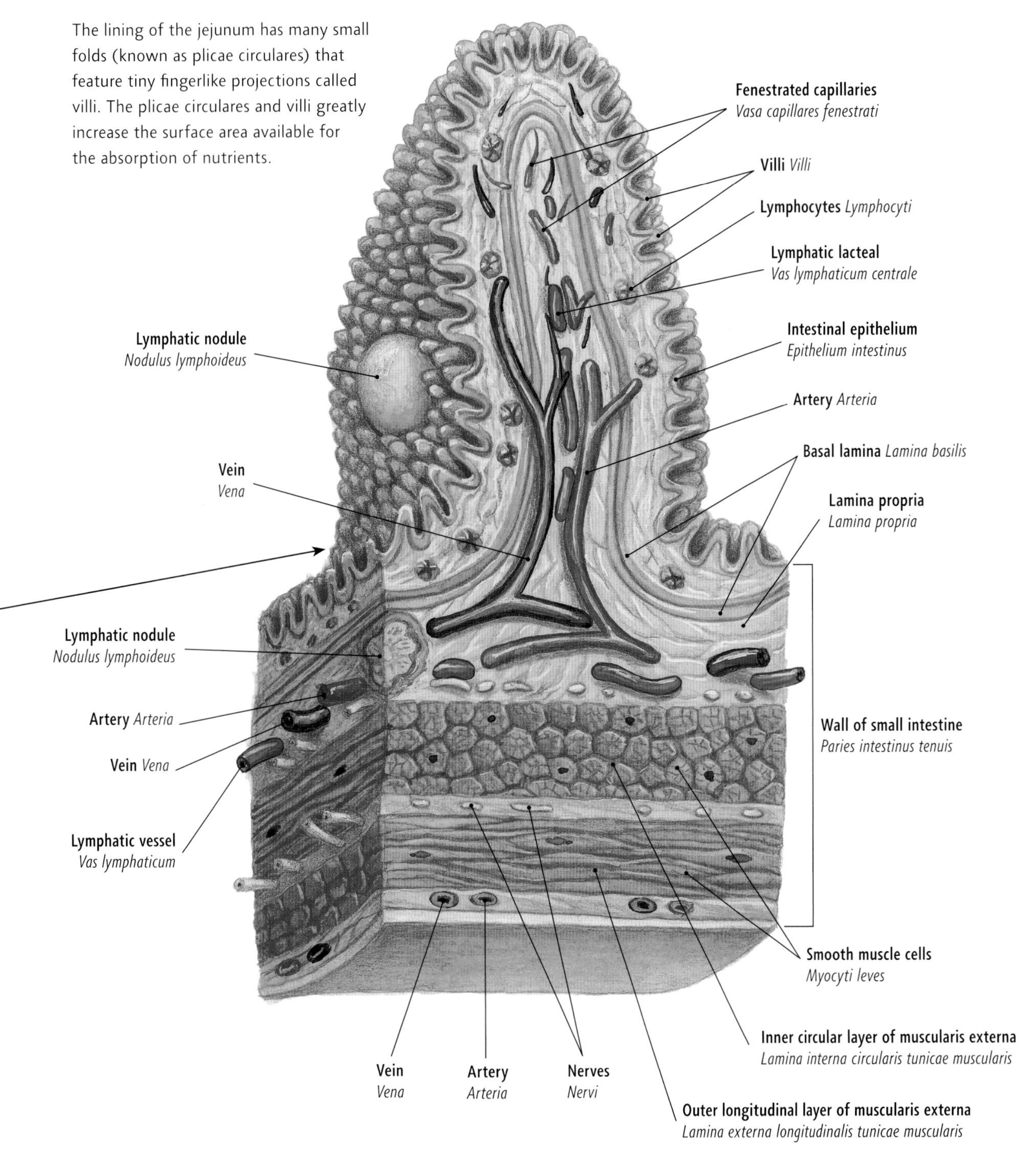

Fenestrated capillaries
Vasa capillares fenestrati
Villi Villi
Lymphocytes Lymphocyti
Lymphatic lacteal
Vas lymphaticum centrale
Intestinal epithelium
Epithelium intestinus
Artery Arteria
Basal lamina Lamina basilis
Lamina propria
Lamina propria
Wall of small intestine
Paries intestinus tenuis
Smooth muscle cells
Myocyti leves
Inner circular layer of muscularis externa
Lamina interna circularis tunicae muscularis
Outer longitudinal layer of muscularis externa
Lamina externa longitudinalis tunicae muscularis
Nerves
Nervi
Artery
Arteria
Vein
Vena
Lymphatic nodule
Nodulus lymphoideus
Vein
Vena
Lymphatic nodule
Nodulus lymphoideus
Artery Arteria
Vein Vena
Lymphatic vessel
Vas lymphaticum

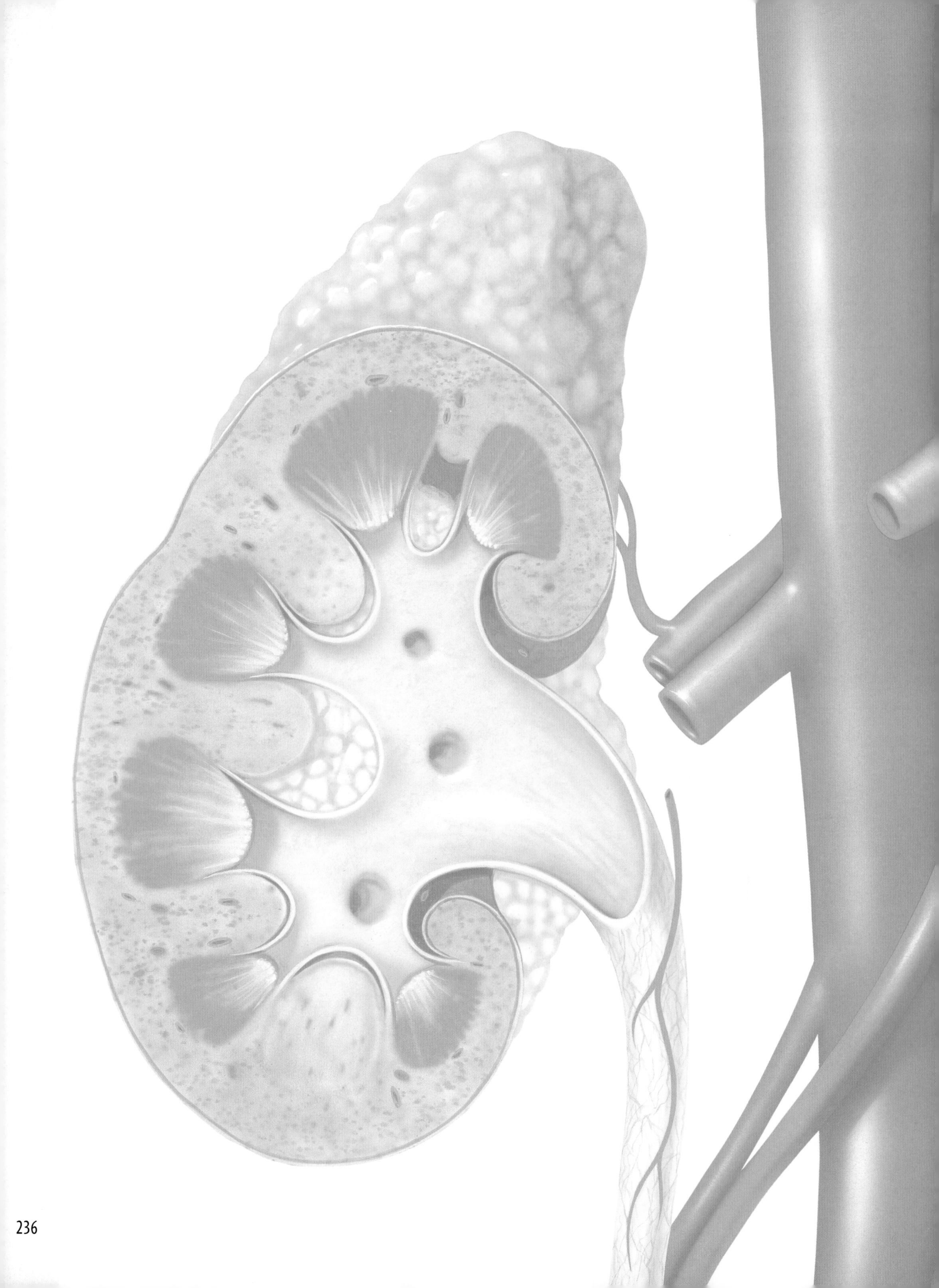

The Urinary System

The Urinary System

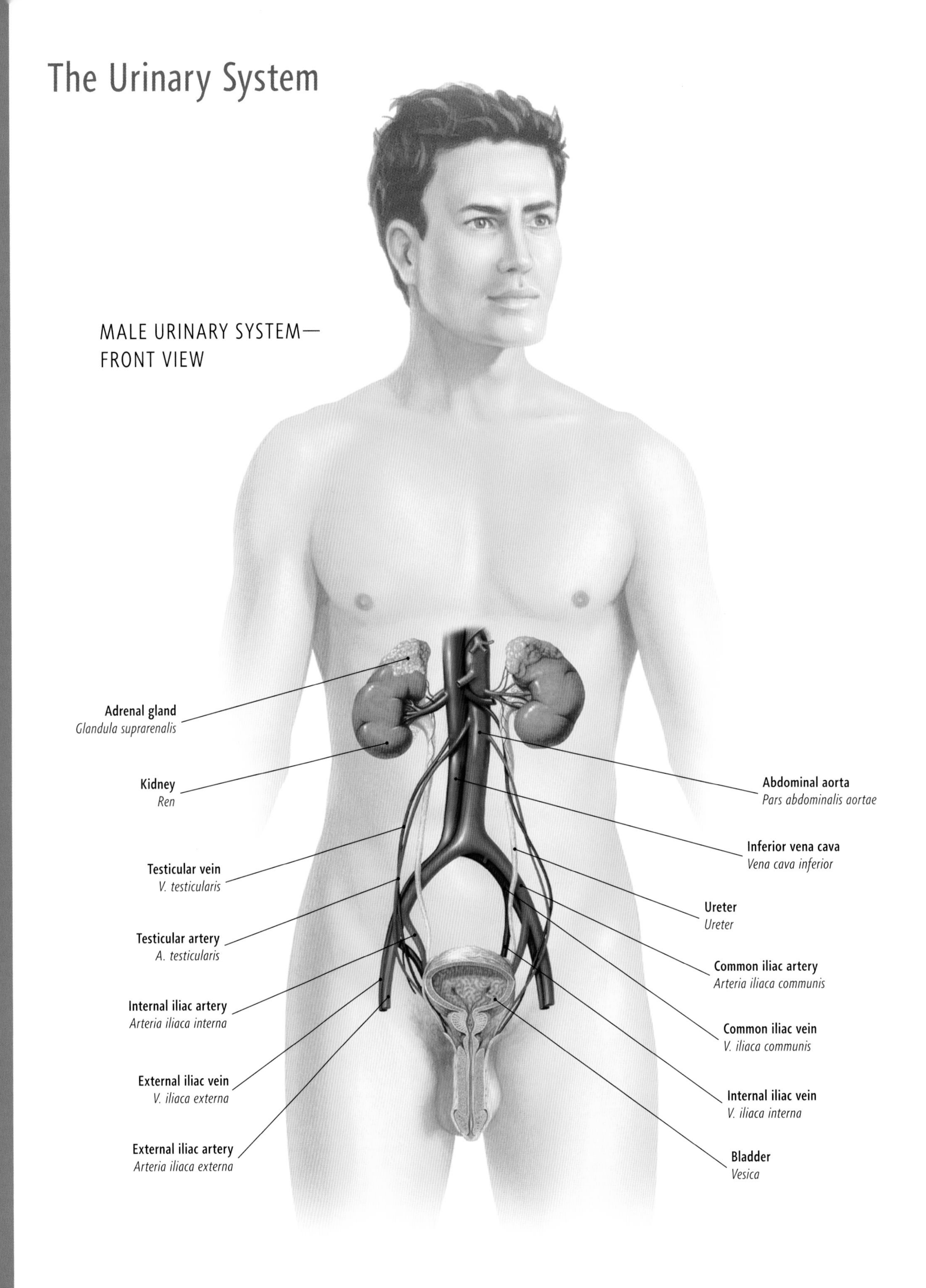
MALE URINARY SYSTEM—
FRONT VIEW
Adrenal gland
Glandula suprarenalis
Kidney
Ren
Testicular vein
V. testicularis
Testicular artery
A. testicularis
Internal iliac artery
Arteria iliaca interna
External iliac vein
V. iliaca externa
External iliac artery
Arteria iliaca externa
Abdominal aorta
Pars abdominalis aortae
Inferior vena cava
Vena cava inferior
Ureter
Ureter
Common iliac artery
Arteria iliaca communis
Common iliac vein
V. iliaca communis
Internal iliac vein
V. iliaca interna
Bladder
Vesica

FEMALE URINARY SYSTEM—
FRONT VIEW

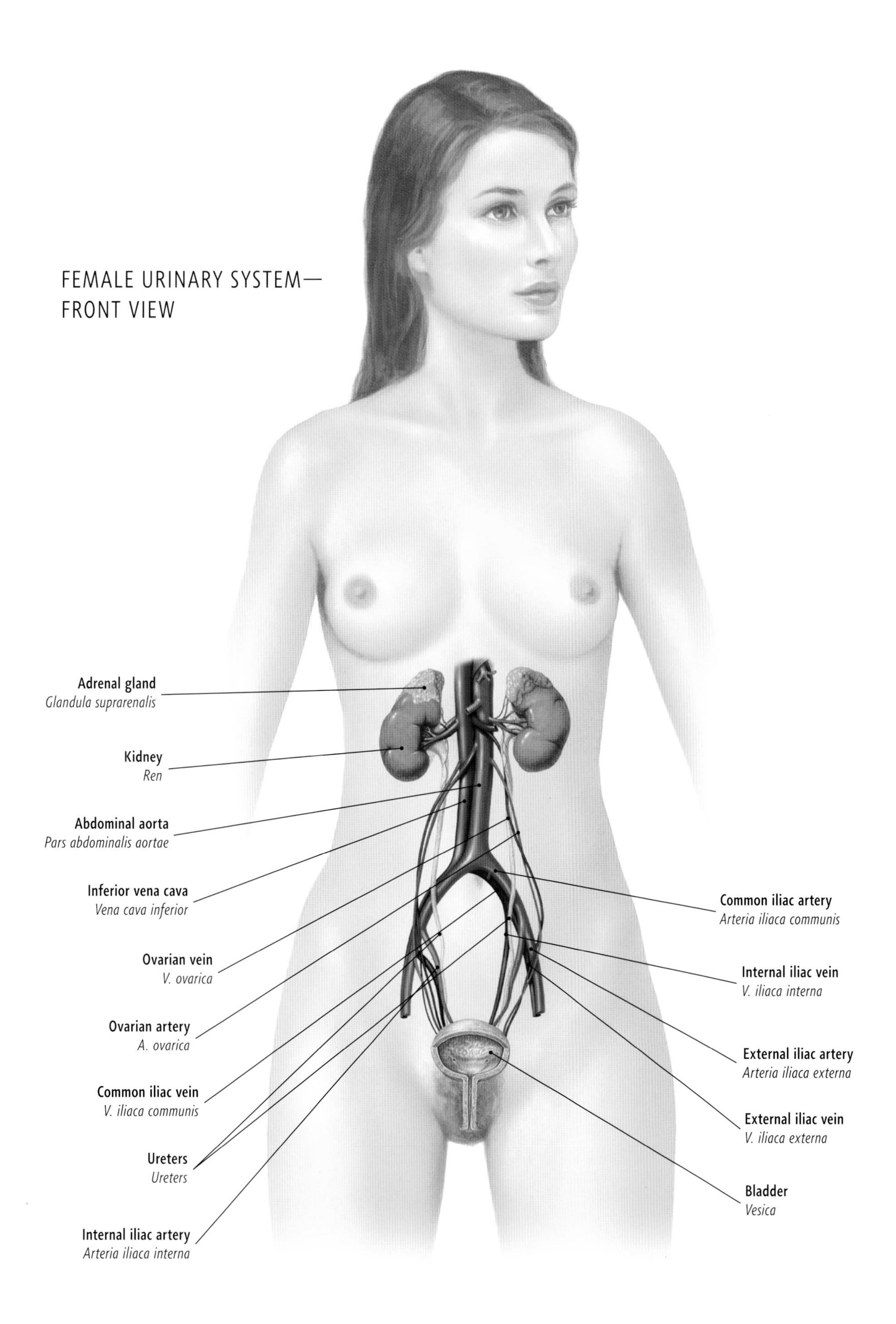
Adrenal gland
Glandula suprarenalis
Kidney
Ren
Abdominal aorta
Pars abdominalis aortae
Inferior vena cava
Vena cava inferior
Ovarian vein
V. ovarica
Ovarian artery
A. ovarica
Common iliac vein
V. iliaca communis
Ureters
Ureters
Internal iliac artery
Arteria iliaca interna
Common iliac artery
Arteria iliaca communis
Internal iliac vein
V. iliaca interna
External iliac artery
Arteria iliaca externa
External iliac vein
V. iliaca externa
Bladder
Vesica

Urinary Tract and Bladder

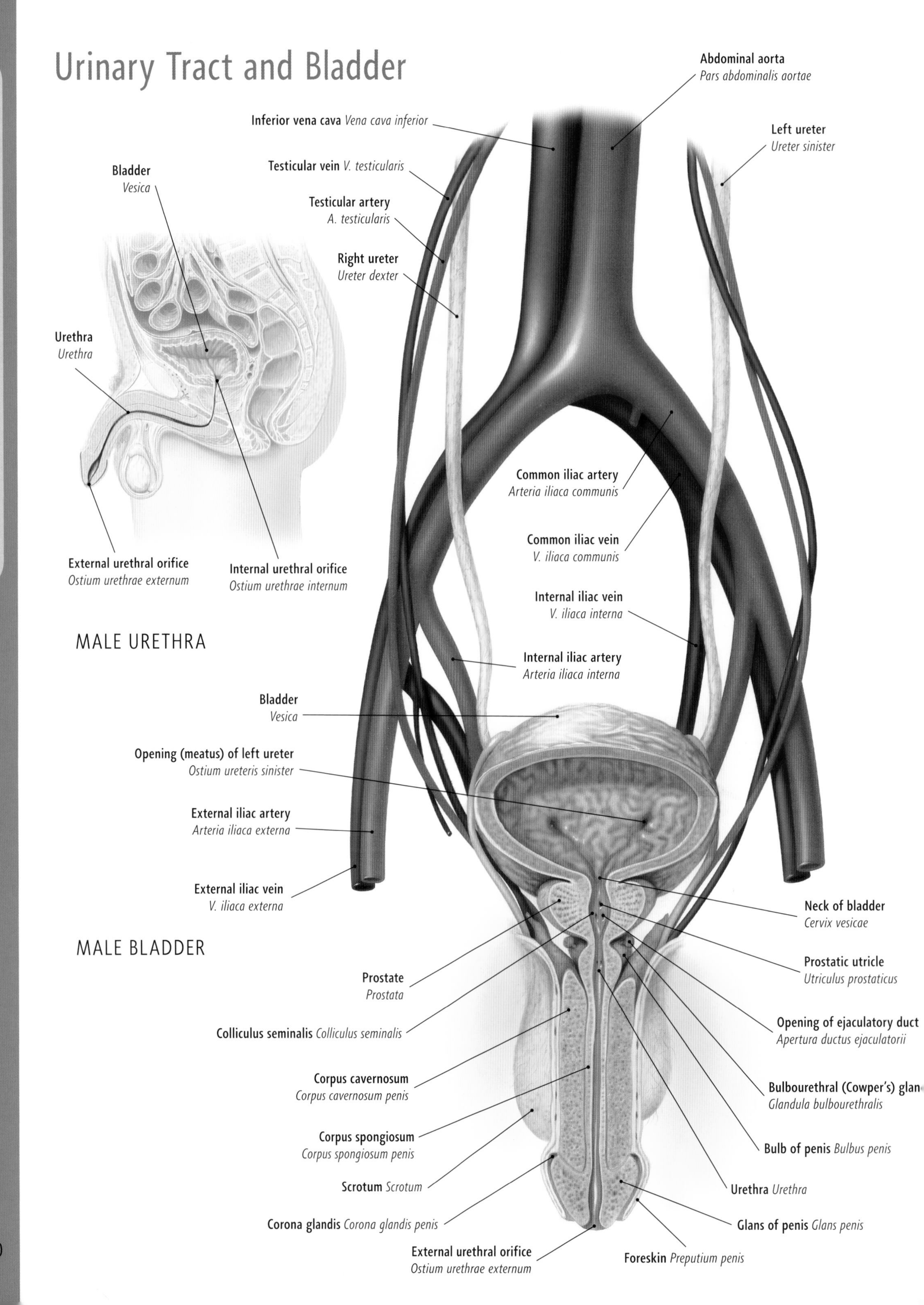

Bladder
Vesica
Urethra
Urethra
External urethral orifice
Ostium urethrae externum
Internal urethral orifice
Ostium urethrae internum
MALE URETHRA
Abdominal aorta
Pars abdominalis aortae
Inferior vena cava Vena cava inferior
Left ureter
Ureter sinister
Testicular vein V. testicularis
Testicular artery
A. testicularis
Right ureter
Ureter dexter
Common iliac artery
Arteria iliaca communis
Common iliac vein
V. iliaca communis
Internal iliac vein
V. iliaca interna
Internal iliac artery
Arteria iliaca interna
Bladder
Vesica
Opening (meatus) of left ureter
Ostium ureteris sinister
External iliac artery
Arteria iliaca externa
External iliac vein
V. iliaca externa
Neck of bladder
Cervix vesicae
MALE BLADDER
Prostatic utricle
Utriculus prostaticus
Prostate
Prostata
Opening of ejaculatory duct
Apertura ductus ejaculatorii
Colliculus seminalis Colliculus seminalis
Corpus cavernosum
Corpus cavernosum penis
Bulbourethral (Cowper's) glan
Glandula bulbourethralis
Corpus spongiosum
Corpus spongiosum penis
Bulb of penis Bulbus penis
Scrotum Scrotum
Urethra Urethra
Corona glandis Corona glandis penis
Glans of penis Glans penis
External urethral orifice
Ostium urethrae externum
Foreskin Preputium penis

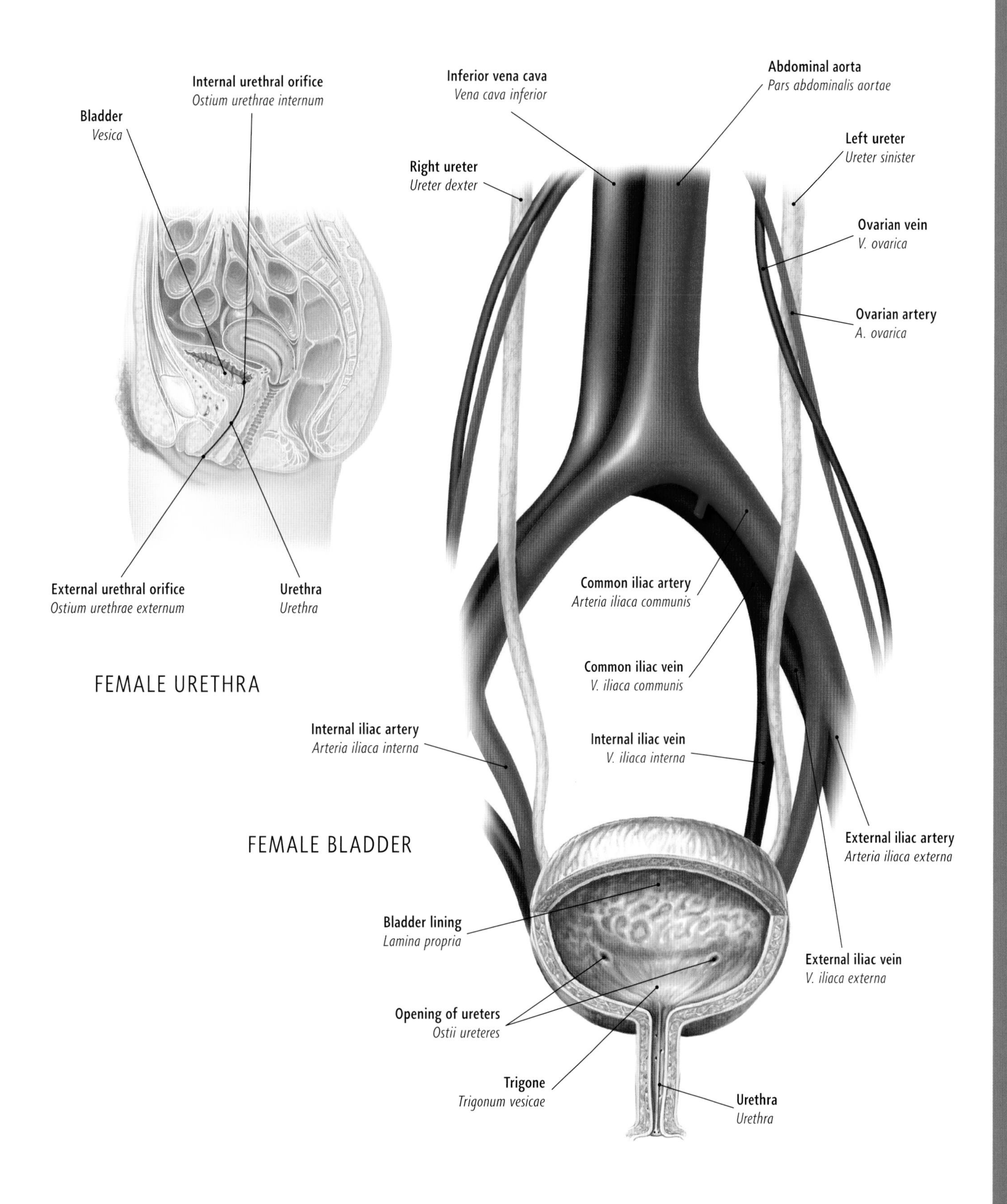
Bladder
Vesica
Internal urethral orifice
Ostium urethrae internum
External urethral orifice
Ostium urethrae externum
Urethra
Urethra
FEMALE URETHRA
Inferior vena cava
Vena cava inferior
Abdominal aorta
Pars abdominalis aortae
Right ureter
Ureter dexter
Left ureter
Ureter sinister
Ovarian vein
V. ovarica
Ovarian artery
A. ovarica
Common iliac artery
Arteria iliaca communis
Common iliac vein
V. iliaca communis
Internal iliac artery
Arteria iliaca interna
Internal iliac vein
V. iliaca interna
External iliac artery
Arteria iliaca externa
FEMALE BLADDER
Bladder lining
Lamina propria
External iliac vein
V. iliaca externa
Opening of ureters
Ostii ureteres
Trigone
Trigonum vesicae
Urethra
Urethra

Kidneys

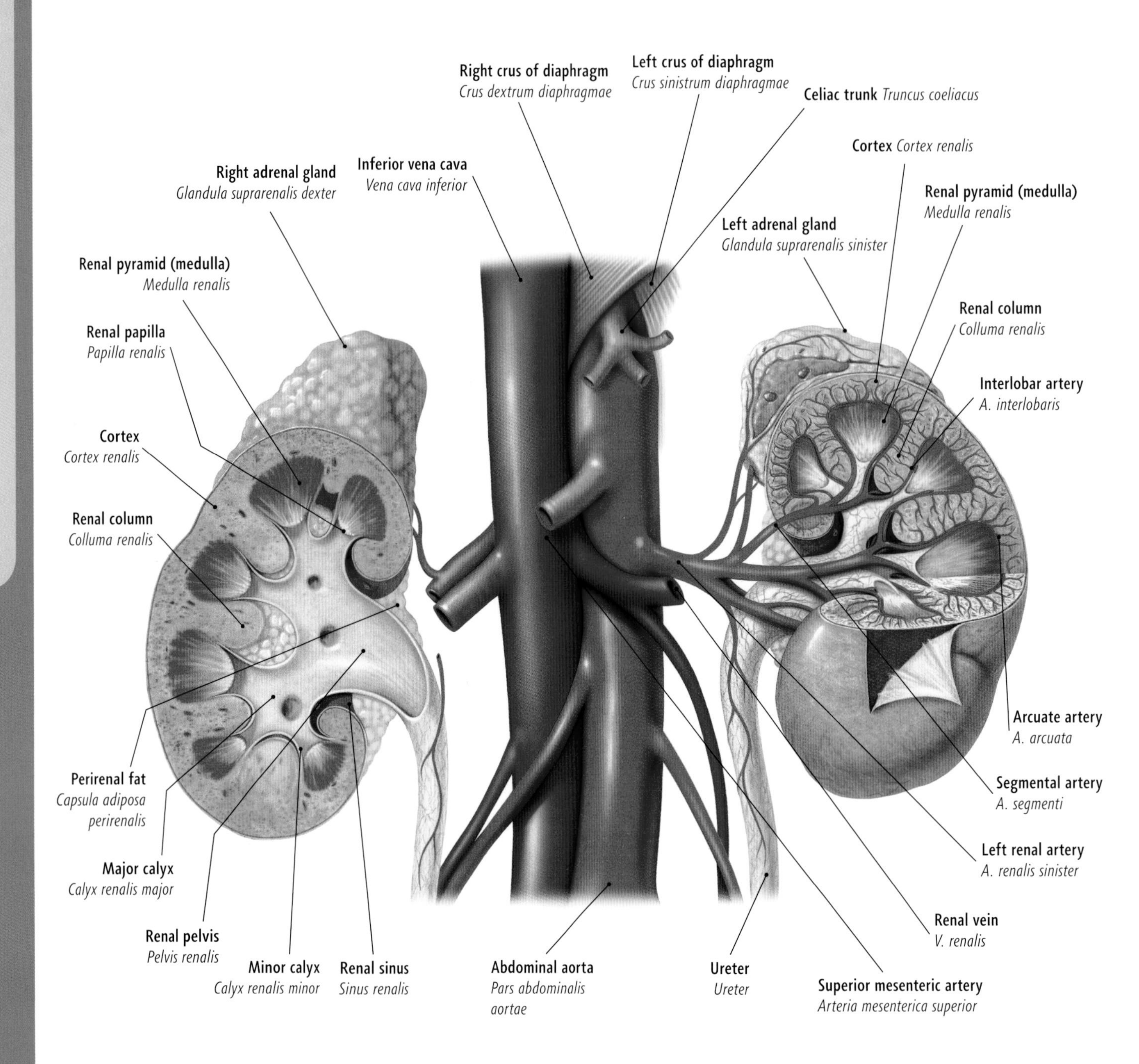
Right crus of diaphragm
Crus dextrum diaphragmae
Left crus of diaphragm
Crus sinistrum diaphragmae
Celiac trunk Truncus coeliacus
Cortex Cortex renalis
Right adrenal gland
Glandula suprarenalis dexter
Inferior vena cava
Vena cava inferior
Renal pyramid (medulla)
Medulla renalis
Left adrenal gland
Glandula suprarenalis sinister
Renal pyramid (medulla)
Medulla renalis
Renal column
Colluma renalis
Renal papilla
Papilla renalis
Interlobar artery
A. interlobaris
Cortex
Cortex renalis
Renal column
Colluma renalis
Arcuate artery
A. arcuata
Perirenal fat
Capsula adiposa perirenalis
Segmental artery
A. segmenti
Major calyx
Calyx renalis major
Left renal artery
A. renalis sinister
Renal pelvis
Pelvis renalis
Renal vein
V. renalis
Minor calyx
Calyx renalis minor
Renal sinus
Sinus renalis
Abdominal aorta
Pars abdominalis aortae
Ureter
Ureter
Superior mesenteric artery
Arteria mesenterica superior

KIDNEYS AND ADRENAL GLANDS

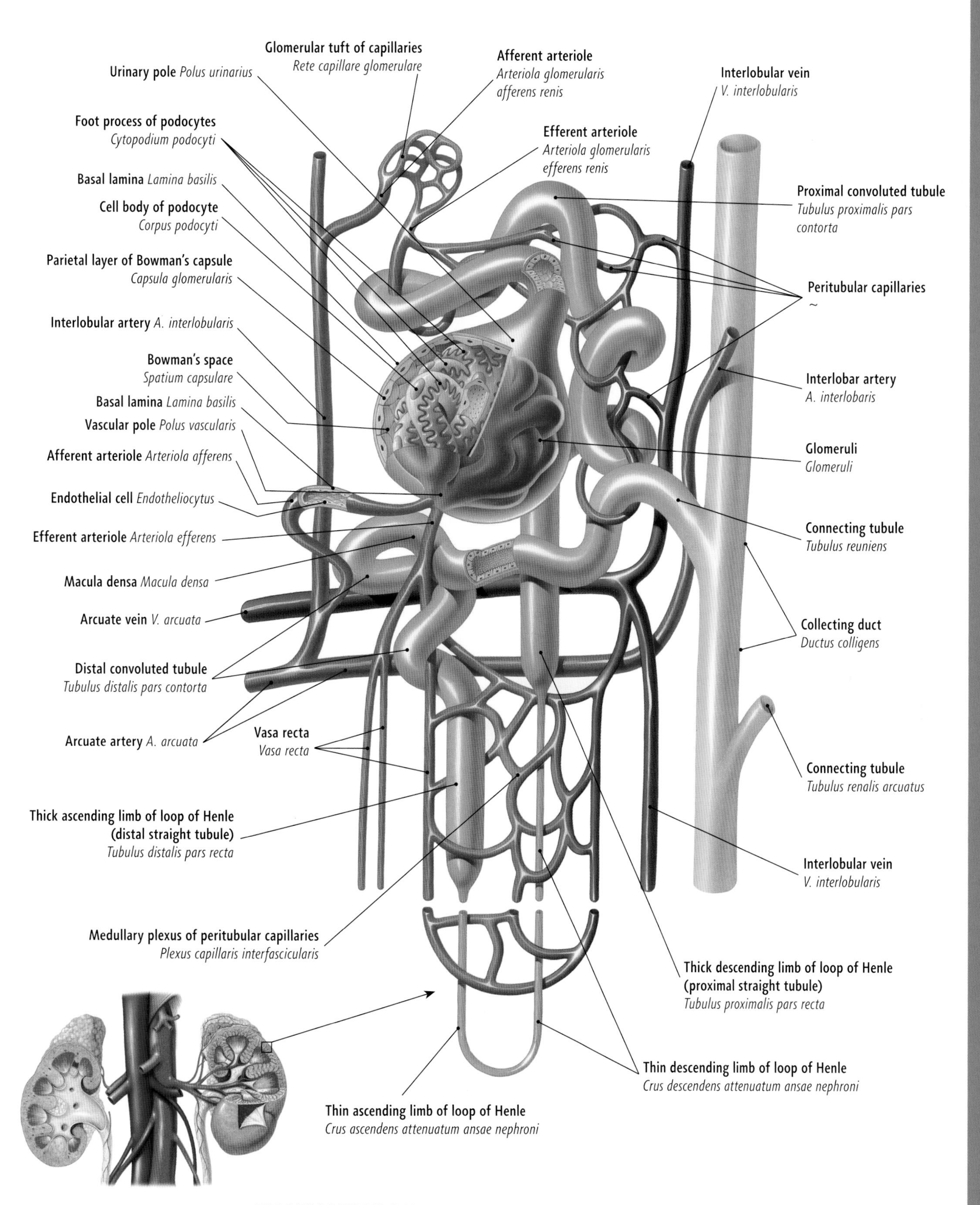
Glomerular tuft of capillaries
Rete capillare glomerulare
Urinary pole Polus urinarius
Afferent arteriole
Arteriola glomerularis afferens renis
Interlobular vein
V. interlobularis
Foot process of podocytes
Cytopodium podocyti
Efferent arteriole
Arteriola glomerularis efferens renis
Basal lamina Lamina basilis
Cell body of podocyte
Corpus podocyti
Proximal convoluted tubule
Tubulus proximalis pars contorta
Parietal layer of Bowman's capsule
Capsula glomerularis
Peritubular capillaries
~
Interlobular artery A. interlobularis
Bowman's space
Spatium capsulare
Interlobar artery
A. interlobaris
Basal lamina Lamina basilis
Vascular pole Polus vascularis
Glomeruli
Glomeruli
Afferent arteriole Arteriola afferens
Endothelial cell Endotheliocytus
Efferent arteriole Arteriola efferens
Connecting tubule
Tubulus reuniens
Macula densa Macula densa
Arcuate vein V. arcuata
Collecting duct
Ductus colligens
Distal convoluted tubule
Tubulus distalis pars contorta
Arcuate artery A. arcuata
Vasa recta
Vasa recta
Connecting tubule
Tubulus renalis arcuatus
Thick ascending limb of loop of Henle
(distal straight tubule)
Tubulus distalis pars recta
Interlobular vein
V. interlobularis
Medullary plexus of peritubular capillaries
Plexus capillaris interfascicularis
Thick descending limb of loop of Henle
(proximal straight tubule)
Tubulus proximalis pars recta
Thin descending limb of loop of Henle
Crus descendens attenuatum ansae nephroni
Thin ascending limb of loop of Henle
Crus ascendens attenuatum ansae nephroni

KIDNEY NEPHRON

The functional units of the kidneys are microscopic structures called nephrons, of which there are estimated to be about 1.2 million in each kidney. The nephrons filter the blood, producing a liquid (called the filtrate) containing minerals, wastes, and water.

[~] = no direct Latin equivalent

Controlling Fluid Balance

MAINTAINING FLUID BALANCE IN THE BODY

When water levels in the blood drop too low, the hypothalamus instructs the posterior pituitary gland to release antidiuretic hormone (ADH). This makes the kidneys retain water and reabsorb it into the blood, and decreases urine production. When the body has returned to a normal state of hydration, the hypothalamus tells the pituitary to slow ADH secretion, the kidneys keep less water in the blood, and the bladder releases urine again.

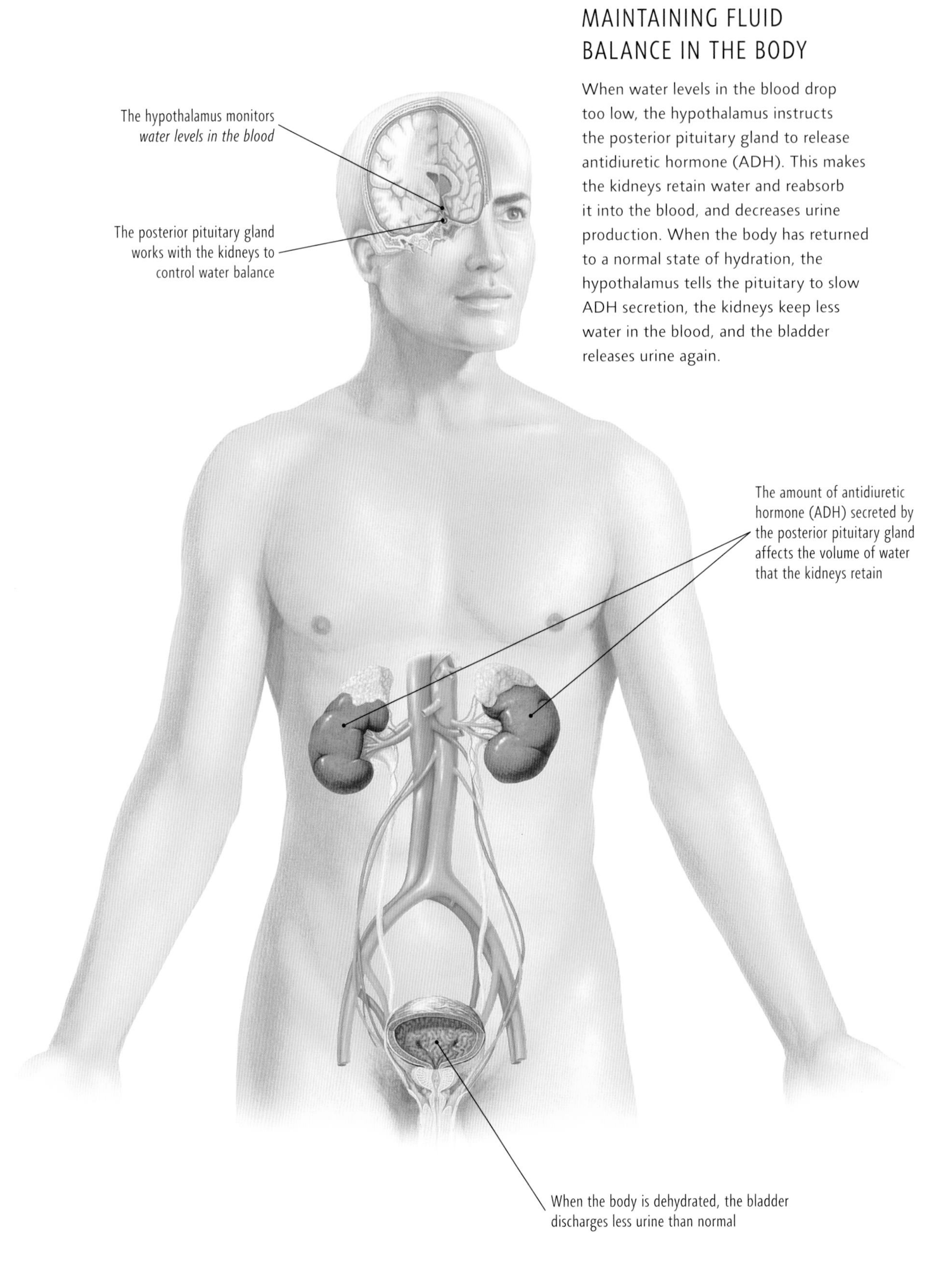
The hypothalamus monitors *water levels in the blood*
The posterior pituitary gland works with the kidneys to control water balance
The amount of antidiuretic hormone (ADH) secreted by the posterior pituitary gland affects the volume of water that the kidneys retain
When the body is dehydrated, the bladder discharges less urine than normal

LOCATION OF THE HYPOTHALAMUS AND POSTERIOR PITUITARY GLAND

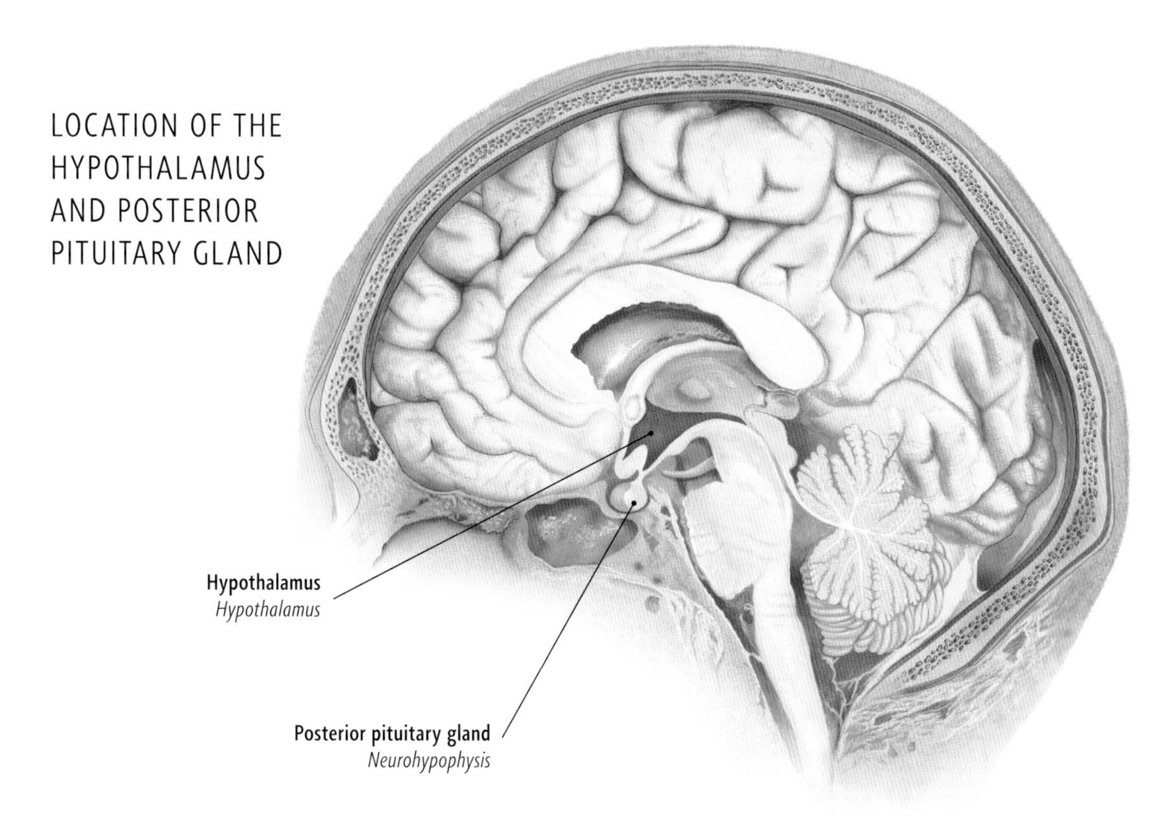
Hypothalamus
Hypothalamus
Posterior pituitary gland
Neurohypophysis

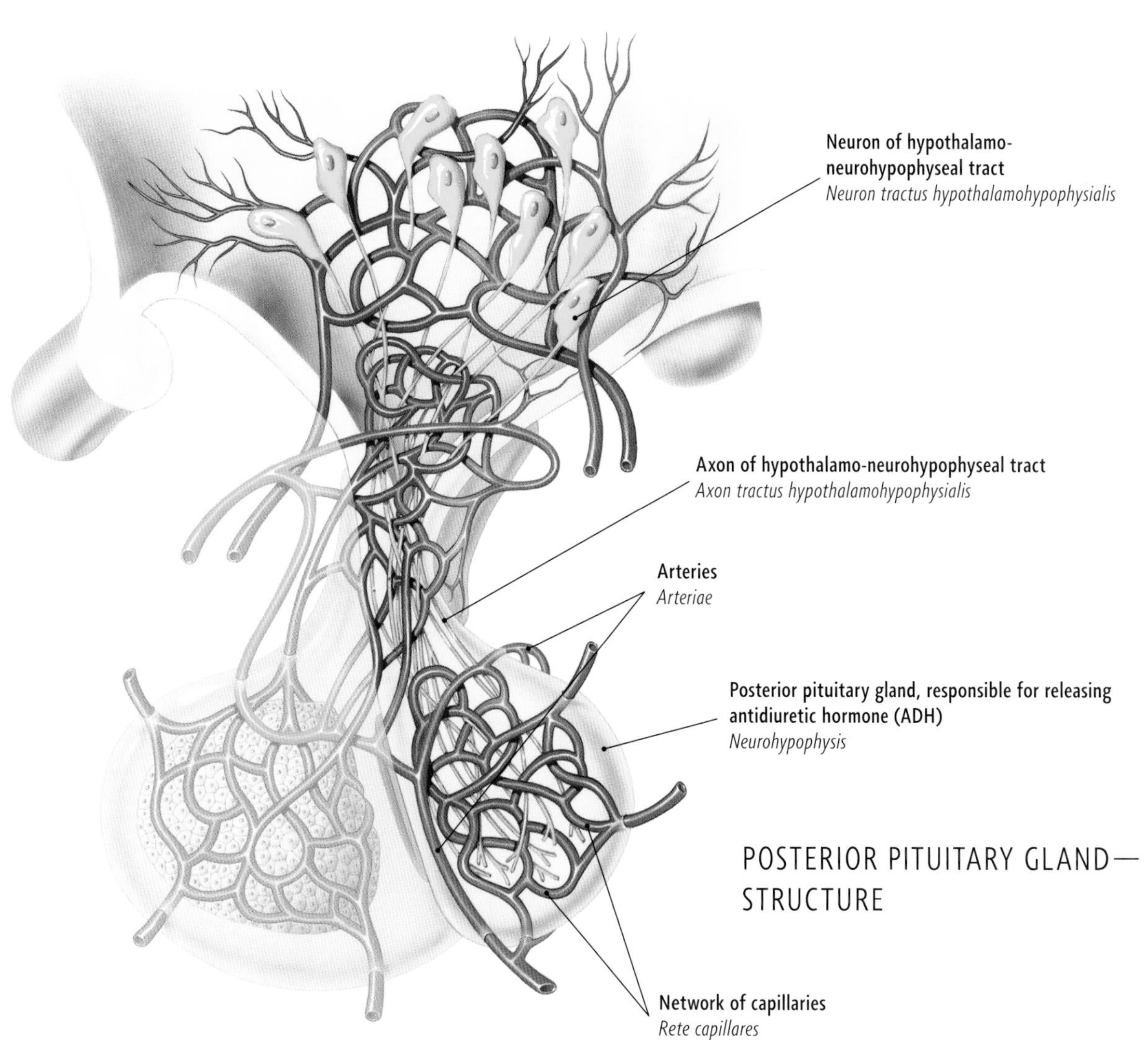
Neuron of hypothalamo-neurohypophyseal tract
Neuron tractus hypothalamohypophysialis
Axon of hypothalamo-neurohypophyseal tract
Axon tractus hypothalamohypophysialis
Arteries
Arteriae
Posterior pituitary gland, responsible for releasing antidiuretic hormone (ADH)
Neurohypophysis
Network of capillaries
Rete capillares

POSTERIOR PITUITARY GLAND—STRUCTURE

[~] = no direct Latin equivalent

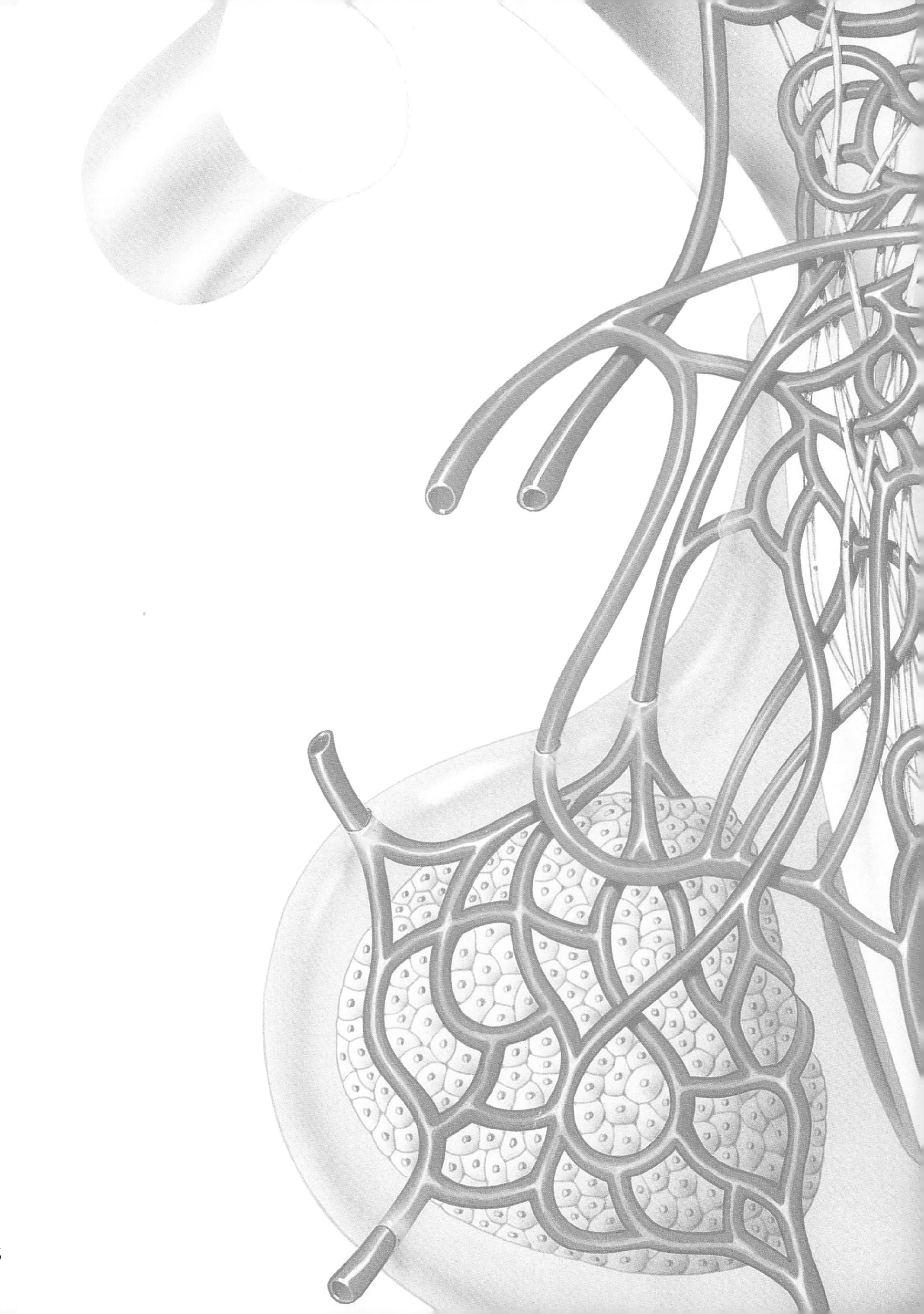

The Endocrine System

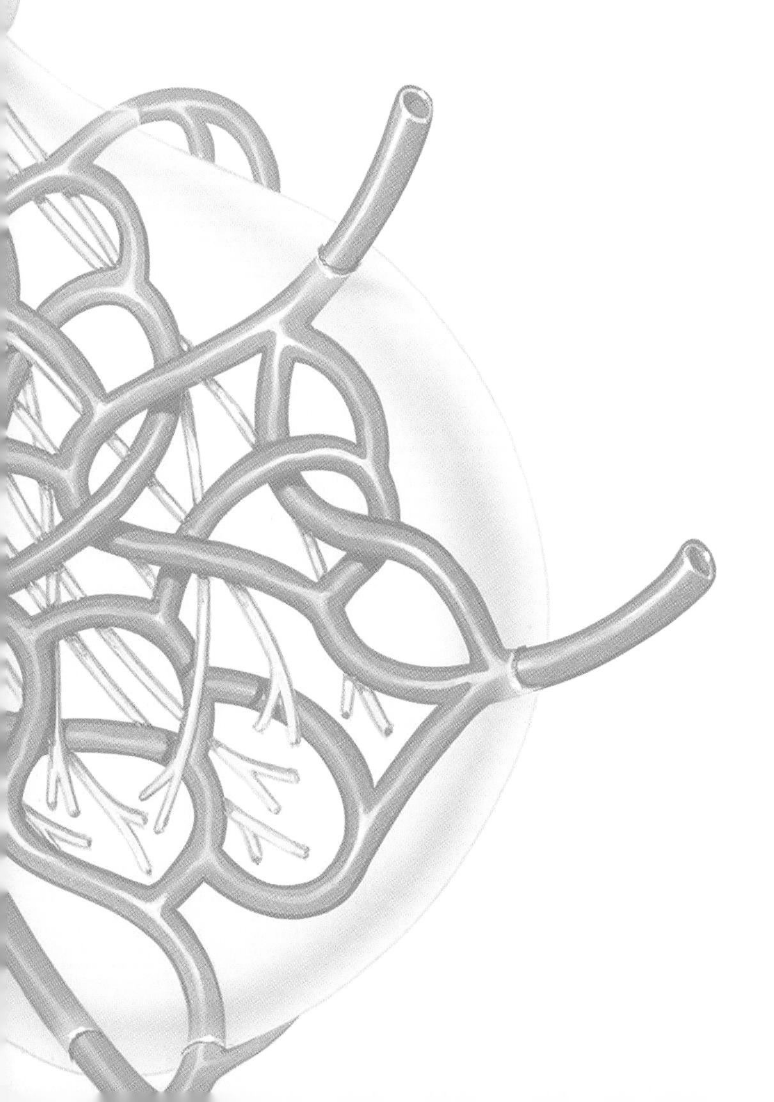

The Endocrine System

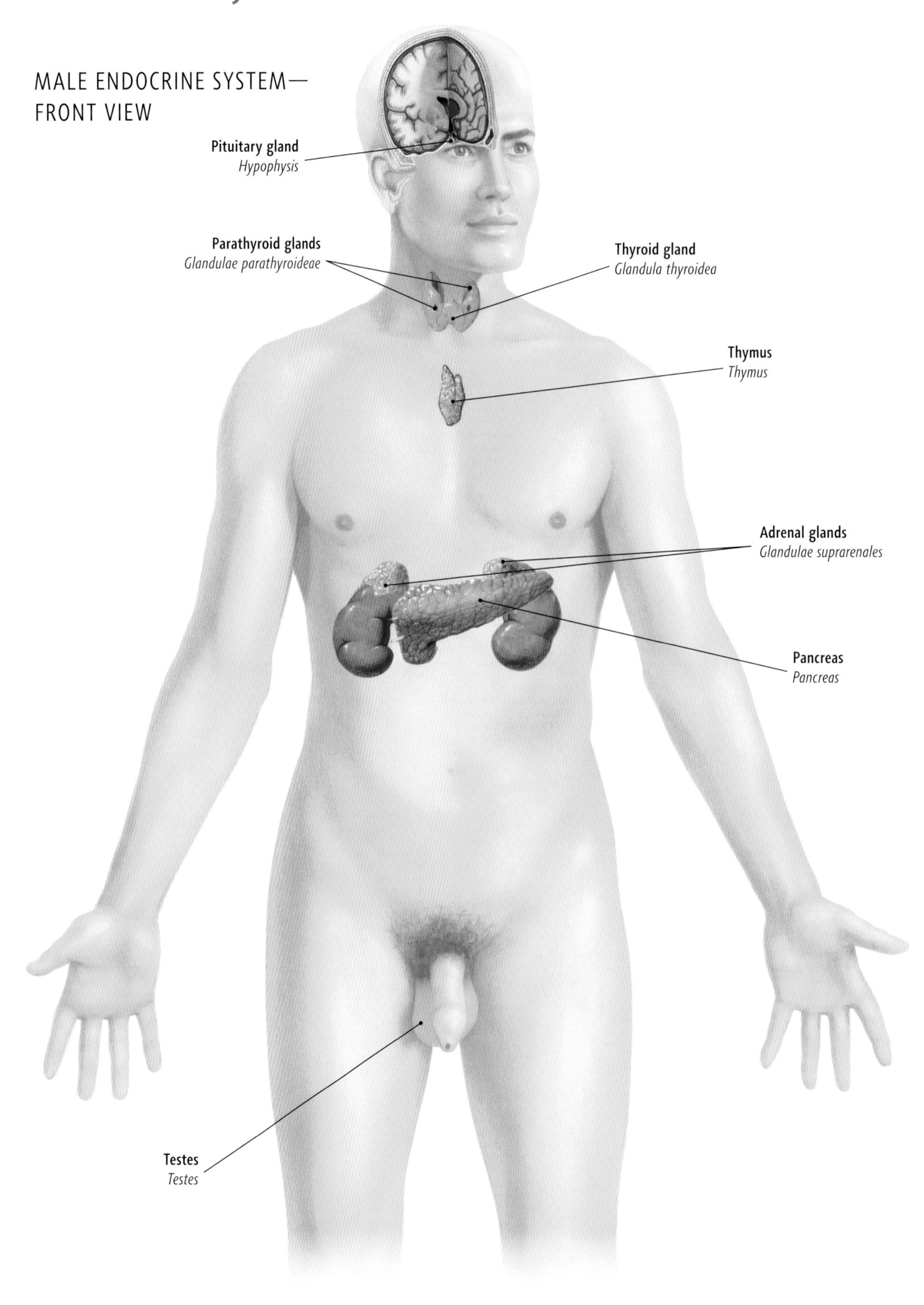
MALE ENDOCRINE SYSTEM—
FRONT VIEW
Pituitary gland
Hypophysis
Parathyroid glands
Glandulae parathyroideae
Thyroid gland
Glandula thyroidea
Thymus
Thymus
Adrenal glands
Glandulae suprarenales
Pancreas
Pancreas
Testes
Testes

FEMALE ENDOCRINE SYSTEM—FRONT VIEW

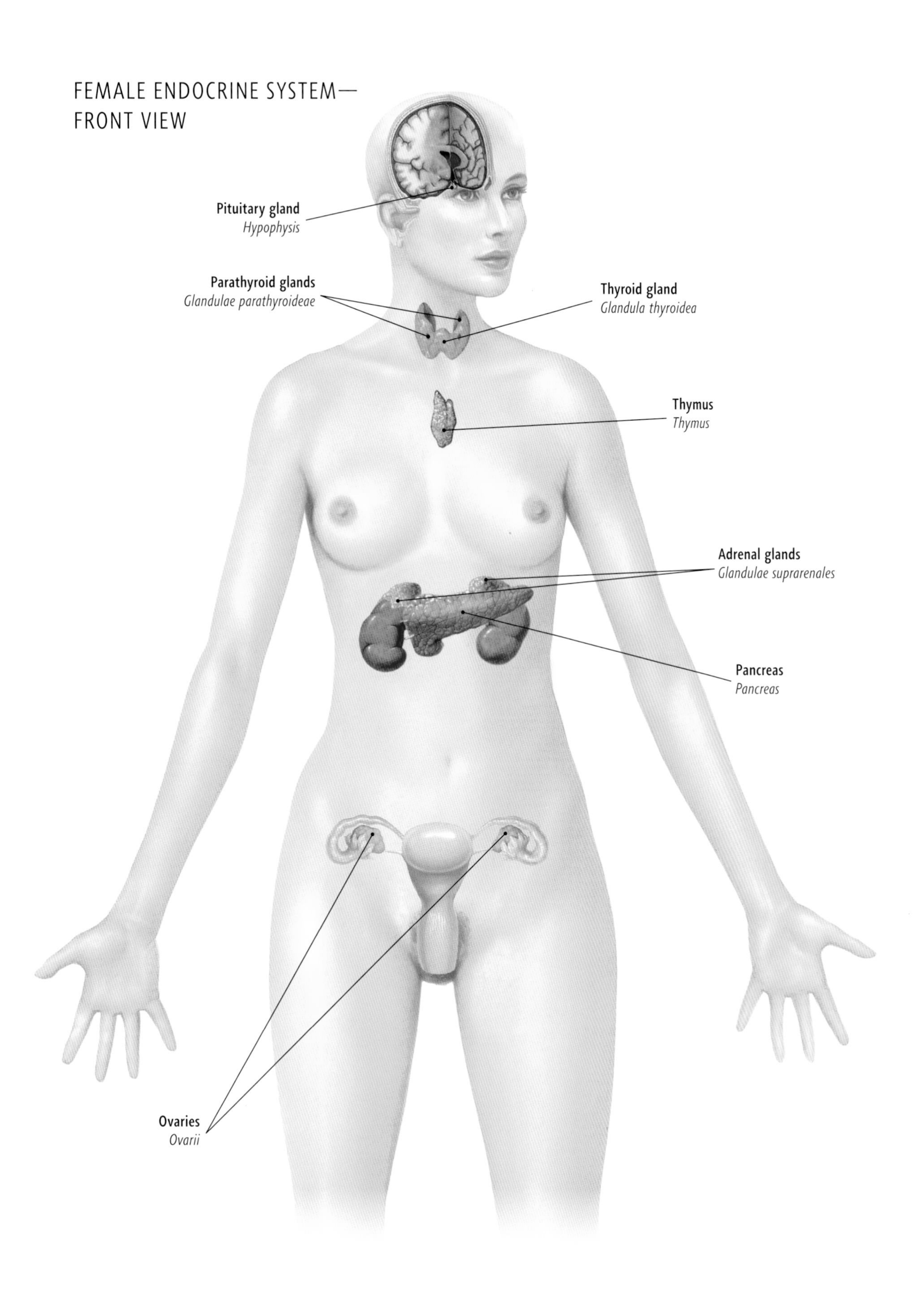
Pituitary gland
Hypophysis
Parathyroid glands
Glandulae parathyroideae
Thyroid gland
Glandula thyroidea
Thymus
Thymus
Adrenal glands
Glandulae suprarenales
Pancreas
Pancreas
Ovaries
Ovarii

Pituitary Gland

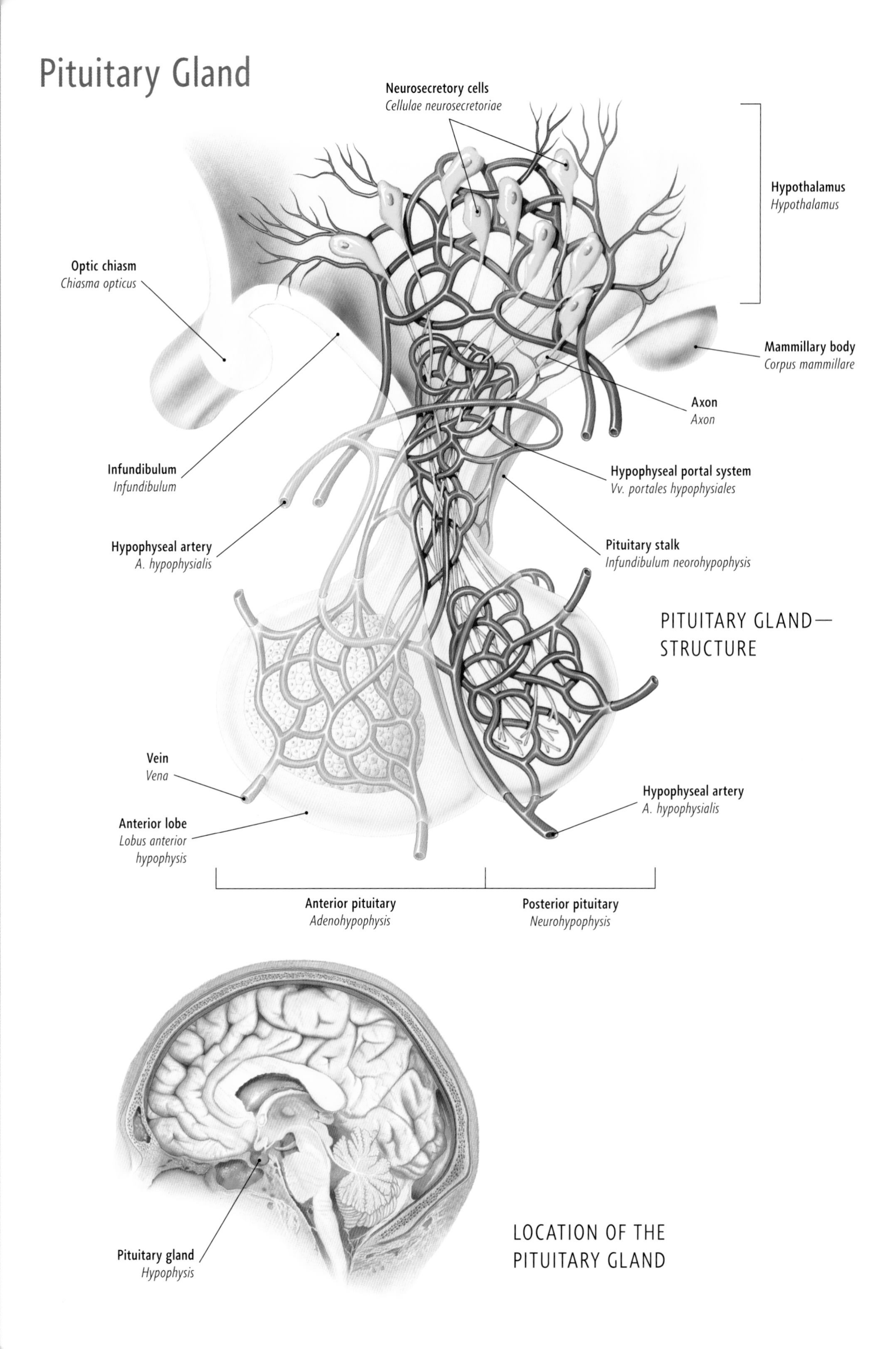
Neurosecretory cells
Cellulae neurosecretoriae
Hypothalamus
Hypothalamus
Optic chiasm
Chiasma opticus
Mammillary body
Corpus mammillare
Axon
Axon
Infundibulum
Infundibulum
Hypophyseal portal system
Vv. portales hypophysiales
Hypophyseal artery
A. hypophysialis
Pituitary stalk
Infundibulum neorohypophysis
PITUITARY GLAND—
STRUCTURE
Vein
Vena
Hypophyseal artery
A. hypophysialis
Anterior lobe
Lobus anterior
hypophysis
Anterior pituitary
Adenohypophysis
Posterior pituitary
Neurohypophysis
Pituitary gland
Hypophysis
LOCATION OF THE
PITUITARY GLAND

GLANDS AND ORGANS AFFECTED BY THE PITUITARY GLAND

The pituitary gland exerts its effects on organs and tissues throughout the body by means of organic chemicals called hormones. This illustration shows the organs and tissues affected by the pituitary gland, and the specific hormones involved.

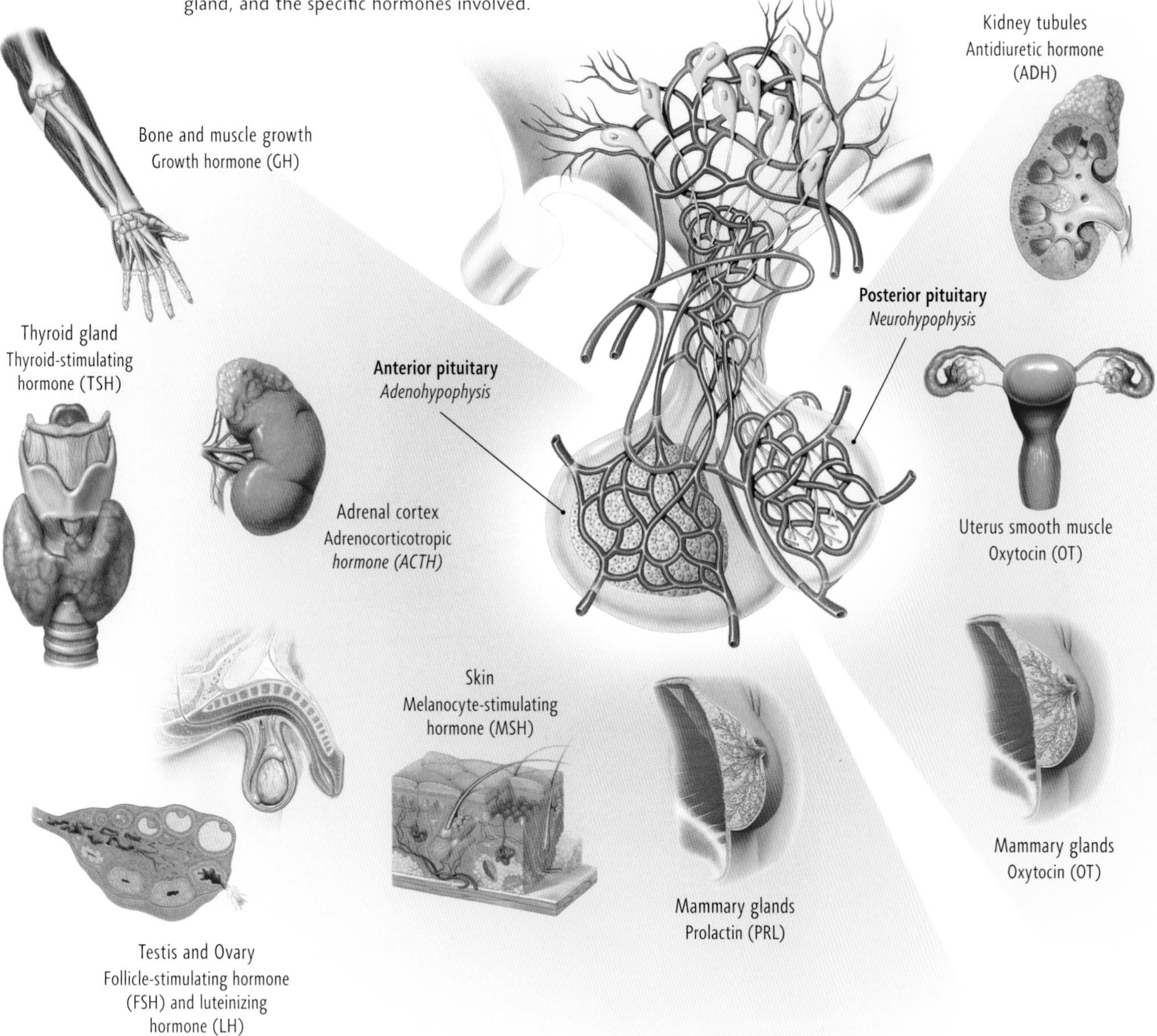

Kidney tubules
Antidiuretic hormone
(ADH)
Bone and muscle growth
Growth hormone (GH)
Posterior pituitary
Neurohypophysis
Thyroid gland
Thyroid-stimulating
hormone (TSH)
Anterior pituitary
Adenohypophysis
Adrenal cortex
Adrenocorticotropic
hormone (ACTH)
Uterus smooth muscle
Oxytocin (OT)
Skin
Melanocyte-stimulating
hormone (MSH)
Mammary glands
Oxytocin (OT)
Mammary glands
Prolactin (PRL)
Testis and Ovary
Follicle-stimulating hormone
(FSH) and luteinizing
hormone (LH)

Endocrine Glands

LOCATION OF THE PINEAL GLAND

The pineal gland is found deep inside the brain. It produces melatonin, thought to be involved in regulating the body's sleep-wake "clock"—melatonin production is highest during a person's normal sleeping hours, and drops off as the body begins to wake.

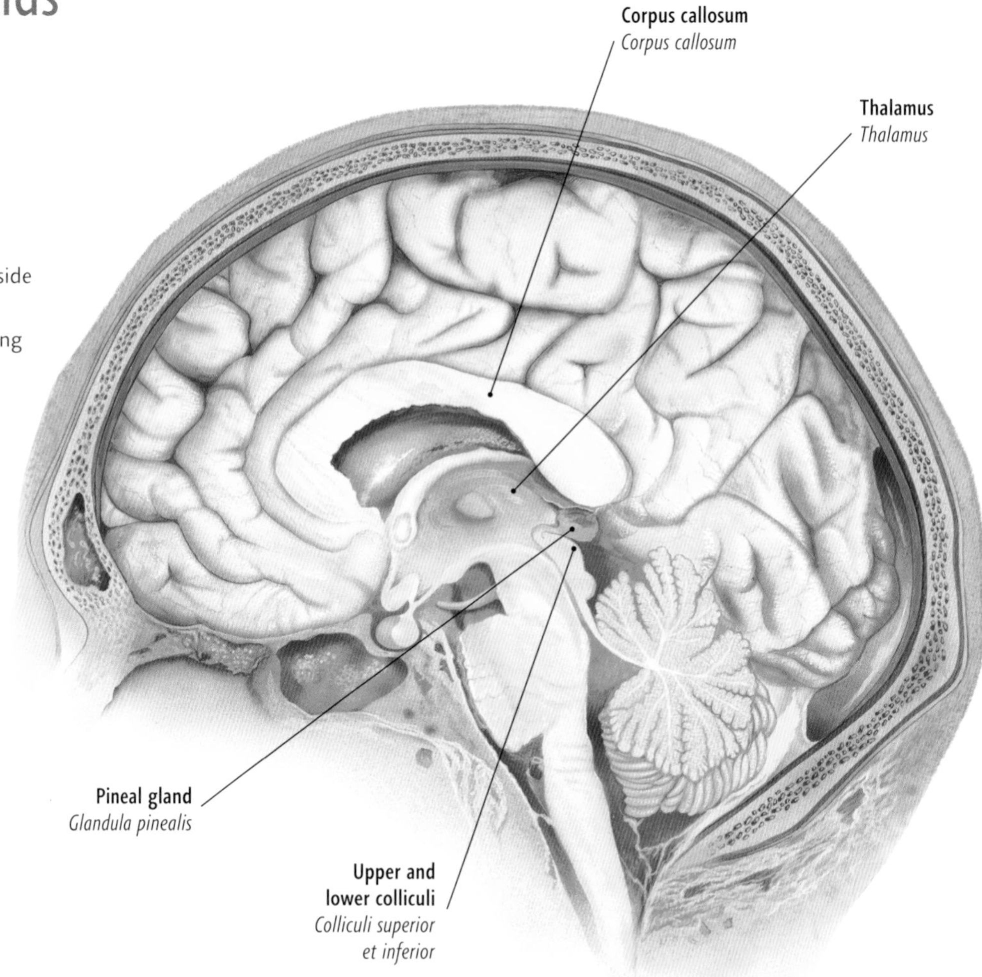
Corpus callosum
Corpus callosum
Thalamus
Thalamus
Pineal gland
Glandula pinealis
Upper and lower colliculi
Colliculi superior et inferior

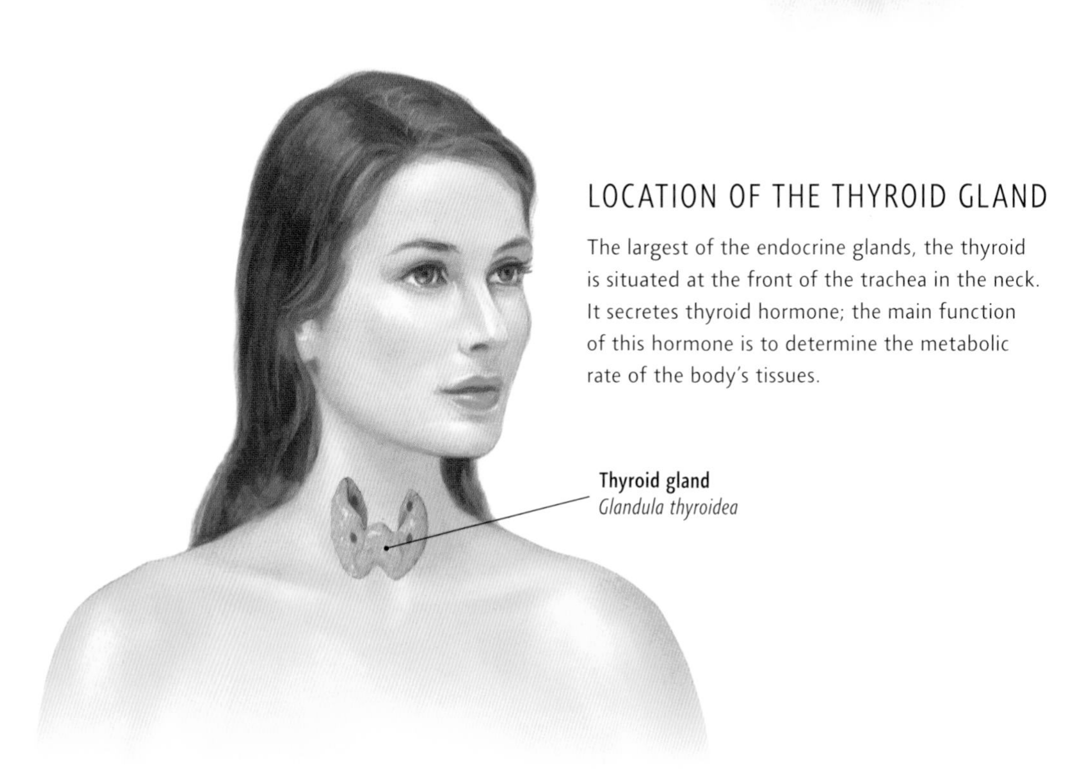
Thyroid gland
Glandula thyroidea

LOCATION OF THE THYROID GLAND

The largest of the endocrine glands, the thyroid is situated at the front of the trachea in the neck. It secretes thyroid hormone; the main function of this hormone is to determine the metabolic rate of the body's tissues.

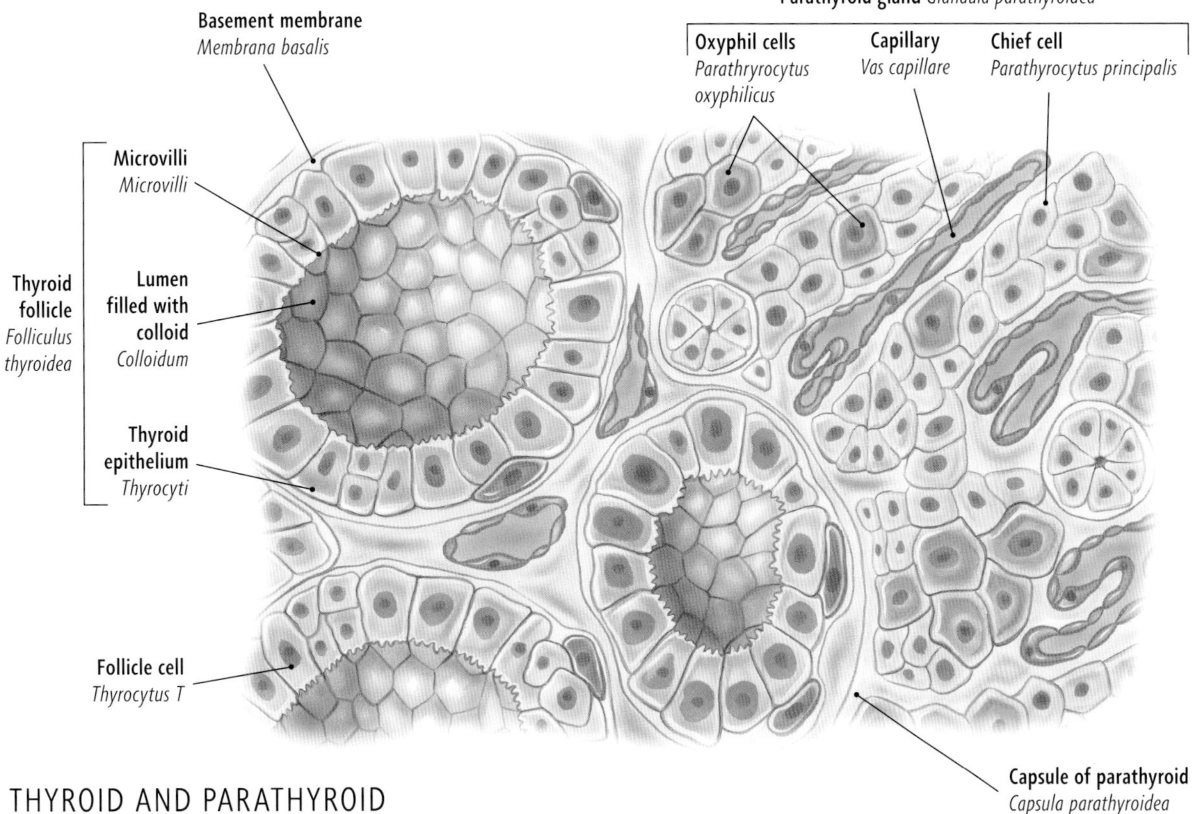
Parathyroid gland Glandula parathyroidea
Basement membrane
Membrana basalis
Oxyphil cells
Parathryrocytus oxyphilicus
Capillary
Vas capillare
Chief cell
Parathyrocytus principalis
Microvilli
Microvilli
Thyroid follicle
Folliculus thyroidea
Lumen filled with colloid
Colloidum
Thyroid epithelium
Thyrocyti
Follicle cell
Thyrocytus T
Capsule of parathyroid
Capsula parathyroidea

THYROID AND PARATHYROID GLANDS—MICROSTRUCTURE

Thyroid gland cells are separated from the cells of the parathyroid glands by a dense capsule of fibers. The thyroid gland comprises many follicles, each consisting of thyroid epithelium cells arranged around a cavity (lumen). The parathyroid glands contain two different cells: chief cells and oxyphil cells.

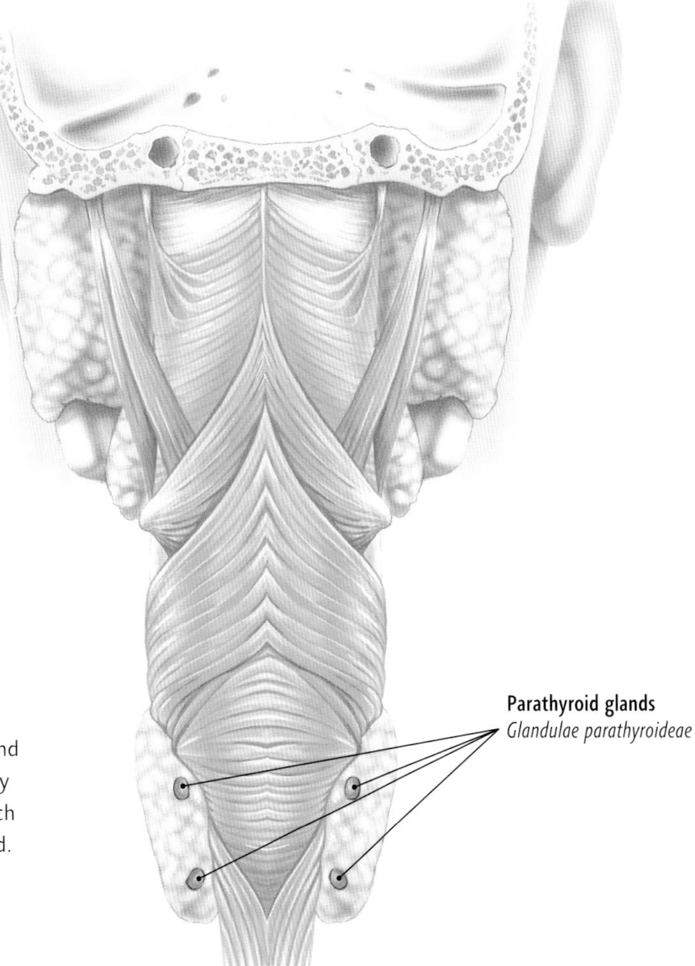
Parathyroid glands
Glandulae parathyroideae

LOCATION OF THE PARATHYROID GLANDS

The parathyroid glands are four (or occasionally three) pea-sized endocrine glands that lie just behind the thyroid gland in the neck. They secrete parathyroid hormone, which controls calcium levels in the blood.

Endocrine Glands

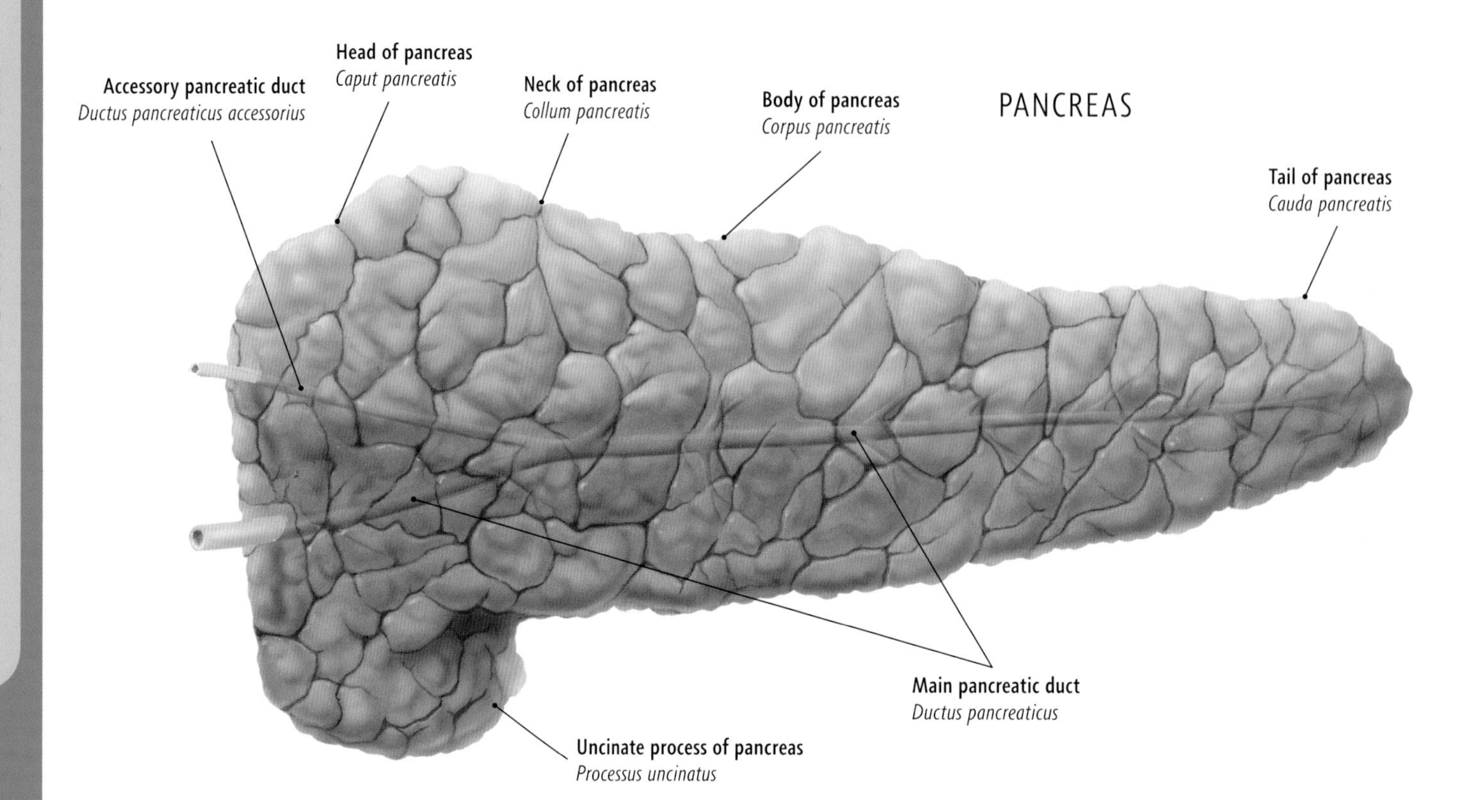
PANCREAS
Head of pancreas
Caput pancreatis
Accessory pancreatic duct
Ductus pancreaticus accessorius
Neck of pancreas
Collum pancreatis
Body of pancreas
Corpus pancreatis
Tail of pancreas
Cauda pancreatis
Main pancreatic duct
Ductus pancreaticus
Uncinate process of pancreas
Processus uncinatus

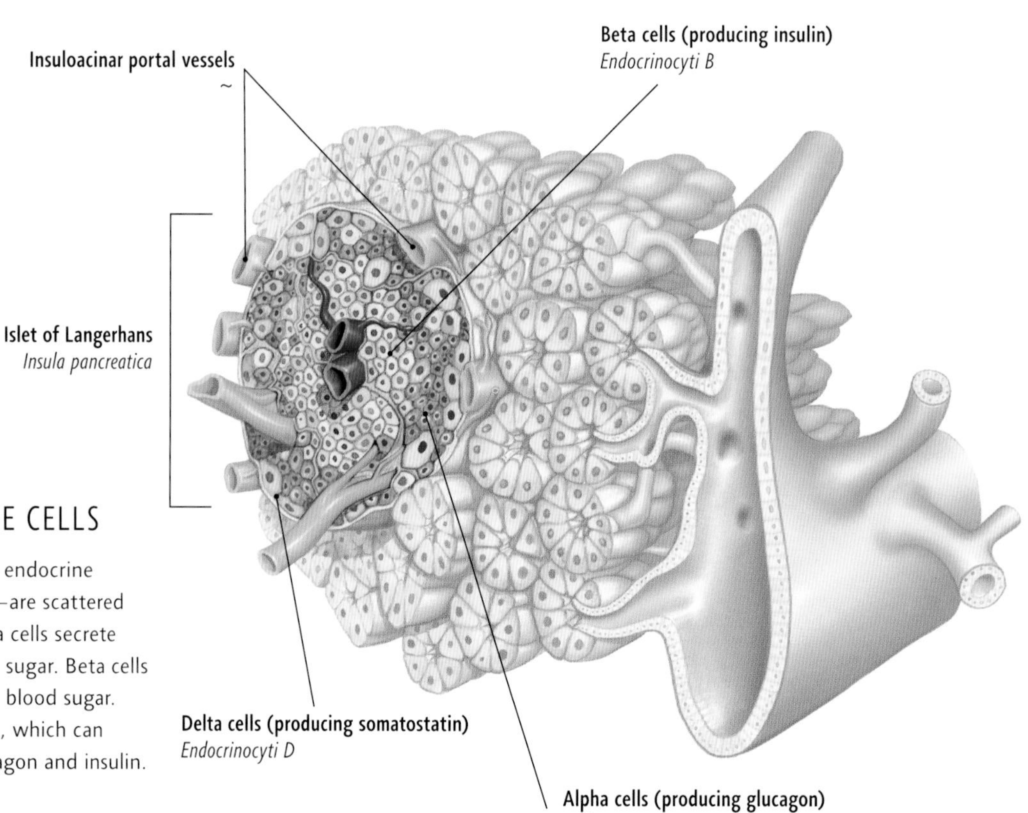
Insuloacinar portal vessels
~
Beta cells (producing insulin)
Endocrinocyti B
Islet of Langerhans
Insula pancreatica
Delta cells (producing somatostatin)
Endocrinocyti D
Alpha cells (producing glucagon)
Endocrinocyti A

PANCREAS: ENDOCRINE CELLS

Clusters of hormone-producing endocrine cells—the islets of Langerhans—are scattered throughout the pancreas. Alpha cells secrete glucagon, which elevates blood sugar. Beta cells secrete insulin, which decreases blood sugar. Delta cells secrete somatostatin, which can inhibit the release of both glucagon and insulin.

[~] = no direct Latin equivalent

ADRENAL GLAND— MICROSTRUCTURE

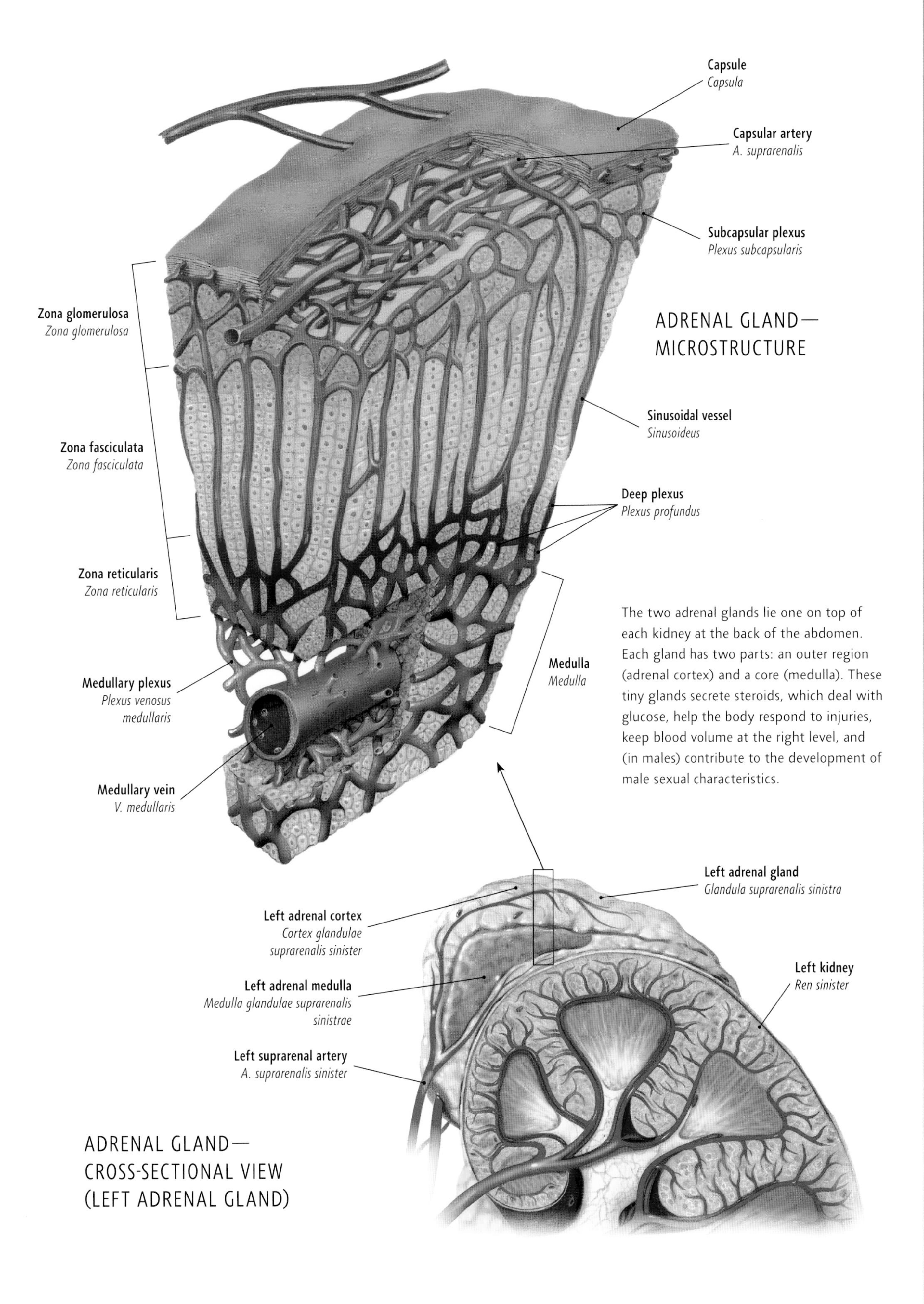
Capsule
Capsula
Capsular artery
A. suprarenalis
Subcapsular plexus
Plexus subcapsularis
Zona glomerulosa
Zona glomerulosa
Sinusoidal vessel
Sinusoideus
Zona fasciculata
Zona fasciculata
Deep plexus
Plexus profundus
Zona reticularis
Zona reticularis
Medulla
Medulla
Medullary plexus
Plexus venosus medullaris
Medullary vein
V. medullaris
Left adrenal gland
Glandula suprarenalis sinistra
Left adrenal cortex
Cortex glandulae suprarenalis sinister
Left kidney
Ren sinister
Left adrenal medulla
Medulla glandulae suprarenalis sinistrae
Left suprarenal artery
A. suprarenalis sinister

The two adrenal glands lie one on top of each kidney at the back of the abdomen. Each gland has two parts: an outer region (adrenal cortex) and a core (medulla). These tiny glands secrete steroids, which deal with glucose, help the body respond to injuries, keep blood volume at the right level, and (in males) contribute to the development of male sexual characteristics.

ADRENAL GLAND— CROSS-SECTIONAL VIEW (LEFT ADRENAL GLAND)

Male and Female Endocrine Glands

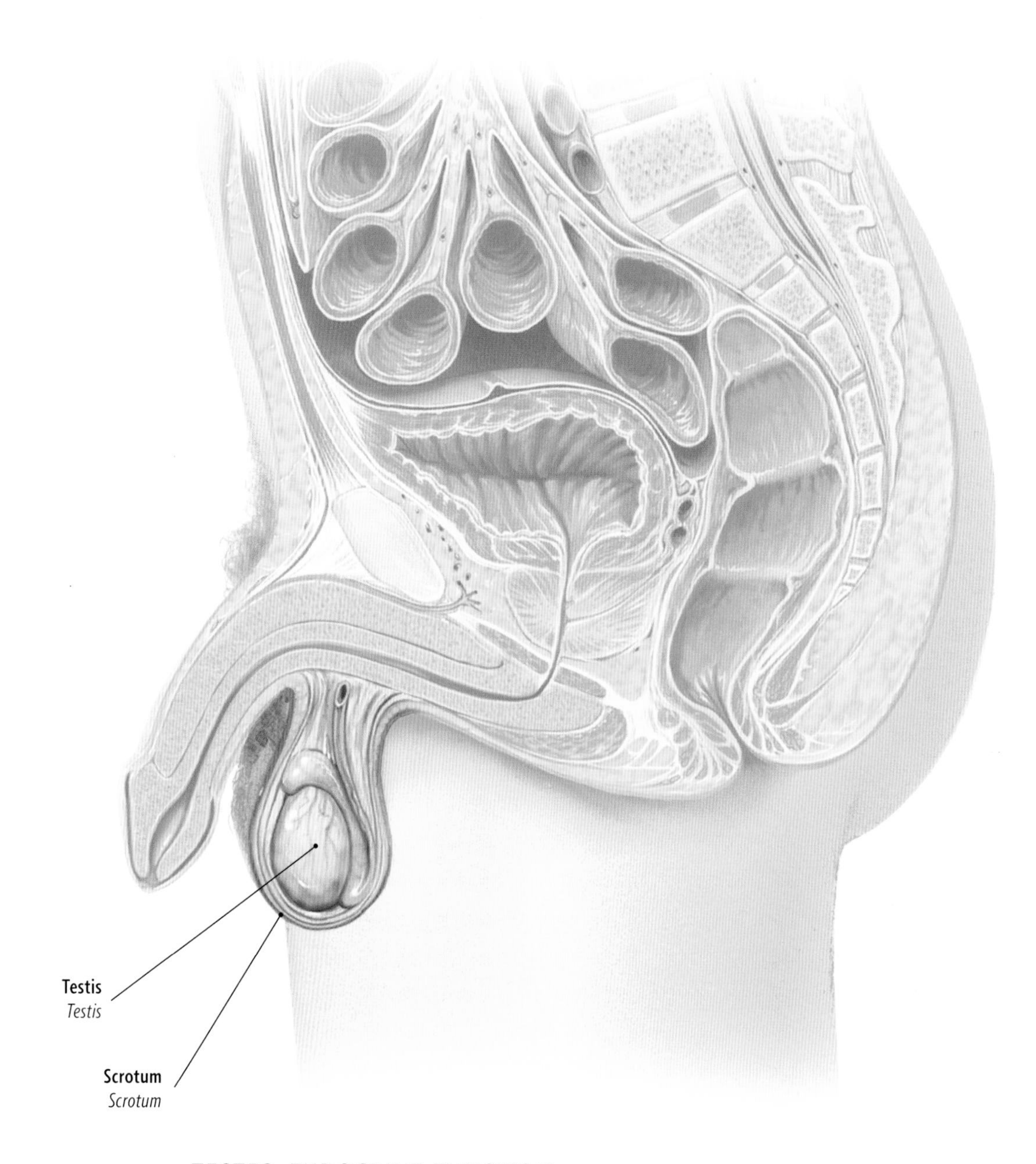
Testis
Testis
Scrotum
Scrotum

TESTES: ENDOCRINE FUNCTION

The testes are two ovoid organs contained in the scrotum, a sac that lies directly behind and beneath the penis. They produce testosterone, which is responsible for the development of secondary sexual characteristics in the male—it stimulates the growth of facial and pubic hair, deepens the voice, and increases muscle tone.

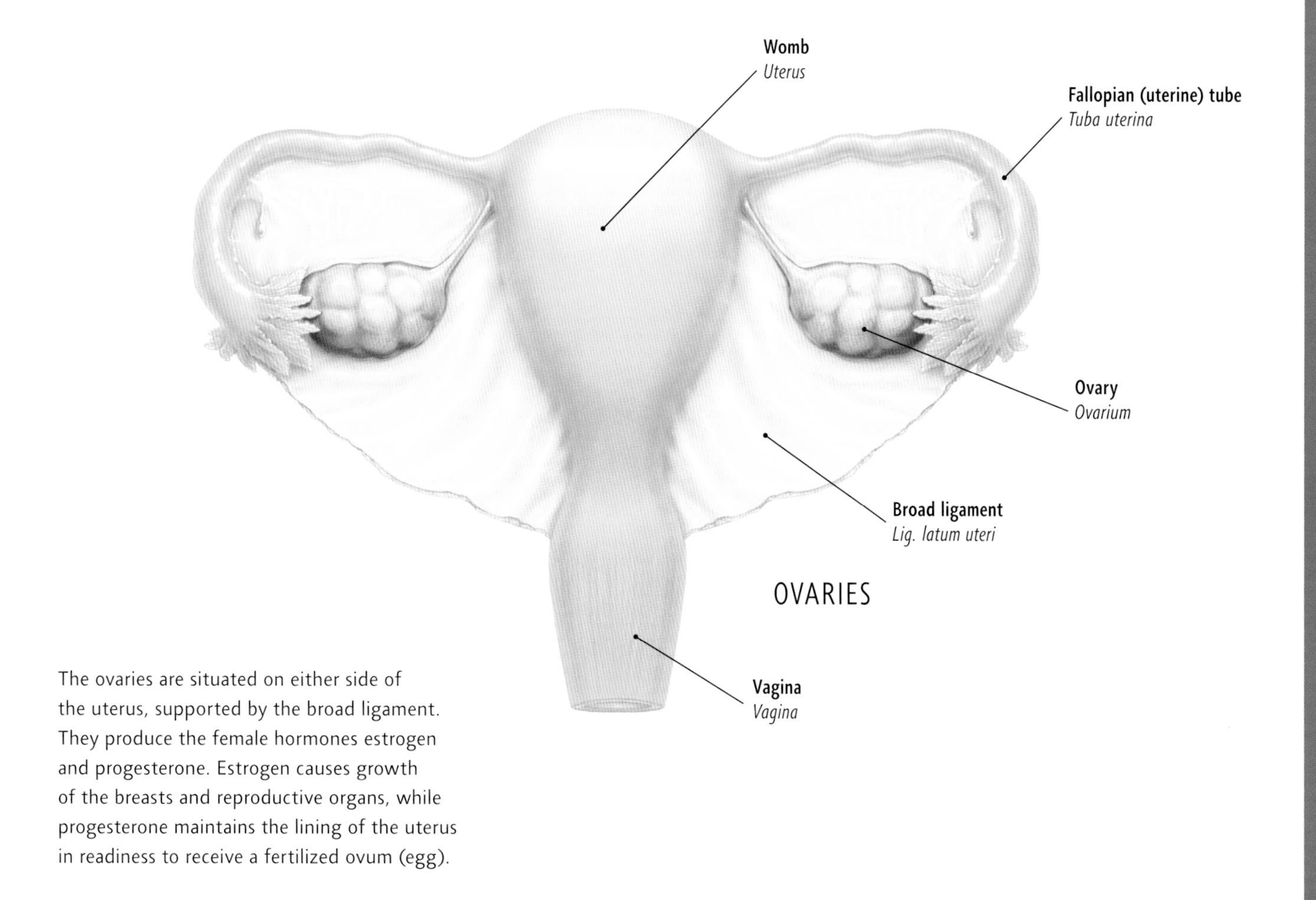
Womb
Uterus
Fallopian (uterine) tube
Tuba uterina
Ovary
Ovarium
Broad ligament
Lig. latum uteri
Vagina
Vagina

OVARIES

The ovaries are situated on either side of the uterus, supported by the broad ligament. They produce the female hormones estrogen and progesterone. Estrogen causes growth of the breasts and reproductive organs, while progesterone maintains the lining of the uterus in readiness to receive a fertilized ovum (egg).

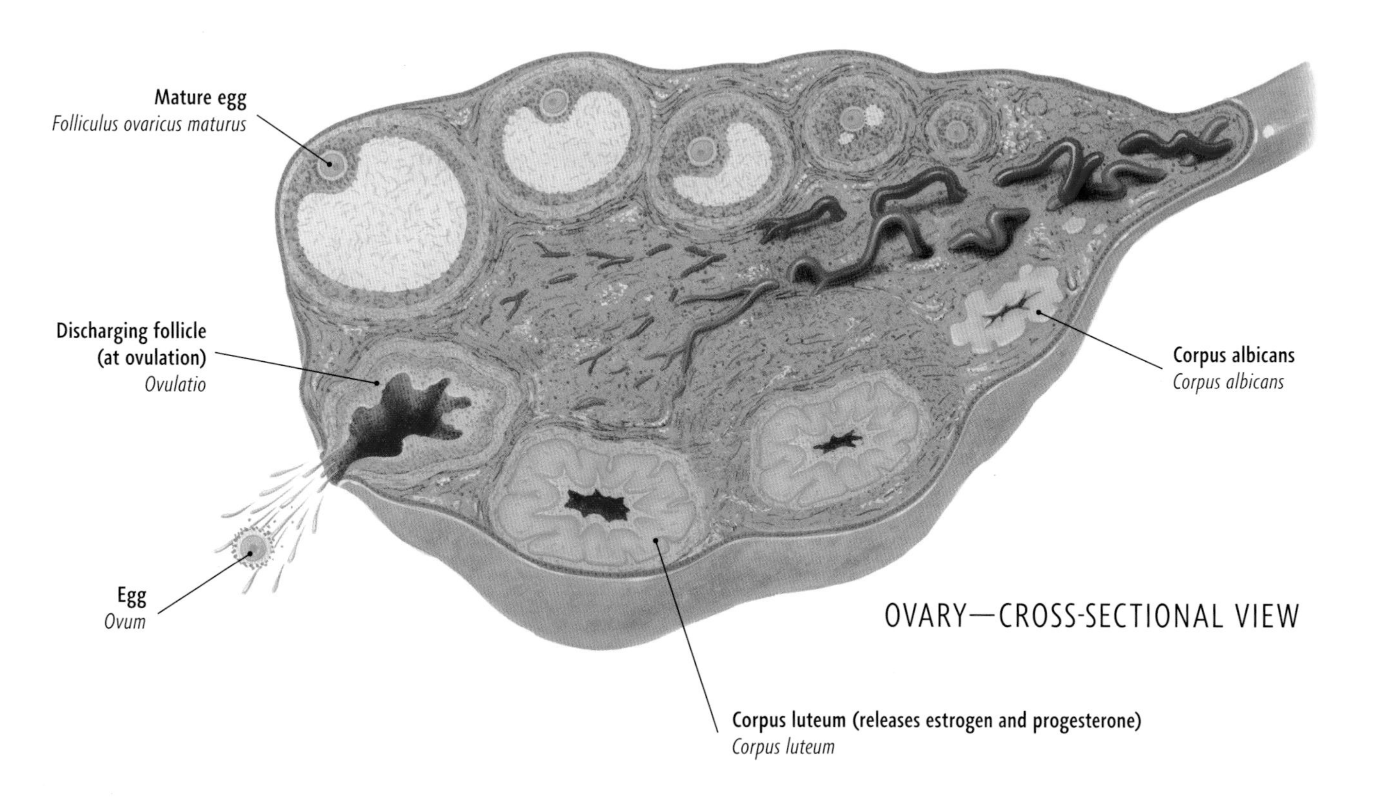
Mature egg
Folliculus ovaricus maturus
Discharging follicle (at ovulation)
Ovulatio
Egg
Ovum
Corpus albicans
Corpus albicans
Corpus luteum (releases estrogen and progesterone)
Corpus luteum

OVARY—CROSS-SECTIONAL VIEW

[~] = no direct Latin equivalent

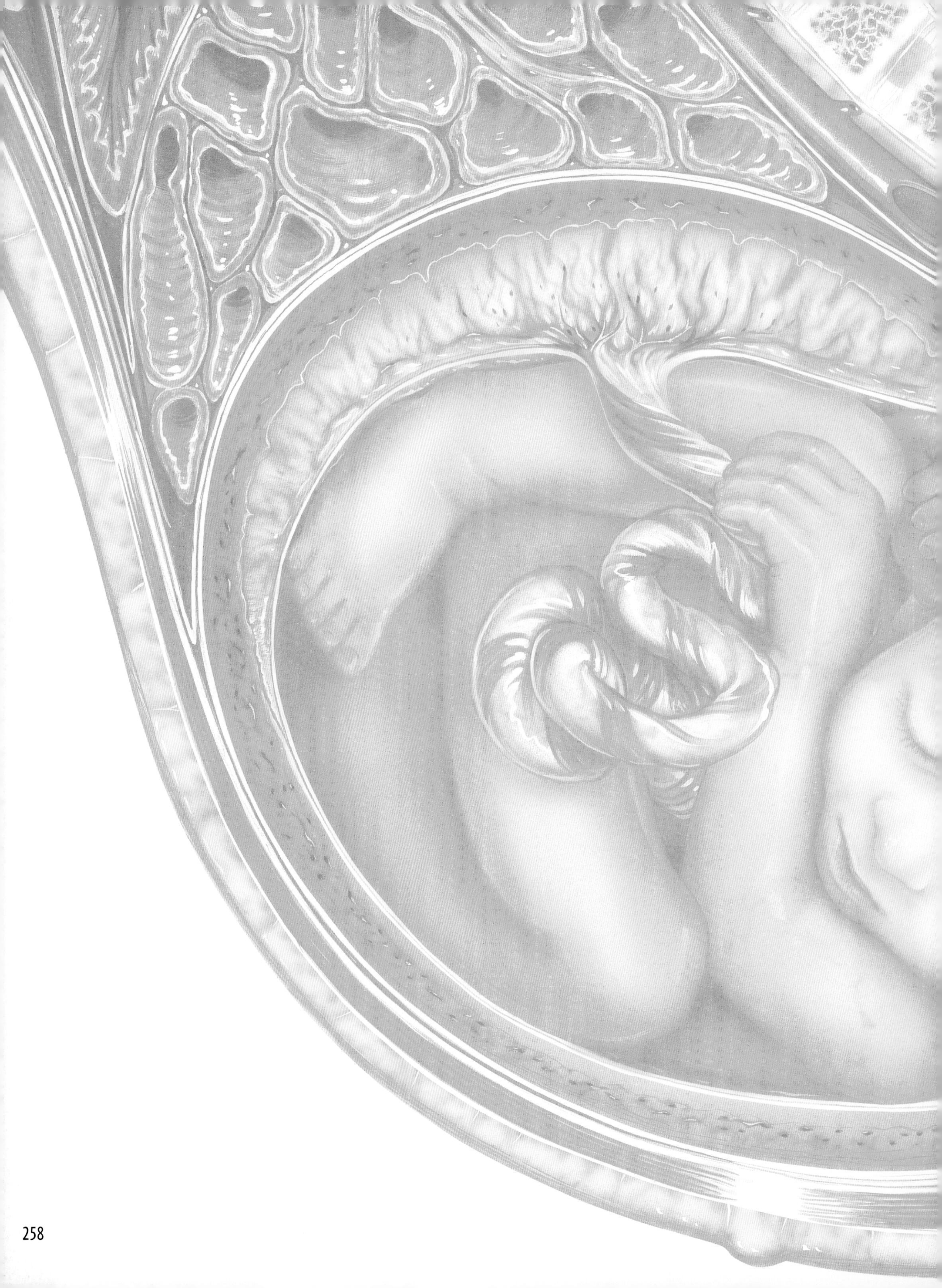

The Reproductive System

Male Reproductive System

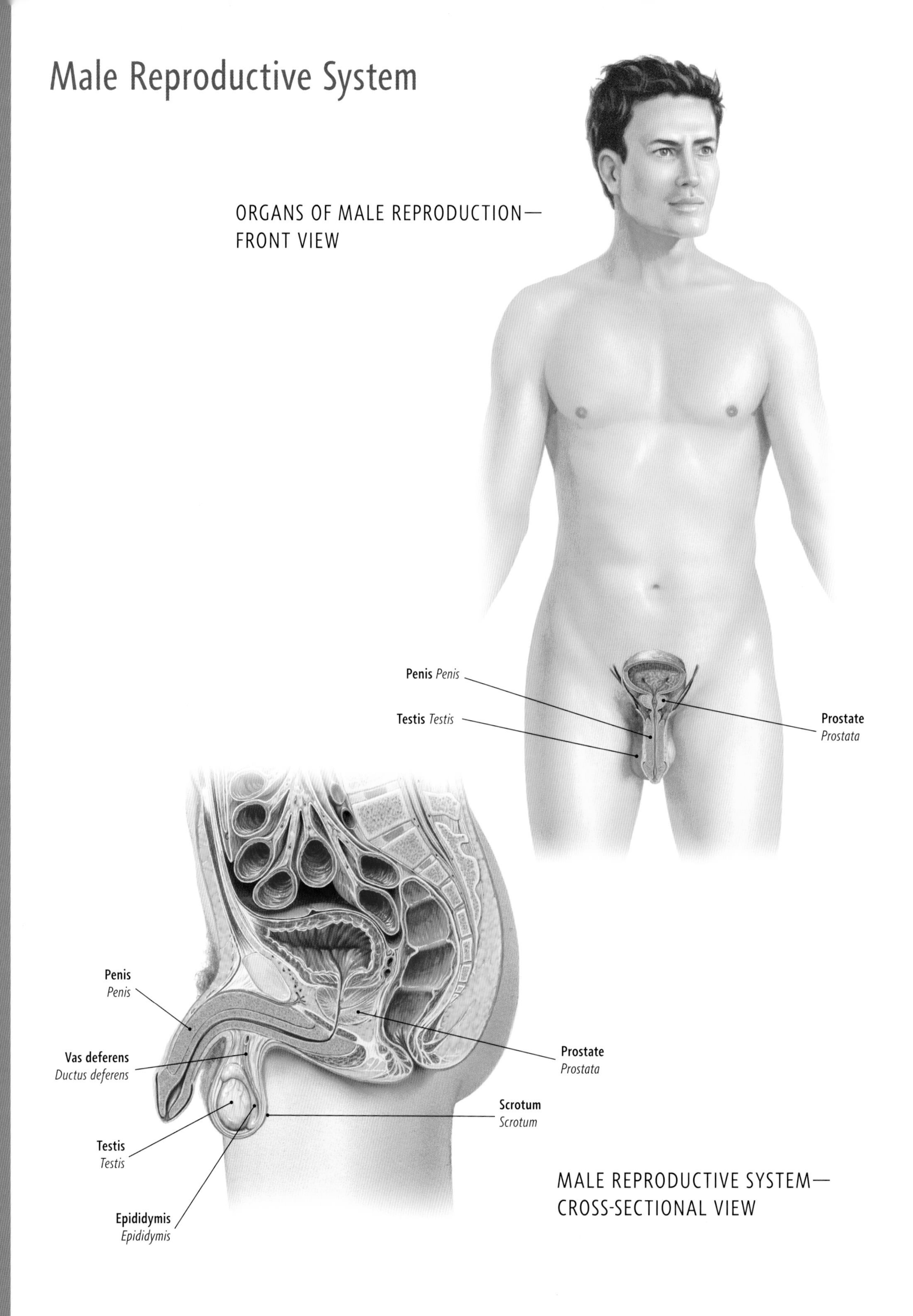
ORGANS OF MALE REPRODUCTION—
FRONT VIEW
Penis *Penis*
Testis *Testis*
Prostate
Prostata
Penis
Penis
Vas deferens
Ductus deferens
Testis
Testis
Epididymis
Epididymis
Prostate
Prostata
Scrotum
Scrotum
MALE REPRODUCTIVE SYSTEM—
CROSS-SECTIONAL VIEW

PROSTATE

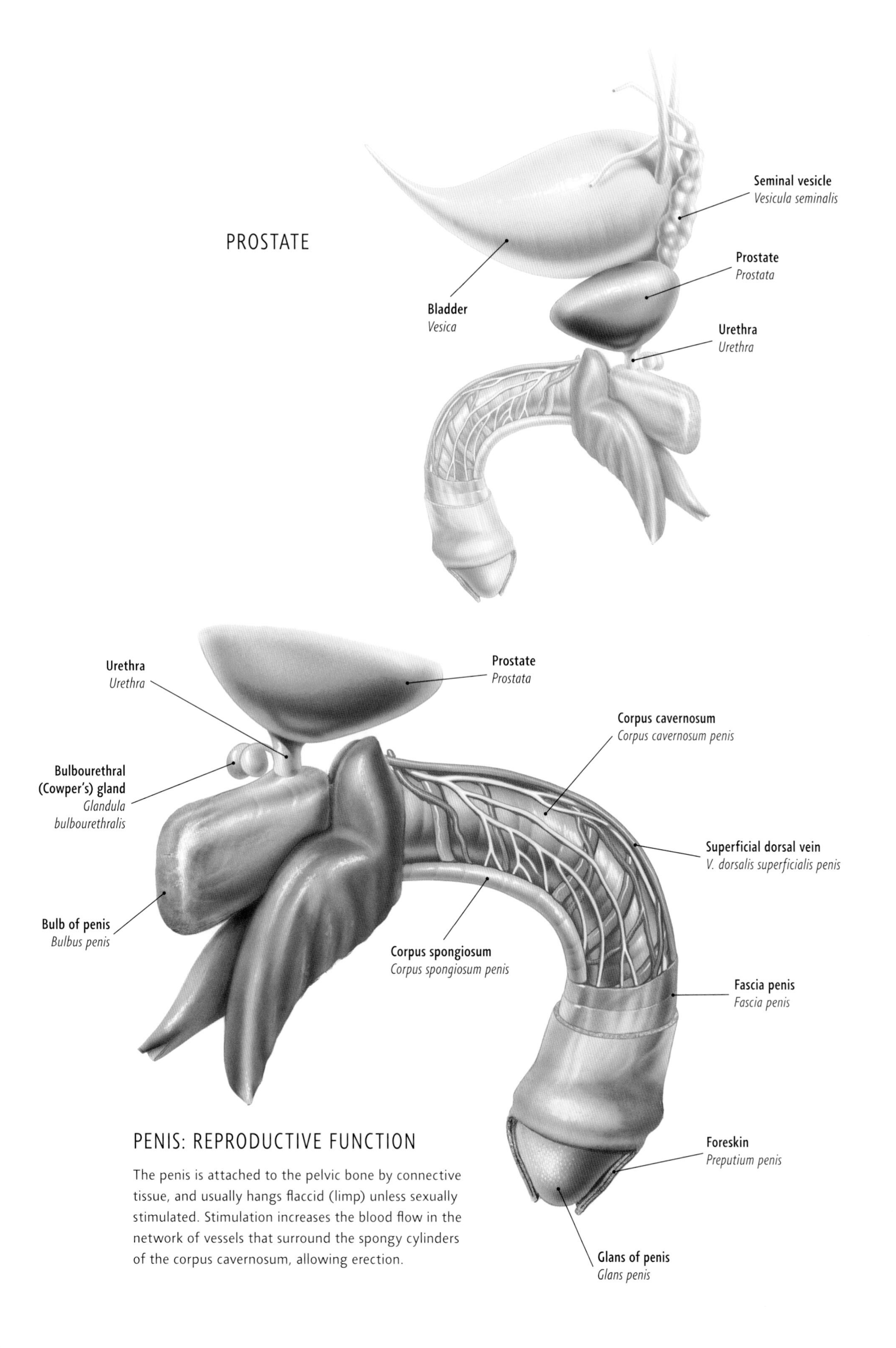
Seminal vesicle
Vesicula seminalis
Bladder
Vesica
Prostate
Prostata
Urethra
Urethra
Urethra
Urethra
Prostate
Prostata
Corpus cavernosum
Corpus cavernosum penis
Bulbourethral
(Cowper's) gland
Glandula
bulbourethralis
Superficial dorsal vein
V. dorsalis superficialis penis
Bulb of penis
Bulbus penis
Corpus spongiosum
Corpus spongiosum penis
Fascia penis
Fascia penis
Foreskin
Preputium penis
Glans of penis
Glans penis

PENIS: REPRODUCTIVE FUNCTION

The penis is attached to the pelvic bone by connective tissue, and usually hangs flaccid (limp) unless sexually stimulated. Stimulation increases the blood flow in the network of vessels that surround the spongy cylinders of the corpus cavernosum, allowing erection.

Male Reproductive System

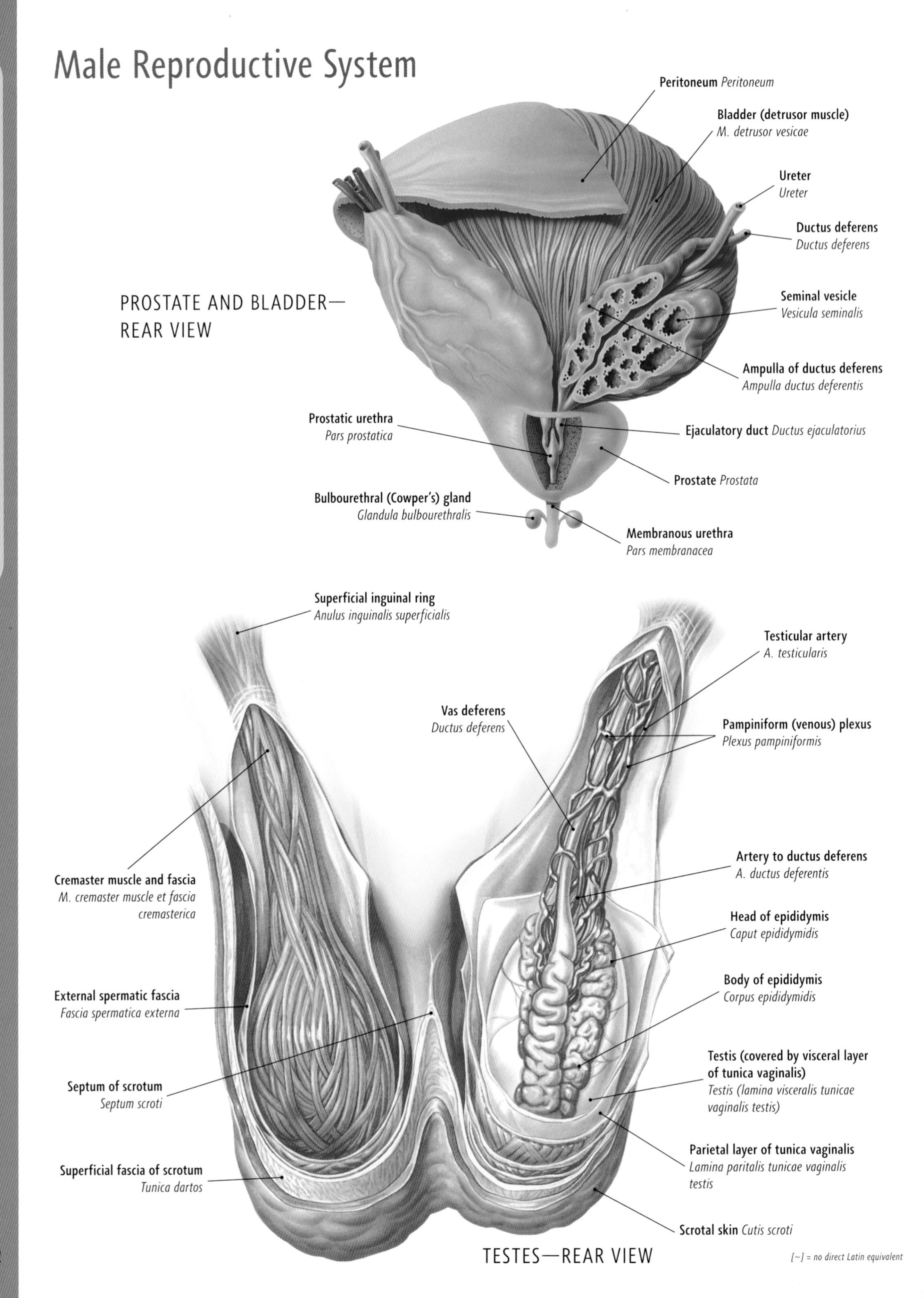

PROSTATE AND BLADDER—REAR VIEW
Peritoneum *Peritoneum*
Bladder (detrusor muscle) *M. detrusor vesicae*
Ureter *Ureter*
Ductus deferens *Ductus deferens*
Seminal vesicle *Vesicula seminalis*
Ampulla of ductus deferens *Ampulla ductus deferentis*
Prostatic urethra *Pars prostatica*
Ejaculatory duct *Ductus ejaculatorius*
Prostate *Prostata*
Bulbourethral (Cowper's) gland *Glandula bulbourethralis*
Membranous urethra *Pars membranacea*
Superficial inguinal ring *Anulus inguinalis superficialis*
Testicular artery *A. testicularis*
Vas deferens *Ductus deferens*
Pampiniform (venous) plexus *Plexus pampiniformis*
Cremaster muscle and fascia *M. cremaster muscle et fascia cremasterica*
Artery to ductus deferens *A. ductus deferentis*
Head of epididymis *Caput epididymidis*
Body of epididymis *Corpus epididymidis*
External spermatic fascia *Fascia spermatica externa*
Septum of scrotum *Septum scroti*
Testis (covered by visceral layer of tunica vaginalis) *Testis (lamina visceralis tunicae vaginalis testis)*
Parietal layer of tunica vaginalis *Lamina paritalis tunicae vaginalis testis*
Superficial fascia of scrotum *Tunica dartos*
Scrotal skin *Cutis scroti*
TESTES—REAR VIEW

[~] = no direct Latin equivalent

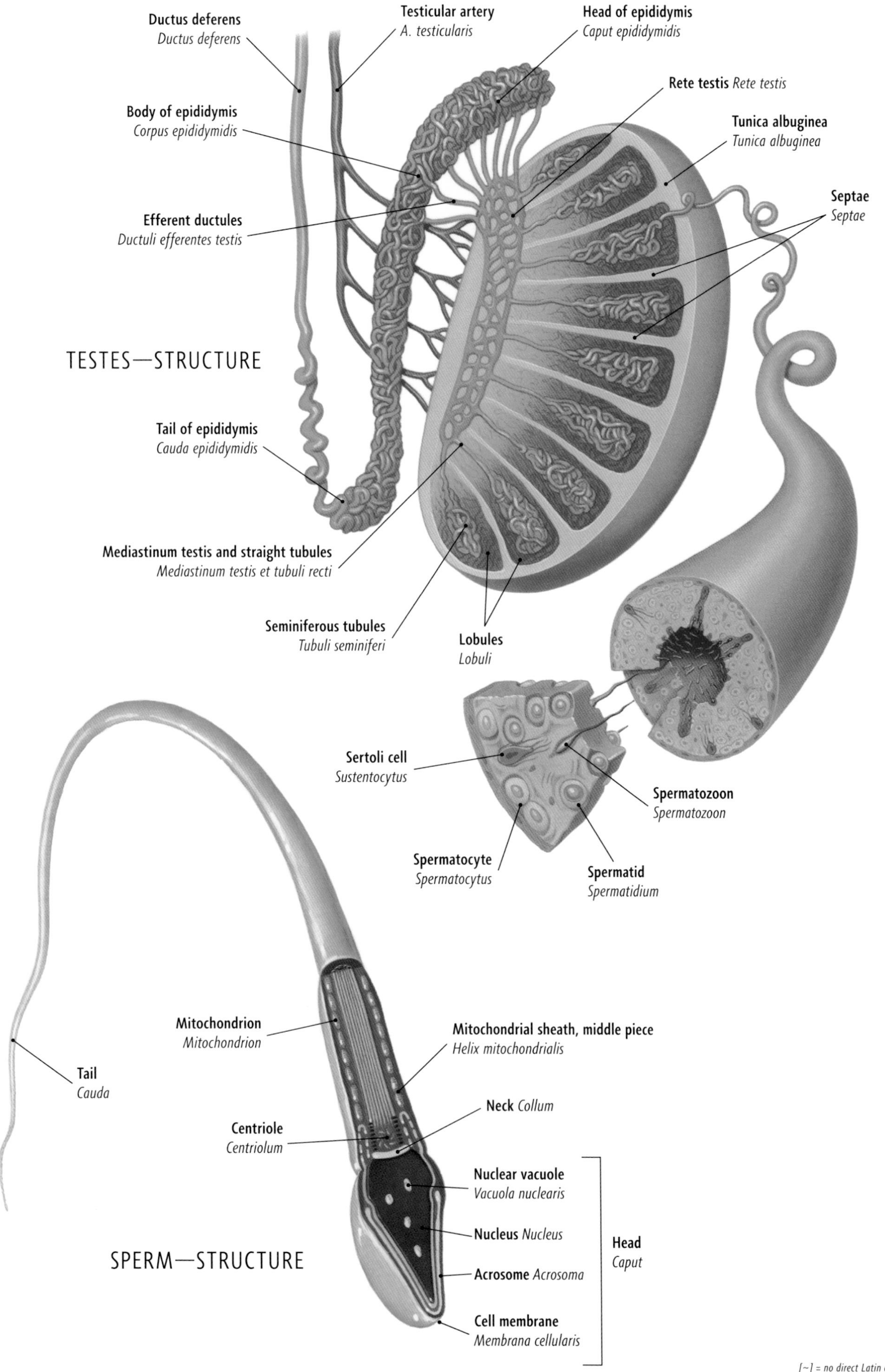
Ductus deferens
Ductus deferens
Testicular artery
A. testicularis
Head of epididymis
Caput epididymidis
Rete testis Rete testis
Body of epididymis
Corpus epididymidis
Tunica albuginea
Tunica albuginea
Septae
Septae
Efferent ductules
Ductuli efferentes testis
TESTES—STRUCTURE
Tail of epididymis
Cauda epididymidis
Mediastinum testis and straight tubules
Mediastinum testis et tubuli recti
Seminiferous tubules
Tubuli seminiferi
Lobules
Lobuli
Sertoli cell
Sustentocytus
Spermatozoon
Spermatozoon
Spermatocyte
Spermatocytus
Spermatid
Spermatidium
Mitochondrion
Mitochondrion
Mitochondrial sheath, middle piece
Helix mitochondrialis
Tail
Cauda
Neck Collum
Centriole
Centriolum
Nuclear vacuole
Vacuola nuclearis
Nucleus Nucleus
Head
Caput
SPERM—STRUCTURE
Acrosome Acrosoma
Cell membrane
Membrana cellularis

[~] = no direct Latin equivalent

Female Reproductive System

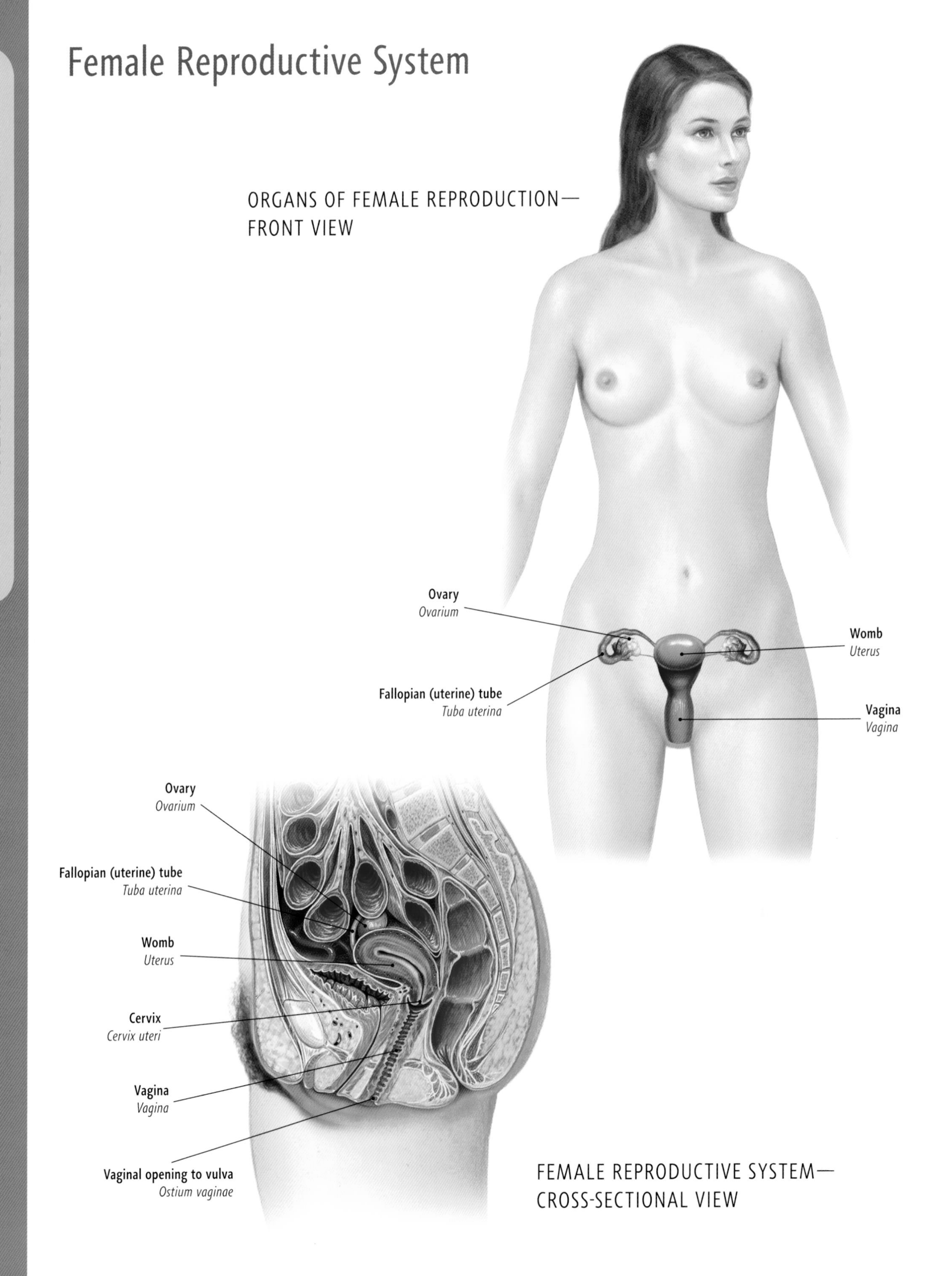
ORGANS OF FEMALE REPRODUCTION—
FRONT VIEW
Ovary
Ovarium
Womb
Uterus
Fallopian (uterine) tube
Tuba uterina
Vagina
Vagina
Ovary
Ovarium
Fallopian (uterine) tube
Tuba uterina
Womb
Uterus
Cervix
Cervix uteri
Vagina
Vagina
Vaginal opening to vulva
Ostium vaginae
FEMALE REPRODUCTIVE SYSTEM—
CROSS-SECTIONAL VIEW

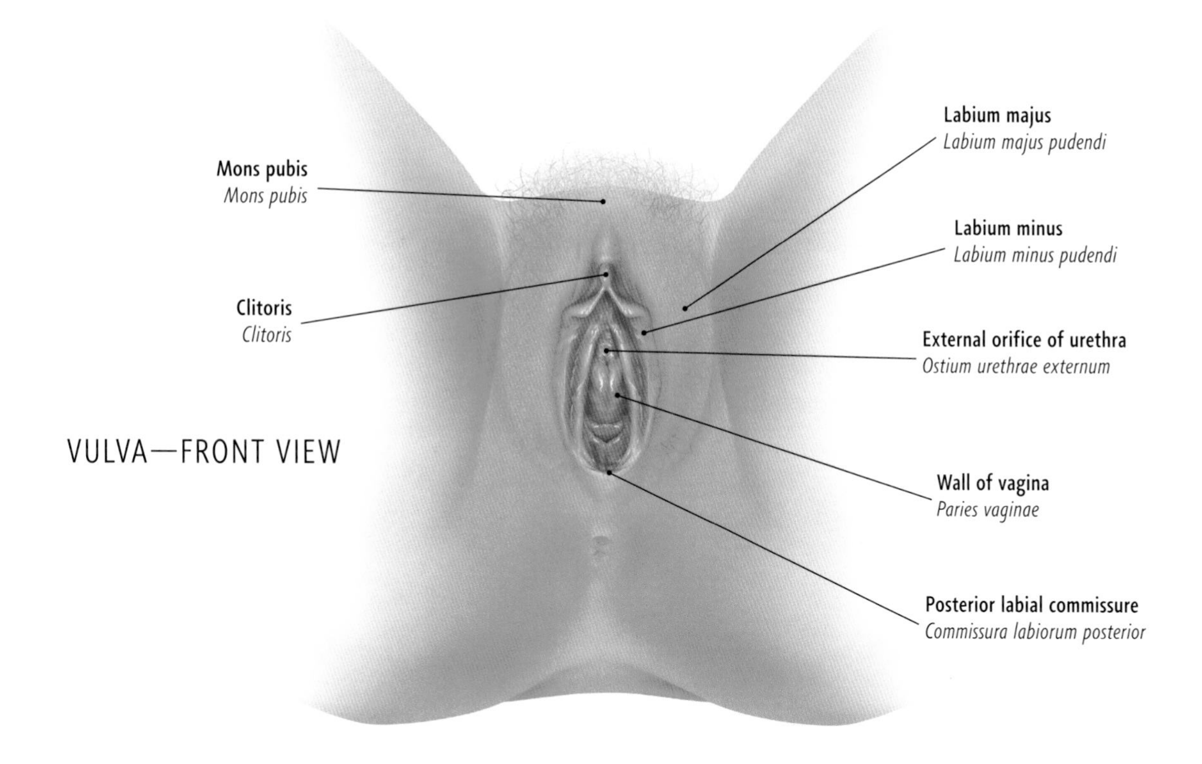
Mons pubis
Mons pubis
Clitoris
Clitoris
Labium majus
Labium majus pudendi
Labium minus
Labium minus pudendi
External orifice of urethra
Ostium urethrae externum
Wall of vagina
Paries vaginae
Posterior labial commissure
Commissura labiorum posterior

VULVA—FRONT VIEW

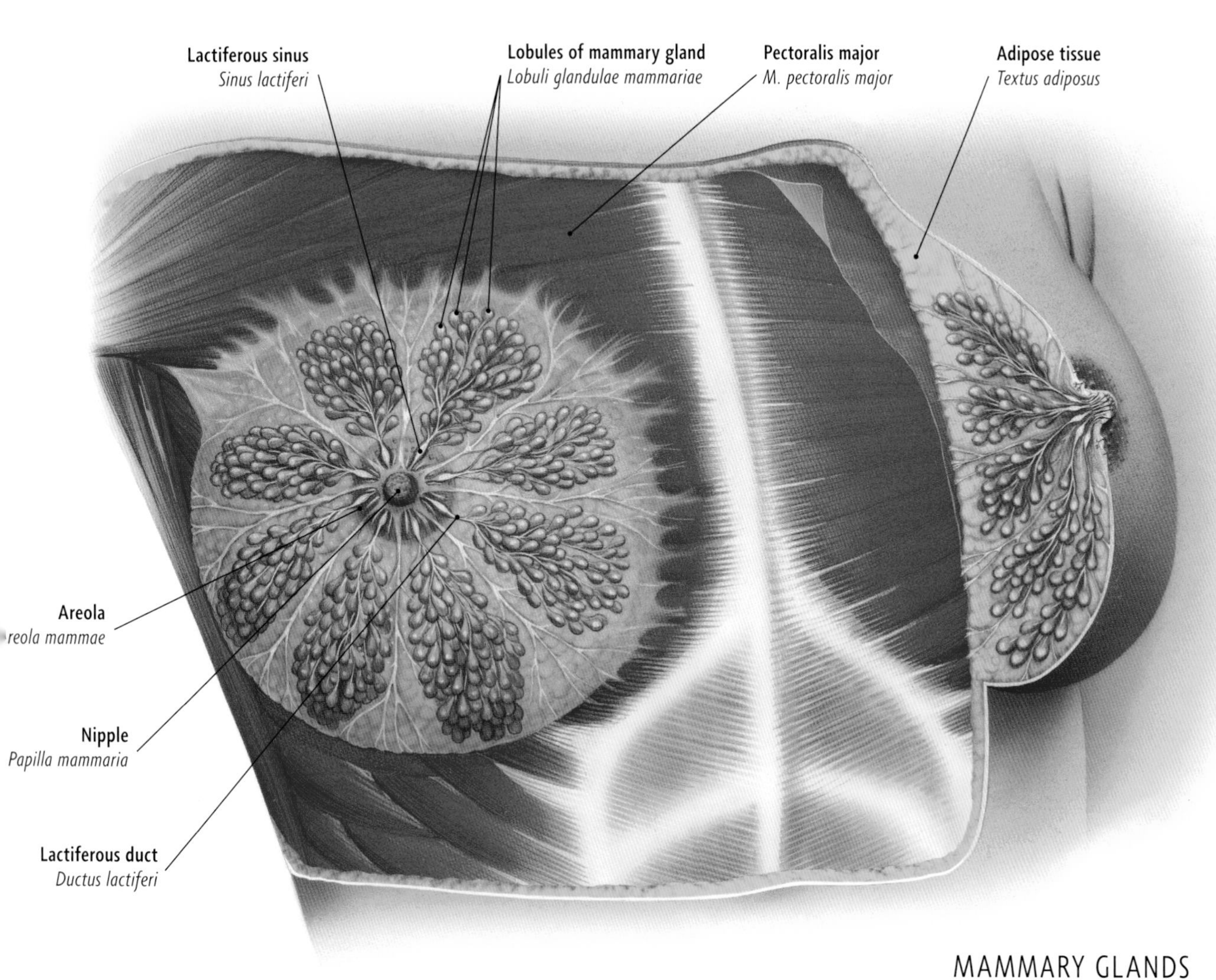
Lactiferous sinus
Sinus lactiferi
Lobules of mammary gland
Lobuli glandulae mammariae
Pectoralis major
M. pectoralis major
Adipose tissue
Textus adiposus
Areola
reola mammae
Nipple
Papilla mammaria
Lactiferous duct
Ductus lactiferi

MAMMARY GLANDS

Menstrual Cycle

CYCLE REGULATION

The hypothalamus and the anterior pituitary gland regulate the menstrual cycle. During each cycle the anterior pituitary gland releases follicle-stimulating hormone (FSH) and luteinizing hormone (LH), which trigger follicles (egg sacs) in the ovaries to mature and release estrogen and progesterone.

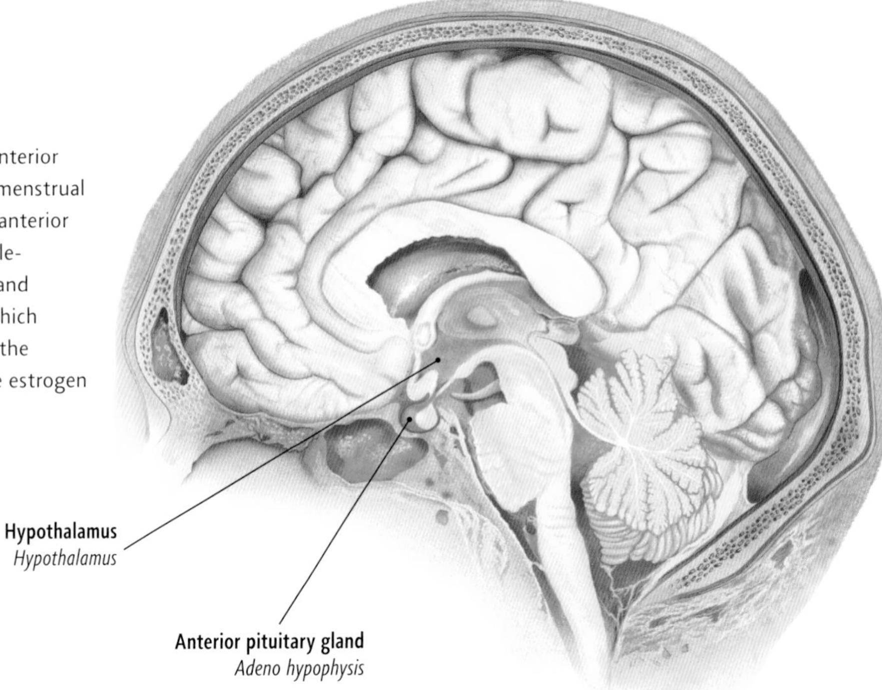

Hypothalamus
Hypothalamus
Anterior pituitary gland
Adeno hypophysis

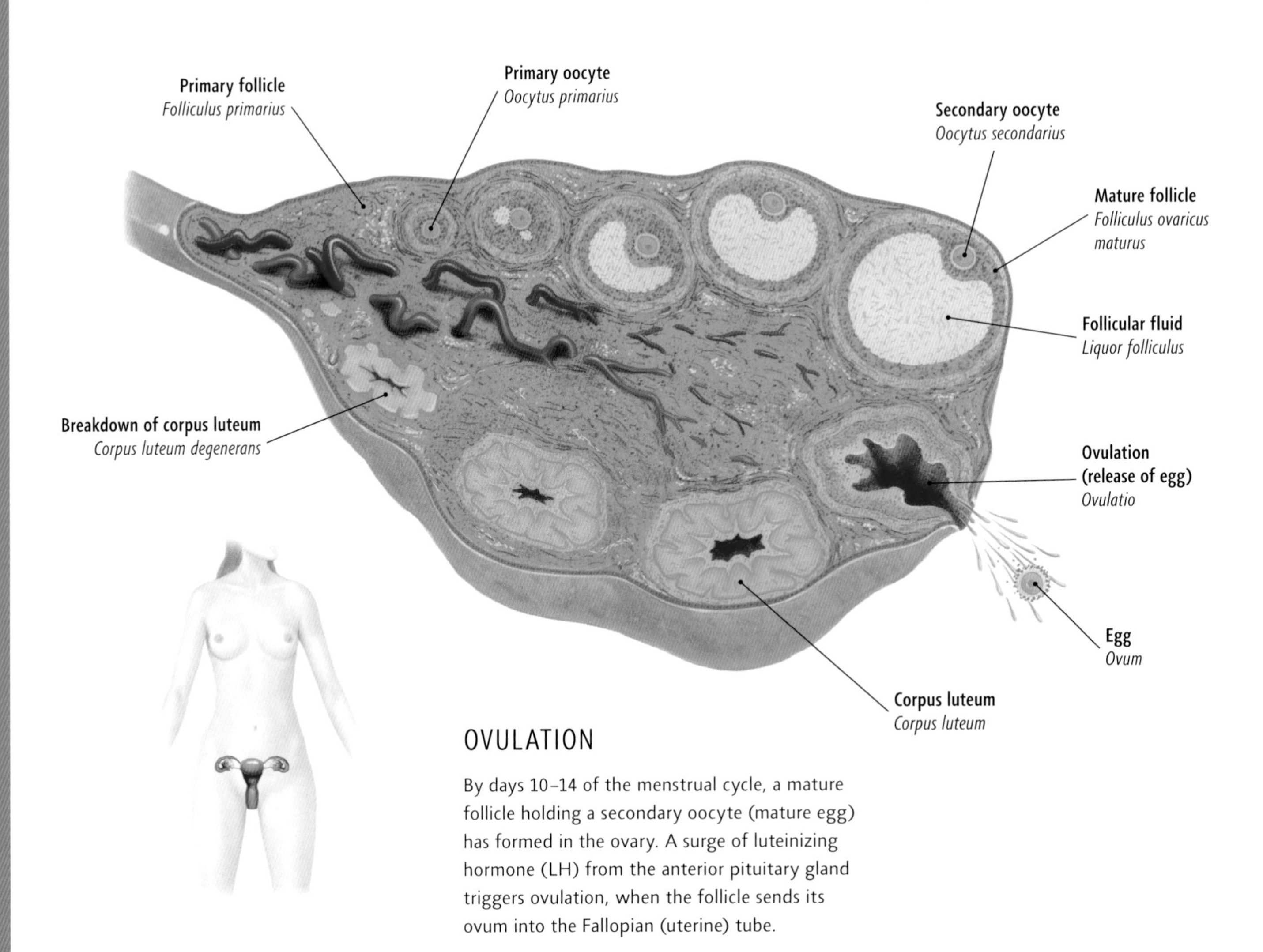

Primary follicle
Folliculus primarius
Primary oocyte
Oocytus primarius
Secondary oocyte
Oocytus secondarius
Mature follicle
Folliculus ovaricus maturus
Follicular fluid
Liquor folliculus
Breakdown of corpus luteum
Corpus luteum degenerans
Ovulation (release of egg)
Ovulatio
Egg
Ovum
Corpus luteum
Corpus luteum

OVULATION

By days 10–14 of the menstrual cycle, a mature follicle holding a secondary oocyte (mature egg) has formed in the ovary. A surge of luteinizing hormone (LH) from the anterior pituitary gland triggers ovulation, when the follicle sends its ovum into the Fallopian (uterine) tube.

[~] = no direct Latin equivalent

a **Days 1-6**

Menstruation

During menstruation, the lining of the uterus (endometrium) breaks down and is discharged.

b **Days 7–13**

Proliferative phase

At the start of the proliferative phase, cells begin to repair the lining of the uterus, and follicles in the ovary begin to form an ovum (egg).

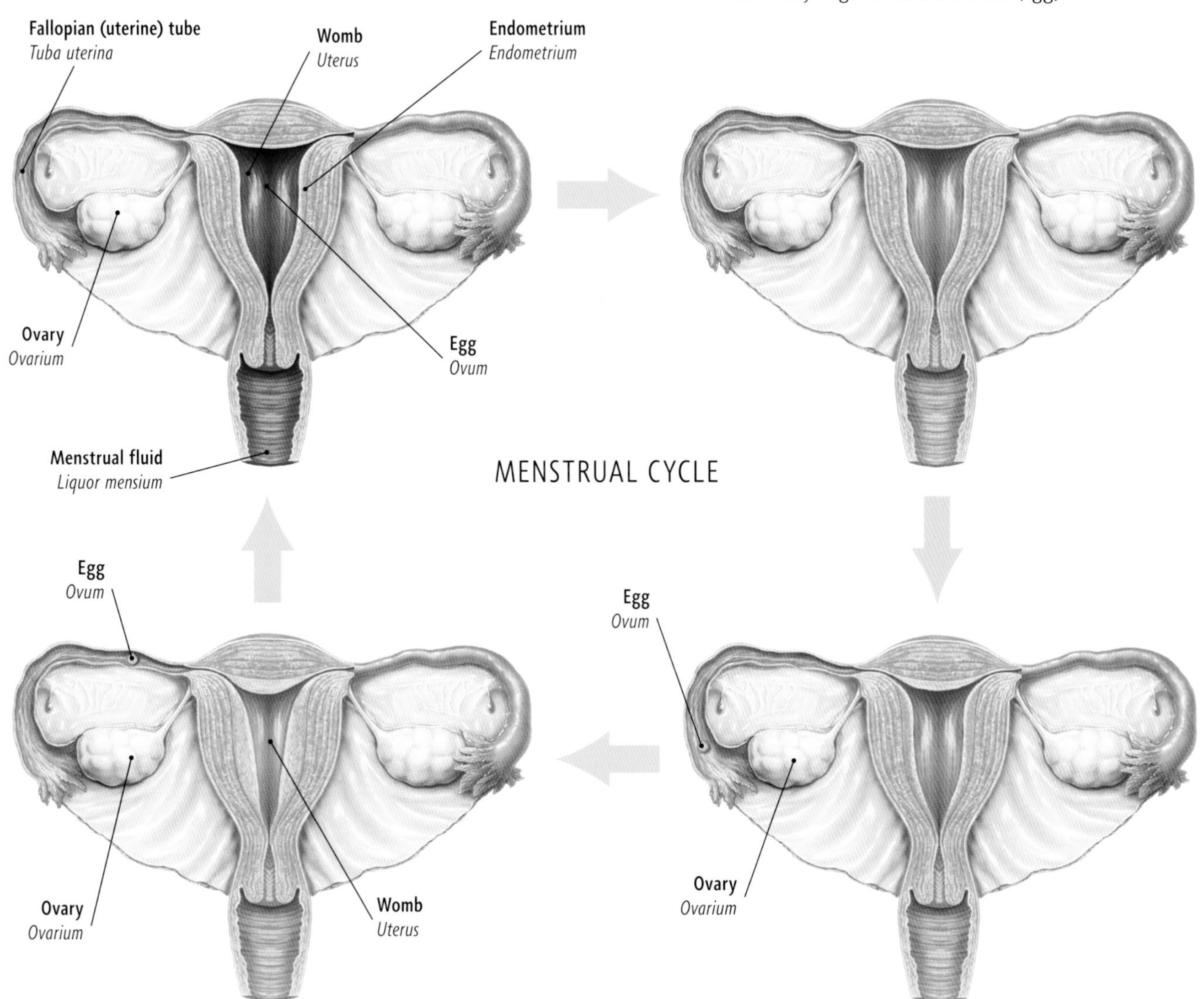
Fallopian (uterine) tube
Tuba uterina
Womb
Uterus
Endometrium
Endometrium
Ovary
Ovarium
Egg
Ovum
Menstrual fluid
Liquor mensium
MENSTRUAL CYCLE
Egg
Ovum
Egg
Ovum
Ovary
Ovarium
Womb
Uterus
Ovary
Ovarium

d **Days 15–28**

Secretory phase

During the secretory phase, the ovum travels along the Fallopian (uterine) tube toward the uterus, and hormones thicken the uterus lining in readiness to support a fertilized ovum. If the ovum is not fertilized, hormone levels fall and blood vessels in the uterus constrict.

c **Day 14**

Ovulation

Ovulation takes place on around day 14 of the cycle, and the ovary releases its ovum into the Fallopian (uterine) tube.

[~] = no direct Latin equivalent

Conception and Early Pregnancy

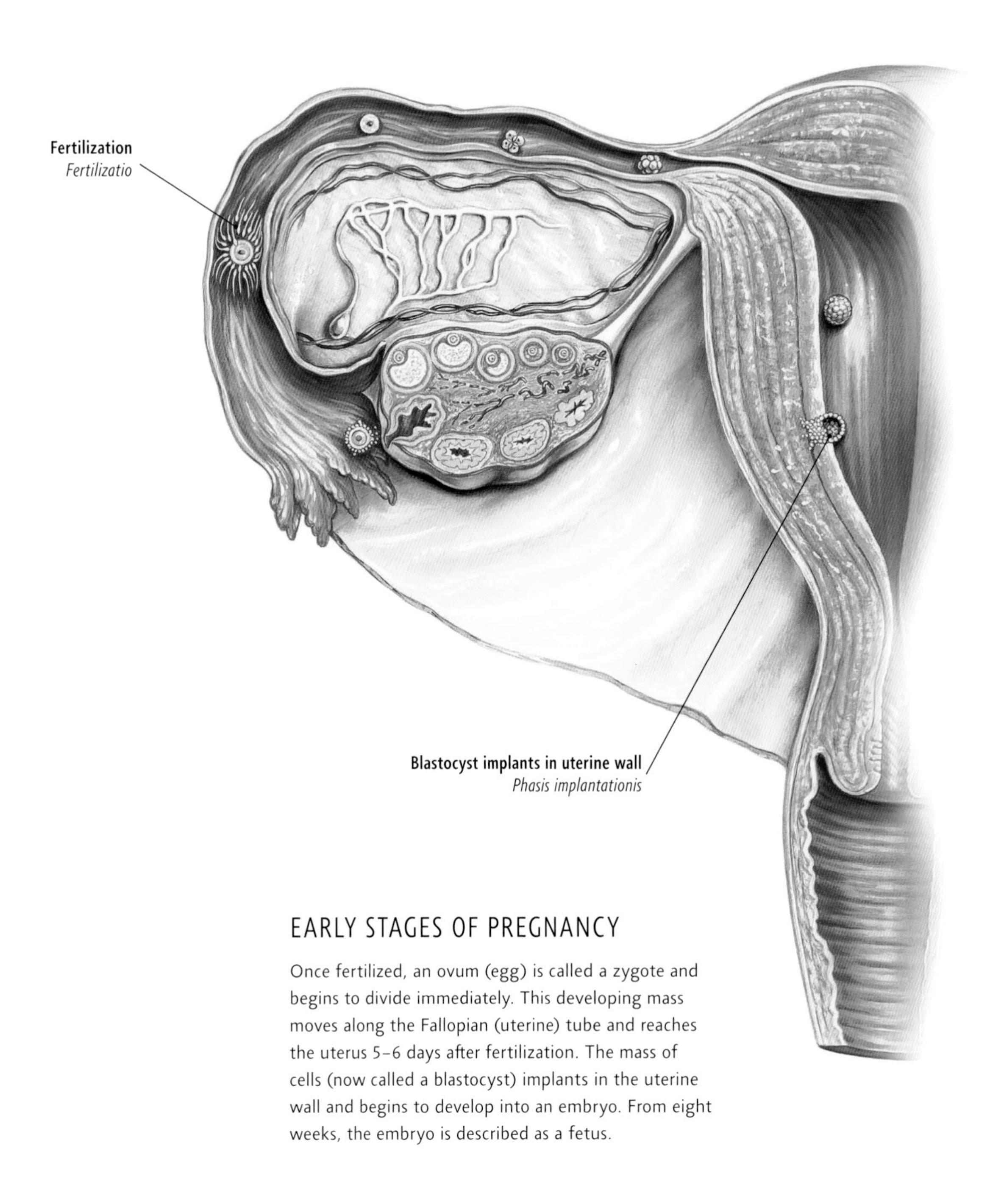
Fertilization
Fertilizatio
Blastocyst implants in uterine wall
Phasis implantationis

EARLY STAGES OF PREGNANCY

Once fertilized, an ovum (egg) is called a zygote and begins to divide immediately. This developing mass moves along the Fallopian (uterine) tube and reaches the uterus 5–6 days after fertilization. The mass of cells (now called a blastocyst) implants in the uterine wall and begins to develop into an embryo. From eight weeks, the embryo is described as a fetus.

[~] = no direct Latin equivalent

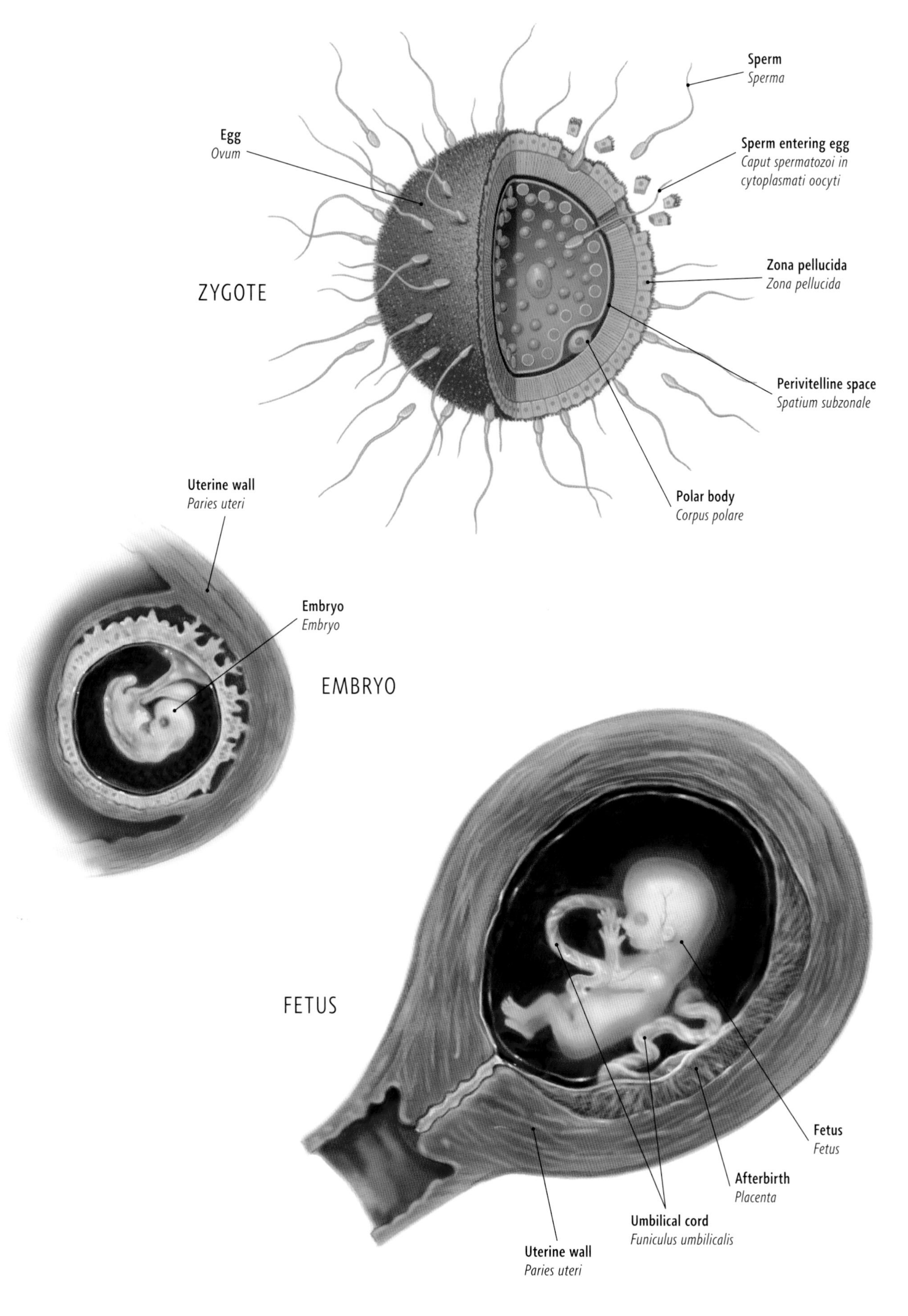
ZYGOTE
Egg
Ovum
Sperm
Sperma
Sperm entering egg
Caput spermatozoi in cytoplasmati oocyti
Zona pellucida
Zona pellucida
Perivitelline space
Spatium subzonale
Polar body
Corpus polare
EMBRYO
Uterine wall
Paries uteri
Embryo
Embryo
FETUS
Fetus
Fetus
Afterbirth
Placenta
Umbilical cord
Funiculus umbilicalis
Uterine wall
Paries uteri

[~] = no direct Latin equivalent

Fetal Development Cycle

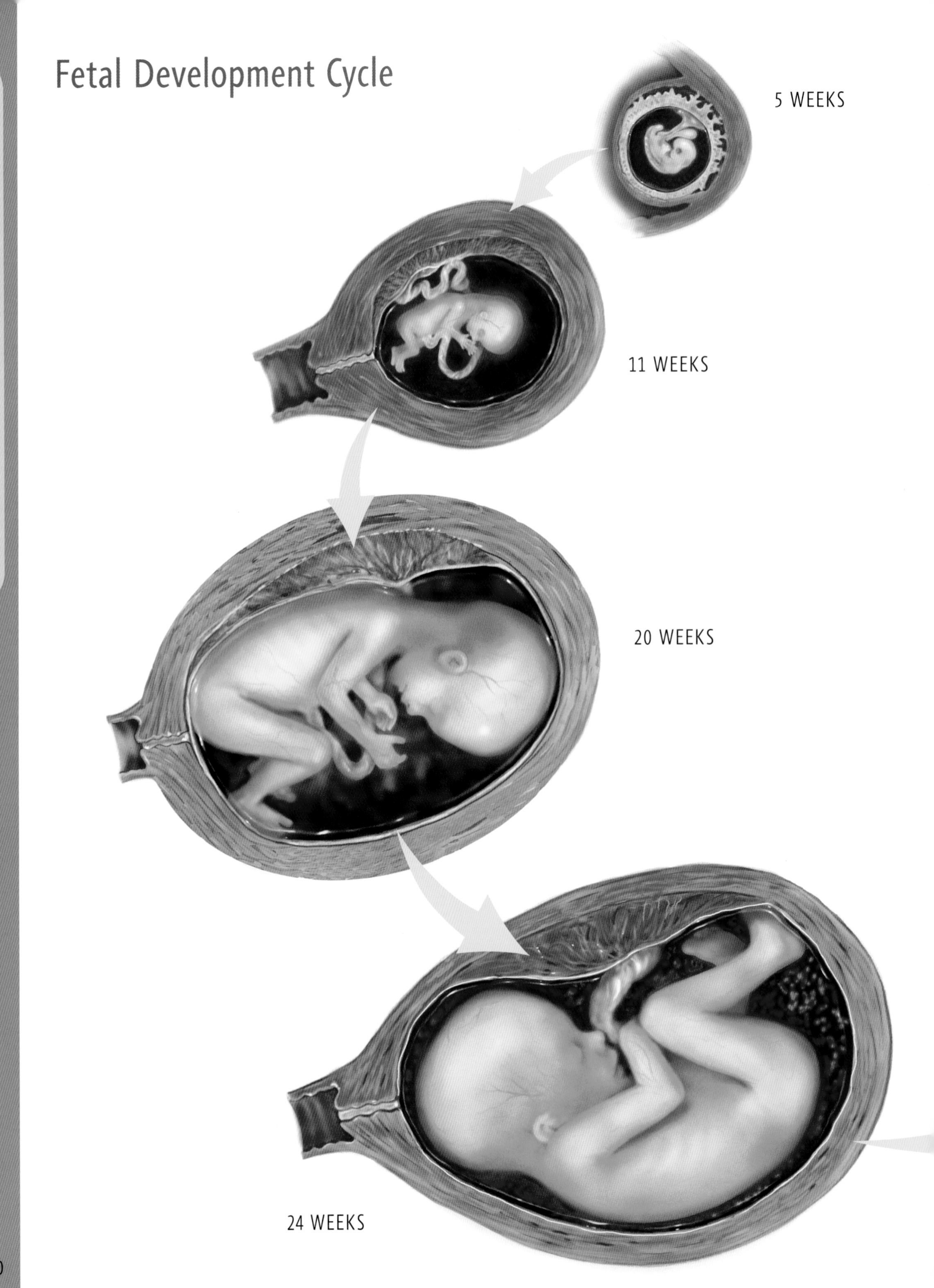
5 WEEKS
11 WEEKS
20 WEEKS
24 WEEKS

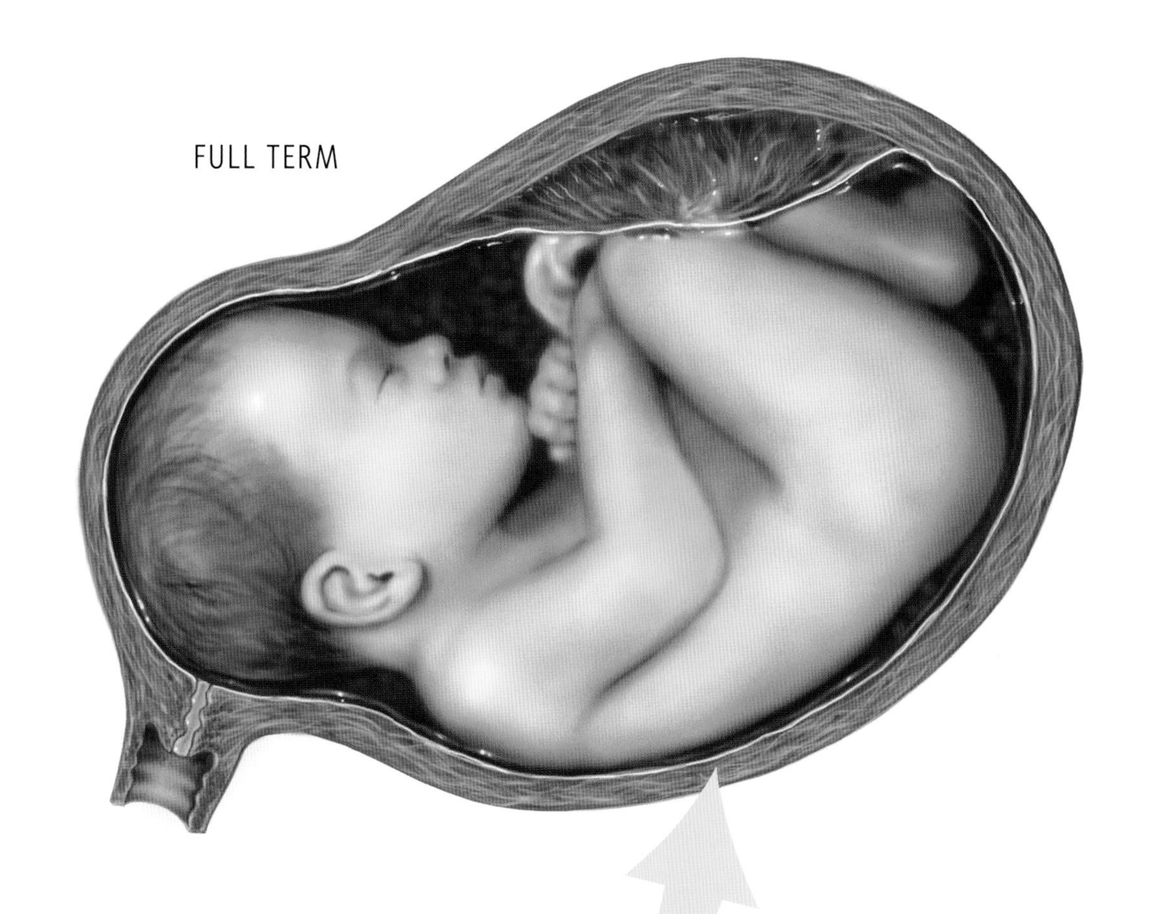
FULL TERM

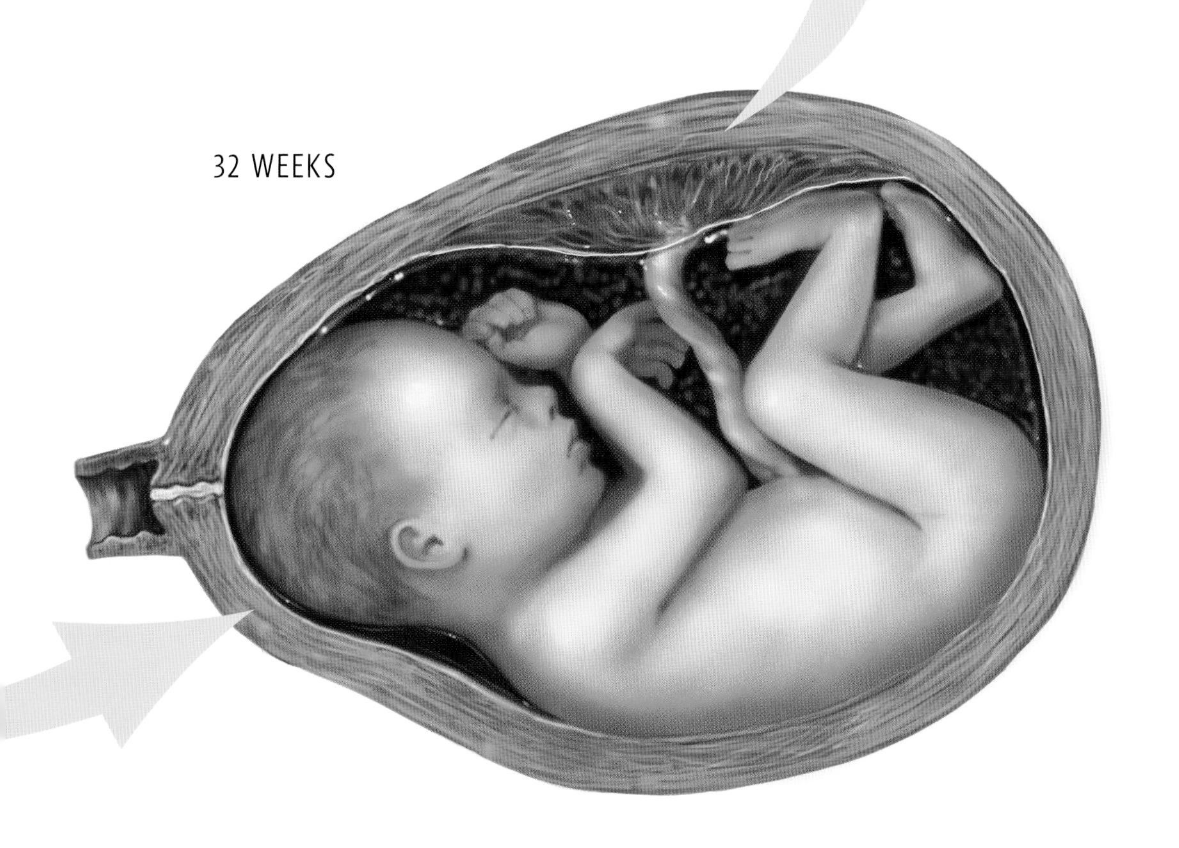
32 WEEKS

Fetal Skull and Bone Development

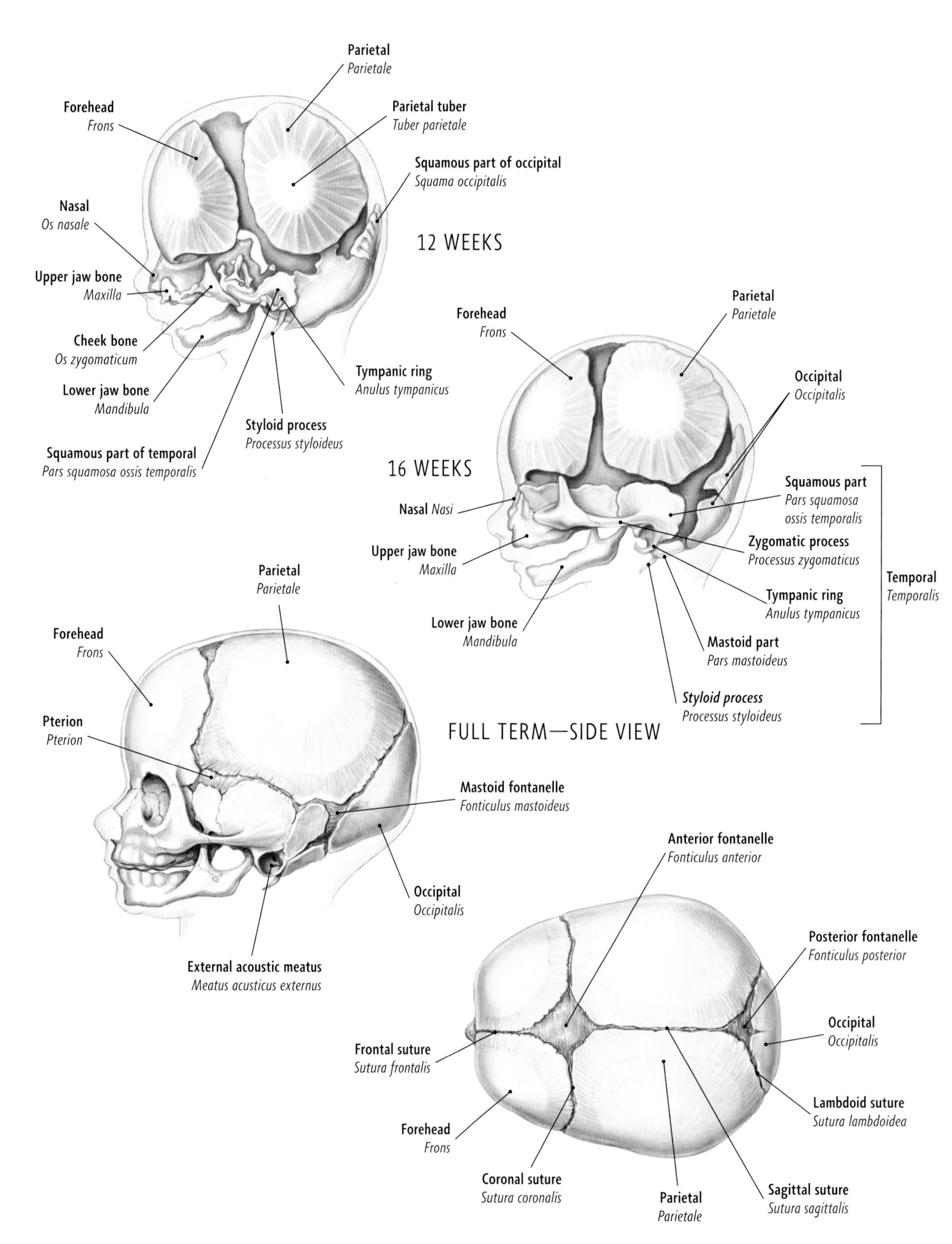
12 WEEKS
Parietal
Parietale
Forehead
Frons
Parietal tuber
Tuber parietale
Squamous part of occipital
Squama occipitalis
Nasal
Os nasale
Upper jaw bone
Maxilla
Cheek bone
Os zygomaticum
Lower jaw bone
Mandibula
Tympanic ring
Anulus tympanicus
Styloid process
Processus styloideus
Squamous part of temporal
Pars squamosa ossis temporalis
16 WEEKS
Forehead
Frons
Parietal
Parietale
Occipital
Occipitalis
Squamous part
Pars squamosa
ossis temporalis
Nasal Nasi
Upper jaw bone
Maxilla
Zygomatic process
Processus zygomaticus
Temporal
Temporalis
Tympanic ring
Anulus tympanicus
Lower jaw bone
Mandibula
Mastoid part
Pars mastoideus
Styloid process
Processus styloideus
FULL TERM—SIDE VIEW
Parietal
Parietale
Forehead
Frons
Pterion
Pterion
Mastoid fontanelle
Fonticulus mastoideus
Occipital
Occipitalis
External acoustic meatus
Meatus acusticus externus
Anterior fontanelle
Fonticulus anterior
Posterior fontanelle
Fonticulus posterior
Occipital
Occipitalis
Frontal suture
Sutura frontalis
Lambdoid suture
Sutura lambdoidea
Forehead
Frons
Coronal suture
Sutura coronalis
Parietal
Parietale
Sagittal suture
Sutura sagittalis
FULL TERM—VIEW FROM ABOVE

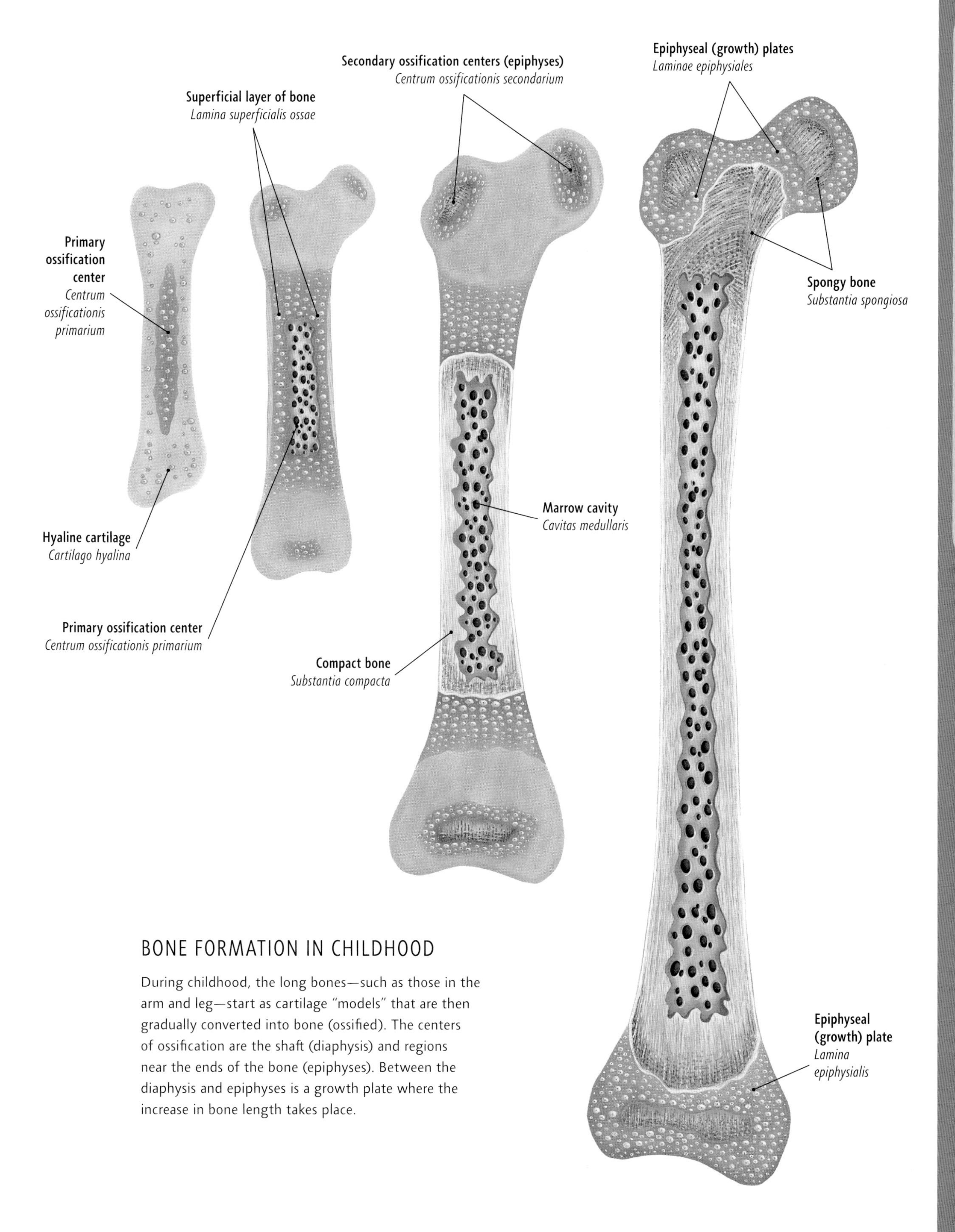
Secondary ossification centers (epiphyses)
Centrum ossificationis secondarium
Epiphyseal (growth) plates
Laminae epiphysiales
Superficial layer of bone
Lamina superficialis ossae
Primary ossification center
Centrum ossificationis primarium
Spongy bone
Substantia spongiosa
Marrow cavity
Cavitas medullaris
Hyaline cartilage
Cartilago hyalina
Primary ossification center
Centrum ossificationis primarium
Compact bone
Substantia compacta
Epiphyseal (growth) plate
Lamina epiphysialis

BONE FORMATION IN CHILDHOOD

During childhood, the long bones—such as those in the arm and leg—start as cartilage "models" that are then gradually converted into bone (ossified). The centers of ossification are the shaft (diaphysis) and regions near the ends of the bone (epiphyses). Between the diaphysis and epiphyses is a growth plate where the increase in bone length takes place.

Fetal Brain Development

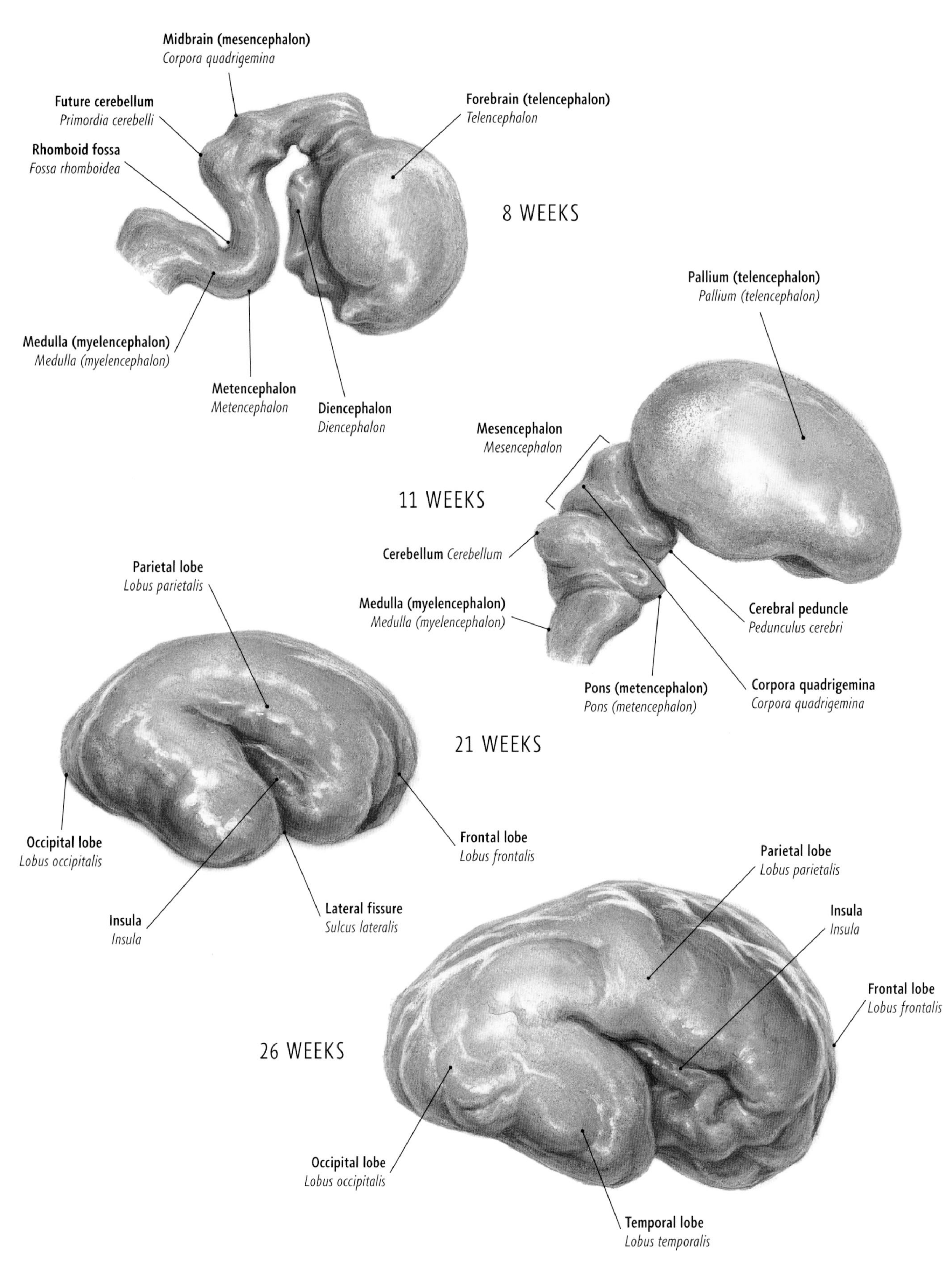
Midbrain (mesencephalon)
Corpora quadrigemina
Future cerebellum
Primordia cerebelli
Rhomboid fossa
Fossa rhomboidea
Forebrain (telencephalon)
Telencephalon
8 WEEKS
Medulla (myelencephalon)
Medulla (myelencephalon)
Metencephalon
Metencephalon
Diencephalon
Diencephalon
Pallium (telencephalon)
Pallium (telencephalon)
Mesencephalon
Mesencephalon
11 WEEKS
Cerebellum Cerebellum
Medulla (myelencephalon)
Medulla (myelencephalon)
Cerebral peduncle
Pedunculus cerebri
Pons (metencephalon)
Pons (metencephalon)
Corpora quadrigemina
Corpora quadrigemina
Parietal lobe
Lobus parietalis
21 WEEKS
Occipital lobe
Lobus occipitalis
Frontal lobe
Lobus frontalis
Insula
Insula
Lateral fissure
Sulcus lateralis
Parietal lobe
Lobus parietalis
Insula
Insula
Frontal lobe
Lobus frontalis
26 WEEKS
Occipital lobe
Lobus occipitalis
Temporal lobe
Lobus temporalis

[~] = no direct Latin equivalent

Parietal lobe
Lobus parietalis

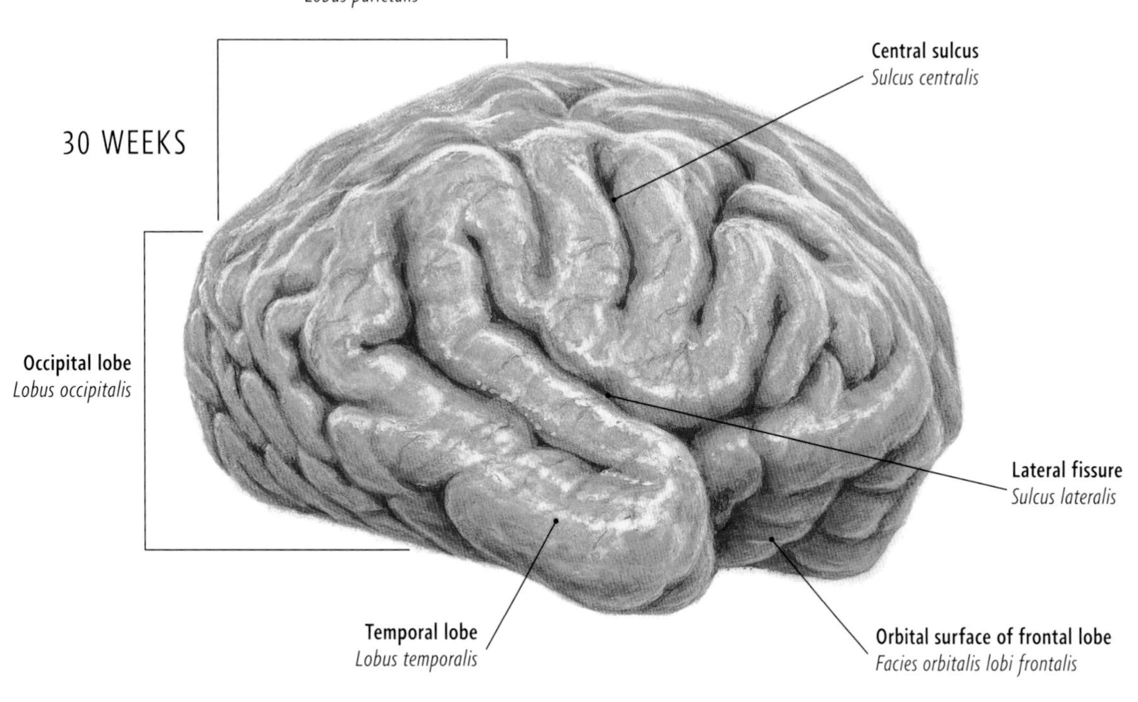
30 WEEKS
Central sulcus
Sulcus centralis
Occipital lobe
Lobus occipitalis
Lateral fissure
Sulcus lateralis
Temporal lobe
Lobus temporalis
Orbital surface of frontal lobe
Facies orbitalis lobi frontalis

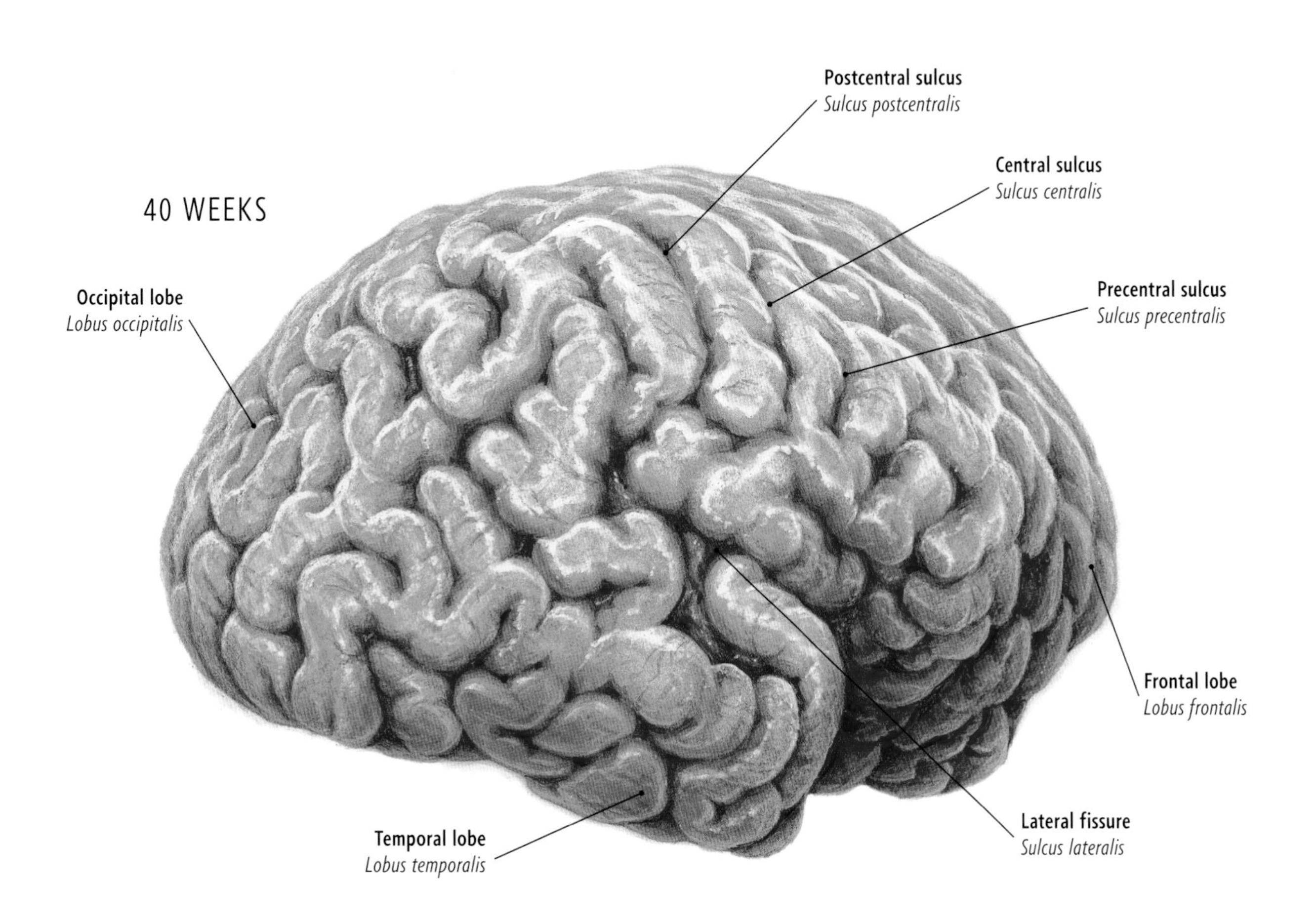
Postcentral sulcus
Sulcus postcentralis
Central sulcus
Sulcus centralis
40 WEEKS
Occipital lobe
Lobus occipitalis
Precentral sulcus
Sulcus precentralis
Frontal lobe
Lobus frontalis
Lateral fissure
Sulcus lateralis
Temporal lobe
Lobus temporalis

Fetal Sex Differentiation

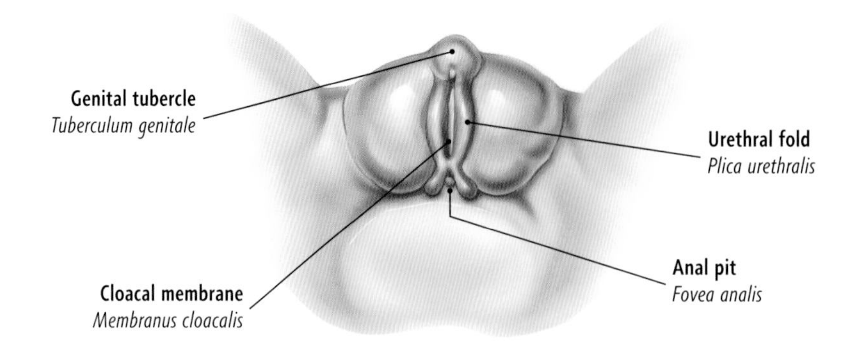
Genital tubercle
Tuberculum genitale
Urethral fold
Plica urethralis
Anal pit
Fovea analis
Cloacal membrane
Membranus cloacalis

UNDIFFERENTIATED (UP TO 12 WEEKS)

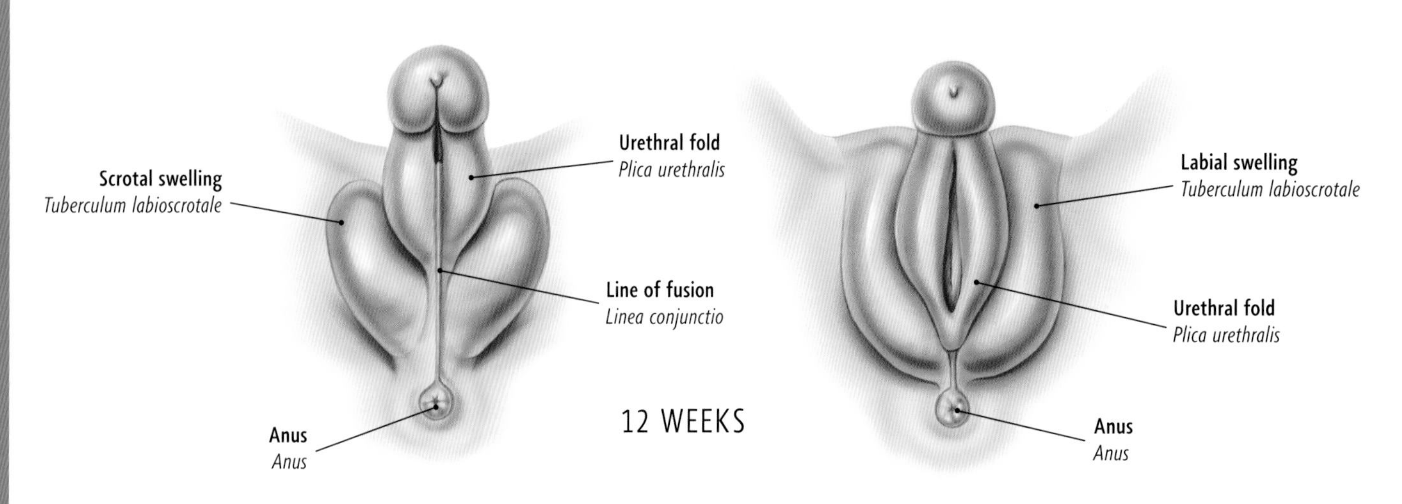
Urethral fold
Plica urethralis
Scrotal swelling
Tuberculum labioscrotale
Line of fusion
Linea conjunctio
Anus
Anus
Labial swelling
Tuberculum labioscrotale
Urethral fold
Plica urethralis
Anus
Anus

12 WEEKS

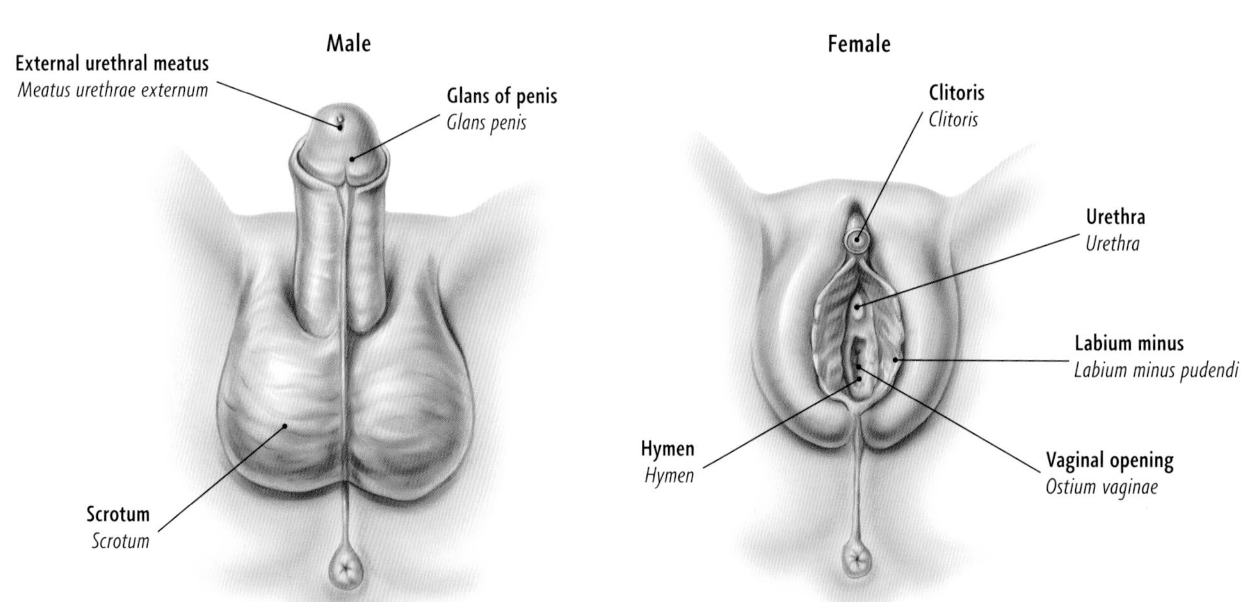
Male
External urethral meatus
Meatus urethrae externum
Glans of penis
Glans penis
Scrotum
Scrotum
Female
Clitoris
Clitoris
Urethra
Urethra
Labium minus
Labium minus pudendi
Hymen
Hymen
Vaginal opening
Ostium vaginae

FULLY DEVELOPED

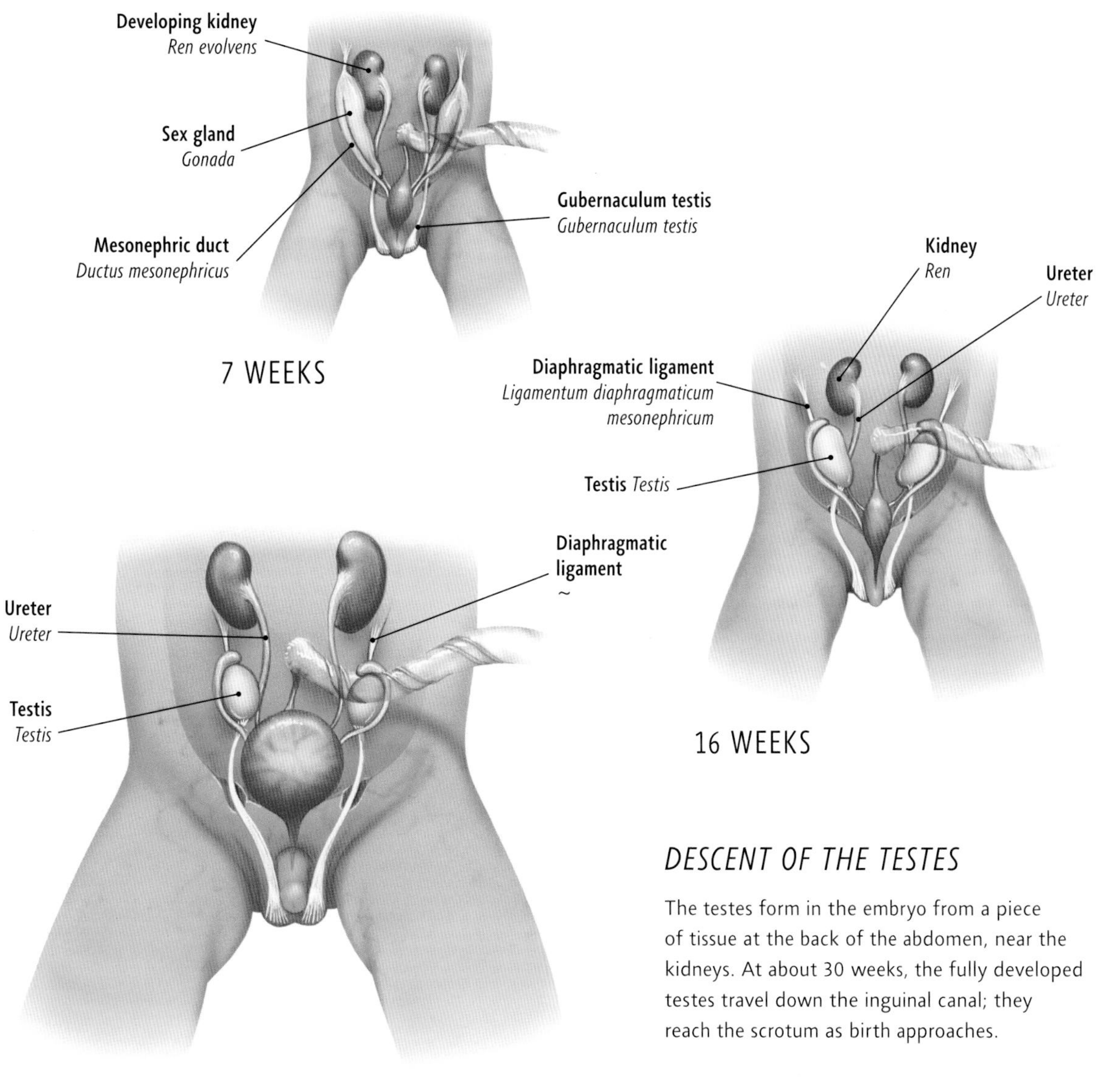
Developing kidney
Ren evolvens
Sex gland
Gonada
Mesonephric duct
Ductus mesonephricus
Gubernaculum testis
Gubernaculum testis
7 WEEKS
Kidney
Ren
Ureter
Ureter
Diaphragmatic ligament
Ligamentum diaphragmaticum mesonephricum
Testis Testis
16 WEEKS
Ureter
Ureter
Testis
Testis
Diaphragmatic ligament
~
30 WEEKS

DESCENT OF THE TESTES

The testes form in the embryo from a piece of tissue at the back of the abdomen, near the kidneys. At about 30 weeks, the fully developed testes travel down the inguinal canal; they reach the scrotum as birth approaches.

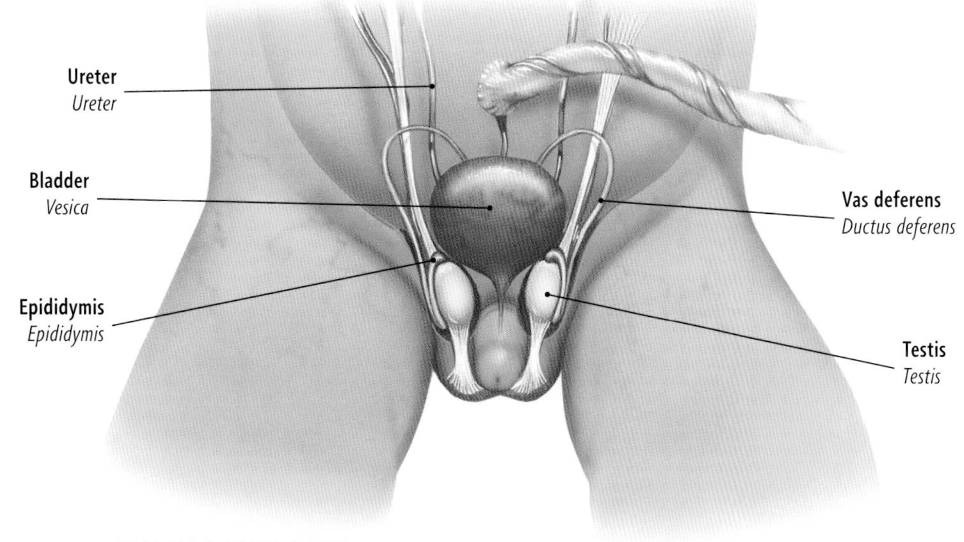
Ureter
Ureter
Bladder
Vesica
Epididymis
Epididymis
Vas deferens
Ductus deferens
Testis
Testis
FULLY DEVELOPED

[~] = no direct Latin equivalent

Late Pregnancy

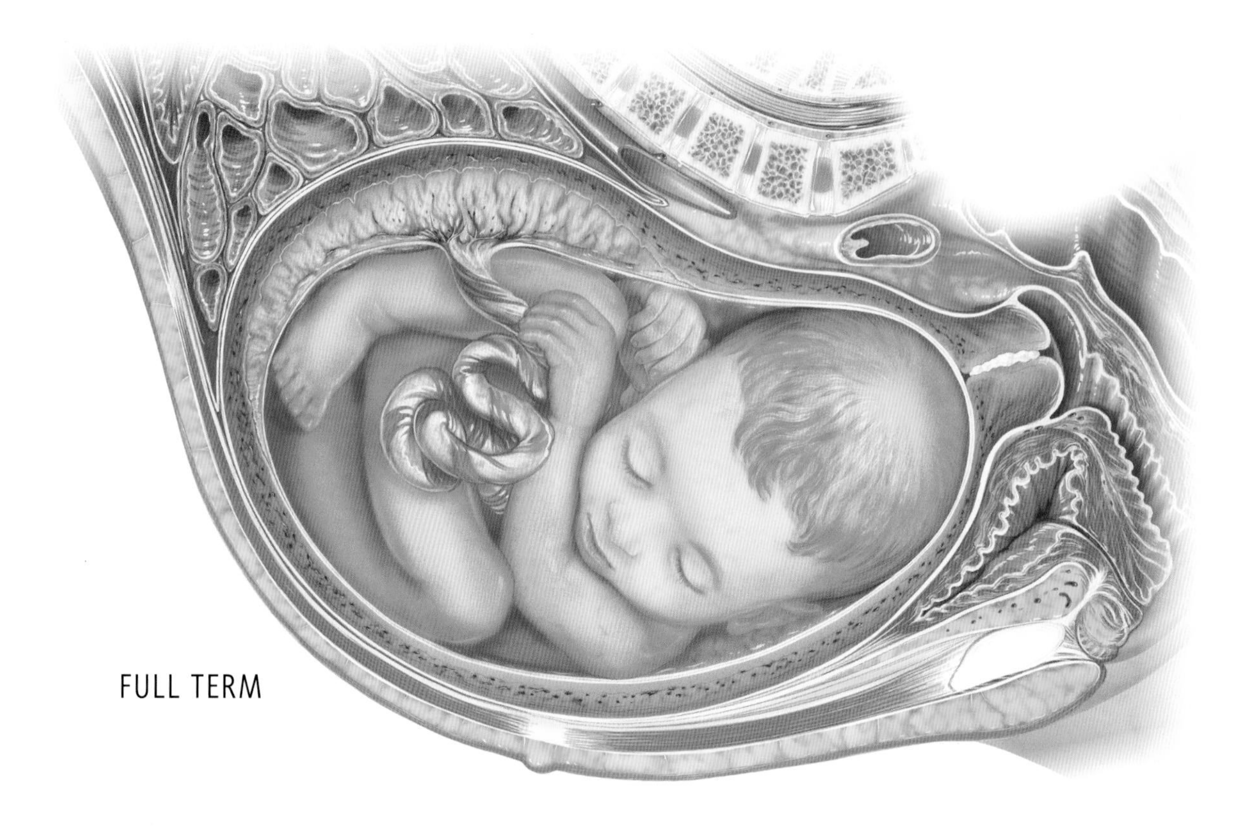

FULL TERM

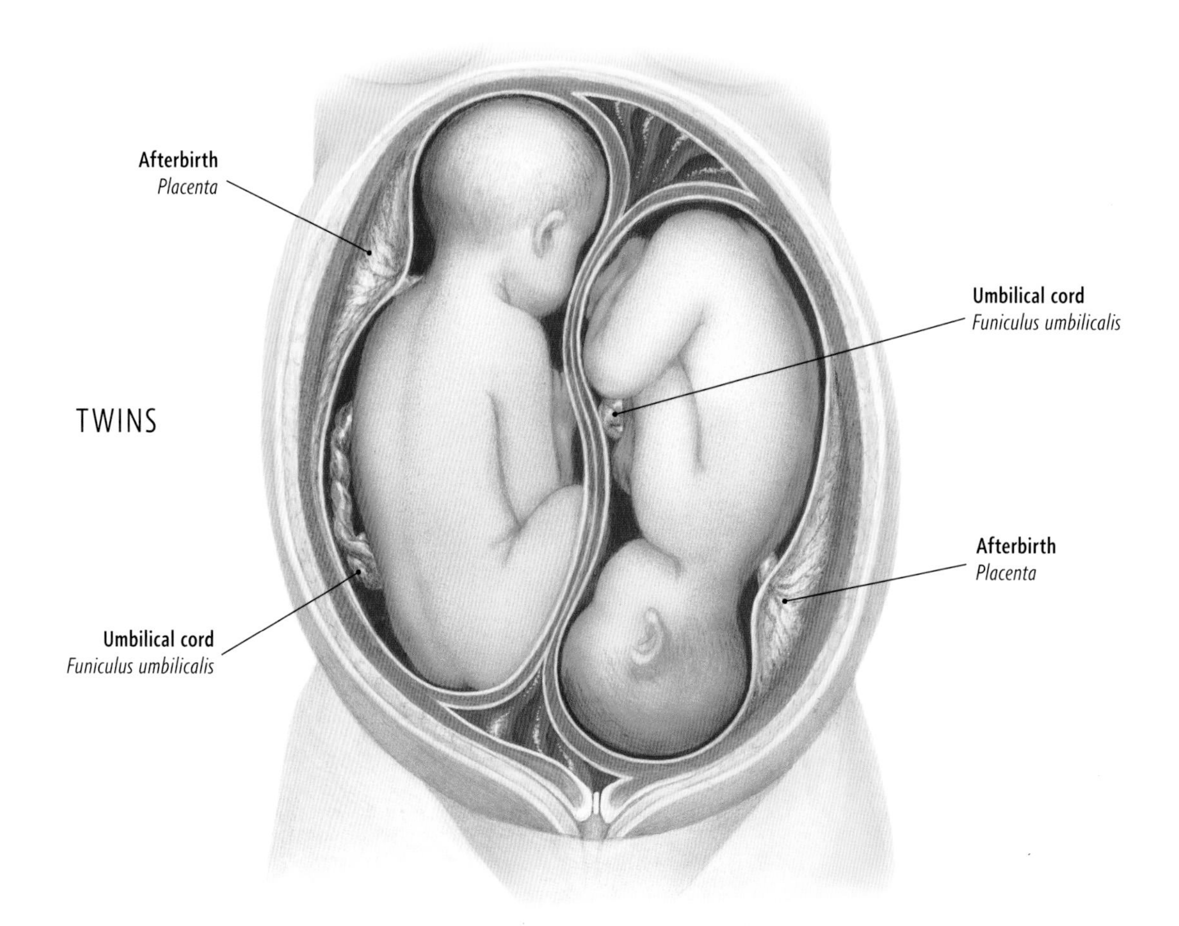
Afterbirth
Placenta
Umbilical cord
Funiculus umbilicalis
Afterbirth
Placenta
Umbilical cord
Funiculus umbilicalis

TWINS

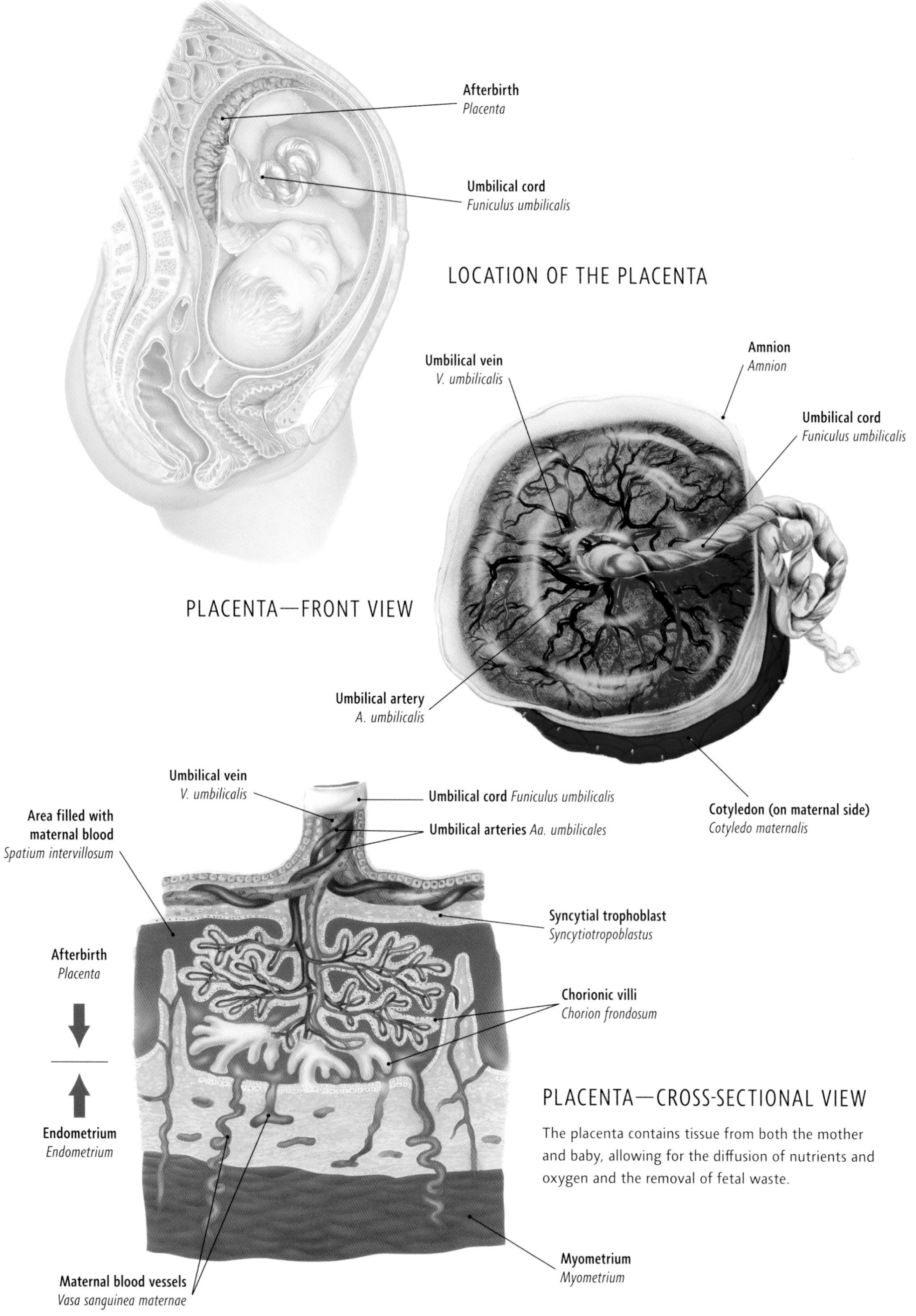
Afterbirth
Placenta
Umbilical cord
Funiculus umbilicalis
LOCATION OF THE PLACENTA
Umbilical vein
V. umbilicalis
Amnion
Amnion
Umbilical cord
Funiculus umbilicalis
PLACENTA—FRONT VIEW
Umbilical artery
A. umbilicalis
Umbilical vein
V. umbilicalis
Umbilical cord Funiculus umbilicalis
Area filled with maternal blood
Spatium intervillosum
Umbilical arteries Aa. umbilicales
Cotyledon (on maternal side)
Cotyledo maternalis
Syncytial trophoblast
Syncytiotropoblastus
Afterbirth
Placenta
Chorionic villi
Chorion frondosum
Endometrium
Endometrium
Myometrium
Myometrium
Maternal blood vessels
Vasa sanguinea maternae

PLACENTA—CROSS-SECTIONAL VIEW

The placenta contains tissue from both the mother and baby, allowing for the diffusion of nutrients and oxygen and the removal of fetal waste.

[~] = no direct Latin equivalent

Childbirth

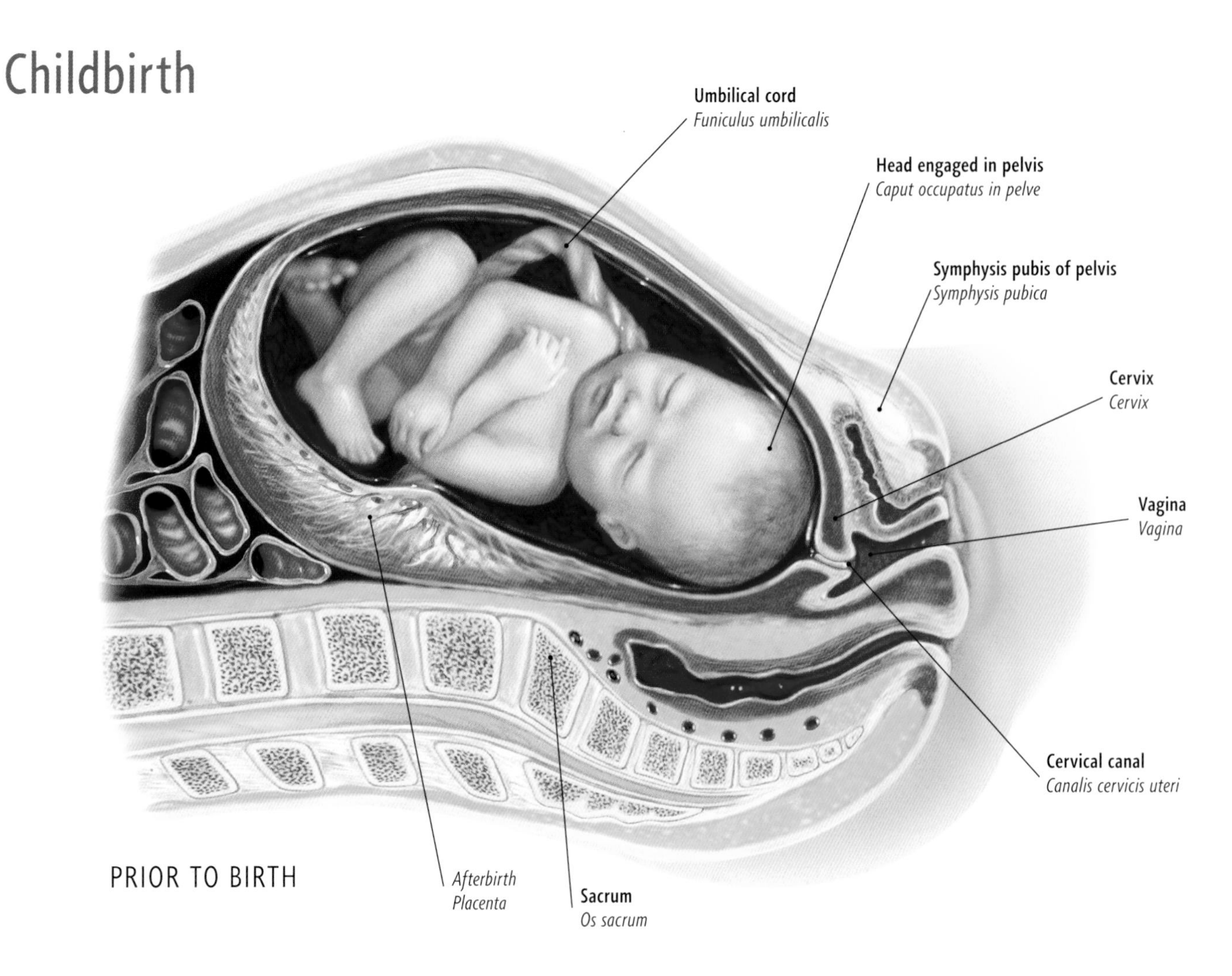
Umbilical cord
Funiculus umbilicalis
Head engaged in pelvis
Caput occupatus in pelve
Symphysis pubis of pelvis
Symphysis pubica
Cervix
Cervix
Vagina
Vagina
Cervical canal
Canalis cervicis uteri
Afterbirth
Placenta
Sacrum
Os sacrum

PRIOR TO BIRTH

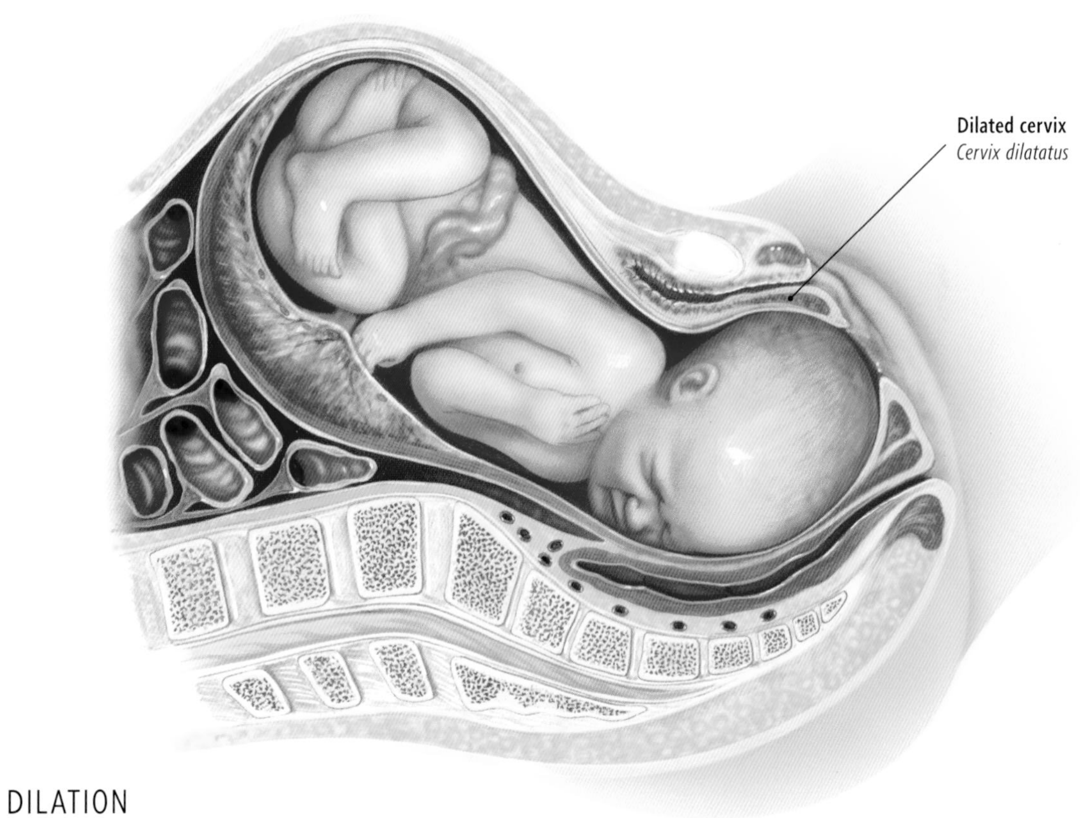
Dilated cervix
Cervix dilatatus

DILATION

[~] = no direct Latin equivalent

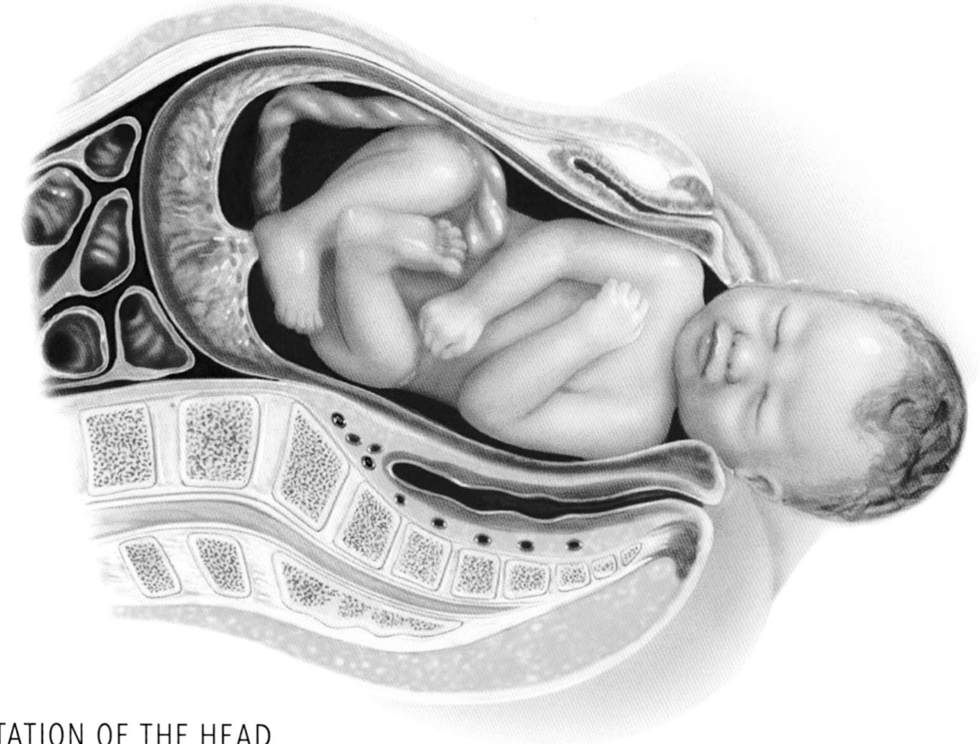

PRESENTATION OF THE HEAD

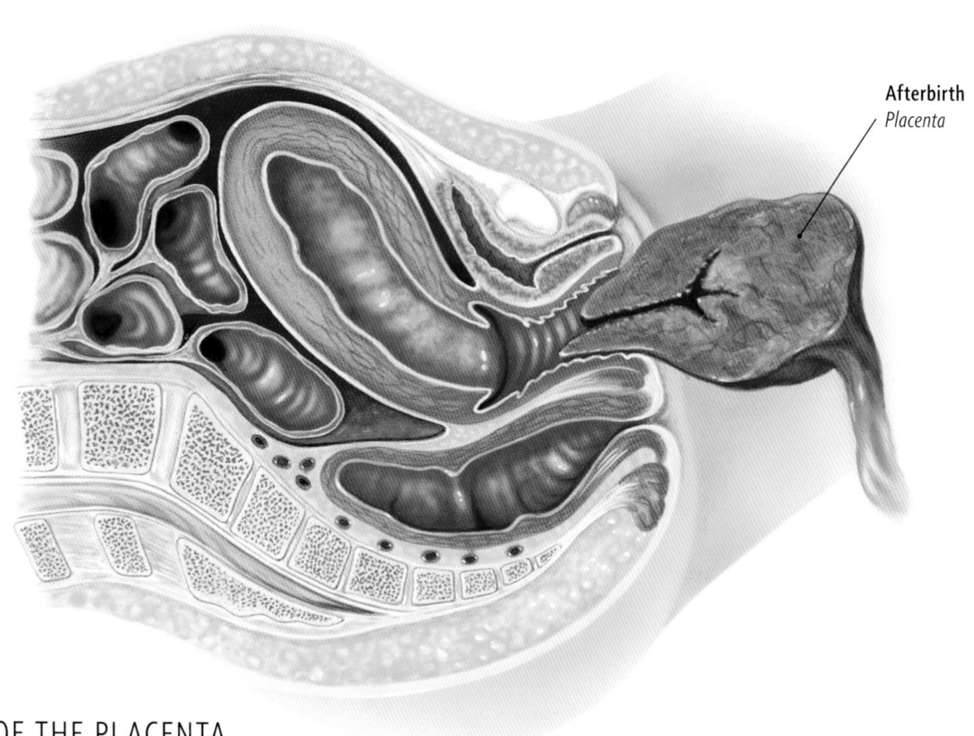
Afterbirth
Placenta

EXPULSION OF THE PLACENTA

Time of Your Life: Female

At about 10 years of age, girls have a growth spurt, and pubic hair and underarm hair start to grow a year or two later. Soon the reproductive system develops, with the first period arriving around the age of 13. By 18 the body is approaching maturity and from there it continues to change, but not grow. From the late 40s the production of eggs and sex hormones decreases, culminating in menopause at around the age of 50. Without the hormonal effects, the breasts and skin lose firmness, fat distribution changes, bones lose minerals, and height decreases.

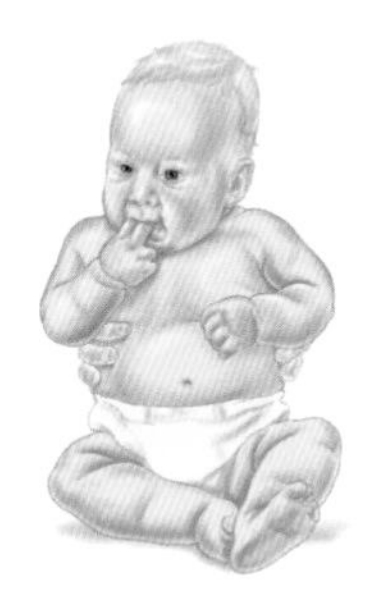

BIRTH–6 MONTHS

6–12 MONTHS

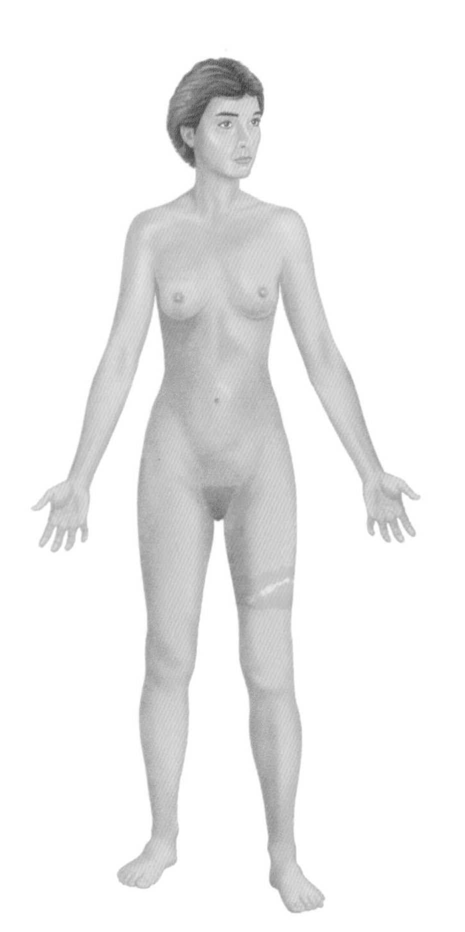

20–30 YEARS

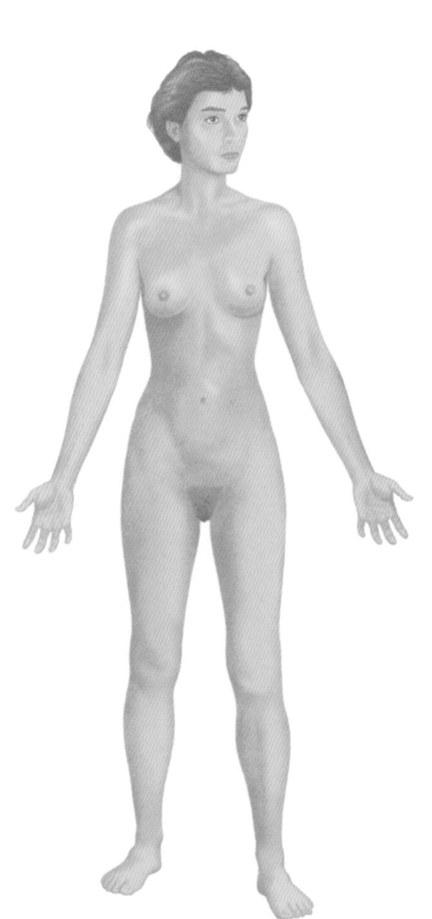

30–40 YEARS

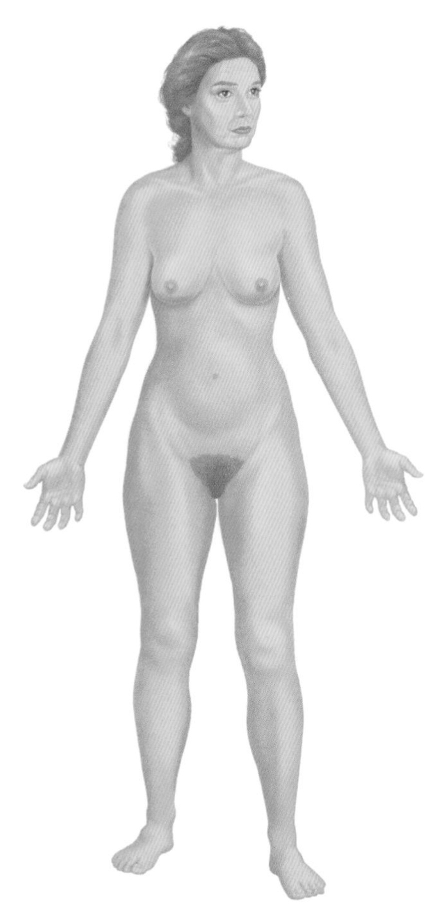

40–50 YEARS

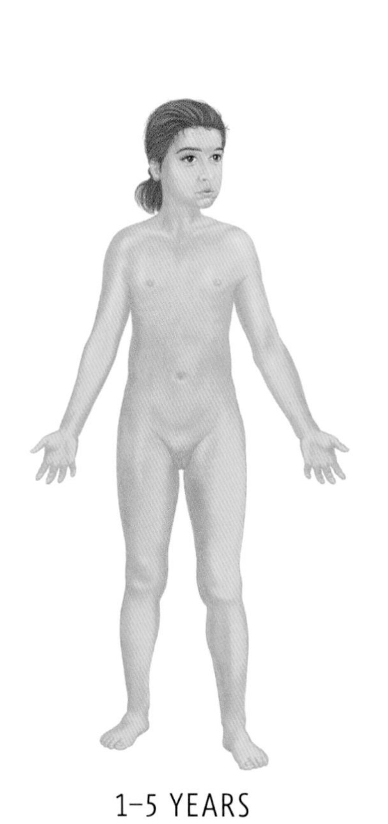

1–5 YEARS

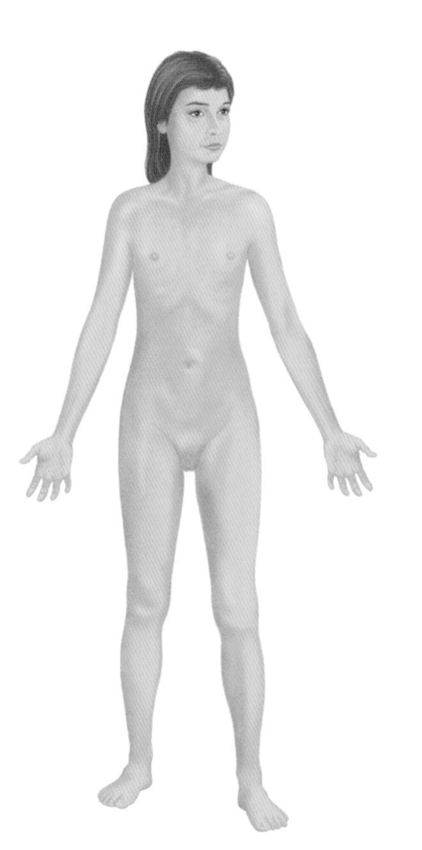

6–11 YEARS

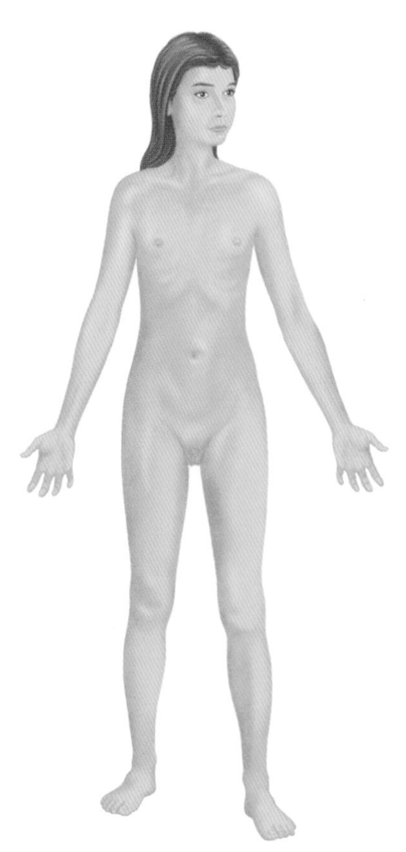

12–19 YEARS

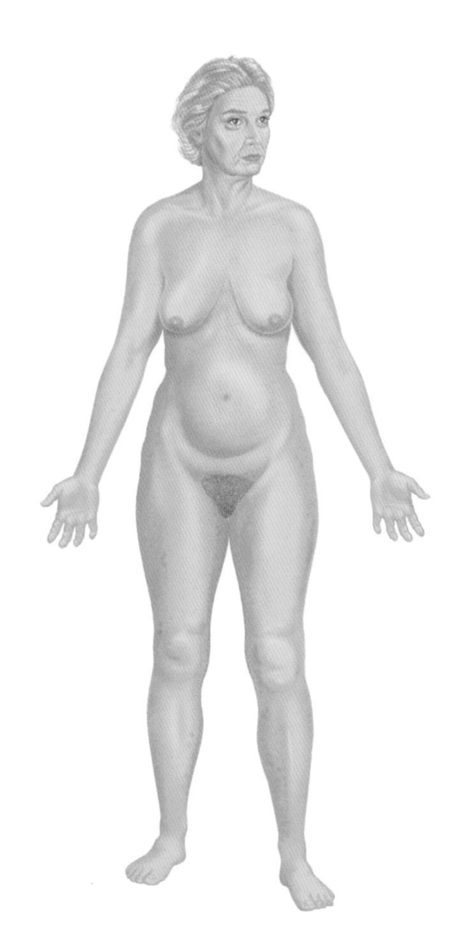

50–65 YEARS

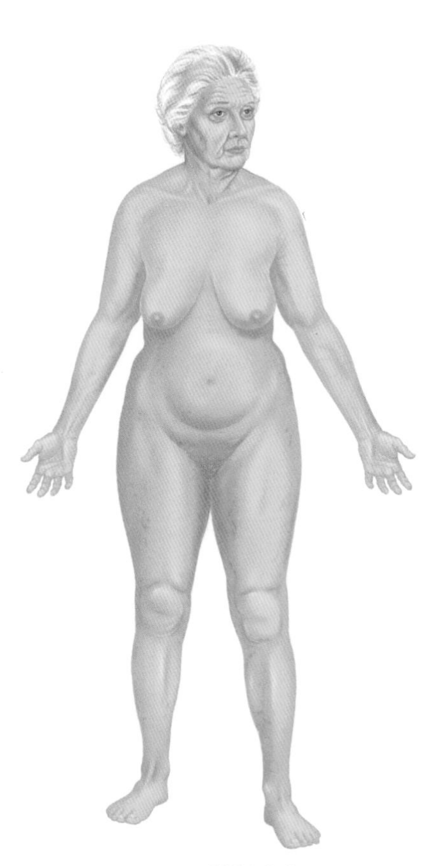

65–85 YEARS

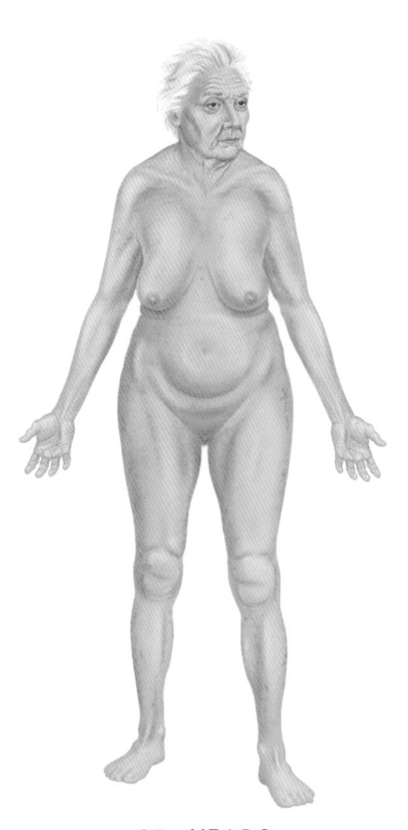

85+ YEARS

Time of Your Life: Male

Boys' growth spurt occurs at around 13 or 14 years of age. Muscles begin to bulk up and body hair grows. The testes grow and produce sperm, and the voice deepens. As men mature the body reaches full muscular development, then starts to decline. The skin and muscles become softer, fat may increase, the prostate enlarges, production of male hormones and sperm reduces, and hair loss occurs.

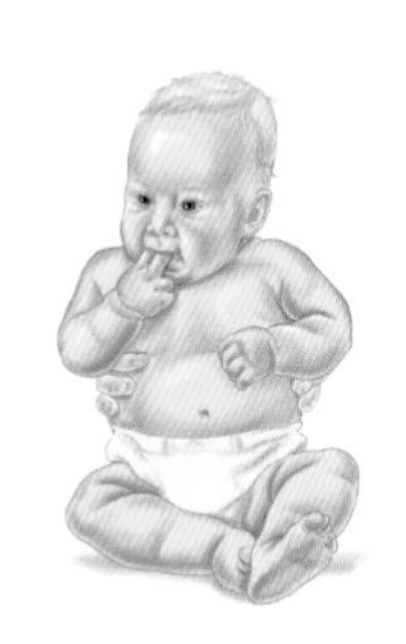

BIRTH–6 MONTHS

6–12 MONTHS

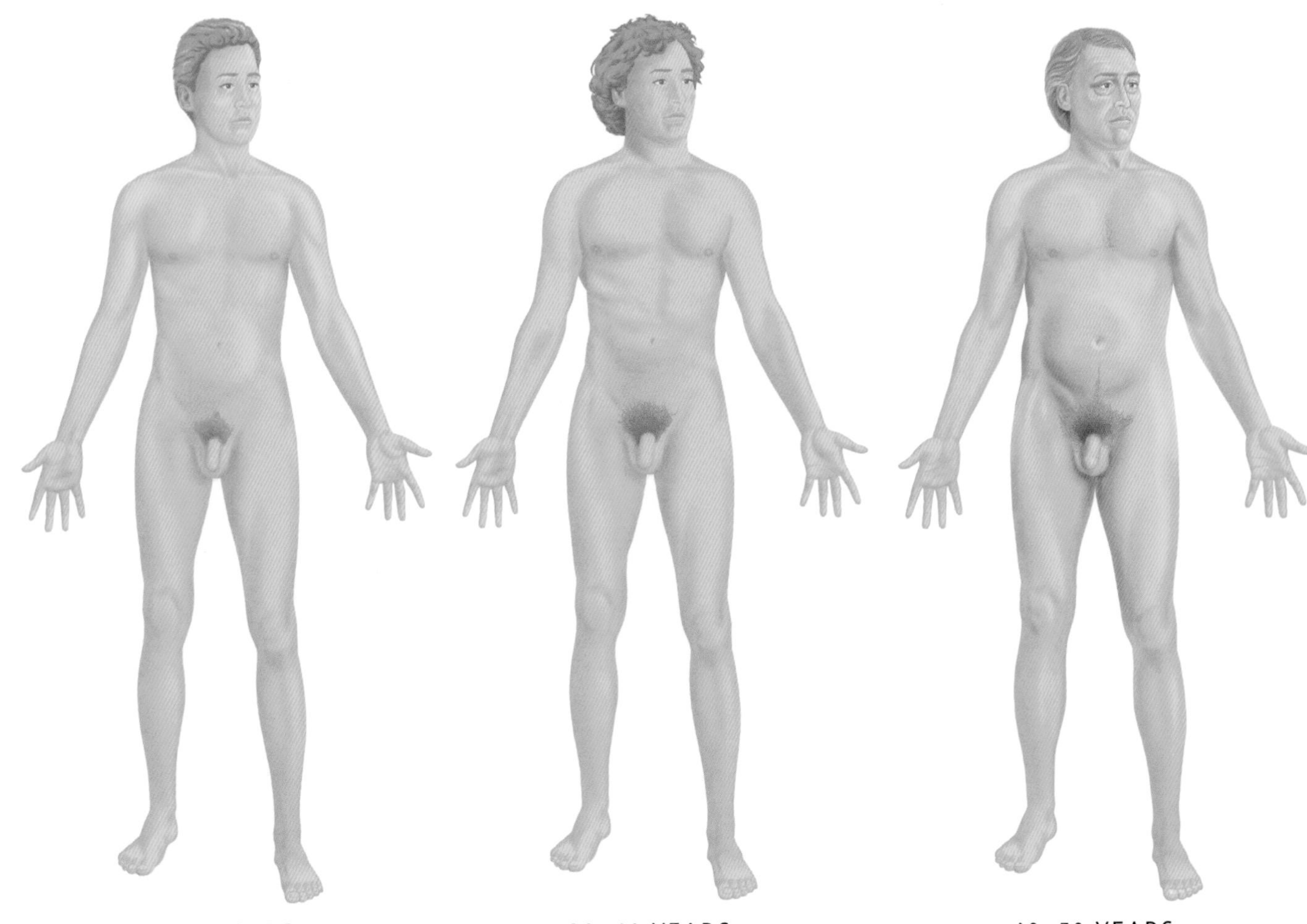

20–30 YEARS

30–40 YEARS

40–50 YEARS

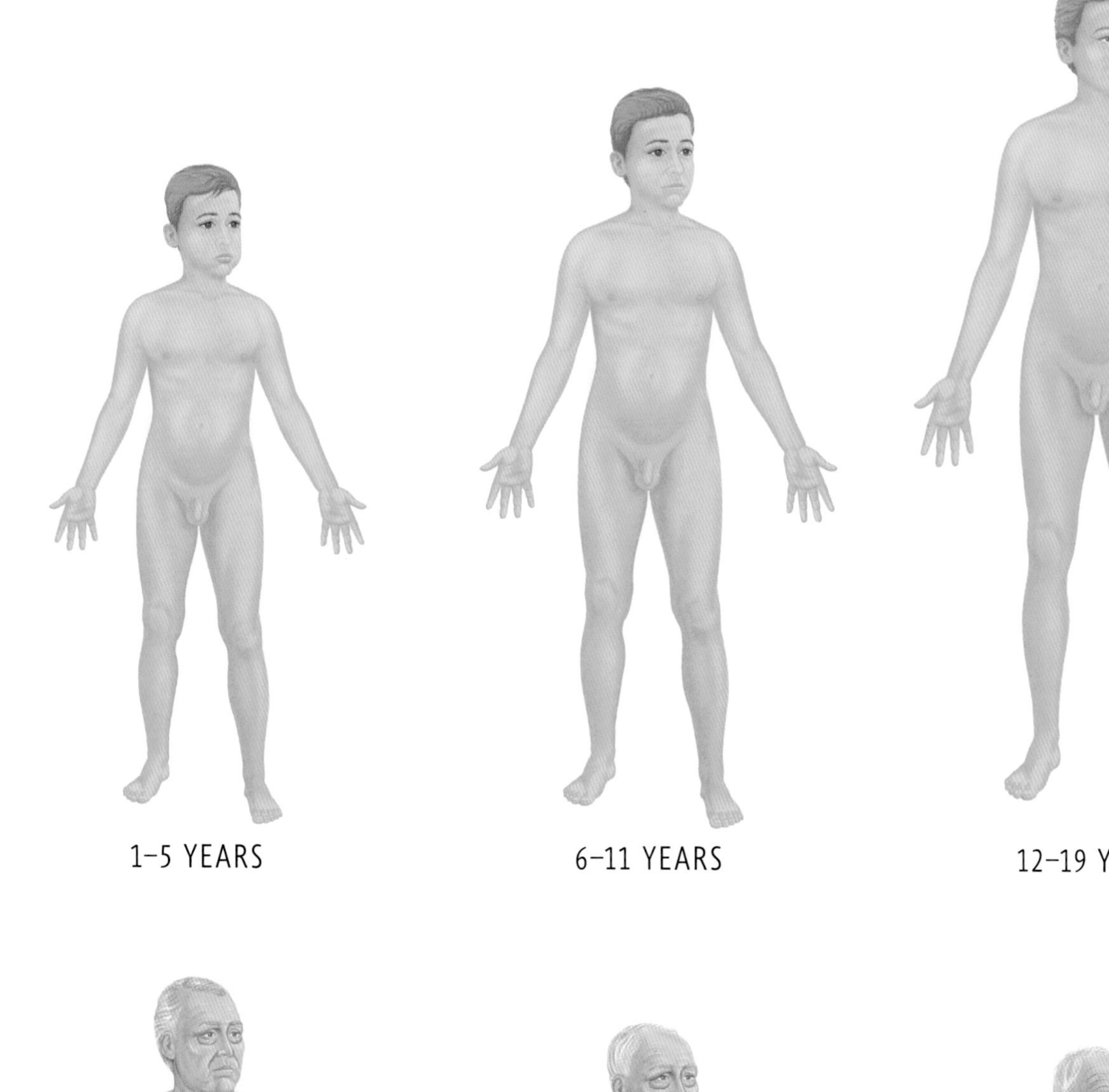

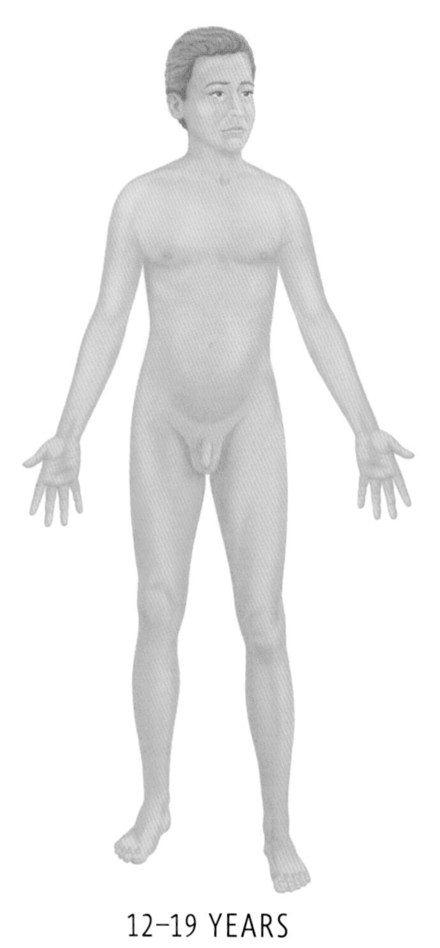

1–5 YEARS

6–11 YEARS

12–19 YEARS

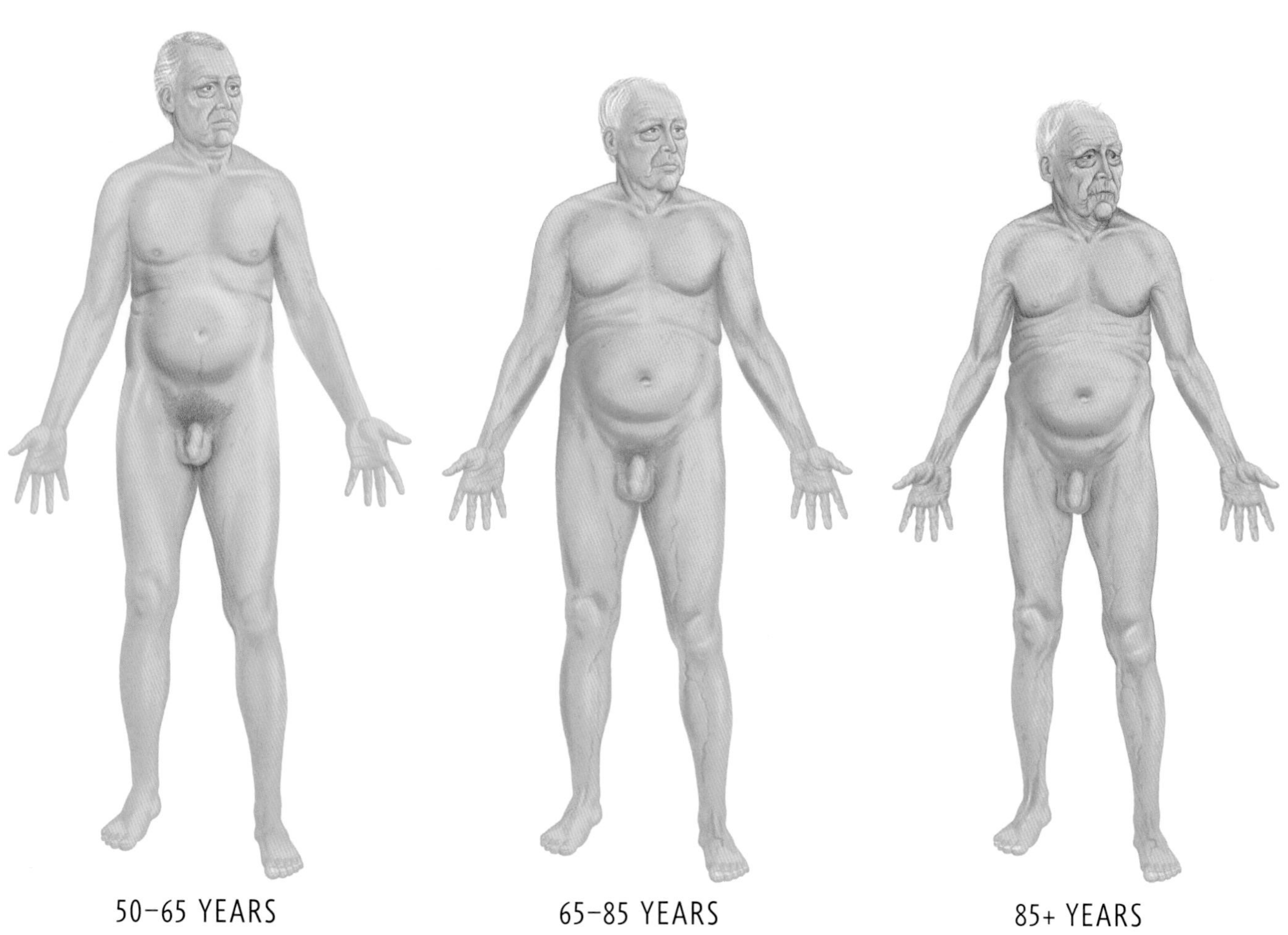

50–65 YEARS

65–85 YEARS

85+ YEARS

Coloring workbook

Muscular System

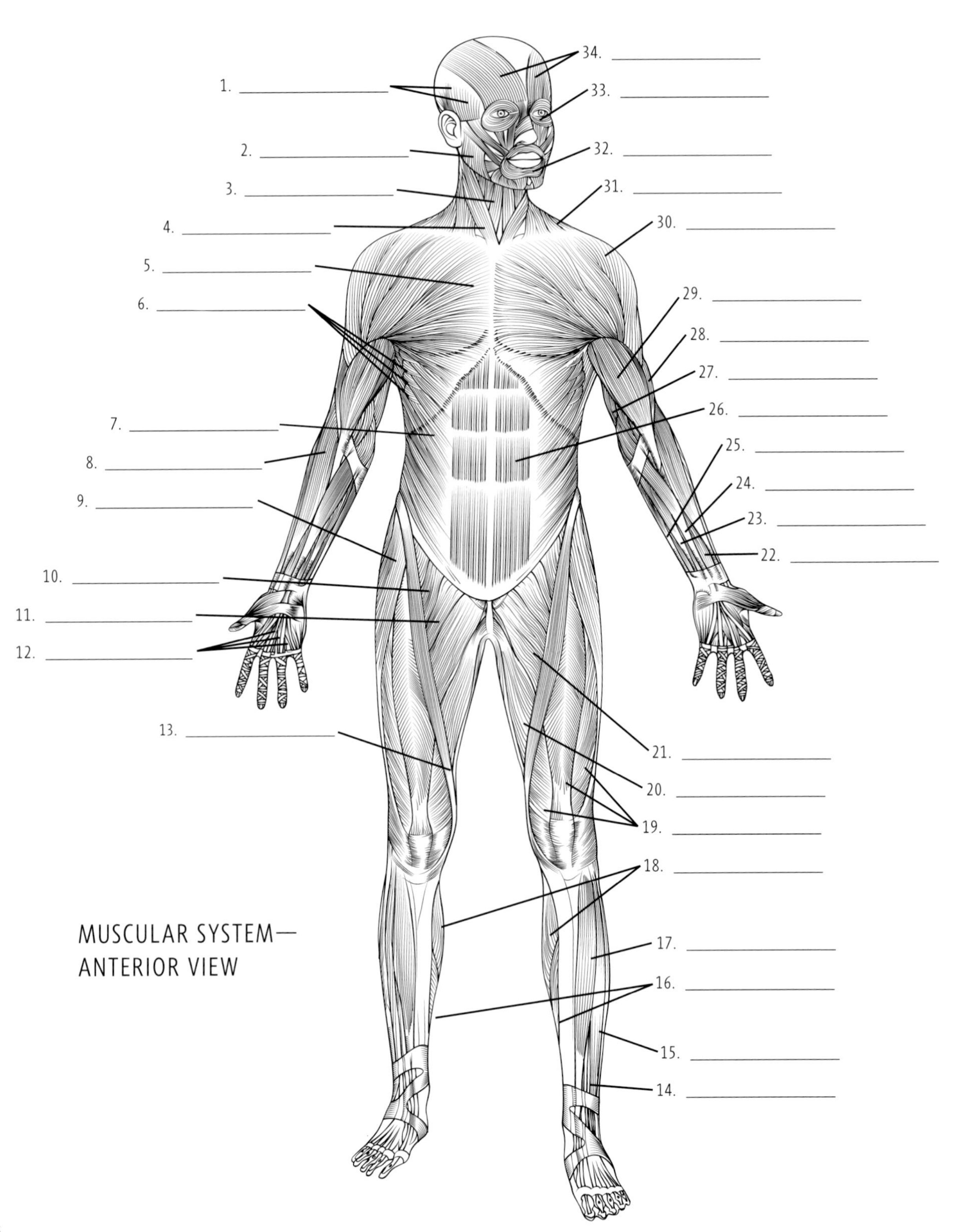
1.
2.
3.
4.
5.
6.
7.
8.
9.
10.
11.
12.
13.
14.
15.
16.
17.
18.
19.
20.
21.
22.
23.
24.
25.
26.
27.
28.
29.
30.
31.
32.
33.
34.

MUSCULAR SYSTEM—
ANTERIOR VIEW

Answers

1. Temporalis, 2. Masseter, 3. Sternohyoid, 4. Sternocleidomastoid, 5. Pectoralis major, 6. Serratus anterior, 7. External abdominal oblique, 8. Brachioradialis, 9. Tensor fasciae latae, 10. Iliopsoas, 11. Pectineus, 12. Lumbricals, 13. Sartorius, 14. Extensor hallucis longus, 15. Extensor digitorum longus, 16. Soleus, 17. Tibialis anterior, 18. Gastrocnemius, 19. Quadriceps femoris, 20. Adductormagnus, 21. Adductor longus, 22. Flexor digitorum superficialis, 23. Tendon of palmaris longus, 24. Tendon of flexor carpi radialis, 25. Tendon of flexor carpi ulnaris, 26. Rectus abdominis, 27. Triceps brachii, 28. Brachialis, 29. Biceps brachii, 30. Deltoid, 31. Trapezius, 32. Orbicularis oris, 33. Orbicularis oculi, 34. Frontalis

MUSCULAR SYSTEM— POSTERIOR VIEW

MUSCULAR SYSTEM— LATERAL VIEW

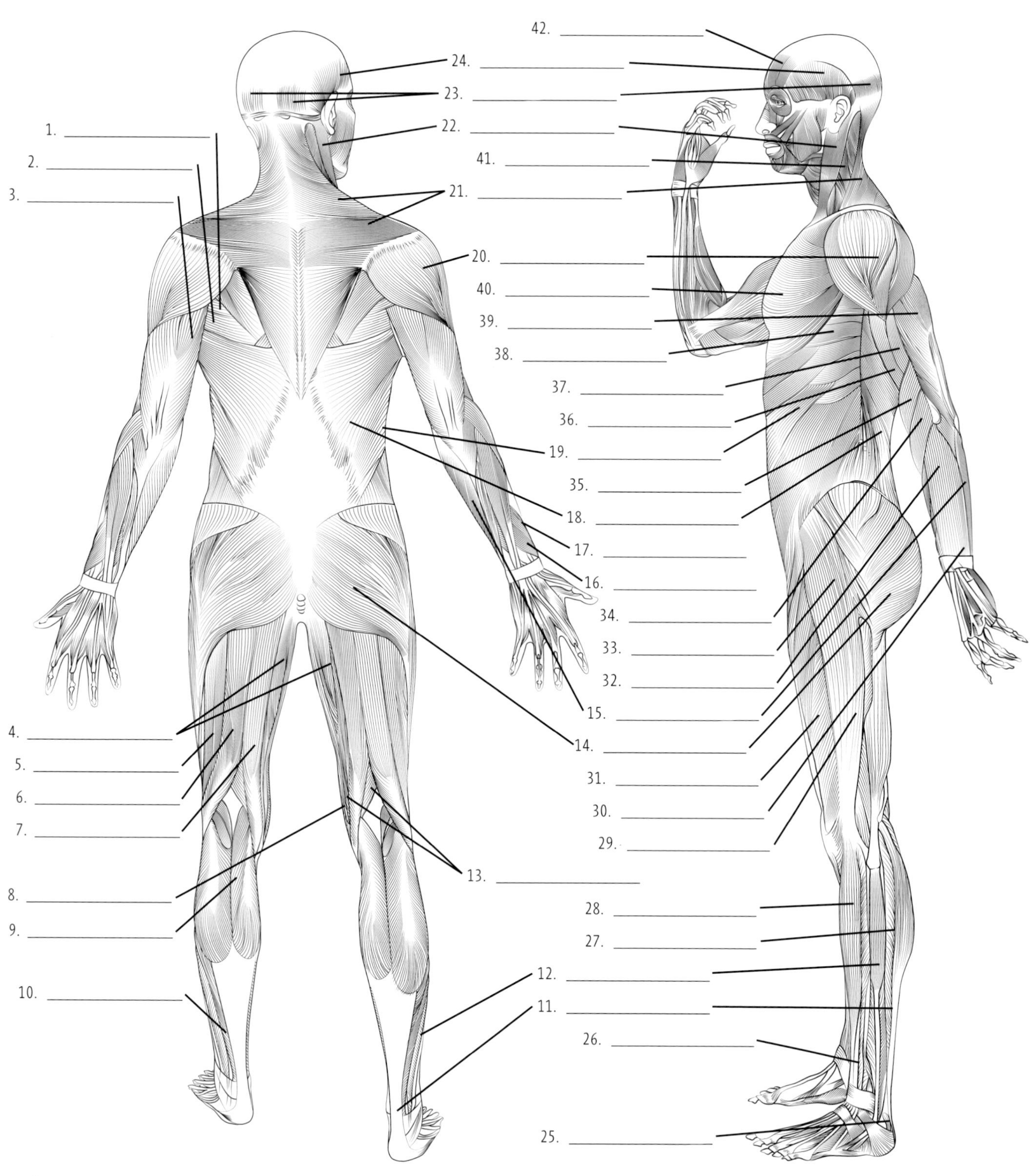
1.
2.
3.
4.
5.
6.
7.
8.
9.
10.
11.
12.
13.
14.
15.
16.
17.
18.
19.
20.
21.
22.
23.
24.
25.
26.
27.
28.
29.
30.
31.
32.
33.
34.
35.
36.
37.
38.
39.
40.
41.
42.

Answers

1. Teres minor, 2. Teres major, 3. Triceps brachii, 4. Adductor magnus, 5. Vastus lateralis, 6. Long head of biceps femoris, 7. Semitendinosus, 8. Gracilis, 9. Gastrocnemius, 10. Soleus, 11. Tendo calcaneus (Achilles tendon), 12. Fibularis (peroneus) longus, 13. Semimembranosus, 14. Gluteus maximus, 15. Flexor carpi ulnaris, 16. Extensor pollicis brevis, 17. Abductor pollicis longus, 18. Latissimus dorsi, 19. External abdominal oblique, 20. Deltoid, 21. Trapezius, 22. Sternocleidomastoid, 23. Occipitalis, 24. Temporalis, 25. Tendo calcaneus (Achilles tendon), 26. Extensor digitorum longus, 27. Lateral head of gastrocnemius, 28. Tibialis anterior, 29. Iliotibial tract, 30. Extensor carpi ulnaris, 31. Quadriceps femoris (vastus lateralis), 32. Extensor digitorum, 33. Tensor fasciae latae, 34. Extensor carpi radialis longus, 35. Brachioradialis, 36. Biceps brachii, 37. Brachialis, 38. Serratus anterior, 39. Lateral head of triceps brachii, 40. Pectoralis major, 41. Levator scapulae, 42. Frontalis

Muscles of the Thorax and Abdomen

SUPERFICIAL AND DEEP MUSCLES OF THE THORAX AND ABDOMEN—ANTERIOR VIEW

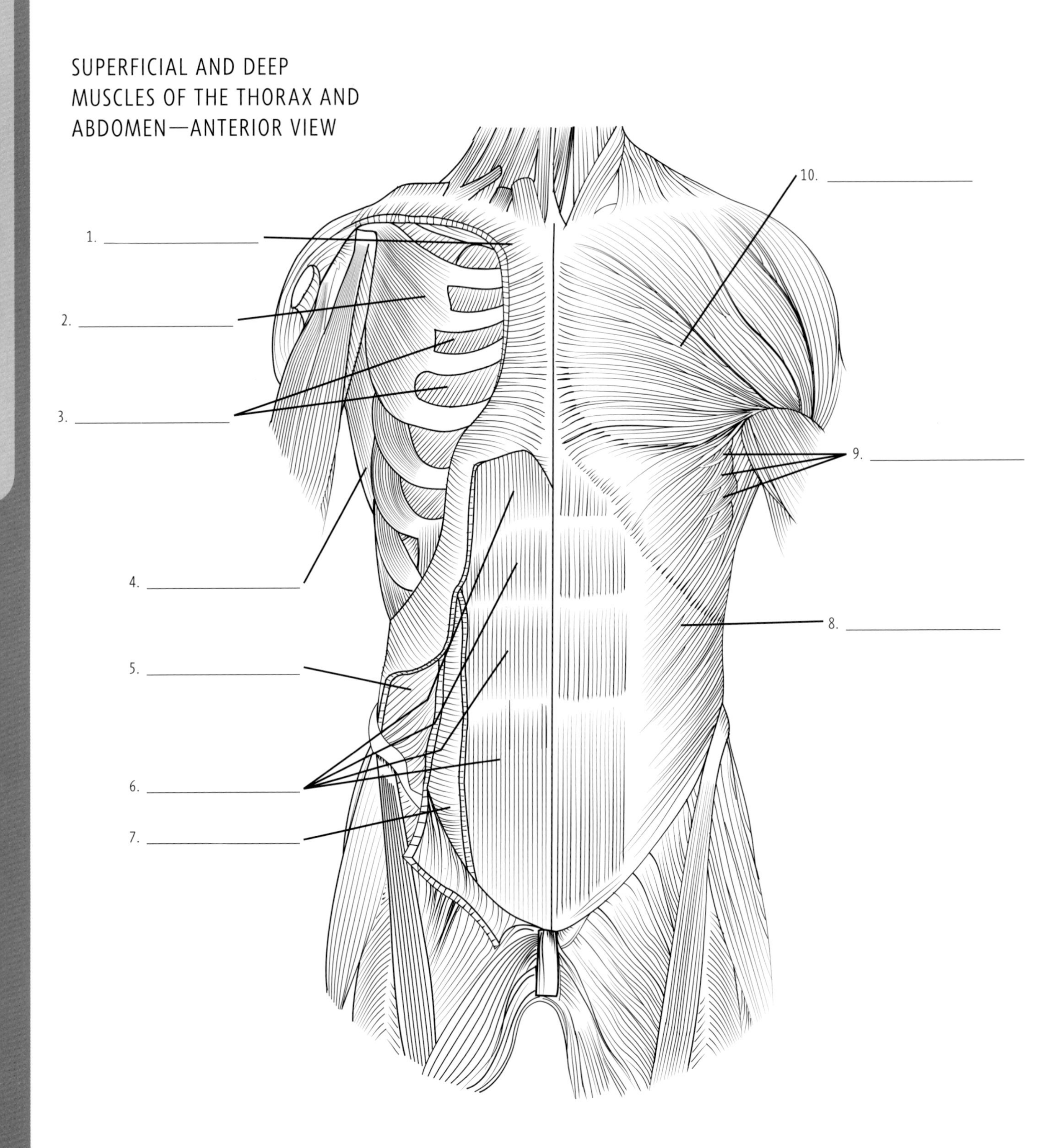
1. ____________
2. ____________
3. ____________
4. ____________
5. ____________
6. ____________
7. ____________
8. ____________
9. ____________
10. ____________

Answers

1. Pectoralis major, 2. Pectoralis minor, 3. Internal intercostals, 4. Latissimus dorsi, 5. Internal abdominal oblique, 6. Rectus abdominis, 7. Transversus abdominis, 8. External abdominal oblique, 9. Serratus anterior, 10. Pectoralis major

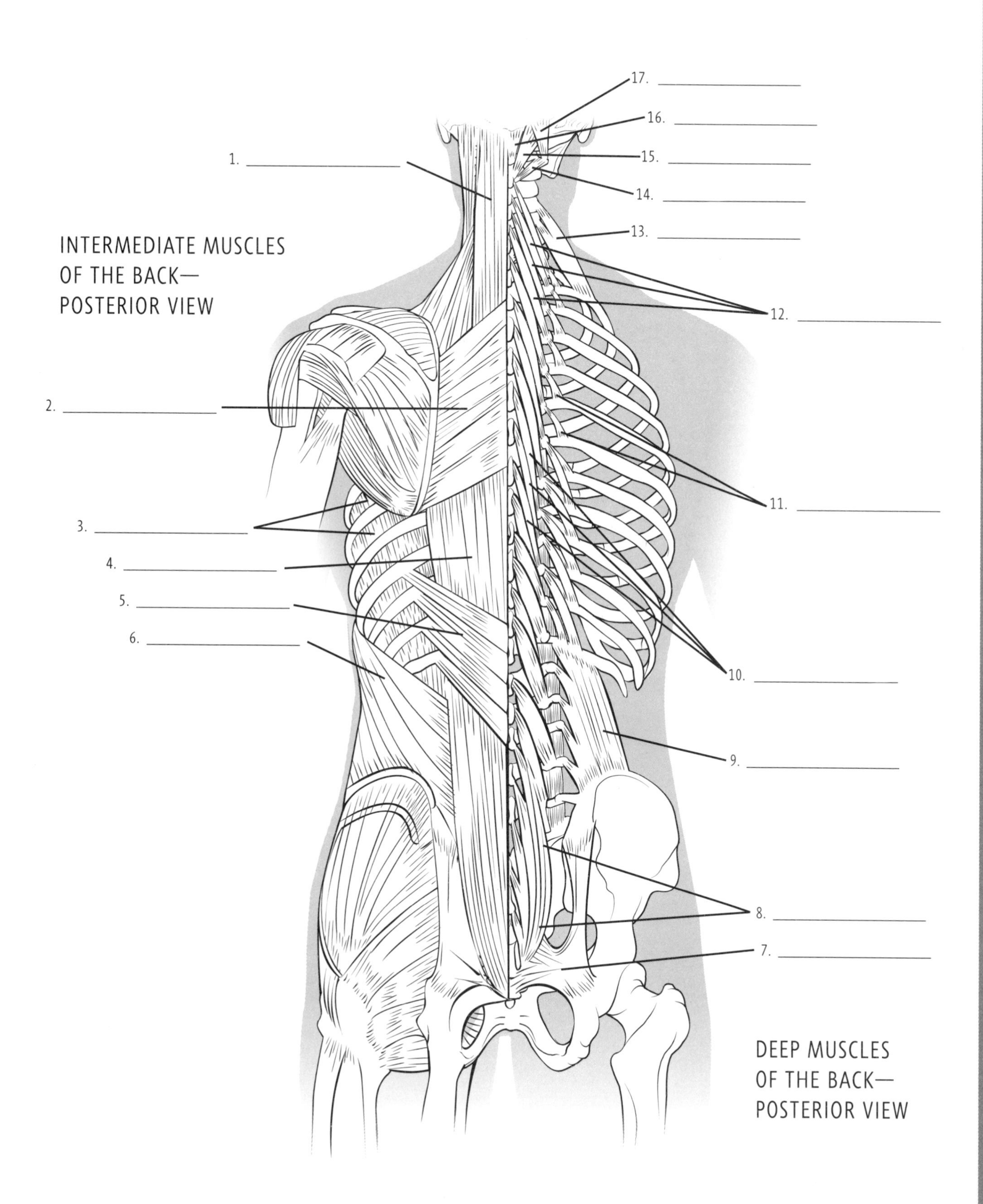
INTERMEDIATE MUSCLES OF THE BACK— POSTERIOR VIEW
DEEP MUSCLES OF THE BACK— POSTERIOR VIEW
1.
2.
3.
4.
5.
6.
7.
8.
9.
10.
11.
12.
13.
14.
15.
16.
17.

Answers

1. Semispinalis capitis, 2. Rhomboid major, 3. External intercostals, 4. Erector spinae, 5. Serratus posterior inferior, 6. Internal oblique, 7. Sacrotuberous ligament, 8. Multifidus, 9. Quadratus lumborum, 10. Semispinalis thoracis, 11. Levatores costarum, 12. Semispinalis cervicis, 13. Scalenus posterior, 14. Obliquus capitis inferior, 15. Rectus capitis posterior major, 16. Rectus capitis posterior minor, 17. Obliquus capitis superior

Muscles of the Upper Limb

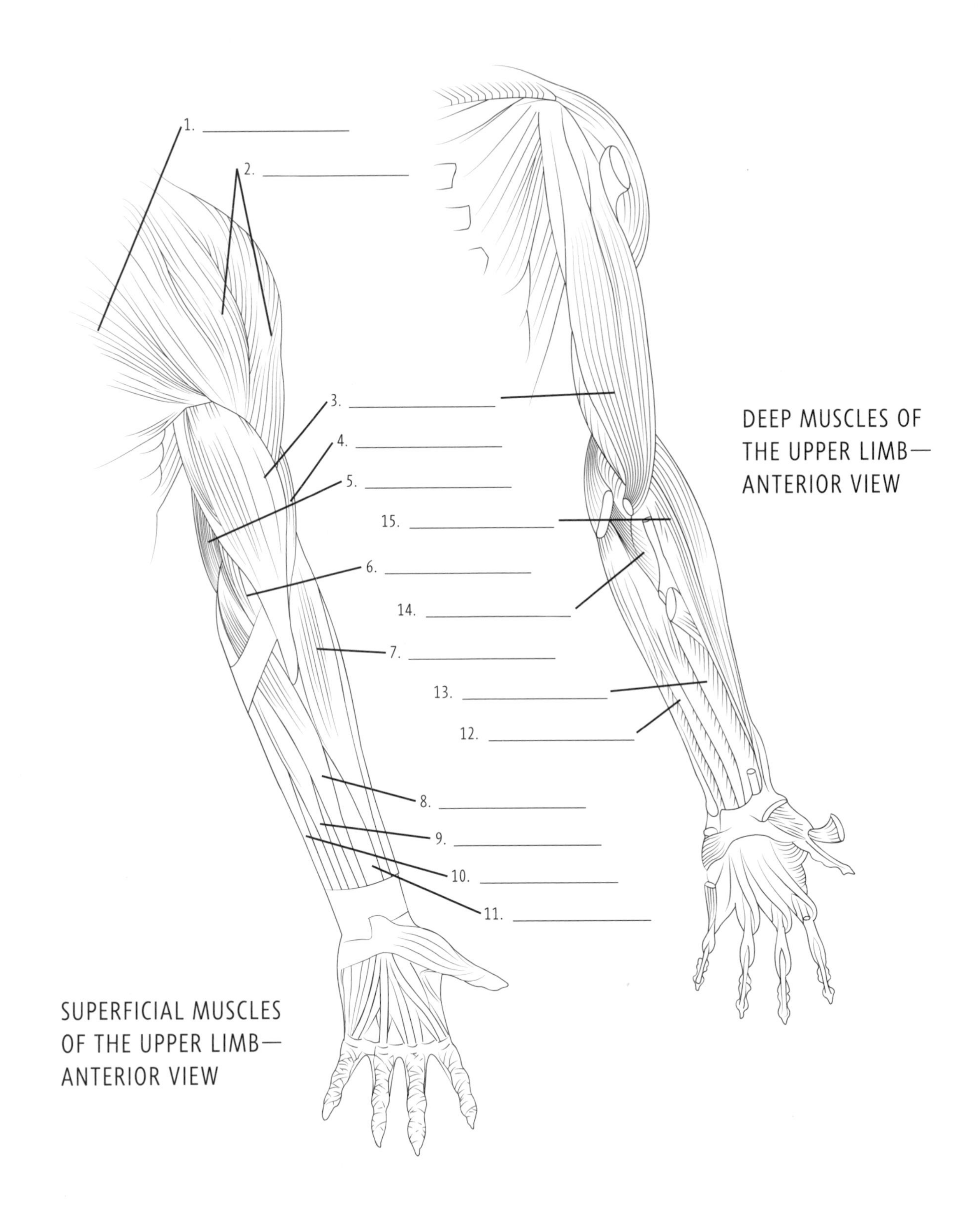
1.
2.
3.
4.
5.
15.
6.
14.
7.
13.
12.
8.
9.
10.
11.
DEEP MUSCLES OF THE UPPER LIMB—ANTERIOR VIEW
SUPERFICIAL MUSCLES OF THE UPPER LIMB—ANTERIOR VIEW

Answers

1. Pectoralis major, 2. Deltoid, 3. Biceps brachii, 4. Brachialis, 5. Triceps brachii, 6. Pronator teres, 7. Brachioradialis, 8. Tendon of flexor carpi radialis, 9. Tendon of palmaris longus, 10. Tendon of flexor carpi ulnaris, 11. Flexor digitorum superficialis, 12. Flexor digitorum profundus, 13. Flexor pollicis longus, 14. Pronator teres, 15. Extensor carpi radialis longus

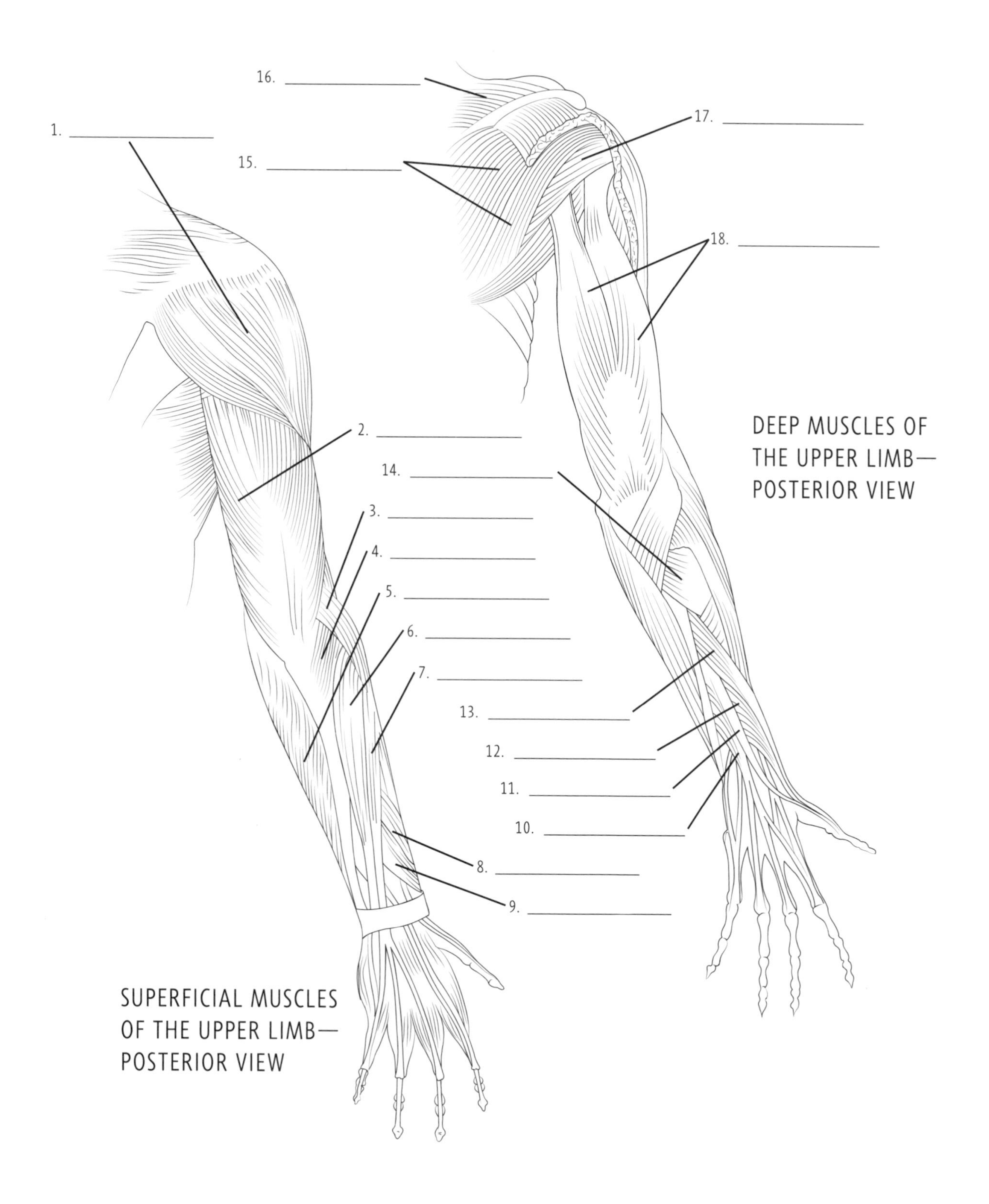
16.
1.
15.
17.
18.
2.
14.
3.
4.
5.
6.
7.
13.
12.
11.
10.
8.
9.
DEEP MUSCLES OF THE UPPER LIMB—POSTERIOR VIEW
SUPERFICIAL MUSCLES OF THE UPPER LIMB—POSTERIOR VIEW

Answers

1. Deltoid, 2. Long head of triceps brachii, 3. Brachioradialis, 4. Anconeus, 5. Flexor carpi ulnaris, 6. Extensor carpi ulnaris, 7. Extensor digiti minimi, 8. Extensor digitorum, 8. Abductor pollicis longus, 9. Extensor pollicis brevis, 10. Extensor indicis, 11. Extensor pollicis longus, 12. Extensor pollicis brevis, 13. Abductor pollicis longus, 14. Supinator, 15. Infraspinatus, 16. Supraspinatus, 17. Teres minor, 18. Triceps brachii

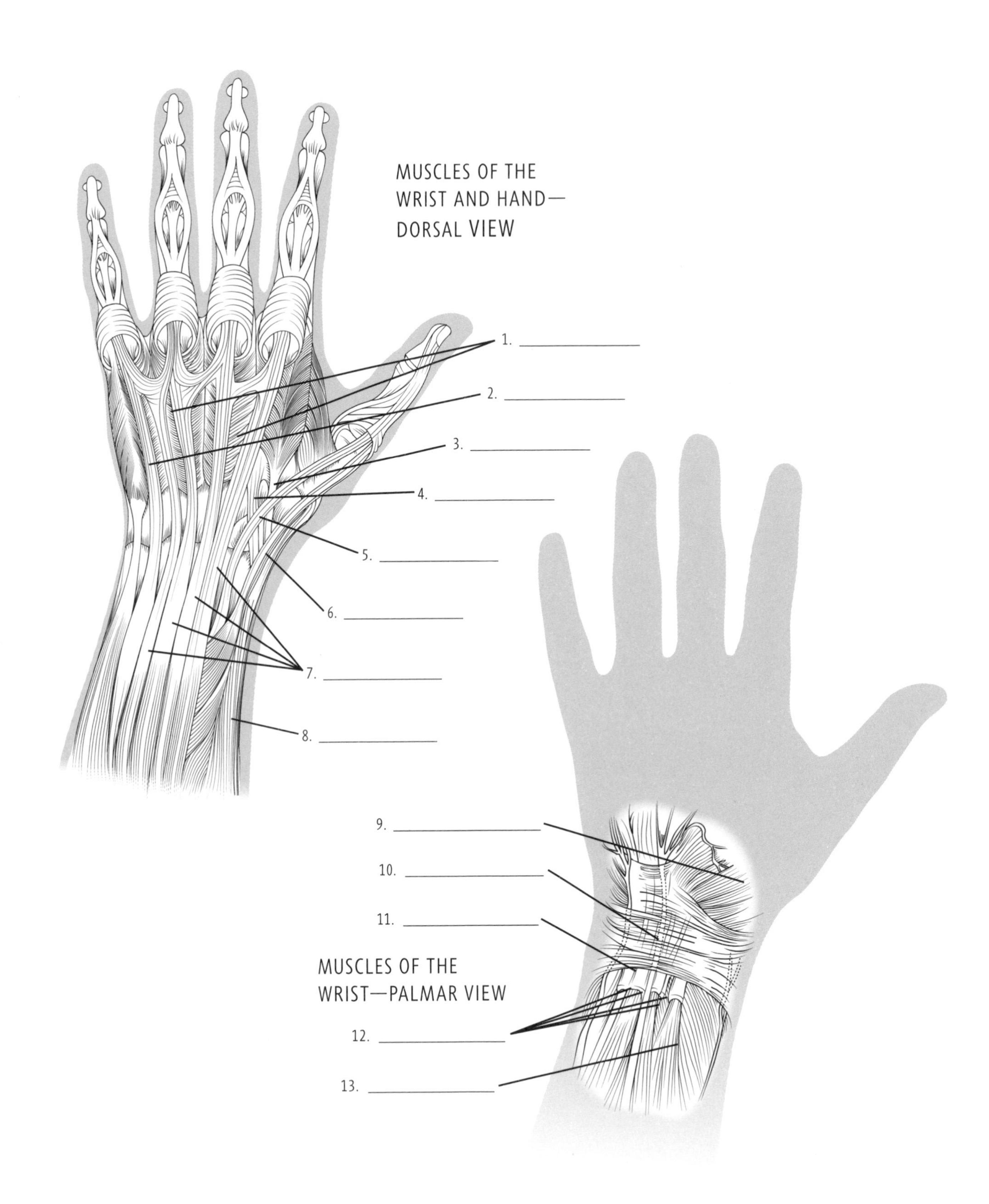
MUSCLES OF THE
WRIST AND HAND—
DORSAL VIEW
1. ______________
2. ______________
3. ______________
4. ______________
5. ______________
6. ______________
7. ______________
8. ______________
9. ______________
10. ______________
11. ______________
MUSCLES OF THE
WRIST—PALMAR VIEW
12. ______________
13. ______________

Answers

1. Dorsal interosseus muscles, 2. Extensor digiti minimi, 3. Extensor carpi radialis longus, 4. Extensor carpi radialis brevis, 5. Extensor pollicis brevis, 6. Extensor pollicis longus, 7. Extensor digitorum, 8. Abductor pollicis longus, 9. Thenar muscles, 10. Flexor retinaculum, 11. Tendinous sheath of flexor digitorum superficialis, 12. Flexor digitorum superficialis tendons, 13. Flexor carpi radialis

Muscles of the Lower Limb

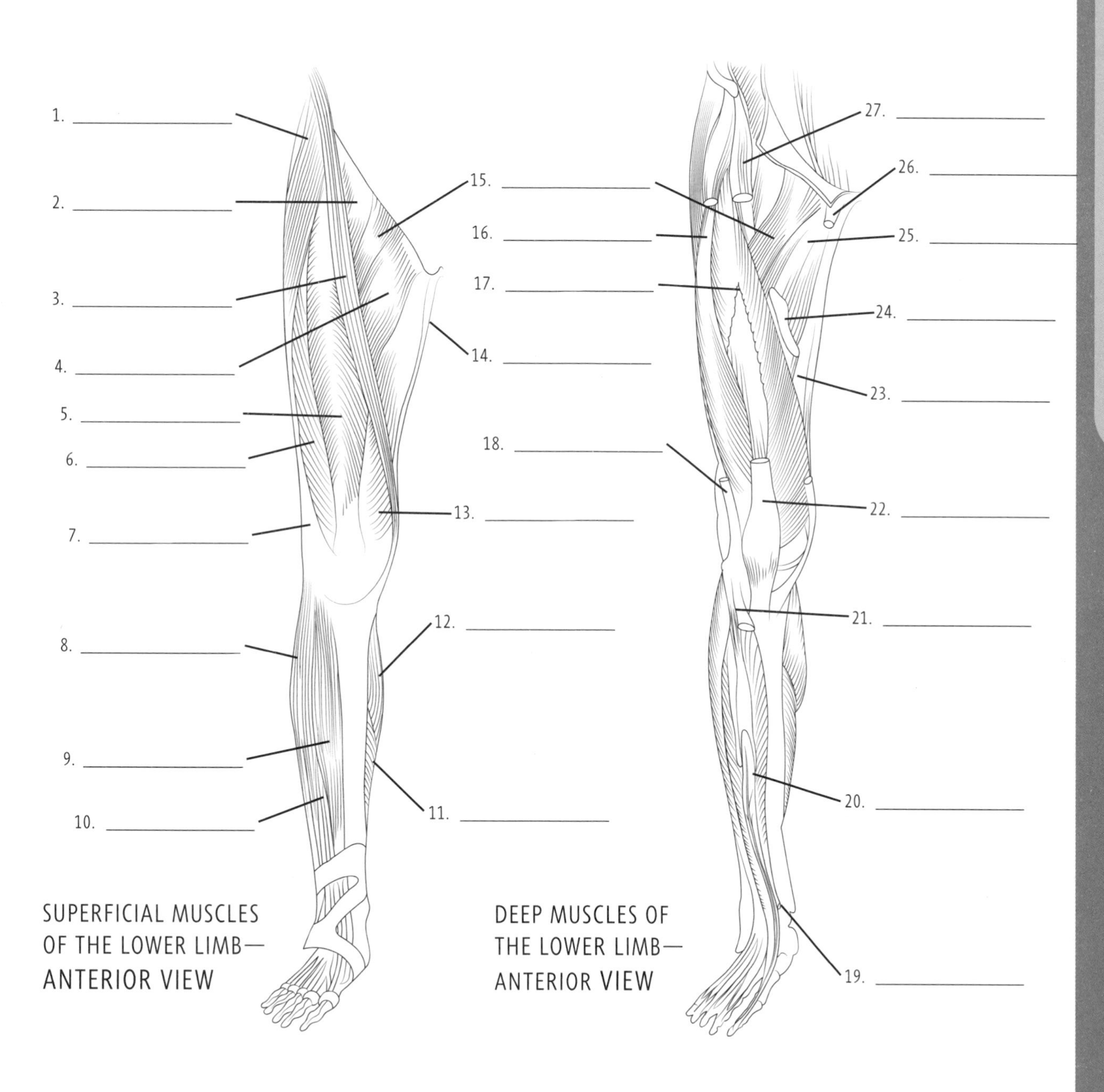
1.
2.
3.
4.
5.
6.
7.
8.
9.
10.
11.
12.
13.
14.
15.
16.
17.
18.
19.
20.
21.
22.
23.
24.
25.
26.
27.

SUPERFICIAL MUSCLES OF THE LOWER LIMB—ANTERIOR VIEW

DEEP MUSCLES OF THE LOWER LIMB—ANTERIOR VIEW

Answers

1. Tensor fasciae latae, 2. Iliopsoas, 3. Sartorius, 4. Adductor longus, 5. Rectus femoris, 6. Vastus lateralis, 7. Iliotibial tract, 8. Fibularis (peroneus) longus, 9. Tibialis anterior, 10. Extensor digitorum longus, 11. Soleus, 12. Gastrocnemius, 13. Vastus medialis, 14. Gracilis, 15. Pectineus, 16. Vastus lateralis, 17. Vastus intermedius, 18. Iliotibial tract (cut), 19. Tibialis anterior (cut), 20. Extensor hallucis longus, 21. Tibialis anterior (cut), 22. Rectus femoris (cut), 23. Adductor longus (cut), 24. Adductor magnus, 25. Adductor longus (cut), 26. Adductor brevis, 27. Sartorius (cut)

Muscles of the Lower Limb

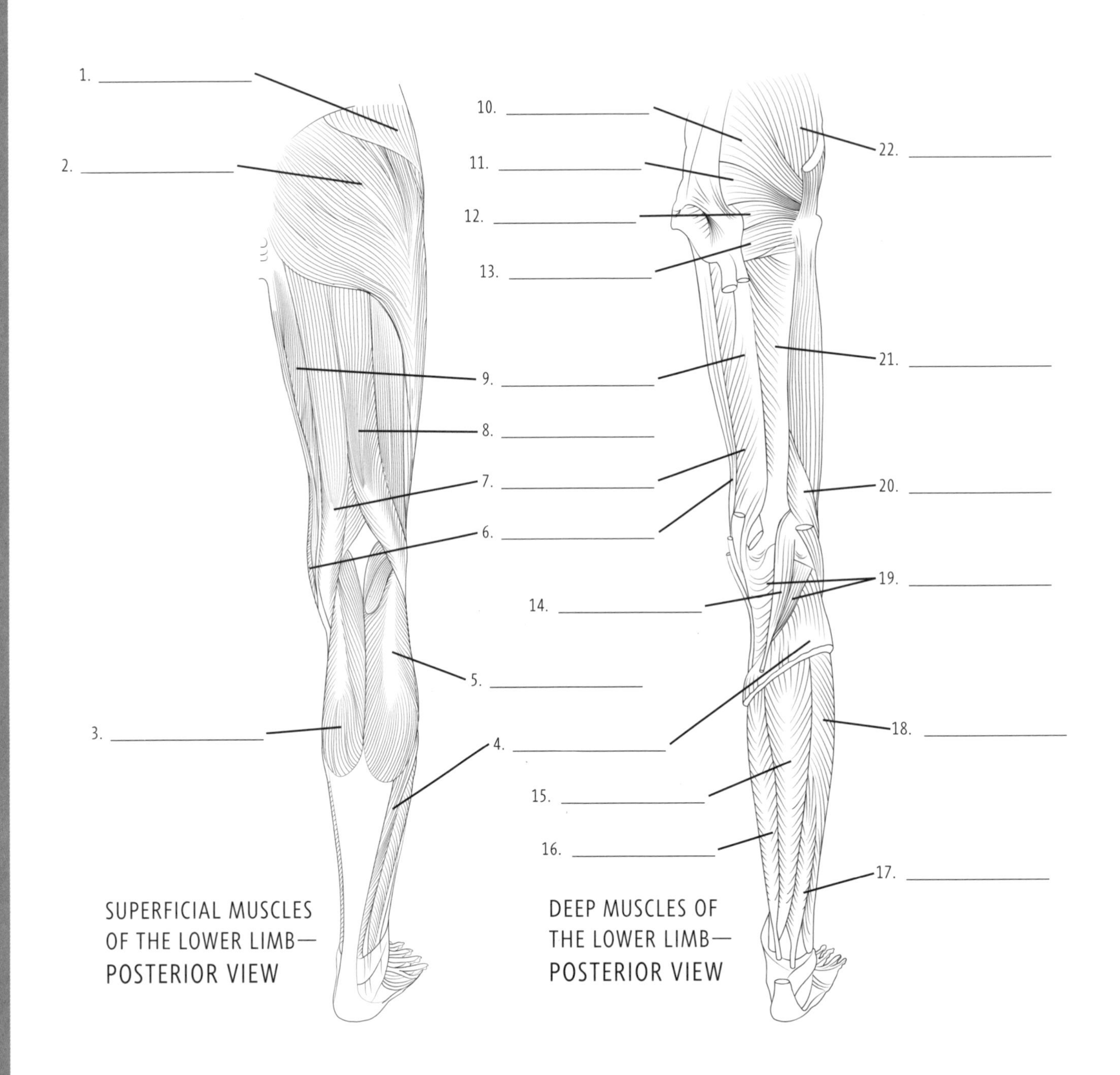
1.
2.
3.
4.
5.
6.
7.
8.
9.
10.
11.
12.
13.
14.
15.
16.
17.
18.
19.
20.
21.
22.
SUPERFICIAL MUSCLES OF THE LOWER LIMB—POSTERIOR VIEW
DEEP MUSCLES OF THE LOWER LIMB—POSTERIOR VIEW

Answers

1. Gluteus medius, 2. Gluteus maximus, 3. Medial head of gastrocnemius, 4. Soleus, 5. Lateral head of gastrocnemius, 6. Gracilis, 7. Semitendinosus, 8. Biceps femoris, 9. Adductor magnus, 10. Piriformis, 11. Superior gemellus, 12. Inferior gemellus, 13. Quadratus femoris, 14. Plantaris, 15. Tibialis posterior, 16. Flexor digitorum longus, 17. Flexor hallucis longus, 18. Fibularis (peroneus) longus, 19. Popliteus, 20. Short head of biceps femoris, 21. Adductor part of adductor magnus, 22. Gluteus minimus

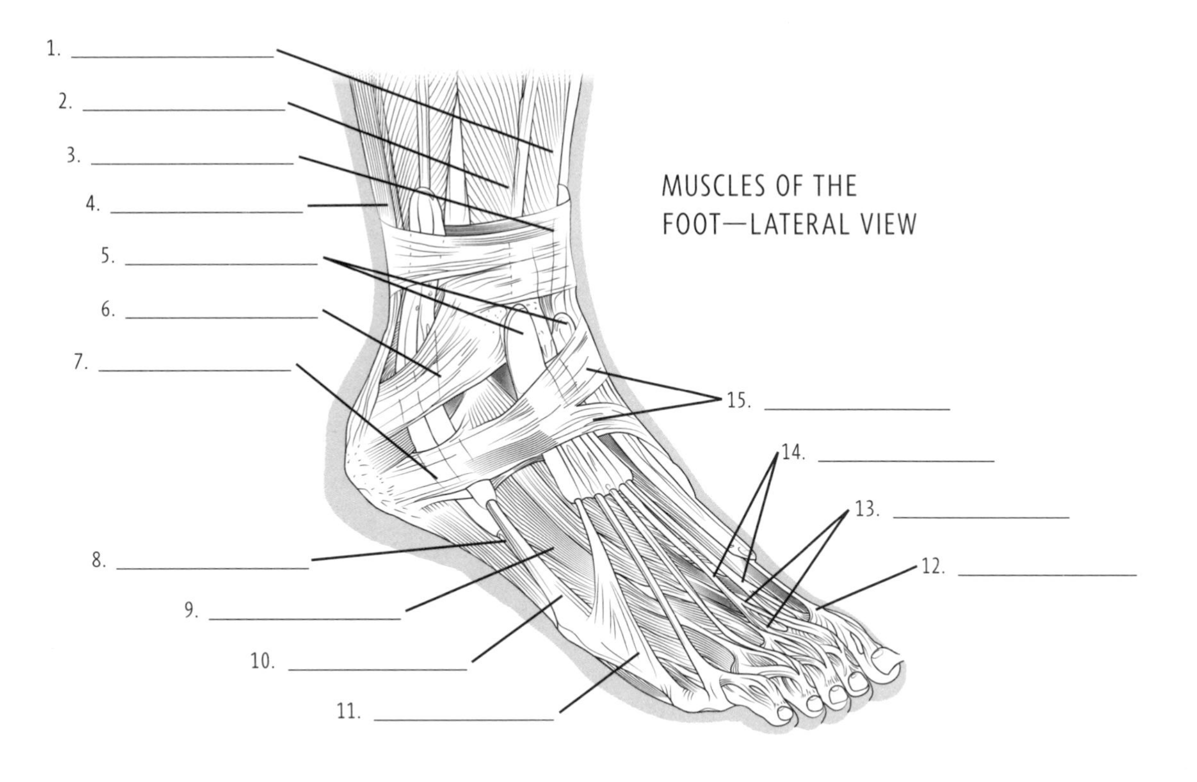
1.
2.
3.
4.
5.
6.
7.
8.
9.
10.
11.
12.
13.
14.
15.
MUSCLES OF THE
FOOT—LATERAL VIEW

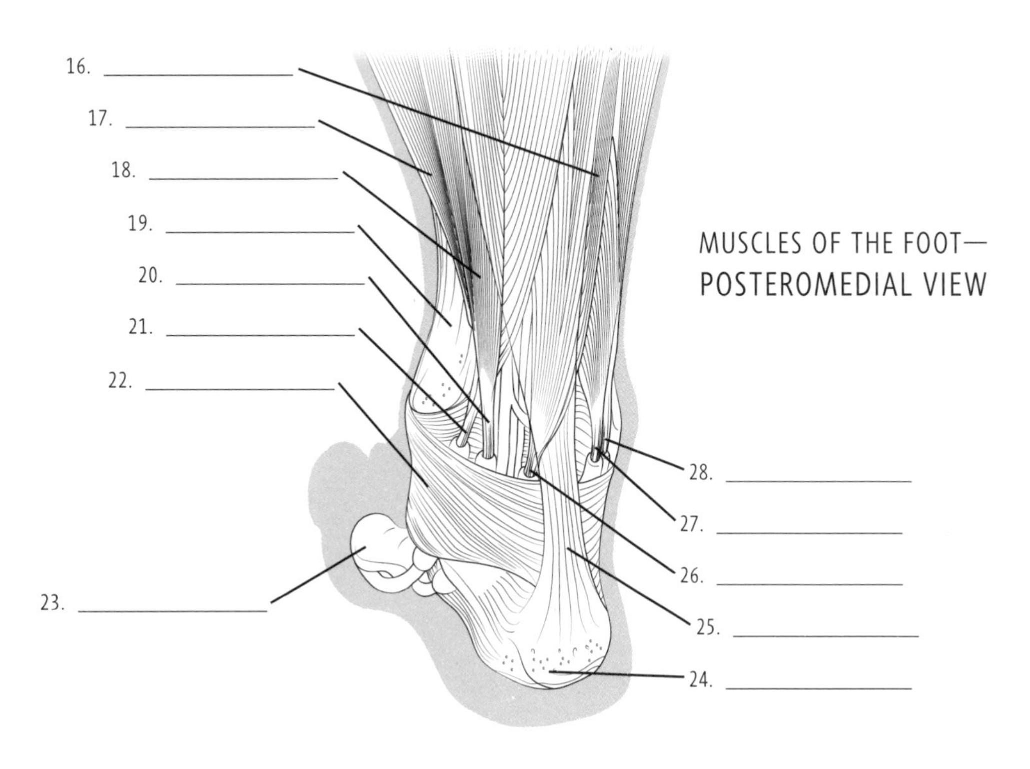
16.
17.
18.
19.
20.
21.
22.
23.
24.
25.
26.
27.
28.
MUSCLES OF THE FOOT—
POSTEROMEDIAL VIEW

Answers

1. Extensor hallucis longus, 2. Extensor digitorum longus, 3. Superior extensor retinaculum, 4. Tendo calcaneus (Achilles tendon), 5. Tendon sheaths, 6. Superior fibular retinaculum, 7. Inferior fibular retinaculum, 8. Fibularis (peroneus) longus tendon, 9. Extensor digitorum brevis, 10. Fibularis (peroneus) brevis tendon, 11. Fibularis (peroneus) tertius, 12. Extensor hallucis longus tendon, 13. Extensor digitorum brevis tendons, 14. Extensor digitorum longus tendons, 15. Inferior extensor retinaculum, 16. Flexor hallucis longus, 17. Tibialis posterior, 18. Flexor digitorum longus, 19. Tibia, 20. Flexor digitorum longus tendon, 21. Tibialis posterior tendon, 22. Flexor retinaculum, 23. First metatarsal, 24. Calcaneal tuberosity, 25. Tendo calcaneus (Achilles tendon), 26. Flexor hallucis longus tendon, 27. Fibularis (peroneus) longus tendon, 28. Fibularis (peroneus) brevis tendon

Skeletal System

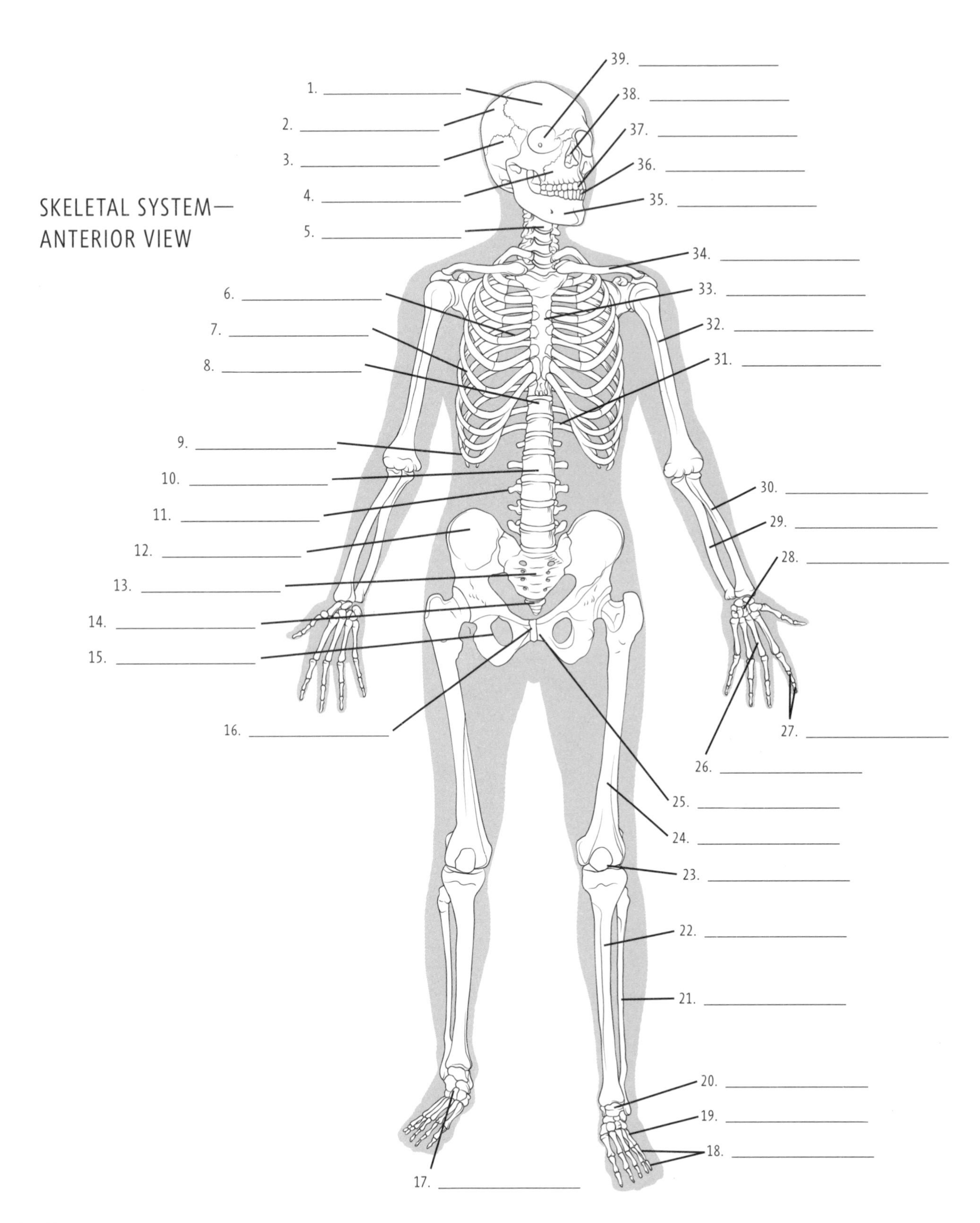
SKELETAL SYSTEM—
ANTERIOR VIEW
1.
2.
3.
4.
5.
6.
7.
8.
9.
10.
11.
12.
13.
14.
15.
16.
17.
18.
19.
20.
21.
22.
23.
24.
25.
26.
27.
28.
29.
30.
31.
32.
33.
34.
35.
36.
37.
38.
39.

Answers

1. Frontal bone, 2. Parietal bone, 3. Temporal bone, 4. Maxilla, 5. Cervical vertebra, 6. Costal cartilage, 7. True rib, 8. Thoracic vertebra, 9. False rib, 10. Lumbar vertebra, 11. Transverse process,12. Ilium, 13. Sacrum, 14. Coccyx, 15. Ischium, 16. Pubic symphysis, 17. Tarsal bones, 18. Phalanges, 19. Metatarsal bones, 20. Talus, 21. Fibula, 22. Tibia, 23. Patella, 24. Femur, 25. Pubis, 26. Metacarpal bones, 27. Phalanges, 28. Carpal bones, 29. Ulna, 30. Radius, 31. Twelfth rib (floating rib), 32. Humerus, 33. Sternum, 34. Clavicle, 35. Mandible, 36. Lower teeth, 37. Upper teeth, 38. Anterior nasal (piriform) aperture, 39. Orbit

SKELETAL SYSTEM—POSTERIOR VIEW

1. ____________
2. ____________
3. ____________
4. ____________
5. ____________
6. ____________
7. ____________
8. ____________
9. ____________
10. ____________
11. ____________
12. ____________
13. ____________
14. ____________
15. ____________
16. ____________
17. ____________
18. ____________
19. ____________
20. ____________
21. ____________
22. ____________
23. ____________
24. ____________
25. ____________
26. ____________
27. ____________
28. ____________
29. ____________
30. ____________
31. ____________
32. ____________
33. ____________
34. ____________
35. ____________

SKELETAL SYSTEM—LATERAL VIEW

36. ____________
37. ____________
38. ____________
39. ____________
40. ____________
41. ____________
42. ____________
43. ____________

Answers

1. Parietal bone, 2. Occipital bone, 3. Atlas (C1), 4. Axis (C2), 5. Spinous process of thoracic vertebra, 6. Thoracic vertebra, 7. Floating ribs (11 & 12), 8. Humerus, 9. Ulna, 10. Radius, 11. Carpal bones, 12. Ischial tuberosity, 13. Phalanges, 14. Metatarsal bones, 15. Calcaneus, 16. Talus, 17. Fibula, 18. Tibia, 19. Femoral condyle, 20. Femur, 21. Phalanges, 22. Metacarpal bones, 23. Pubis, 24. Coccyx, 25. Sacrum, 26. Ilium, 27. Lumbar vertebra, 28. False rib, 29. True rib, 30. Scapula, 31. Acromion, 32. Spine of the scapula, 33. Clavicle, 34. Mandible, 35. Zygomatic bone, 36. Metatarsal bones, 37. Phalanges, 38. Navicular, 39. Patella, 40. Ischium, 41. Iliac crest, 42. Intervertebral disk, 43. Humerus

Vertebral Column

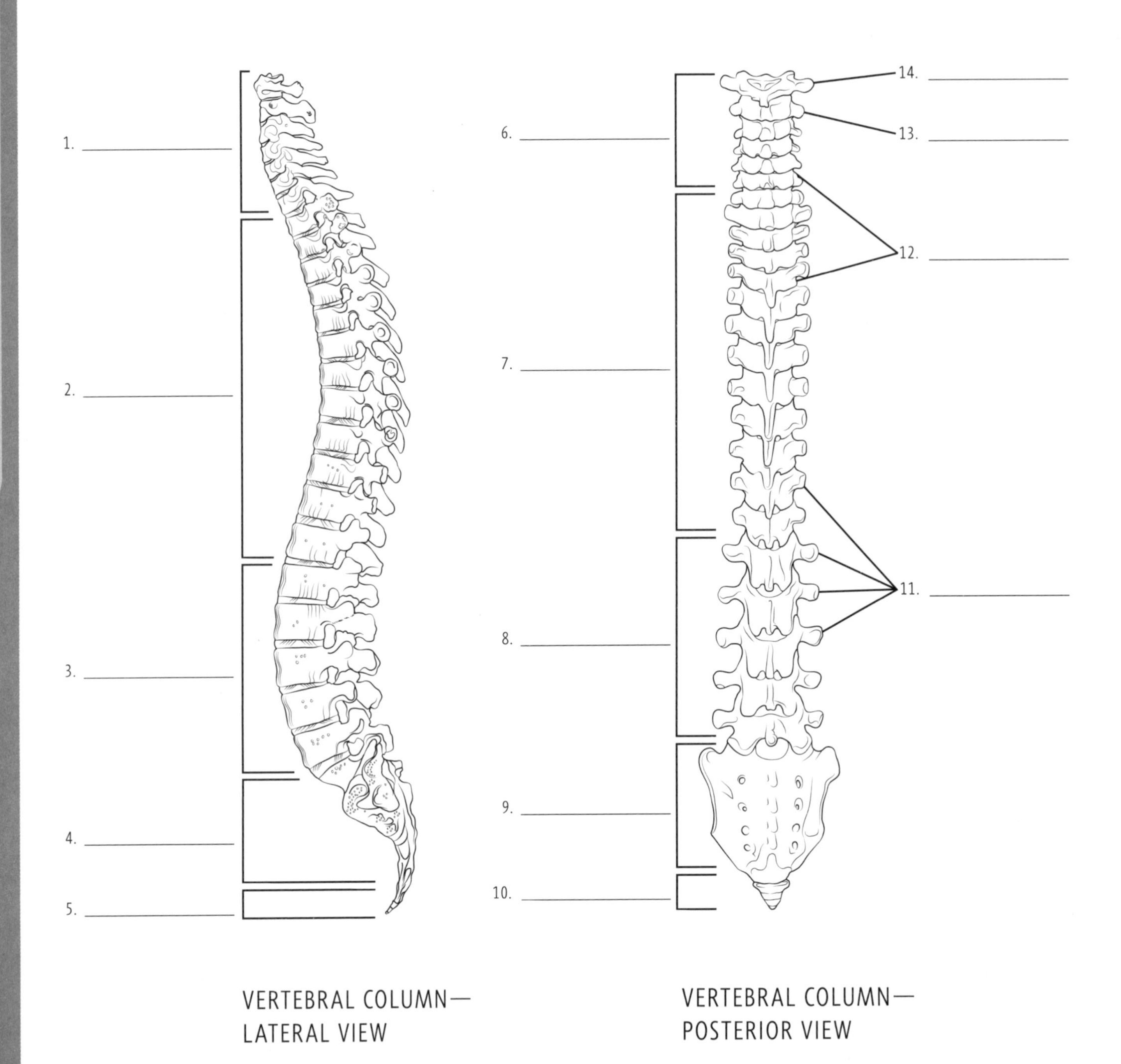
1.
2.
3.
4.
5.
6.
7.
8.
9.
10.
11.
12.
13.
14.

VERTEBRAL COLUMN—
LATERAL VIEW

VERTEBRAL COLUMN—
POSTERIOR VIEW

Answers

1. Cervical vertebrae (C1 – C7), 2. Thoracic vertebrae (T1–T12), 3. Lumbar vertebrae (L1–L5), 4. Sacrum, 5. Coccyx, 6. Cervical region (C1–C7), 7. Thoracic region (T1–T12), 8. Lumbar region (L1–L5), 9. Sacral region (S1–S5), 10. Coccygeal region, 11. Transverse processes, 12. Spinous processes, 13. Axis (C2), 14. Atlas (C1)

Index of English medical terms

Bold numbers indicate chapter names and subject titles. Plain numbers indicate illustration names and labels. *Italicized* numbers indicate references in captions.

A

B

C

D

E

F

G

H

M

N

O

P

Q

R

S

T

U

V

W

X

Z

Index of Latin medical terms

A

B

C

D

E

F

G

H

I

J

L

[lig. = ligamentum]

M

[m. = musculus]

N

[n. = nervus]

O

P

R

[r. = ramus]

S

T

U

V

[v. = vena]

Z

ACKNOWLEDGMENTS

The publisher would like to thank Professor Ian Whitmore for his work on the Latin terms.

The Latin anatomical terms used in this book are based on *Terminologia Anatomica* (2011, Georg Thieme Verlag), which represents the international standard on anatomical terminology developed by the Federative Committee on Anatomical Terminology (FCAT) and the International Federation of Associations of Anatomists (IFAA) in 1998.